Series in BioEngineering

The Series in Bioengineering serves as an information source for a professional audience in science and technology as well as for advanced students. It covers all applications of the physical sciences and technology to medicine and the life sciences. Its scope ranges from bioengineering, biomedical and clinical engineering to biophysics, biomechanics, biomaterials, and bioinformatics.

Indexed by WTI Frankfurt eG, zbMATH.

Neetu Talreja · Divya Chauhan ·
Mohammad Ashfaq
Editors

Microneedles (MNs)-Based Technology

Springer

Editors
Neetu Talreja
Department of Science
Alliance University
Bengaluru, Karnataka, India

Divya Chauhan
Ernst and Young Global LLP
New Delhi, Delhi, India

Mohammad Ashfaq
UCRD
Chandigarh University
Mohali, Punjab, India

ISSN 2196-8861 ISSN 2196-887X (electronic)
Series in BioEngineering
ISBN 978-981-96-3915-1 ISBN 978-981-96-3916-8 (eBook)
https://doi.org/10.1007/978-981-96-3916-8

This Springer imprint is published by the registered company Springer Nature Singapore Pte Ltd.
The registered company address is: 152 Beach Road, #21-01/04 Gateway East, Singapore 189721, Singapore

Contents

Chapter 1
Microneedles: Fabrication, Characterization and Translational Potential

Antonio José Guillot, Miquel Martínez-Navarrete, José Alejandro Bernabeu-Martínez, Ana Sara Cordeiro, and Ana Melero

1.1 Overview of the Current Landscape of Transdermal Drug Delivery

Among the most common drug administration routes, the transdermal one is gaining relevance for the treatment of certain diseases that were traditionally treated via parenteral or oral delivery. Clinicians and healthcare professionals agree that the oral route is most frequently used because of its accessibility and non-invasiveness [1]. In addition, advances in dosage forms for oral administration—such as tablets, capsules, etc., made it stable, very economic and easy-to-use for patients through self-administration, improving adherence to treatment [2]. However, not all drugs can be orally administered since the effectiveness of this route depends on the drug's bioavailability and its stability through the gastrointestinal tract [3, 4]. When drugs are ingested, they face several challenges before they can be absorbed into the systemic circulation in therapeutic doses. The main challenges drugs must overcome are gastric degradation and hepatic first-pass metabolism [4–7].

Other administration routes have been explored as an alternative to oral and parenteral ones [8, 9], including vaginal [10], rectal [11], and sublingual [12], although their clinical usage in certain conditions and diseases is still quite limited [13]. Topical administration routes, including the ophthalmic and intravitreal routes [14, 15], seem very promising, particularly due to being painless and allowing self-administration. This is an ideal approach for diseases locally affecting the skin

A. J. Guillot · M. Martínez-Navarrete · J. A. Bernabeu-Martínez · A. Melero
Department of Pharmacy and Pharmaceutical Technology and Parasitology, Faculty of Pharmacy, University of Valencia, Burjassot, Spain

A. S. Cordeiro (✉)
Leicester Institute for Pharmaceutical, Health and Social Care Innovations, De Montfort University, Leicester, UK
e-mail: ana.melero@uv.es

N. Talreja et al. (eds.), *Microneedles (MNs)-Based Technology*, Series in BioEngineering, https://doi.org/10.1007/978-981-96-3916-8_1

tissue—wounds, burns, infections, melanoma [16]—but also for systemic purposes [17], including vaccination and immunotherapy [18]. The deeper skin layers (dermis and hypodermis) are in contact with small capillary vessels that can absorb the drug deposited in the skin to the systemic circulation [19]. This dual strategy makes the topical and transdermal routes very versatile.

Although this route provides several advantages over other ones, the main function of the skin is to protect the entrance of xenobiotics into the body, limiting drug penetration. This function is performed by the *stratum corneum* (SC), the outermost layer of the skin constituted mainly by corneocytes [20]. The SC is very difficult to penetrate for some molecules because of its high lipophilicity and cell cohesion. Therefore, the main challenge in transdermal delivery is for drug molecules to penetrate the SC and to partition and diffuse though the skin layers [21]. Drugs can enter the skin by passive diffusion [22] provided that they possess some specific characteristics: (i) low degree of ionization [23], (ii) specific distribution coefficient and (iii) low molecular weight (under 500 Da) [24]. Absorption is also determined by other factors, such as skin conditions (hydration, cell aggregation index, etc.). From a mechanical point of view, skin drug absorbance is determined by Fick's diffusion law, which determines that the drug concentration in the skin and in the vehicle, as well as the partition coefficient between vehicle and skin, are also important factors determining the absorption process [25].

As a result of this opposition of the skin to molecule permeation, some strategies have been widely investigated to improve transdermal drug delivery. Delivery systems designed for this route aim to temporarily break the barrier function of the SC or to improve the physicochemical properties of the drug. They can be classified into chemical or physical approaches [26]. The use of chemical enhancement, prodrugs and micro/nanocarriers can be considered the most frequent, as they can easily be included into other supramolecular systems such as patches, hydrogels, or emulsions. However, this administration route is often slow and not adequate for drugs that require rapid action.

Some active strategies have also been developed to increase the spectrum of drugs that can be delivered through the skin. Most of them use external energy as the driving force of compounds through the skin and generally allow faster drug access to the systemic circulation [27]. However, these strategies usually present important drawbacks associated with their aggressiveness against the skin barrier, as they are based on mechanical, electrical (iontophoresis and electroporation), ultrasonic (sonophoresis) or thermal (laser and radiofrequency) methods and velocity-based devices (jet injectors) to overcome the SC [28–35].

Microneedle technology is a physical method for overcoming the skin barrier. Microarray patches (MAPs) contain numerous microscopic needle-shaped projections that can mechanically pierce the outer skin layers and easily reach the dermis without touching the nerve endings [36]. Microneedle length in these patches is usually within the range of 100–1000 μm with a tip radius ranging between 30 and 100 μm [37]. These devices create physical pathways through which the drug can easily diffuse to the deeper skin layers, allowing the administration of drugs independently of their limiting characteristics, including proteins or nanoparticles

[36, 38]. MAPs are also easy to manipulate, apply and wear, minimally invasive and painless [37, 39], avoiding injection-associated costs [40, 41]. Advances in the development of new materials have given MAPs an economic advantage for both topical and transdermal drug administration. This chapter aims to collect and analyze the current information on this growing technology to offer an overview of MAPs-based available approaches, the most convenient materials for MAPs fabrication and current fabrication and characterization methods for MAPs design and development in academia, basic research, and industry translation.

1.2 Microneedle Types and Applications

1.2.1 "Poke and Patch" Strategy Using Solid Microneedles

Solid MAPs are used for transdermal delivery with the following two-step strategy: first, solid MAPs are applied onto the skin and removed after piercing. The drug is then applied to the treated area incorporated in a conventional topical dosage form (hydrogel, cream, patch, solution, etc.) [42] (Fig. 1.1). The created microchannels enhance drug absorption to the deeper skin layers since the SC is no longer a limiting factor [43]. This strategy is very attractive because of its simplicity and translational opportunities. Still, the microchannels close after approximately 3 h [44], hindering drug absorption after this period, thus being an important drawback for long-term sustained drug release. Furthermore, if the microchannels would remain unclosed, the risk of infection increases [45].

An important improvement using this strategy is the currently commercially-available Dermaroller® [46]. This device is a metallic roller-like microneedle system

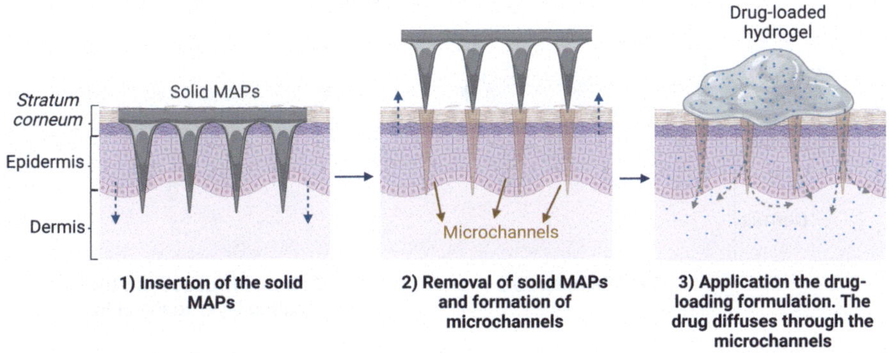

Fig. 1.1 Illustration of "poke and patch" strategy. Solid MAPs are used to pierce the skin and create microchannels to enhance drug absorption from an external pharmaceutical formulation. This illustration has been created using BioRender.com under an active license

that can perforate broader surfaces of skin thanks to its wheel-like mechanism [47]. Its benefits are being explored in the cosmetology and pharmacology fields [48, 49].

1.2.2 "Coat and Poke" Strategy Using Coated Microneedles

Solid MAPs can also be used for the "coat and poke" approach. In this strategy, the surface of solid MAPs is coated with a drug-loaded formulation [50]. When the microneedle device is applied, the coated tips penetrate the skin, inserting the coating material in the deeper skin layers (Fig. 1.2). In addition, this strategy has been successfully used to co-deliver multiple compounds with different properties [51]. Nevertheless, the use of coated MAPs requires highly specialized fabrication techniques and only low drug amounts can be loaded due to surface area limitations. The coating thickness can be increased to increase drug loading, although it often compromises the skin insertion properties of the arrays. Coated MAPs became popular for transdermal vaccination due to the generally low doses needed, their high efficiency and the high presence of immune cells in the skin, which can trigger an immune response after exposure to low antigen doses [18, 52].

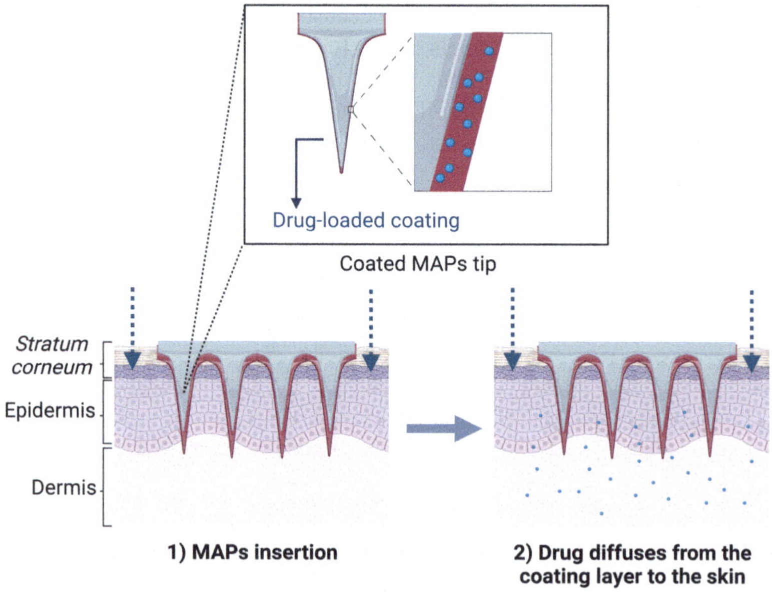

Fig. 1.2 Illustration of "coat and poke" strategy. Solid MAPs are coated with a drug-loaded layer that releases the cargo only when it is dissolved by the interstitial fluid after skin insertion. This illustration has been created using BioRender.com under an active license

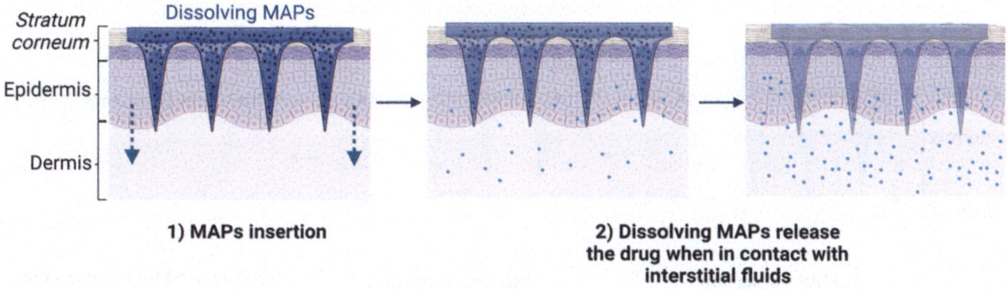

Fig. 1.3 Illustration of "poke and release" strategy. Dissolving MAPs are loaded with the drug, which is released after polymer dissolution when inserted into the skin. This illustration has been created using BioRender.com under an active license

1.2.3 "Poke and Release" Strategy Using Dissolving Microneedles

Dissolving MAPs are fabricated with biocompatible polymers that dissolve in the interstitial skin fluid (ISF). This strategy is an important variation of the previous two since the drug is loaded into the polymeric matrix and released while the MAPs dissolve [53]. Dissolving MAPs allow an increase in drug loading in comparison with coated ones and can guarantee sustained release controlled by the polymer characteristics. These MAPs are applied in one step and remain in the skin until complete dissolution [54] (Fig. 1.3), improving patient comfort [55] and reducing infection risk. The main drawback of this approach is that the polymer matrix may not have sufficient sharpness and mechanical strength for appropriate skin penetration [56]. Additionally, the dose that these systems can administer is limited by the microneedle dimensions [57], narrowing their application down to highly potent molecules.

1.2.4 "Poke and Break" Strategy Using Rapidly Separable Microneedles

As a variation of the dissolving MAPs, separable MAPs are composed of an insoluble inert matrix and a dissolving polymer loading [58]. This strategy involves two steps: the separable MAPs pierce the skin, and the microneedle tips remain inside the skin after MAPs removal [59]. When the polymeric tips start to degrade, the drug is released [60] (Fig. 1.4).

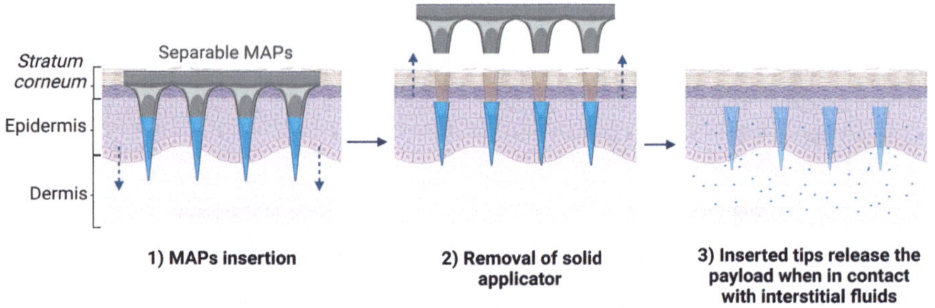

Fig. 1.4 Illustration of "poke and break" strategy. Separable MAPs are constituted by a solid part that acts as an applicator, and a separable dissolving part that contains the drug and remains inside the skin after insertion. This illustration has been created using BioRender.com under an active license

1.2.5 "Poke and Flow" Strategy Using Hollow Microneedles

Hollow MAPs have needle-shaped microchannel projections, which can enhance the absorption of liquids into the skin. A drug-loaded liquid is pushed through this type of MAPs with some type of pressure applicator, which enters the skin through the channels after insertion [61, 62] (Fig. 1.5). This strategy imitates parenteral injection but at a non-invasive, microscopic scale [63]. Although it has the potential to increase patient acceptance and allow self-administration [40], the fabrication costs for this microneedle structure are high, given the high accuracy and specialization needed [64].

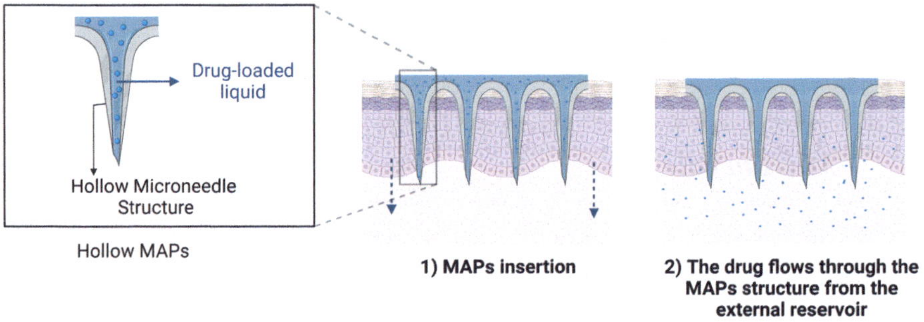

Fig. 1.5 Illustration of "poke and flow" strategy. Hollow MAPs present a cavity in their structure that is used for pumping a drug-loaded liquid. This illustration has been created using BioRender.com under an active license

1.2.6 "Swelling and Release" Strategy Using Swellable Microneedles

Swellable MAPs can absorb ISF and swell rapidly, releasing the drug contained in the baseplate or in an external reservoir [65]. When swelling MAPs absorb ISF, a repulsion between the polymer chains is produced, leading to a volume increase [66], which enhances drug diffusion through the baseplate to the swollen tips and drags the drug through the microneedle structure due to the concentration gradient [67] (Fig. 1.6). This mechanism presents some benefits, such as an easy application (one-step administration), low infection risk (the microchannels created in the skin are always in contact with the microneedle), increased drug dose (through the use of an external reservoir), and minimal polymer residues left in the patient's body [68, 69].

1.3 Materials Used for MAPs Fabrication

A wide range of materials has been used for MAP fabrication, including metals, silicon, ceramics, glass, sugars, and polymers [69]. The materials need to fulfil some requirements for this purpose, such as being inert, producing no toxicity or immunogenicity, having high tensile strength for appropriate skin piercing, being fracture-resistant, corrosion-resistant, and showing good biocompatibility and long-term stability. However, depending on the specific approach [70], the most adequate fabrication material should be selected according to the desired properties of the final device [71, 72].

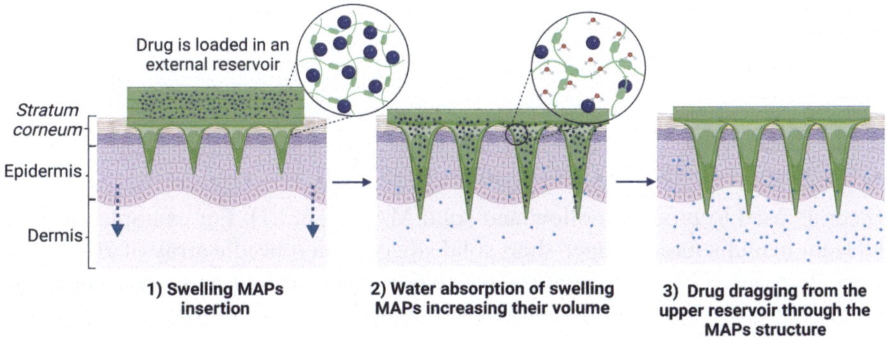

Fig. 1.6 Illustration of "swelling and release" strategy. Swelling or hydrogel-forming MAPs are constituted by polymers that can absorb water, increasing their volume and dragging the drug through the microneedle structure into the skin. This illustration has been created using BioRender.com under an active license

1.3.1 Metals

Solid, coated, and hollow MAPs have conventionally been produced using different metallic materials. Stainless steel and titanium are the most common ones [73–77]. Both, as a consequence of their high tensile strength and low fragility, allow easy penetration through the external skin layers and, subsequently, increase transdermal drug absorption. Other less-used metals include gold and nickel [78, 79]. Dermaroller® has appeared in the field as one of the main applications of metallic microneedles. Nevertheless, Dermaroller® performance is limited to cosmetic approaches [80–82] and has not yet been explored in the drug delivery field.

Despite its excellent mechanical properties, metallic materials have low biocompatibility, and their usage in microneedles has been declining in recent years. Currently, metallic microneedles are used in combination with other more biocompatible materials, mainly polymers, for coated microneedle devices. Metallic microneedles provide a rigid structure with excellent mechanical properties, which can overcome the main limitations of some materials that are not rigid enough to perforate the skin. Ulla et al. used this strategy to address the fragility of porous polymers. The authors designed stainless-steel microneedles coated with porous poly (lactic-co-glycolic acid) (PLGA). This way, PLGA achieved high drug loadings while stainless steel provided sufficient mechanical strength to pierce the skin [83]. On the other hand, Narayanan and collaborators designed a silicon microneedle coated with a gold layer [84]. The metallic layer enhanced the insertion properties of the microneedle using a minimum gold amount (layer thickness of 1 μm). Gold was used in this case because of its excellent biocompatibility. Metallic matrices are also used for sensor fabrication. Tabrizi et al. designed a glutamate transdermal sensor based on platinum using electrochemical analysis due to the conduciveness of the metal. Glutamate-responsive enzymes were then covalently linked onto the device surface. This metal-based biosensor was able to capture the electrochemical signal and detect glutamate within a good sensitivity range [85].

1.3.2 Silicones

Silicon is used to produce hollow and solid MAPs [86, 87]. For example, Wei-Ze and team manufactured a super-short solid silicone microneedle array of 70–80 μm in needle height [42]. This strategy was efficient at piercing and overcoming the *stratum corneum*, as shown in histological analysis. This prototype also increased drug permeation through the skin, compared with the free-drug solution. In some other cases, hollow silicon MAPs were made with certain success in terms of sharpness and pore size. Li et al. developed a silicone-based hollow microneedle device and proved its efficacy for piercing the skin and creating microchannels in pig skin [88]. Bolton and coworkers also fabricated silicone hollow microneedles [89]. Silicon was used in this study due to the low manufacturing costs compared with other materials, such as metals since the fabrication of hollow microneedles is complex. In this study, again, silicon hollow microneedles created deep microchannels of 400 μm, which were strong enough to resist the application stress without fracture. Additionally, transdermal drug flux was achieved using the hollow microneedle device. These studies demonstrate the feasibility of silicone for providing micro-scaled complex structures. However, despite its availability, low cost, properties and ease of manufacturing, some authors report problems related to scalability and ineffective skin insertion [90].

1.3.3 Ceramics

Ceramic MAPs were designed to follow the "poke and flow" approach. However, in this case, the coated layer does not exist, and water-soluble drugs are loaded by capillarity into the porous structure of the device [91]. Therefore, the structure of the device acts as a drug reservoir, allowing higher doses, which is especially useful for vaccination purposes [92, 93]. Even though Alumina presents optimal biocompatibility, some issues related to its brittleness have been detected. Zirconia has improved fracture toughness, bending strength, and wear resistance compared to alumina [94, 95]. These two ceramic materials are considered attractive for research purposes since ceramic-based hollow MAPs have been showed highly effective in administering large doses of medications [96]. Gholami et al. used ultraporous alumina to study the ability of this clay to incorporate large biomolecules—from 15 to 30 μm. In addition, the formulation presented excellent biocompatibility, skin insertion and drug release performance [97]. To follow on, Schepens et al. used porous ceramic microneedles for influenza vaccination. This study demonstrated the effectiveness of these microneedles for vaccine delivery, triggering an antibody-specific immune response in vaccinated mice [98]. Despite these promising advances in the field, ceramic microneedles face some downsides, like the high cooking temperatures for the ceramic sintering (1200–1500 °C), and no ceramic microneedles for the delivery of nanoparticles have been reported yet.

1.3.4 Glass

Glass has also been used to manufacture MAPs [99]. Different studies have demonstrated that glass microneedles offer a simple method to infuse fluid into the skin that can be carried out with little to no pain [62, 100]. For instance, Hu et al. studied the suitability of glass microneedles for manufacturing production. In this study, they demonstrated that the thermoplastic properties of metallic glasses make this material suitable for complex microneedle structures, such as hollow microneedles, and opened new horizons for scalability [101]. However, other studies have manifested concerns due to their potential to break off, producing tissue injuries [102].

1.3.5 Sugars

Rapid-dissolving sugars and polysaccharides such as maltose [103], trehalose [104], mannitol [105] and galactose have also been explored to prepare dissolvable MAPs with appropriate mechanical properties. Puttaswamy and team fabricated a maltose-based microneedle device, which was further coated with PLGA [106]. Maltose provided biocompatibility to the device and desired mechanical strength. Nonetheless, sugars are not only appealing because of their biocompatibility. Maltose is a promising material for microneedle-based biosensors because of its osmotic properties. Following this idea, Zheng et al. developed an osmotic microneedle patch composed of hyaluronic acid and maltose [107]. The presence of maltose in the formulation increased proportionally the amounts of interstitial skin fluid that the microneedle could extract. Additionally, maltose did not affect the crosslinking of the polymeric matrix or the mechanical properties of the microneedle array. Trehalose was used by Tian et al. in combination with polysaccharides to achieve some specific mechanical properties that are vital for antigen delivery. Firstly, the use of sugars made the manufacturing process more suitable for antibody stability. Secondly, trehalose-based microneedles achieved good mechanical resistance, were stable at storage conditions for 4 weeks and could dissolve within 15 min into the skin. Moreover, the microneedle device was tested in vivo and successfully protected the animals against influenza virus infection. In this experiment, microneedles were tested in different storage conditions (freshly made and after being stored for 4 weeks), and no difference in vaccination effectiveness was observed [108]. Nevertheless, problems in handling these materials, such as caramelization or hydrolysis due to high temperatures, can occur. Moreover, stability problems have been detected due to sugar MAPs being hygroscopic, which suggests associated storage difficulties [109, 110].

1.3.6 Polymers

Coating layers and dissolvable MAPs are made of a wide variety of natural and synthetic polymers exported from other fields of pharmaceutical research. For this reason, these materials usually present biodegradable properties and, therefore, excellent biocompatibility. Some typical examples include polyvinylpyrrolidone [111], poly methylvinylether-co-maleic acid [112], polyethylene glycol [113], polycarbonate [114], polyvinyl alcohol [115], SU-8 [116], hydroxypropyl methylcellulose [117], chitosan, hyaluronic acid (HA), carboxymethylcellulose (CMC)

[118], polylactic acid and PLGA [119, 120]. Due to their wide range of properties and characteristics, polymeric materials can be adapted for microneedle fabrication using several approaches, namely controlled release, swelling microneedles, fast dissolving microneedles, and responsive delivery, among others. Albadr et al. designed a polymer-based fast-dissolving microneedle for ocular delivery. In this work, the authors loaded the polymeric matrix consisting of HA and PVP–with antifungal drugs [121]. Results showed that the polymeric microneedles not only scored excellent piercing performance but also dissolved within seconds and delivered the entire cargo. A microneedle patch consisting of PVA and PVP and incorporating Cyclosporine A-loaded lipid vesicles for controlled release of the drug for the treatment of inflammatory skin diseases was also developed [122]. The lipid vesicles allowed an enhanced drug loading, added a step in drug release and presented intrinsic emollient properties, which are beneficial for topical applications. With this combination of biodegradable polymers, the authors demonstrated that microneedles could deliver the intact cargo in a controlled manner for 24 h with a single application. This study also reported that PVA/PVP microneedles retrieved skin inflammation in an in vivo murine model.

Polymeric materials have been used for other approaches that do not necessarily involve degradation of the microneedles. As an example, Tekko et al. designed a hydrogel-forming microneedle device for controlled and non-invasive methotrexate delivery [123]. The polymeric matrix of these microneedle tips swells in contact with the interstitial fluid, thus slowly releasing the cargo. Once swollen, microneedles remained attached to the skin and the cargo was delivered from the reservoir to the microneedle tips. Finally, thanks to the new advances in polymer design, some authors have developed responsive microneedles using "smart" polymers that react to specific stimuli. For example, Wang et al. developed a glucose-responsive microneedle array for insulin delivery. The microneedle matrix consisted of a mixture of crosslinked PVA/HA/methacrylated PVA, able to release insulin in hypoglycemic conditions [124].

In other fields, Yao and the team developed a dissolving microneedle patch for wound healing. For this purpose, microneedles were fabricated, loading imidazole nanoparticles in a photo-crosslinked HA matrix [125]. The aim was to use a soft material to avoid risks of extra damage to the wound tissue. Still, HA was also selected because it is very biodegradable—it is hydrolyzed by endogenous enzymes—and has long been used for ophthalmic and wound healing operations. Finally, HA-based imidazole-loaded microneedles showed good mechanical properties to reach the deeper wound, often covered by blood or fat. They effectively reduced the wound area when tested in vivo. The authors also reported that the HA microneedle surface interaction with the wound tissue was beneficial for nutrient exchange and cell growth.

1.4 Microneedle Manufacturing

The above-mentioned MAP types can be manufactured through different techniques according to the materials selected, as shown in Table 1.1.

The different methods described for MAPs fabrication are depicted in Fig. 1.7.

1.4.1 Laser Cutting

Laser cutting is a technique frequently used to produce stainless steel solid MAPs [126, 127]. First, the array designs are created using different graphic software programs. Then, a laser removes part of the material block through smelting or highly localized vaporization (Fig. 1.7a). Generally, the MAPs are in the same plane

Table 1.1 Summary of the most common combinations of materials and fabrication methods for MAPs manufacturing

MAP type		Material	Fabrication
Solid MAPs		Metal	Laser cutting Laser ablation Electrodeposition
		Silicones	Microetching
		Glass	Glass drawing
Hollow MAPs		Metal	Laser ablation
		Silicones	Microstereolithography Microetching
		Glass	Glass drawing
Coated MAPs	Microneedle	Metal	Laser cutting Laser ablation Electrodeposition
		Silicones	Microstereolithography Microetching
		Ceramic	Ceramic sintering
	Coating	Polymers	Dipping Spraying
Dissolving MAPs	Master template	Metal	Laser cutting Laser ablation Electrodeposition
		Silicones	Microstereolithography Microetching
	Microneedle matrix	Polymers	Solvent casting
		Sugars	
Swelling MAPs	Polymers	Solvent casting	

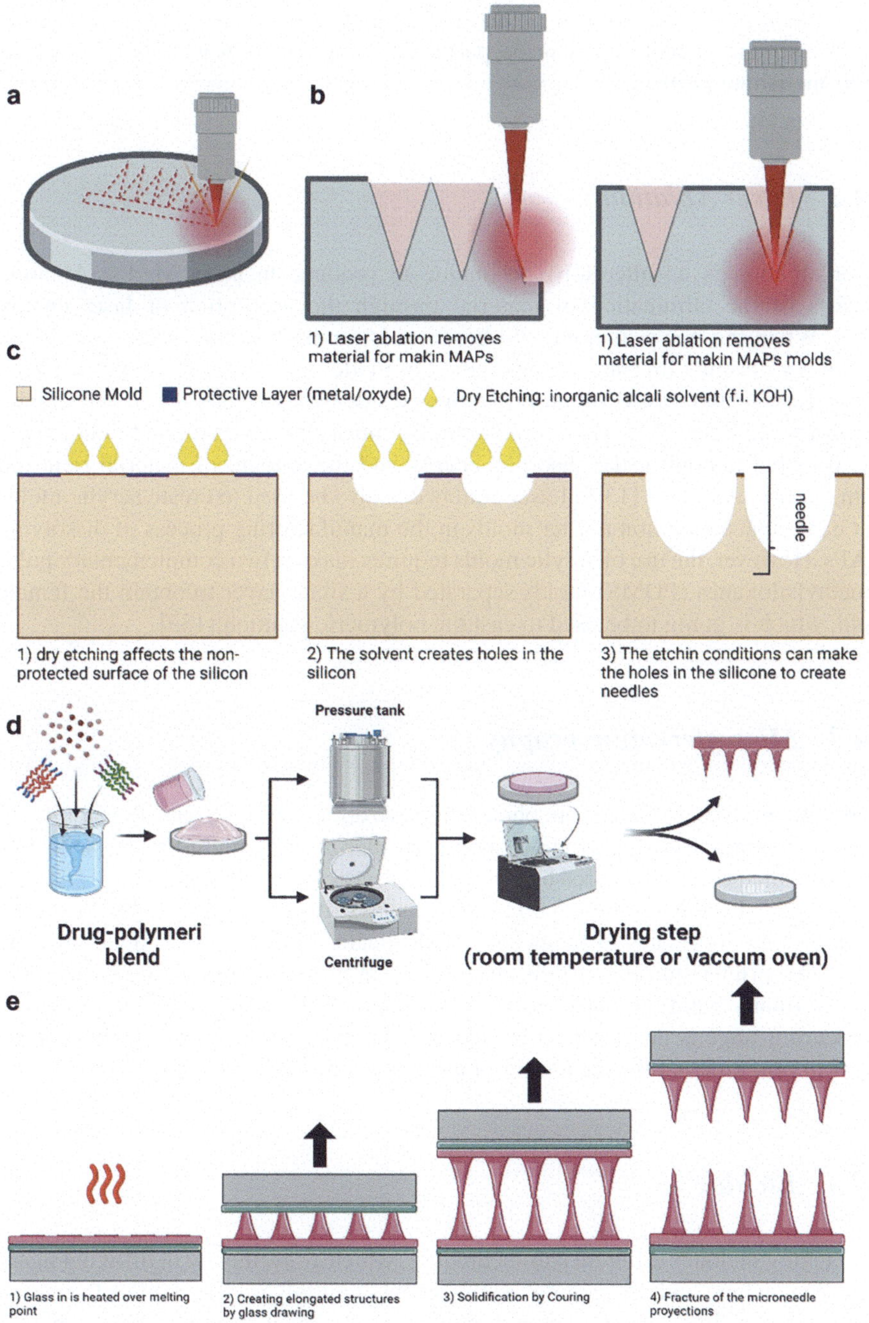

Fig. 1.7 Schematic illustrations of the most common methods for MAPs fabrication: **a** Laser cutting, **b** laser ablation, **c** etching, **d** solvent-casting and **e** glass drawing. This illustration has been created using BioRender.com under an active license

as the metal piece and are bent to the desired angle after cleaning [128]. Also, the sharpness of the needle tips can be perfected using different techniques, such as electropolishing [129].

1.4.2 Laser Ablation

Laser ablation is an alternative technique to produce metallic MAPs. Ablation consists of the elimination of material through the absorption of laser energy [130]. When the power is high, and the laser is focused on the surface, the metal becomes a plasma (4th state of aggregation of matter with different properties from solids, liquids and gases) because the energy's density exceeds the material ablation threshold [130, 131]. Therefore, through control and modulation of light pulses, it is possible to control the absorbed energy and the amount of material removed from the original piece [132]. Laser cutters can also be used to create acrylic molds that can replace common master molds in the manufacturing process of dissolving MAPs. However, the use of acrylic molds requires making two complementary polydimethylsiloxanes (PDMS) molds separated by a silane layer to obtain the female mold, which is going to be used to cast the polymeric solution [133].

1.4.3 Microstereolithography

Stereolithography is based on photopolymerization and presents the ability to produce micrometric shapes [134]. Different thermostable liquid resins with low density, which are locally cured by UV radiation, are employed [135]. The 3D models are virtually sliced into multiple thin layers, using software interpreted by the printer in Cartesian coordinates to focus a UV light beam. This process is done layer by layer. The printer introduces thin layers of resin with the appropriate surface per layer to finally build the object under construction [135, 136]. After the process, a post-curing stage is necessary to increase the hardness of the pieces and boost the physical properties of the resin or even introduce a metallic covering [135].

1.4.4 Etching

The etching includes different sub-techniques, which allow the use of different materials. Wet etching and dry etching are the most commonly used and differ from each other in the etching agent. Wet etching uses a chemical solution [137], whereas ions and electrons generated by a plasma are used for dry etching [138]. Etching processes can be classified in two types, depending on the etch direction. In the case of anisotropic etching, the material is sculpted vertically [139]. In the case of isotropic etching, the etch is produced in all directions [140]. Anisotropic etching is particularly interesting in microneedle manufacturing, as it allows the production of microneedle tips with the desired properties regarding size, length, morphology, etc. Ideal materials to etch with this technique are silicon or metals, such as aluminum or copper [141]. In terms of the etchants, the most widely used are potassium hydroxide (KOH) [142], tetramethylammonium hydroxide (TMAH) [143], and hydrogen fluoride (HF) [144]. Besides, the process requires the use of a mask to protect certain areas of the material from the etchant's effects. These masks are usually made of silicon oxide or nitride [145, 146].

1.4.5 Lithography

Lithography presents three different sub-techniques, which differ in the source of radiation. Photolithography is probably more widespread than electron beam lithography, ion beam lithography or X-ray lithography [147, 148]. Like the etching technique, it also uses a mask and photoresist. Still, in this case, UV light is used as a radiation source, changing the properties of polymers, such as their solubility, and allowing the development of the piece [149, 150].

1.4.6 Solvent Casting or Micromolding

The solvent casting method has been used to prepare polymeric films that are applied in other fields, such as tissue engineering or coating procedures. It has been transferred to manufacture polymeric MAPs [151–161]. This technique is based on the use of polymeric dispersions, which are deposited in a mold to create a polymeric structure with a negative shape of the mold after solvent evaporation. Then, the polymeric structure is peeled off from the mold by the application of a mechanical force or using a chemical solvent, which disintegrates only the mold [162] (Fig. 1.7d). Master molds are made from plastics, metals or even simple polymers [163], through different techniques, being lithography the most common one [163–165]. They are used to create a second (female) mold, which is then used for solvent casting. PDMS is the most widely used polymer to make female molds, due to its low adhesion and flexibility, which facilitates the separation of the MAPs [165]. In addition, it also presents high stability after the drying process [166]. When the polymeric dispersion is loaded in the female mold, before solvent removal, it is common to centrifuge the molds to ensure homogeneity in the dispersion concentration and complete filling of the microneedle tips [56]. Crosslinking can then be carried out to improve the mechanical properties of the MAP, and a baseplate made of the same or a different polymer or mixture can be added to the final device [69, 167].

As mentioned previously, polymeric MAPs present certain interesting advantages, such as biocompatibility and the ability to load relatively high quantities of drugs in their structure [168, 169]. These advantages, linked with the fact that micro molding is a low-cost, quick and probably the simplest manufacturing process from a technological point of view, led to dissolving MAPs becoming the most used and extended type of MAPs in basic research [163, 170]. However, this technique presents the disadvantage of requiring the previous use of other techniques, such as lithography or, more recently, 3D printing [171], for manufacturing of the master molds. Luckily, these molds can be used repeatedly to produce a considerable number of subsequent female molds. Therefore, micromolding is an increasingly selected method in several laboratories to produce MAPs at a low scale and real perspectives to achieve reliable scalability [162, 172].

1.4.7 Glass Drawing

Pulling pipettes is the most commonly used glass drawing method to make hollow glass MAPs with micrometric inner and outer diameters. This technique requires the use of specialized pulling machines which can control different parameters during the process, such as heat, force, movement and cooling time, to achieve a reproducible production of glass structures [173, 174]. Glass tubes, generally borosilicate, with around 1–1.5 mm of outer diameter and 0.5–1 mm of inner diameter, are also needed for the procedure [75, 175]. Tubes are heated and when the temperature is adequate, the glass becomes soft and the pipette puller system tractions from both sides, separating the tube into two pieces with the microneedle tips, which are then inserted in a plate, setting up the arrays [75].

1.4.8 Ceramic Sintering

Ceramic MAPs have been made mainly using ceramic sintering. Other techniques are required to prepare a female mold and, therefore, a previous master mold [92, 176]. Commonly, female molds are made from PDMS, and master molds are produced by other techniques reviewed above, such as lithography. Slurry is the base product used to prepare ceramic MAPs, containing several components, including the essential ceramic powder, plasticizers, binders, and solvents. Dried slurry, which can be prepared by spray-drying [177], is used to fill the PDMS molds creating a non-sintered ceramic item. Then, through the application of high temperatures, the sintering process is completed, and MAPs are ready after a cool-down period [176].

1.4.9 Coating Methods

The various methods used to coat MAPs are summarized in Fig. 1.8.

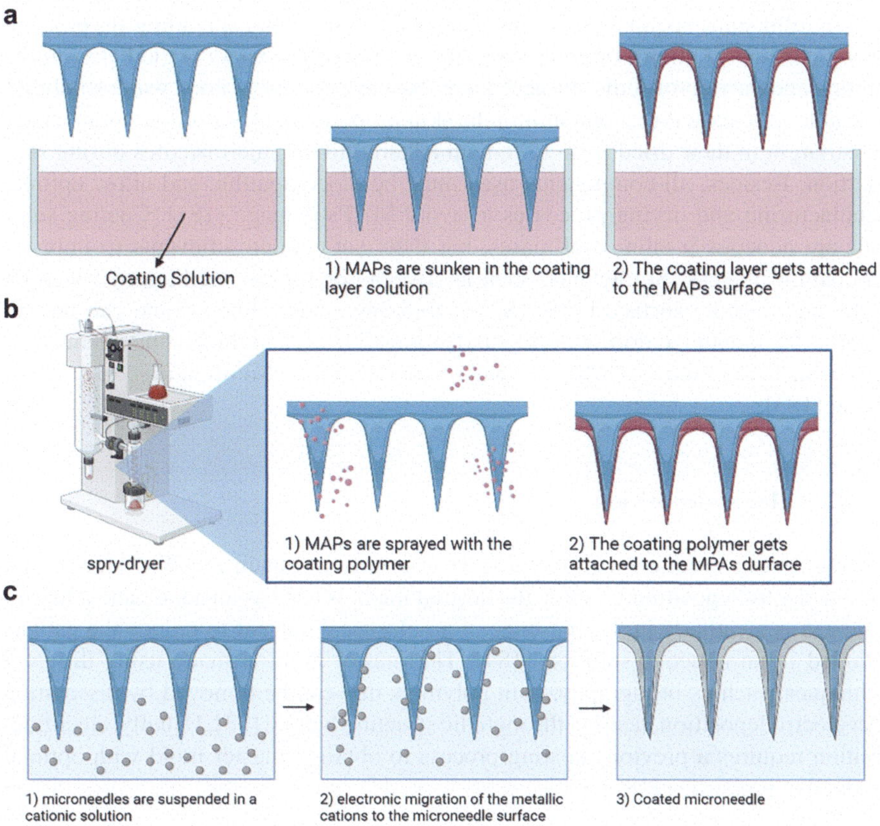

Fig. 1.8 Schematic illustrations of the most common methods for coating MAPs: **a** Dipping, **b** Spraying and **c** electrodeposition. This illustration has been created using BioRender.com under an active license

1.4.9.1 Dipping and Spraying

Different techniques have been described to coat MAPs, with dipping and spraying being the most common ones. These techniques are similar but differ in the application of the coating solution/dispersion. In the case of spraying, MAPs are coated using atomized sprays containing the coating solution [178] (Fig. 1.8a). On the other hand, dipping consists in simply sinking the microneedle tips into a coating solution once or repeatedly. If possible, dipping is done in micro-positioning machines, which introduce each tip into microwells containing the coating solution in a very controlled manner [179]. After coating, the devices are left to dry at room temperature, generally for several hours (Fig. 1.8b). The layer-by-layer technique has also been used to coat MAPs. In this case, the microneedles are introduced alternatively into a solution charged with the molecule of interest and a dispersion of an oppositely charged polymer [180]. As described previously in the literature, coating solutions need to fulfil some requirements, like controlled wetting and spreading (to produce a homogeneous coating), optimal viscosity and reduced contact angle between the solution and the microneedle surface. They also have to present good water solubility to be able to release their content into the skin properly and have sufficient mechanical strength in their dried form to remain adhered to the microneedles during skin insertion. Besides, all components used must be biocompatible and allow optimal manufacturing and drying processes to avoid MAPs damage [181]. Coating solutions are aqueous in almost all cases, but they can contain adjuvants to improve the coating process. Surfactants, such as Tweens or Lutrol, facilitate the wetting of the microneedle surface [178, 182]; stabilizing agents, like sugars, can protect the drug from damage during drying and storage [182]; and thickening agents e.g., carboxymethylcellulose (CMC) or hyaluronan (HA), are also often used to improve drying [183].

1.4.9.2 Electrodeposition

Coating a base structure with metal is another way to obtain metallic MAPs. In a normal electrodeposition process, the desired metal is used as an anode and a master mold act as a cathode in the presence of an electrical current to reduce the cations obtained in an aqueous solution [184]. The master mold is made using different techniques, such as lithography, with polymers that can be removed by dissolution after electrodeposition, leaving the metallic structure behind [63]. Usually, electrodeposition requires a previous coating process to obtain a master mold with optimal conductive properties [185].

1.5 Common Techniques to Characterize Microneedle Devices

The optimal properties and characteristics of MAPs depend on the final aim of these devices and their specific intended use. In November 2020, the US Food and Drugs Administration (FDA) published a brief guidance for clarifying when a product should be considered a microneedling device and therefore subjected to specific requirements [186]. In this report, several aspects such as needle length and arrangement to achieve penetration into living skin layers, needle sharpness and degree of control of manual or automatic microneedling products are reviewed. However, this guidance is merely informative and no compliance is required. For academic and basic research purposes, the properties which must be thoroughly determined and analyzed include: needle length, geometry, mechanical resistance to compression, insertion performance, drug loading, drug release, residual water content, structural composition and stability, ex vivo drug absorption and safety profile.

1.5.1 Needle Length, Geometry and Density Per Array

Optical microscopy and scanning electron microscopy (SEM) are accurate techniques that can be used to determine the needle length and geometry properties of MAPs. These properties are crucial since they can help to overcome skin elasticity and achieve proper insertion, which can be challenging, particularly when developing soft microneedle prototypes, such as dissolving MAPs. For SEM imaging, MAPs must be coated with a metal (commonly gold or palladium) and, depending on the nature of this layer, different voltages are applied to obtain the images, ranging from 600 V to 30 kV [187–189]. It has been observed that geometry can affect the mechanical properties and dissolution profile of MAPs. Li et al. compared different shapes of dextran-based MAPs, including cone, cone-cylinder, rectangular pyramids and hexagonal pyramids, proving that there is a huge difference in dissolution rates among these prototypes under in vivo conditions [190]. Particularly, cone-shaped MAPs showed a twofold higher dissolution rate when compared to cone-cylindrical MAPs, which only achieved 40% dissolution. In fact, cone-cylinder MAPs also showed to be the geometry with the worst insertion properties (38% of the needle height). Sillankorva et al. also demonstrated the impact of geometry on MAP mechanical properties since the force required for needle failure with PVA-based conic MAPs was threefold higher than with cylindrical-shaped MAPs [191].

Microneedle density also plays a crucial role in MAP efficacy for drug delivery, as shown by Yan et al. Solid silicon MAPs with different needle lengths (100–1100 μm) and densities (400–11900 needles/cm^2) were tested. Longer needles (>600 μm) were more effective in creating channels through the skin and enhancing acyclovir flux. MAPs with a lower number of needles (<2000 needles/cm^2) were more effective if the needles were larger than 600 μm. Altogether, these results show how important these physical properties and their characterization are to minimize the usual incomplete insertion of MAPs in the skin [192].

1.5.2 Mechanical Resistance to Compression

As mentioned before, the mechanical properties of MAPs have a huge impact on their performance. However, they do not depend solely on the length, geometry and density of needles per array. The materials used for MAP production have an enormous influence on their mechanical properties. Solid MAPs made of metals or ceramics and hollow MAPs manufactured with metals or glass have good tensile strength but high brittleness, which can lead to easy fracture and, therefore, undesired skin reactions [76, 96, 176]. Mechanical resistance is usually studied using texture analyzers that compress MAPs against a stainless-steel block to force the deformation of the MAPs tips. Needle length reduction can be calculated by measuring this parameter before and after compression [193]. These studies are more interesting when polymeric MAPs are developed since the type of polymers and their concentrations used in manufacturing may lead to different properties. For instance, Permana et al. observed that PVP-based MAPs show high deformation (up to 30% of needle height reduction) [194], and Teodorescu et al. noticed that the combination of polymers with hydroxyl and carbonyl groups could significantly improve the mechanical resistance of MAPs due to the interaction between those groups [195].

1.5.3 Insertion Performance

Insertion performance is usually assessed ex vivo before in vivo studies. For that purpose, a texture analyzer can be used to compress dye-loaded MAPs into a piece of excised skin. Larrañeta et al. reported that for these studies, a compressing force of 20–32 N is the most appropriate since patients tend to apply an equivalent thumb pressure when inserting MAPs by themselves [196]. As a gold standard, human, porcine, or even rodent skin is appropriate for conducting an insertion test. However, to reduce costs and avoid ethical concerns and restricted availability, different surrogate models have been proposed using Parafilm M® layers, wax or polysaccharide-based gels [197]. Gelatin and agarose gels are highly accepted, but the Parafilm M® model is more widely used, as it is very easy to determine insertion percentage and depth by recording the perforation of consecutive Parafilm M® layers (each measuring 127 μm in thickness) [198]. This material also exhibits an excellent correlation with the results obtained using porcine skin [199]. Moreover, optical coherence tomography can be used for imaging MAP insertion in all kinds of models: in vivo [200], ex vivo [201] and in surrogate models.

1.5.4 Drug Loading

Drug loading constitutes one of the main challenges in MAP fabrication. It depends on different parameters, including the size/volume of MAP tips and the physicochemical properties of the drug. Water-soluble drugs are not highly affected by this hurdle since they can be more concentrated in MAPs matrix. However, the concentration of hydrophobic drugs that can be loaded onto MAPs is usually very limited, which becomes problematic when high doses of active ingredients are required [202]. To solve this hurdle, MAPs have been loaded with different types of micro- and nanoparticles containing poorly water-soluble drugs. This also contributes to improving the stability of the drug and providing a controlled release profile [38]. Nevertheless, it makes it more difficult to determine the actual drug concentration in MAP tips. For this, an estimation can be calculated based on the density and concentration of drug formulation and the number, height (length), and base width of the needle tips (Eq. 1.1) [203]:

$$\text{Drug in MAPs tips} = \text{N}\frac{((\text{h}\alpha^2)\rho[\text{drug}])}{3} \tag{1.1}$$

where N is the number of needle tips, h is the height of the tips, α is the base width of the tips, ρ is the density of the dry formulation, and [drug] is the concentration of the drug in the dry formulation.

1.5.5 Drug Release and Swelling Behavior

In vitro drug release studies are regularly used in the optimization process of pharmaceutical dosage forms. The use of different materials to produce MAPs leads to different mechanical and chemical properties that influence the drug release process considerably. Different models have been proposed to study drug release from MAPs, and those which use biological tissues for in vitro assessment present obvious difficulties due to the inherent variability of those tissues. Larrañeta et al. reported a simple method using artificial membranes in a sink-dissolution or Franz-diffusion cell setup [204]. Even though the release of active ingredients from MAPs is mainly governed by diffusion in coated MAPs or dissolution of the matrix in dissolving MAPs, other mechanisms are possible, for example, when polymers are insoluble in biological fluids, or MAPs are coupled with controlled release systems. Different kinetic models can be used to model the release process. However, those which consider more than just passive drug diffusion from MAPs are particularly interesting. For instance, the Korsmeyer-Peppas, Kim or Peppas-Sahlin models, which consider the erosion or relaxation process of the MAPs matrix [205] (Eqs. 1.2–1.4 respectively) are among these examples.

$$\frac{M_t}{M_\infty} = \text{k}t^n \tag{1.2}$$

$$\frac{M_t}{M_\infty} = \text{k}t^n + b \tag{1.3}$$

In these equations, M_t is the drug amount released at time t, M_∞ is the maximum amount of drug released, M_t/M_∞ is the drug fraction released at time t, k is the constant that stands for the release rate, n is the release exponent that explains the mechanism governing the process, and b is the parameter associated to the existence of a burst release. Low values of release exponent ($n < 0.5$) indicate that the release occurs by polymer diffusion/dissolution. On the other hand, high values ($n > 0.85$) indicate that the process is mainly governed by relaxation/swelling, whereas intermediate values ($0.5 < n < 0.85$) indicate the co-existence of both [206–208].

$$\frac{M_t}{M_\infty} = k_1 t^n + k_2 t^{2n} \tag{1.4}$$

In the Peppas-Sahlin model indicated above, M_t is the amount released at time t, M_∞ is the maximum amount of drug released, M_t/M_∞ is the drug fraction released at time t, k_1 and k_2 are the constants associated with diffusion and relaxation, respectively, and n is the release exponent [209].

A especial behaviour after insertion is shown by swelling MAPs, which tend to uptake ISF. The swelling percentage of MAPs can be easily monitored by calculating the mass gain at pre-determined time points after insertion, according to the following formula (1.5):

$$W\ (\%) = \left[\frac{W_s - W_d}{W_d}\right] \times 100 \tag{1.5}$$

where W (%) is the swelling percentage, W_s is the mass of the swollen MAPs after liquid absorption, and W_d is the mass of the MAPs in a dry state [69].

1.5.6 Residual Water Content

Moisture content plays a key role in the properties of MAPs, including rigidity, flexibility, stability, efficacy and dissolution kinetics [210]. In particular, a high content of water will impair MAPs ability to penetrate the skin. Thermogravimetric analysis (TGA) is widely used to monitor mass loss as a function of temperature. In this sense, the moisture content of a sample can be easily calculated by TGA since water mass corresponds to weight loss between room temperature and 100 °C [211]. In general, microneedles can contain approximately 10% of residual water after their production [210]; however, it has also been observed that the storage period, relative humidity and temperature drastically affect the drying process and the residual water content. Kolluru et al. studied the water content of gelatin-based dissolvable MAPs at different storage conditions [212]. The authors observed that, after 1 week at 25 and 40 °C (0% relative humidity), the residual moisture content decreased from 6.25% to 3.75% and 2.5%, respectively. Thus, it seems that MAPs drying is an important process that requires attention since, depending on their composition, the balance between manufacturing aspects and storage conditions necessary to guarantee their quality can change.

1.5.7 Composition and Structural Changes After Thermal Treatment

Enthalpic changes in the constituent materials of MAPs can be measured by Differential Scanning Calorimetry (DSC). This thermal analysis can provide a formulation fingerprint, which is useful for quality control, traceability and scaling-up processes

[213]. DSC analyzes the differences in heat flow between the sample and reference material, required to maintain both at the same temperature since the sample experiences physical transformations triggered by heating, including phase transitions (T_m and T_g for crystalline and amorphous materials, respectively), dehydration, relaxation, or crystallization, among others [214]. For instance, Arshad et al. demonstrated the crystalline drug nature of cetirizine based on a sharp endothermic peak at 220 °C [215]. In addition, the dehydration of alginate was proved by an endothermic peak at 100 °C, while an exothermic peak at 240–60 °C was observed due to a pyrolysis reaction. The melting process of cetirizine was indicated by a broadened endothermic peak at 195 °C that could be coupled with a possible decomposition of the polymer since another thermal event was also observed.

Chitosan was the other polymeric component under study, which showed a broad endothermic peak at 80 °C, related to the release of water, while an exothermic peak at 300 °C was derived from the decomposition of amine groups of the polymer [215]. Qiang et al. studied the properties of PVP and CMC-based MAPs, which showed melting peaks at 70 and 105 °C, respectively [57]. The melting temperature of the composite (PVP-CMC combined) MAPs was significantly higher in comparison to bare PVP but lower than that of CMC-Na. The melting temperature was considerably influenced by the polymer's proportion, with the prototypes containing higher concentrations of CMC showing higher endothermic signals. However, the melting peak distribution of MAPs was wide, meaning that the crystallization of the composite MAPs was not uniform and, proving that the mass percentage of the mixture of the different polymers affects the crystallization process when preparing dissolving MAPs. Another example of investigating the physical stability of dissolving PVP/PVA-based MAPs using DSC was reported by Ramadon et al. [216]. Based on the obtained results, it could be concluded that the drug (ketoprofen) can show improved solubility when formulated in a nanosuspension, but crystal traces of the drug were also detected, as a sharp endothermic peak was still found with a lower intensity when compared with the pure drug at 100 °C. However, when the proportion of components was optimized, the endothermic peak was not seen, meaning that no traces of crystals could be detected. This is likely due to the physical interaction between the components, which may change the solid crystalline form of ketoprofen to an amorphous form.

1.5.8 Ex Vivo Permeability Studies

Diffusion cells are the reference method for estimating transdermal drug permeability [217]. Specifically, the Franz-diffusion cell (FDC) setup is widely used in a huge number of studies to assess transdermal absorption facilitated by MAPs [68, 218]. FDCs are composed of two coupled compartments, with a membrane or tissue placed in between. The receptor chamber is filled with a medium to receive the drug after its passage through the studied barrier. Regarding skin barrier function, although human skin is considered the gold standard, porcine or rodent skin is also accepted as a surrogate model [219]. To mimic in vivo conditions, the receptor compartment is stirred and warmed to the physiological temperature of the skin (32–37 °C) [220] and includes a sampling port.

1.5.9 Safety Profile

The safety of microneedles could be evaluated in various aspects, including pore closure, potential for infection, local reactions, and cell toxicity. Transepidermal water loss (TEWL) measurements were used by Ogunjimi et al. to confirm micropore formation after stainless-steel solid MAPs insertion [221]. Changes in TEWL and impedance were significant in all skin types under study, confirming the microchannel formation regardless of skin type. However, it was observed that microchannels lifetime was significantly longer in Afro-American skin (66.5 h) in comparison to Asiatic (44.1 h), multi-racial (48.0 h), and Caucasic population skin types (50.2 h). Calcein imaging can be used to observe pore closure directly, as Kalluri et al. study demonstrates [43]. Solid MAPs made with metal (770 μm long, base width of 140 μm) successfully created microchannels in the skin (average depth of 152.5 ± 9.6 μm and diameter of 70.7 ± 9.9 μm), which were closed after 18 h. However, TEWL measurements (as an indirect indicator of the recovery of skin barrier function) pointed out that skin regains its barrier function approximately 4–5 h after MAP application.

An important concern linked to the microchannel's lifetime is the risk of infections. Nevertheless, this has been extensively studied. Donnelly et al. concluded that silicon solid MAPs puncture resulted in significantly less microbial penetration than 21G hypodermic needle puncture. In this case, the number of microorganisms (*Candida albicans*, *Pseudomonas aeruginosa* and *Staphylococcus epidermidis*) detected in tape-stripped *stratum corneum*, viable tissue and FDC receptor solution after 24 h was an order of magnitude lower than what was found with hypodermic needle punctures [222]. Although irritation produced by MAPs use has been studied by skin redness and local skin blood flow, local reactions and inflammation triggered by MAPs use are better monitored by different biomarkers, including TNF-α, IL-6 or IgG. However, no significant increases were detected in different studies to consider that MAPs induce a humoral response or intense local and irritation reactions [223]. Finally, if the biocompatibility of any MAP components needs to be tested, the most accurate cell lines to assess this parameter by MTT or ATP/luciferase assays are usually fibroblasts (NHF) and keratinocytes (HaCat), as these are characteristic cells of the skin tissue [224, 225].

1.6 Translational Potential

Since the early 70s, when the first microneedle patent was registered and the technology made its entrance into the drug delivery field, its evolution has been exponential. In the year 2000, when dissolving MAPs were developed, they brought to light new and relevant microneedle applications in the fields of cosmetics, vaccination, diagnostics, and gene delivery [109, 226]. Since then, this technology has gathered a big interest in the scientific community, as the increasing number of publications in scientific databases demonstrates (Fig. 1.9).

Autoregressive statistical models also confirm this permanent growth [227]. A similar trend can be seen in the number of patents registered recently, which also evidences the clinical and industrial interest in this technology [228–230]. To date, 3925 patents have been registered in Espacenet (European Patent Office) under the term "microneedle", and 143 clinical trials (96 completed and 47 ongoing, mainly in North America and Europe) are currently included in the ClinicalTrials.gov database. However, MAPs have still to achieve clinical translation.

Some of the barriers that academic and industrial researchers have encountered in the "bench-to-bedside" development of MAP-based approaches are related to the inherent characteristics of these devices. Concerns over the time of micropore closure with solid MAPs, as well as the minor risk of infection associated with the application of various MAP types, have been discussed by various researchers [102]. Overall, it is accepted that the use of microneedles is a relatively safe procedure despite the occurrence of some minor adverse reactions, particularly at the site of application. However, the long-term effects of repeated application must be thoroughly assessed, particularly for dissolving MAPs, which leave their components behind in the skin

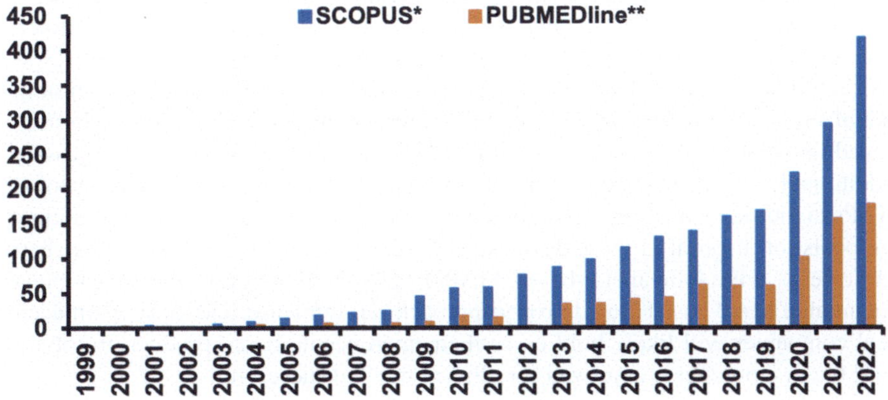

Fig. 1.9 Evolution of scientific publications containing the "microneedle" word in PUBMEDline and SCOPUS databases from 1999 to 2022. *Scopus search Boolean operators: ((TITLE(microneedle*) AND KEY(microneedle*)) AND PUBYEAR > 1999 AND PUBYEAR < 2022 AND (LIMIT-TO (DOCTYPE, "ar"))). **PUBMEDline search command: microneedle*[Title/Abstract]

[231]. Besides this, there is a difficult balance to achieve between the dosing needs for certain active ingredients and the patch size that is acceptable and manageable for patients. In these cases, a dosing schedule that aims at weekly MAP applications for long-acting formulations, for example, may be an appropriate compromise [232]. Finally, the sterility of MAP-based products is a key aspect of discussion in this field, as it is a common regulatory requirement for products that pierce the most external layer of the skin. In the case of MAPs, choosing an appropriate terminal sterilization method which does not cause any damage to the MAP itself and/or the drug cargo is the most difficult yet most important step. Aseptic manufacturing, which is the alternative to terminal sterilization, is likely to increase production costs to a level that may not be compatible with the commercial potential of the product [233].

As mentioned in relation to patch sizes, patient acceptability is key in the journey to translate a MAP-based drug delivery approach into clinical practice. In this sense, researchers must endeavour to include patient opinions and views in the most initial steps of their design process, ensuring that aspects such as patch size, easiness of application and confirmation of successful MAP insertion through some form of feedback mechanism are covered in their product development pathway [234, 235]. Similarly, marketing choices for these products will be crucial in achieving patient acceptability and true translation into the clinic [231].

When aiming for industrial development and clinical application of MAP-based approaches, the scalability of the manufacturing processes is a central point. Collaboration between academic, industrial, and regulatory entities in this process can largely contribute to achieving this, particularly in terms of the cost-efficiency of MAP manufacturing and turnaround time [231]. However, the lack of official regulatory standards is still a hurdle for MAP developers to overcome. In 2020, a guideline document was published by the FDA for Industry and Food and Drug Administration staff guidance ("Regulatory Considerations for Microneedling Products") [186]. Nevertheless, it is still uncertain if regulatory agencies will accept MAPs as drug delivery systems or medical devices. Additionally, the standardization of microneedle products in terms of defined standards and quality control tests is fundamental to prompt manufacturers to invest in large-scale MAP production [102]. Experts have hypothesized on what these tests may look like [231], but further work is still necessary. Additionally, it will also be crucial to achieving consistently reproducible insertion of MAPs in the skin, ensuring standardized delivery efficiency, as well as well-defined protocols for disposal of these devices, particularly when they still constitute sharp waste (e.g., with solid metal-based MAPs) [233, 235]. Overall, the translational potential of MAP-based drug delivery approaches is immense, and most pharmaceutical companies will likely hold on to their microneedle-based product prototypes until the first product is approved for clinical use and successful.

1.7 Conclusions

MAPs technology is a growing field of research in both academic and industrial environments, and it has well-established benefits for patients and healthcare professionals. There is still significant room for innovation in the use of MAPs in drug delivery, diagnostics and biomedicine. The use of stimuli-responsive materials for producing smart MAPs [236] or the association between MAPs and nanoparticles pave the way for new innovations and therapeutic strategies [38]. In the future, a deeper assessment of the long-term safety of these devices, alongside the standardization of their manufacturing and characterization through the development of regulatory guidelines, will be a significant step towards the introduction of the first microneedle-based product in clinical practice.

References

1. Anselmo, A.C., Mitragotri, S.: An overview of clinical and commercial impact of drug delivery systems. J. Control. Release **190**, 15–28 (2014). https://doi.org/10.1016/j.jconrel.2014.03.053
2. Darji, M.A., Lalge, R.M., Marathe, S.P., Mulay, T.D., Fatima, T., Alshammari, A., Lee, H.K., Repka, M.A., Narasimha Murthy, S.: Excipient stability in oral solid dosage forms: a review. AAPS PharmSciTech **19**, 12–26 (2018). https://doi.org/10.1208/s12249-017-0864-4
3. Davis, S.S., Hardy, J.G., Fara, J.W.: Transit of pharmaceutical dosage forms through the small intestine. Gut **27**, 886–892 (1986). https://doi.org/10.1136/gut.27.8.886
4. Maderuelo, C., Lanao, J.M., Zarzuelo, A.: Enteric coating of oral solid dosage forms as a tool to improve drug bioavailability. Eur. J. Pharm. Sci. **138**, 105019 (2019). https://doi.org/10.1016/j.ejps.2019.105019
5. Parsons, R.L.: Drug absorption in gastrointestinal disease with particular reference to malabsorption syndromes. Clin. Pharmacokinet. **2**, 45–60 (1977). https://doi.org/10.2165/00003088-197702010-00004
6. Pond, S.M., Tozer, T.N.: First-pass elimination basic concepts and clinical consequences. Clin. Pharmacokinet. **9**, 1–25 (1984)
7. Varma, M.V.S., Obach, R.S., Rotter, C., Miller, H.R., Chang, G., Steyn, S.J., El-Kattan, A., Troutman, M.D.: Physicochemical space for optimum oral bioavailability: contribution of human intestinal absorption and first-pass elimination. J. Med. Chem. **53**, 1098–1108 (2010). https://doi.org/10.1021/jm901371v
8. AMERICAN ACADEMY OF PEDIATRICS Alternative Routes of Drug Administration-Advantages and Disadvantages (Subject Review) (1997). http://publications.aap.org/pediatrics/article-pdf/100/1/143/889898/pe0797001430.pdf
9. Mignani, S., El Kazzouli, S., Bousmina, M., Majoral, J.P.: Expand classical drug administration ways by emerging routes using dendrimer drug delivery systems: a concise overview. Adv. Drug Deliv. Rev. **65**, 1316–1330 (2013). https://doi.org/10.1016/j.addr.2013.01.001
10. Alexander, N.J., Baker, E., Kaptein, M., Karck, U., Miller, L., Zampaglione, E.: Why consider vaginal drug administration? Fertil. Steril. **82**, 1–12 (2004). https://doi.org/10.1016/j.fertnstert.2004.01.025
11. Bialik, M., Kuras, M., Sobczak, M., Oledzka, E.: Achievements in thermosensitive gelling systems for rectal administration. Int. J. Mol. Sci. **22** (2021). https://doi.org/10.3390/ijms22115500
12. Frati, F., La Grutta, S., Bernardini, R., Zampogna, S., Scurati, S., Puccinelli, P., Riario-Sforza, G.G., Incorvaia, C.: Sublingual immunotherapy: administration, dosages, use. Int. J. Immunopathol. Pharmacol. **22**, 13–16 (2009)

13. Hua, S.: Physiological and pharmaceutical considerations for rectal drug formulations. Front. Pharmacol. **10** (2019). https://doi.org/10.3389/fphar.2019.01196
14. Arroyo, C.M., Quinteros, D., Cózar-Bernal, M.J., Palma, S.D., Rabasco, A.M., González-Rodríguez, M.L.: Ophthalmic administration of a 10-fold-lower dose of conventional nano-liposome formulations caused levels of intraocular pressure similar to those induced by marketed eye drops. Eur. J. Pharm. Sci. **111**, 186–194 (2018). https://doi.org/10.1016/j.ejps.2017.09.024
15. Koulich, E., Roland, P.S., Pawlowski, K.S.: Comparison of systemic and otic administration of ofloxacin. Laryngoscope **120**, 2083–2088 (2010). https://doi.org/10.1002/lary.21088
16. Sun, J., Carr, M.J., Khushalani, N.I.: Principles of targeted therapy for melanoma. Surg. Clin. North Am. **100**, 175–188 (2020). https://doi.org/10.1016/j.suc.2019.09.013
17. Paudel, K.S., Milewski, M., Swadley, C.L., Brogden, N.K., Ghosh, P., Stinchcomb, A.L.: Challenges and opportunities in dermal/transdermal delivery. Ther. Deliv. **1**, 109–131 (2010). https://doi.org/10.4155/tde.10.16
18. Matejuk, A.: Skin immunity. Arch. Immunol. Ther. Exp. (Warsz.) **66**, 45–54 (2018). https://doi.org/10.1007/s00005-017-0477-3
19. Hern, S., Mortimer, P.S.: Visualization of dermal blood vessels–capillaroscopy. Clin. Exp. Dermatol. **24**, 473–478 (1999). https://doi.org/10.1046/j.1365-2230.1999.00537.x
20. Wilknson, J.B., Moore, R.J., Ship, A.G.: Harry's cosmeticology. Seventh Edition. Plast Reconstr Surg **73**, 320 (1984). https://doi.org/10.1097/00006534-198402000-00035
21. Beck, R., Guterres, S., Pohlmann, A.: Nanocosmetics and Nanomedicines: new Approaches for Skin Care. Springer (2011)
22. Fox, L.T., Gerber, M., Du Plessis, J., Hamman, J.H.: Transdermal drug delivery enhancement by compounds of natural origin. Molecules **16**, 10507–10540 (2011). https://doi.org/10.3390/molecules161210507
23. Singh, S., Singh, J.: Transdermal drug delivery by passive diffusion and iontophoresis: a review. Med. Res. Rev. **13**, 569–621 (1993). https://doi.org/10.1002/med.2610130504
24. Bos, J.D., Meinardi, M.M.: The 500 Dalton rule for the skin penetration of chemical compounds and drugs. Exp. Dermatol. **9**, 165–169 (2000). https://doi.org/10.1034/j.1600-0625.2000.009003165.x
25. Mitragotri, S., Anissimov, Y.G., Bunge, A.L., Frasch, H.F., Guy, R.H., Hadgraft, J., Kasting, G.B., Lane, M.E., Roberts, M.S.: Mathematical models of skin permeability: an overview. Int. J. Pharm. **418**, 115–129 (2011). https://doi.org/10.1016/j.ijpharm.2011.02.023
26. Asbill, C.S., El-Kattan, A.F., Michniak, B.: Enhancement of transdermal drug delivery: chemical and physical approaches. Crit. Rev. Ther. Drug Carrier Syst. **17**, 621–658 (2000)
27. Arora, A., Prausnitz, M.R., Mitragotri, S.: Micro-scale devices for transdermal drug delivery. Int. J. Pharm. **364**, 227–236 (2008). https://doi.org/10.1016/j.ijpharm.2008.08.032
28. Rao, R., Nanda, S.: Sonophoresis: recent advancements and future trends. J. Pharm. Pharmacol. **61**, 689–705 (2009). https://doi.org/10.1211/jpp.61.06.0001
29. Rich, K.T., Hoerig, C.L., Rao, M.B., Mast, T.D.: Relations between acoustic cavitation and skin resistance during intermediate- and high-frequency sonophoresis. J. Control. Release **194**, 266–277 (2014). https://doi.org/10.1016/j.jconrel.2014.08.007
30. Dhote, V., Bhatnagar, P., Mishra, P.K., Mahajan, S.C., Mishra, D.K.: Iontophoresis: a potential emergence of a transdermal drug delivery system. Sci. Pharm. **80**, 1–28 (2012). https://doi.org/10.3797/scipharm.1108-20
31. Nagar, R., Sengar, S.S.: A simple user-made iontophoresis device for palmoplantar hyperhidrosis. J. Cutan. Aesthet. Surg. **9**, 32–33 (2016). https://doi.org/10.4103/0974-2077.178542
32. Zhu, Q., Shen, Y., Zhang, A., Xu, L.X.: Numerical study of the influence of water evaporation on radiofrequency ablation. Biomed. Eng. Online **12**, 127 (2013). https://doi.org/10.1186/1475-925X-12-127
33. Lin, C.-H., Aljuffali, I.A., Fang, J.-Y.: Lasers as an approach for promoting drug delivery via skin. Expert Opin. Drug Deliv. **11**, 599–614 (2014). https://doi.org/10.1517/17425247.2014.885501

34. Miyazaki, H., Atobe, S., Suzuki, T., Iga, H., Terai, K.: Development of pyro-drive jet injector with controllable jet pressure. J. Pharm. Sci. **108**, 2415–2420 (2019). https://doi.org/10.1016/j.xphs.2019.02.021
35. Kim, D.H., Kim, T.H., Lee, S.H., Lee, A.Y.: Treatment of palmar hyperhidrosis with tap water iontophoresis: a randomized, sham-controlled, single-blind, and parallel-designed clinical trial. Ann. Dermatol. **29**, 728–734 (2017). https://doi.org/10.5021/ad.2017.29.6.728
36. Rzhevskiy, A.S., Singh, T.R.R., Donnelly, R.F., Anissimov, Y.G.: Microneedles as the technique of drug delivery enhancement in diverse organs and tissues. J. Control. Release **270**, 184–202 (2018). https://doi.org/10.1016/j.jconrel.2017.11.048
37. Gill, H.S., Denson, D.D., Burris, B.A., Prausnitz, M.R.: Effect of microneedle design on pain in human volunteers. Clin. J. Pain **24**, 585–594 (2008). https://doi.org/10.1097/AJP.0b013e31816778f9
38. Guillot, A.J., Martínez-navarrete, M., Melero, A., Zinchuk-mironova, V.: Microneedle-assisted transdermal delivery of nanoparticles: recent insights and prospects, 1–30 (2023). https://doi.org/10.1002/wnan.1884
39. Escobar-Chávez, J.J., Bonilla-Martínez, D., Villegas-González, M.A., Molina-Trinidad, E., Casas-Alancaster, N., Revilla-Vázquez, A.L.: Microneedles: a valuable physical enhancer to increase transdermal drug delivery. J. Clin. Pharmacol. **51**, 964–977 (2011). https://doi.org/10.1177/0091270010378859
40. Aronson, R.: The role of comfort and discomfort in insulin therapy. Diabetes Technol. Ther. **14**, 741–747 (2012). https://doi.org/10.1089/dia.2012.0038
41. Kruger, D.F., LaRue, S., Estepa, P.: Recognition of and steps to mitigate anxiety and fear of pain in injectable diabetes treatment. Diabetes Metab Syndr Obes **8**, 49–56 (2015). https://doi.org/10.2147/DMSO.S71923
42. Wei-Ze, L., Mei-Rong, H., Jian-Ping, Z., Yong-Qiang, Z., Bao-Hua, H., Ting, L., Yong, Z.: Super-short solid silicon microneedles for transdermal drug delivery applications. Int. J. Pharm. **389**, 122–129 (2010). https://doi.org/10.1016/j.ijpharm.2010.01.024
43. Kalluri, H., Kolli, C.S., Banga, A.K.: Characterization of microchannels created by metal microneedles: formation and closure. AAPS J. **13**, 473–481 (2011). https://doi.org/10.1208/s12248-011-9288-3
44. Gupta, J., Gill, H.S., Andrews, S.N., Prausnitz, M.R.: Kinetics of skin resealing after insertion of microneedles in human subjects. J. Control. Release **154**, 148–155 (2011). https://doi.org/10.1016/j.jconrel.2011.05.021
45. Donnelly, R.F., Singh, T.R.R., Alkilani, A.Z., McCrudden, M.T.C., O'Neill, S., O'Mahony, C., Armstrong, K., McLoone, N., Kole, P., Woolfson, A.D.: Hydrogel-forming microneedle arrays exhibit antimicrobial properties: potential for enhanced patient safety. Int. J. Pharm. **451**, 76–91 (2013). https://doi.org/10.1016/j.ijpharm.2013.04.045
46. Kumar, A., Bharti, R., Agarwal, S.: Microneedling with Dermaroller 192 needles along with 5-fluorouracil solution in the treatment of stable vitiligo. J. Am. Acad. Dermatol. **81**, e67–e69 (2019). https://doi.org/10.1016/j.jaad.2018.12.066
47. Badran, M.M., Kuntsche, J., Fahr, A.: Skin penetration enhancement by a microneedle device (Dermaroller®) in vitro: dependency on needle size and applied formulation. Euro. J. Pharm. Sci. **36**, 511–523 (2009). https://doi.org/10.1016/j.ejps.2008.12.008
48. Permana, A.D., Elim, D., Ananda, P.W.R., Zaman, H.S., Muslimin, W., Tunggeng, M.G.R.: Enhanced and sustained transdermal delivery of primaquine from polymeric thermoresponsive hydrogels in combination with Dermarollers®, Colloids Surf. B Biointerfaces **219** (2022). https://doi.org/10.1016/j.colsurfb.2022.112805
49. Ahad, A., Al-Saleh, A.A., Al-Mohizea, A.M., Al-Jenoobi, F.I., Raish, M., Yassin, A.E.B., Alam, M.A.: Pharmacodynamic study of eprosartan mesylate-loaded transfersomes Carbopol® gel under Dermaroller® on rats with methyl prednisolone acetate-induced hypertension. Biomed. Pharmacother. **89**, 177–184 (2017). https://doi.org/10.1016/j.biopha.2017.01.164
50. Chen, Y., Chen, B.Z., Wang, Q.L., Jin, X., Guo, X.D.: Fabrication of coated polymer microneedles for transdermal drug delivery. J. Control. Release **265**, 14–21 (2017). https://doi.org/10.1016/j.jconrel.2017.03.383

51. Li, S., Li, W., Prausnitz, M.: Individually coated microneedles for co-delivery of multiple compounds with different properties. Drug Deliv. Transl. Res. **8**, 1043–1052 (2018). https://doi.org/10.1007/s13346-018-0549-x
52. Koutsonanos, D.G., del Pilar Martin, M., Zarnitsyn, V.G., Sullivan, S.P., Compans, R.W., Prausnitz, M.R., Skountzou, I.: Transdermal influenza immunization with vaccine-coated microneedle arrays. PLoS One **4**, e4773 (2009). https://doi.org/10.1371/journal.pone.0004773
53. Lee, W.-J., Han, M.-R., Kim, J.-S., Park, J.-H.: A tearable dissolving microneedle system for shortening application time. Expert Opin. Drug Deliv. **16**, 199–206 (2019). https://doi.org/10.1080/17425247.2019.1583645
54. Rodgers, A.M., McCrudden, M.T.C., Vincente-Perez, E.M., Dubois, A.V., Ingram, R.J., Larrañeta, E., Kissenpfennig, A., Donnelly, R.F.: Design and characterisation of a dissolving microneedle patch for intradermal vaccination with heat-inactivated bacteria: a proof of concept study. Int. J. Pharm. **549**, 87–95 (2018). https://doi.org/10.1016/j.ijpharm.2018.07.049
55. Ito, Y., Yoshimitsu, J.-I., Shiroyama, K., Sugioka, N., Takada, K.: Self-dissolving microneedles for the percutaneous absorption of EPO in mice. J. Drug Target. **14**, 255–261 (2006). https://doi.org/10.1080/10611860600785080
56. Hong, X., Wei, L., Wu, F., Wu, Z., Chen, L., Liu, Z., Yuan, W.: Dissolving and biodegradable microneedle technologies for transdermal sustained delivery of drug and vaccine. Drug Des Devel Ther **7**, 945–952 (2013). https://doi.org/10.2147/DDDT.S44401
57. Qiang, N., Liu, Z., Lu, M., Yang, Y., Liao, F., Feng, Y., Liu, G., Qiu, S.: Preparation and properties of polyvinylpyrrolidone/sodium carboxymethyl cellulose soluble microneedles. Materials **16** (2023). https://doi.org/10.3390/ma16093417
58. Wang, C., Jiang, X., Zeng, Y., Terry, R.N., Li, W.: Rapidly separable microneedle patches for controlled release of therapeutics for long-acting therapies. Med Drug Discov **13**, 100118 (2022). https://doi.org/10.1016/j.medidd.2021.100118
59. Zhu, D.D., Zhang, X.P., Shen, C.B., Cui, Y., Guo, X.D.: The maximum possible amount of drug in rapidly separating microneedles. Drug Deliv Transl. Res. **9**, 1133–1142 (2019). https://doi.org/10.1007/s13346-019-00658-7
60. Chen, M.-C., Huang, S.-F., Lai, K.-Y., Ling, M.-H.: Fully embeddable chitosan microneedles as a sustained release depot for intradermal vaccination. Biomaterials **34**, 3077–3086 (2013). https://doi.org/10.1016/j.biomaterials.2012.12.041
61. Demartis, S., Rassu, G., Anjani, Q.K., Volpe-Zanutto, F., Hutton, A.R.J., Sabri, A.B., McCarthy, H.O., Giunchedi, P., Donnelly, R.F., Gavini, E.: Improved pharmacokinetic and lymphatic uptake of Rose Bengal after transfersome intradermal deposition using hollow microneedles. J. Controll. Release **369**, 363–375 (2024). https://doi.org/10.1016/j.jconrel.2024.03.048
62. Martanto, W., Moore, J.S., Kashlan, O., Kamath, R., Wang, P.M., O'Neal, J.M., Prausnitz, M.R.: Microinfusion using hollow microneedles. Pharm. Res. **23**, 104–113 (2006). https://doi.org/10.1007/s11095-005-8498-8
63. Norman, J.J., Choi, S.-O., Tong, N.T., Aiyar, A.R., Patel, S.R., Prausnitz, M.R., Allen, M.G.: Hollow microneedles for intradermal injection fabricated by sacrificial micromolding and selective electrodeposition. Biomed. Microdevices **15**, 203–210 (2013). https://doi.org/10.1007/s10544-012-9717-9
64. He, X., Sun, J., Zhuang, J., Xu, H., Liu, Y., Wu, D.: Microneedle system for transdermal drug and vaccine delivery: devices, safety, and prospects. Dose Response **17**, 1559325819878585 (2019). https://doi.org/10.1177/1559325819878585
65. Seeni, R.Z., Yu, X., Chang, H., Chen, P., Liu, L., Xu, C.: Iron oxide nanoparticle-powered micro-optical coherence tomography for in situ imaging the penetration and swelling of polymeric microneedles in the skin. ACS Appl. Mater. Interfaces **9**, 20340–20347 (2017). https://doi.org/10.1021/acsami.7b00481
66. Donnelly, R.F., Mooney, K., McCrudden, M.T.C., Vicente-Pérez, E.M., Belaid, L., González-Vázquez, P., McElnay, J.C., Woolfson, A.D.: Hydrogel-forming microneedles increase in volume during swelling in skin, but skin barrier function recovery is unaffected. J. Pharm. Sci. **103**, 1478–1486 (2014). https://doi.org/10.1002/jps.23921

67. Yadav, P.R., Dobson, L.J., Pattanayek, S.K., Das, D.B.: Swellable microneedles based transdermal drug delivery: mathematical model development and numerical experiments. Chem. Eng. Sci. **247**, 117005 (2022). https://doi.org/10.1016/j.ces.2021.117005
68. Donnelly, R.F., Singh, T.R.R., Garland, M.J., Migalska, K., Majithiya, R., McCrudden, C.M., Kole, P.L., Mahmood, T.M.T., McCarthy, H.O., Woolfson, A.D.: Hydrogel-forming microneedle arrays for enhanced transdermal drug delivery. Adv. Funct. Mater. **22**, 4879–4890 (2012). https://doi.org/10.1002/adfm.201200864
69. Donnelly, R.F., McCrudden, M.T.C., Zaid Alkilani, A., Larrañeta, E., McAlister, E., Courtenay, A.J., Kearney, M.-C., Singh, T.R.R., McCarthy, H.O., Kett, V.L., Caffarel-Salvador, E., Al-Zahrani, S., Woolfson, A.D.: Hydrogel-forming microneedles prepared from "super swelling" polymers combined with lyophilised wafers for transdermal drug delivery. PLoS One **9**, e111547 (2014). https://doi.org/10.1371/journal.pone.0111547
70. Dharadhar, S., Majumdar, A., Dhoble, S., Patravale, V.: Microneedles for transdermal drug delivery: a systematic review. Drug Dev. Ind. Pharm. **45**, 188–201 (2019). https://doi.org/10.1080/03639045.2018.1539497
71. Mistilis, M.J., Joyce, J.C., Esser, E.S., Skountzou, I., Compans, R.W., Bommarius, A.S., Prausnitz, M.R.: Long-term stability of influenza vaccine in a dissolving microneedle patch. Drug Deliv Transl. Res. **7**, 195–205 (2017). https://doi.org/10.1007/s13346-016-0282-2
72. Wu, Y., Qiu, Y., Zhang, S., Qin, G., Gao, Y.: Microneedle-based drug delivery: Studies on delivery parameters and biocompatibility. Biomed. Microdevices **10**, 601–610 (2008). https://doi.org/10.1007/s10544-008-9171-x
73. Rajabi, M., Roxhed, N., Shafagh, R.Z., Haraldson, T., Fischer, A.C., van der Wijngaart, W., Stemme, G., Niklaus, F.: Flexible and stretchable microneedle patches with integrated rigid stainless steel microneedles for transdermal biointerfacing. PLoS One **11**, e0166330 (2016). https://doi.org/10.1371/journal.pone.0166330
74. Verbaan, F.J., Bal, S.M., van den Berg, D.J., Groenink, W.H.H., Verpoorten, H., Lüttge, R., Bouwstra, J.A.: Assembled microneedle arrays enhance the transport of compounds varying over a large range of molecular weight across human dermatomed skin. J. Control. Release **117**, 238–245 (2007). https://doi.org/10.1016/j.jconrel.2006.11.009
75. Jiang, J., Gill, H.S., Ghate, D., McCarey, B.E., Patel, S.R., Edelhauser, H.F., Prausnitz, M.R.: Coated microneedles for drug delivery to the eye. Invest. Ophthalmol. Vis. Sci. **48**, 4038–4043 (2007). https://doi.org/10.1167/iovs.07-0066
76. Li, J., Liu, B., Zhou, Y., Chen, Z., Jiang, L., Yuan, W., Liang, L.: Fabrication of a Ti porous microneedle array by metal injection molding for transdermal drug delivery. PLoS One **12**, e0172043 (2017). https://doi.org/10.1371/journal.pone.0172043
77. Parker, E.R., Rao, M.P., Turner, K.L., Meinhart, C.D., MacDonald, N.C.: Bulk micromachined titanium microneedles. J. Microelectromech. Syst. **16**, 289–295 (2007)
78. Senel, M., Dervisevic, M., Voelcker, N.H.: Gold microneedles fabricated by casting of gold ink used for urea sensing. Mater. Lett. **243**, 50–53 (2019). https://doi.org/10.1016/j.matlet.2019.02.014
79. Davis, S.P., Martanto, W., Allen, M.G., Prausnitz, M.R.: Hollow metal microneedles for insulin delivery to diabetic rats. IEEE Trans. Biomed. Eng. **52**, 909–915 (2005). https://doi.org/10.1109/TBME.2005.845240
80. El-Domyati, M., Abdel-Wahab, H., Hossam, A.: Combining microneedling with other minimally invasive procedures for facial rejuvenation: a split-face comparative study. Int. J. Dermatol. **57**, 1324–1334 (2018). https://doi.org/10.1111/ijd.14172
81. Vijaya Lakshmi, Y., Swetha Reddy, L., Naga Neelima Devi, K., Phani Kumar, K., Guru Karthik, G., Srinivas Chakravarthy, P., Nageswar Rao, K.: Evaluation of microneedling therapy in management of facial scars. J. Craniofac. Surg. **31**, e214–e217 (2020). https://doi.org/10.1097/SCS.0000000000006145
82. Elfar, N.N., Hasby, E.A.: Efficacy and safety of plasma gel as a new modality in treatment of atrophic acne scars. Int. J. Dermatol. **59**, 620–626 (2020). https://doi.org/10.1111/ijd.14815
83. Ullah, A., Kim, C.M., Kim, G.M.: Porous polymer coatings on metal microneedles for enhanced drug delivery. R. Soc. Open Sci. **5** (2018). https://doi.org/10.1098/rsos.171609

84. Pradeep Narayanan, S., Raghavan, S.: Fabrication and characterization of gold-coated solid silicon microneedles with improved biocompatibility. Int. J. Adv. Manuf. Technol. **104**, 3327–3333 (2019). https://doi.org/10.1007/s00170-018-2596-3
85. Amouzadeh Tabrizi, M.: A facile method for the fabrication of the microneedle electrode and its application in the enzymatic determination of glutamate. Biosensors (Basel) **13** (2023). https://doi.org/10.3390/bios13080828
86. Wilke, N., Mulcahy, A., Ye, S.-R., Morrissey, A.: Process optimization and characterization of silicon microneedles fabricated by wet etch technology. Microelectron. J. **36**, 650–656 (2005). https://doi.org/10.1016/j.mejo.2005.04.044
87. Das, A., Singha, C., Bhattacharyya, A.: Development of silicon microneedle arrays with spontaneously generated micro-cavity ring for transdermal drug delivery. Microelectron Eng **210**, 14–18 (2019). https://doi.org/10.1016/j.mee.2019.03.019
88. Li, Y., Zhang, H., Yang, R., Laffitte, Y., Schmill, U., Hu, W., Kaddoura, M., Blondeel, E.J.M., Cui, B.: Fabrication of sharp silicon hollow microneedles by deep-reactive ion etching towards minimally invasive diagnostics. Microsyst. Nanoeng. **5** (2019). https://doi.org/10.1038/s41378-019-0077-y
89. Bolton, C.J.W., Howells, O., Blayney, G.J., Eng, P.F., Birchall, J.C., Gualeni, B., Roberts, K., Ashraf, H., Guy, O.J.: Hollow silicon microneedle fabrication using advanced plasma etch technologies for applications in transdermal drug delivery. Lab Chip **20**, 2788–2795 (2020). https://doi.org/10.1039/d0lc00567c
90. Kim, H., Theogarajan, L.S., Pennathur, S.: A repeatable and scalable fabrication method for sharp, hollow silicon microneedles. J. Micromech. Microeng. **28**, 35007 (2018). https://doi.org/10.1088/1361-6439/aaa6a8
91. Indermun, S., Luttge, R., Choonara, Y.E., Kumar, P., du Toit, L.C., Modi, G., Pillay, V.: Current advances in the fabrication of microneedles for transdermal delivery. J. Control. Release **185**, 130–138 (2014). https://doi.org/10.1016/j.jconrel.2014.04.052
92. Verhoeven, S. Bystrova, L. Winnubst, H. Qureshi, T.D. de Gruijl, R.J. Scheper, R. Luttge, Applying ceramic nanoporous microneedle arrays as a transport interface in egg plants and an ex-vivo human skin model. Microelectron. Eng. **98**, 659–662 (2012). https://doi.org/10.1016/j.mee.2012.07.022
93. van der Maaden, K., Luttge, R., Vos, P.J., Bouwstra, J., Kersten, G., Ploemen, I.: Microneedle-based drug and vaccine delivery via nanoporous microneedle arrays. Drug Deliv Transl. Res. **5**, 397–406 (2015). https://doi.org/10.1007/s13346-015-0238-y
94. Sommer, F., Landfried, R., Kern, F., Gadow, R.: Mechanical properties of zirconia toughened alumina with 10–24vol.% 1.5mol% Y-TZP reinforcement. J. Eur. Ceram. Soc. **32**, 3905–3910 (2012). https://doi.org/10.1016/j.jeurceramsoc.2012.02.057
95. De Aza, A.H., Chevalier, J., Fantozzi, G., Schehl, M., Torrecillas, R.: Crack growth resistance of alumina, zirconia and zirconia toughened alumina ceramics for joint prostheses. Biomaterials **23**, 937–945 (2002). https://doi.org/10.1016/S0142-9612(01)00206-X
96. Ita, K.: Ceramic microneedles and hollow microneedles for transdermal drug delivery: two decades of research. J. Drug Deliv. Sci. Technol. **44**, 314–322 (2018). https://doi.org/10.1016/j.jddst.2018.01.004
97. Gholami, S., Mohebi, M.M., Hajizadeh-Saffar, E., Ghanian, M.H., Zarkesh, I., Baharvand, H.: Fabrication of microporous inorganic microneedles by centrifugal casting method for transdermal extraction and delivery. Int. J. Pharm. **558**, 299–310 (2019). https://doi.org/10.1016/j.ijpharm.2018.12.089
98. Schepens, B., Vos, P.J., Saelens, X., van der Maaden, K.: Vaccination with influenza hemagglutinin-loaded ceramic nanoporous microneedle arrays induces protective immune responses. Eur. J. Pharm. Biopharm. **136**, 259–266 (2019). https://doi.org/10.1016/j.ejpb.2019.02.002
99. McAllister, D.V., Wang, P.M., Davis, S.P., Park, J.-H., Canatella, P.J., Allen, M.G., Prausnitz, M.R.: Microfabricated needles for transdermal delivery of macromolecules and nanoparticles: fabrication methods and transport studies. Proc. Natl. Acad. Sci. **100**, 13755–13760 (2003). https://doi.org/10.1073/pnas.2331316100

100. Gupta, J., Park, S.S., Bondy, B., Felner, E.I., Prausnitz, M.R.: Infusion pressure and pain during microneedle injection into skin of human subjects. Biomaterials **32**, 6823–6831 (2011). https://doi.org/10.1016/j.biomaterials.2011.05.061
101. Hu, Z., Meduri, C.S., Ingrole, R.S.J., Gill, H.S., Kumar, G.: Solid and hollow metallic glass microneedles for transdermal drug-delivery. Appl. Phys. Lett. **116** (2020). https://doi.org/10.1063/5.0008983
102. Sivamani, R.K., Liepmann, D., Maibach, H.I.: Microneedles and transdermal applications. Expert Opin. Drug Deliv. **4**, 19–25 (2007). https://doi.org/10.1517/17425247.4.1.19
103. Miyano, T., Tobinaga, Y., Kanno, T., Matsuzaki, Y., Takeda, H., Wakui, M., Hanada, K.: Sugar micro needles as transdermic drug delivery system. Biomed. Microdev. **7**, 185–188 (2005). https://doi.org/10.1007/s10544-005-3024-7
104. Lee, J.W., Choi, S.-O., Felner, E.I., Prausnitz, M.R.: Dissolving microneedle patch for transdermal delivery of human growth hormone. Small **7**, 531–539 (2011). https://doi.org/10.1002/smll.201001091
105. Raphael, A.P., Crichton, M.L., Falconer, R.J., Meliga, S., Chen, X., Fernando, G.J.P., Huang, H., Kendall, M.A.F.: Formulations for microprojection/microneedle vaccine delivery: structure, strength and release profiles. J. Control. Release **225**, 40–52 (2016). https://doi.org/10.1016/j.jconrel.2016.01.027
106. Puttaswamy, S.V., Lubarsky, G.V., Kelsey, C., Zhang, X., Finlay, D., McLaughlin, J.A., Bhalla, N.: Nanophotonic-carbohydrate lab-on-a-microneedle for rapid detection of human cystatin c in finger-prick blood. ACS Nano **14**, 11939–11949 (2020). https://doi.org/10.1021/acsnano.0c05074
107. Zheng, M., Wang, Z., Chang, H., Wang, L., Chew, S.W.T., Lio, D.C.S., Cui, M., Liu, L., Tee, B.C.K., Xu, C.: Osmosis-powered hydrogel microneedles for microliters of skin interstitial fluid extraction within minutes. Adv. Healthc. Mater. **9** (2020). https://doi.org/10.1002/adhm.201901683
108. Tian, Y., Lee, J., van der Maaden, K., Bhide, Y., de Vries-Idema, J.J., Akkerman, R., O'Mahony, C., Jiskoot, W., Frijlink, H.W., Huckriede, A.L.W., Hinrichs, W.L.J., Bouwstra, J.A., Beukema, M.: Intradermal administration of influenza vaccine with trehalose and pullulan-based dissolving microneedle arrays. J. Pharm. Sci. **111**, 1070–1080 (2022). https://doi.org/10.1016/j.xphs.2022.01.033
109. Bariya, S.H., Gohel, M.C., Mehta, T.A., Sharma, O.P.: Microneedles: an emerging transdermal drug delivery system. J. Pharm. Pharmacol. **64**, 11–29 (2012). https://doi.org/10.1111/j.2042-7158.2011.01369.x
110. Donnelly, R.F., Morrow, D.I.J., Singh, T.R.R., Migalska, K., McCarron, P.A., O'Mahony, C., Woolfson, A.D.: Processing difficulties and instability of carbohydrate microneedle arrays. Drug Dev. Ind. Pharm. **35**, 1242–1254 (2009). https://doi.org/10.1080/03639040902882280
111. Sun, W., Inayathullah, M., Manoukian, M.A.C., Malkovskiy, A.V., Manickam, S., Marinkovich, M.P., Lane, A.T., Tayebi, L., Seifalian, A.M., Rajadas, J.: Transdermal delivery of functional collagen via polyvinylpyrrolidone microneedles. Ann. Biomed. Eng. **43**, 2978–2990 (2015). https://doi.org/10.1007/s10439-015-1353-0
112. Castañeda, P.S., Domínguez Delgado, C.L., Cruz, I.M.R., Contreras, L.M.M., Trinidad, E.M.M., Cervantes, M.L., Escobar-Chávez, J.J.: Development of Poly (Methyl vinyl ether-alt-maleic acid) microneedles for transdermal delivery of atorvastatin calcium. Curr. Pharm. Biotechnol. **21**, 852–861 (2020). https://doi.org/10.2174/1389201021666200217103302
113. Gittard, S.D., Ovsianikov, A., Akar, H., Chichkov, B., Monteiro-Riviere, N.A., Stafslien, S., Chisholm, B., Shin, C.-C., Shih, C.-M., Lin, S.-J., Su, Y.-Y., Narayan, R.J.: Two photon polymerization-micromolding of polyethylene glycol-gentamicin sulfate microneedles. Adv. Eng. Mater. **12**, B77–B82 (2010). https://doi.org/10.1002/adem.200980012
114. Oh, J.-H., Park, H.-H., Do, K.-Y., Han, M., Hyun, D.-H., Kim, C.-G., Kim, C.-H., Lee, S.S., Hwang, S.-J., Shin, S.-C., Cho, C.-W.: Influence of the delivery systems using a microneedle array on the permeation of a hydrophilic molecule, calcein. Eur. J. Pharm. Biopharm. **69**, 1040–1045 (2008). https://doi.org/10.1016/j.ejpb.2008.02.009

115. Nguyen, H.X., Bozorg, B.D., Kim, Y., Wieber, A., Birk, G., Lubda, D., Banga, A.K.: Poly (vinyl alcohol) microneedles: fabrication, characterization, and application for transdermal drug delivery of doxorubicin. Eur. J. Pharm. Biopharm. **129**, 88–103 (2018). https://doi.org/10.1016/j.ejpb.2018.05.017
116. Xiang, Z., Wang, H., Pant, A., Pastorin, G., Lee, C.: Development of vertical SU-8 microneedles for transdermal drug delivery by double drawing lithography technology. Biomicrofluidics **7**, 66501 (2013). https://doi.org/10.1063/1.4843475
117. Kim, J.-Y., Han, M.-R., Kim, Y.-H., Shin, S.-W., Nam, S.-Y., Park, J.-H.: Tip-loaded dissolving microneedles for transdermal delivery of donepezil hydrochloride for treatment of Alzheimer's disease. Eur. J. Pharm. Biopharm. **105**, 148–155 (2016). https://doi.org/10.1016/j.ejpb.2016.06.006
118. Park, Y.H., Ha, S.K., Choi, I., Kim, K.S., Park, J., Choi, N., Kim, B., Sung, J.H.: Fabrication of degradable carboxymethyl cellulose (CMC) microneedle with laser writing and replica molding process for enhancement of transdermal drug delivery. Biotechnol. Bioprocess Eng. **21**, 110–118 (2016). https://doi.org/10.1007/s12257-015-0634-7
119. Nguyen, H.X., Banga, A.K.: Delivery of methotrexate and characterization of skin treated by fabricated PLGA microneedles and fractional ablative laser. Pharm. Res. **35**, 68 (2018). https://doi.org/10.1007/s11095-018-2369-6
120. Vora, L.K., Donnelly, R.F., Larrañeta, E., González-vázquez, P., Raj, R., Thakur, S., Vavia, P.R.: Novel bilayer dissolving microneedle arrays with concentrated PLGA nano-microparticles for targeted intradermal delivery: proof of concept. J. Control. Release **265**, 93–101 (2017). https://doi.org/10.1016/j.jconrel.2017.10.005
121. Albadr, A.A., Tekko, I.A., Vora, L.K., Ali, A.A., Laverty, G., Donnelly, R.F.: Rapidly dissolving microneedle patch of amphotericin B for intracorneal fungal infections 931–943 (2022)
122. Martínez-Navarrete, M., Guillot, A.J., Lobita, M.C., Recio, M.C., Giner, R., Aparicio-Blanco, J., Montesinos, M.C., Santos, H.A., Melero, A.: Cyclosporin A-loaded dissolving microneedles for dermatitis therapy: development, characterisation and efficacy in a delayed-type hypersensitivity in vivo model. Drug Deliv. Transl. Res. (2024). https://doi.org/10.1007/s13346-024-01542-9
123. Tekko, I.A., Chen, G., Domínguez-Robles, J., Thakur, R.R.S., Hamdan, I.M.N., Vora, L., Larrañeta, E., McElnay, J.C., McCarthy, H.O., Rooney, M., Donnelly, R.F.: Development and characterisation of novel poly (vinyl alcohol)/poly (vinyl pyrrolidone)-based hydrogel-forming microneedle arrays for enhanced and sustained transdermal delivery of methotrexate. Int. J. Pharm. **586**, 119580 (2020). https://doi.org/10.1016/j.ijpharm.2020.119580
124. Luo, F.-Q., Chen, G., Xu, W., Zhou, D., Li, J.-X., Huang, Y.-C., Lin, R., Gu, Z., Du, J.-Z.: Microneedle-array patch with pH-sensitive formulation for glucose-responsive insulin delivery. Nano Res. **14**, 2689–2696 (2021). https://doi.org/10.1007/s12274-020-3273-z
125. Yao, S., Chi, J., Wang, Y., Zhao, Y., Luo, Y., Wang, Y.: Zn-MOF Encapsulated antibacterial and degradable microneedles array for promoting wound healing. Adv. Healthc. Mater. **10**, e2100056 (2021). https://doi.org/10.1002/adhm.202100056
126. Kim, B.H., Seo, Y.H.: Transdermal drug delivery devices based on microneedles: a review. J. Mucopolysacch. Rare Dis. **1**, 5–14 (2015). https://doi.org/10.19125/jmrd.2015.1.1.5
127. Luo, X., Yang, L., Cui, Y.: Microneedles: materials, fabrication, and biomedical applications. Biomed. Microdev. **25**, 20 (2023). https://doi.org/10.1007/s10544-023-00658-y
128. Hara, Y., Yamada, M., Tatsukawa, C., Takahashi, T., Suzuki, M., Aoyagi, S.: Fabrication of stainless steel microneedle with laser-cut sharp tip and its penetration and blood sampling performance. Int. J. Autom. Technol. **10**, 950–957 (2016)
129. Gill, H.S., Söderholm, J., Prausnitz, M.R., Sällberg, M.: Cutaneous vaccination using microneedles coated with hepatitis C DNA vaccine. Gene Ther. **17**, 811–814 (2010). https://doi.org/10.1038/gt.2010.22
130. Musazzi, S., Perini, U.: Laser-Induced Breakdown Spectroscopy: theory and Applications. Springer Berlin Heidelberg (2014). https://books.google.es/books?id=I68TnwEACAAJ

131. Kang, S.U., Seo, S.J., Kim, Y.S., Shin, Y.S., Koh, Y.W., Lee, C.M., Yang, S.S., Lee, J.S., Moon, E., Kang, H., Ryeo, J.B., Lee, Y., Kim, C.H.: Comparative effects of non-thermal atmospheric pressure plasma on migration and invasion in oral squamous cell cancer, by gas type. Yonsei Med. J. **58**, 272–281 (2017). https://doi.org/10.3349/ymj.2017.58.2.272
132. Omatsu, T., Chujo, K., Miyamoto, K., Okida, M., Nakamura, K., Aoki, N., Morita, R.: Metal microneedle fabrication using twisted light with spin. Opt. Express **18**, 17967–17973 (2010). https://doi.org/10.1364/OE.18.017967
133. Nejad, H.R., Sadeqi, A., Kiaee, G., Sonkusale, S.: Low-cost and cleanroom-free fabrication of microneedles. Microsyst. Nanoeng. **4**, 1–7 (2018). https://doi.org/10.1038/MICRONANO.2017.73
134. Farias, C., Lyman, R., Hemingway, C., Chau, H., Mahacek, A., Bouzos, E., Mobed-Miremadi, M.: Three-dimensional (3D) printed microneedles for microencapsulated cell extrusion. Bioengineering (Basel) **5** (2018). https://doi.org/10.3390/bioengineering5030059
135. Raman, R., Bhaduri, B., Mir, M., Shkumatov, A., Lee, M.K., Popescu, G., Kong, H., Bashir, R.: High-resolution projection microstereolithography for patterning of neovasculature. Adv Healthc Mater **5**, 610–619 (2016). https://doi.org/10.1002/adhm.201500721
136. Lu, Y., Mantha, S.N., Crowder, D.C., Chinchilla, S., Shah, K.N., Yun, Y.H., Wicker, R.B., Choi, J.-W.: Microstereolithography and characterization of poly(propylene fumarate)-based drug-loaded microneedle arrays. Biofabrication **7**, 45001 (2015). https://doi.org/10.1088/1758-5090/7/4/045001
137. Baraissov, Z., Pacco, A., Koneti, S., Bisht, G., Panciera, F., Holsteyns, F., Mirsaidov, U.: Selective wet etching of silicon germanium in composite vertical nanowires. ACS Appl. Mater. Interfaces **11**, 36839–36846 (2019). https://doi.org/10.1021/acsami.9b11934
138. Ohmi, H., Sato, J., Shirasu, Y., Hirano, T., Kakiuchi, H., Yasutake, K.: Significant improvement of copper dry etching property of a high-pressure hydrogen-based plasma by nitrogen gas addition. ACS Omega **4**, 4360–4366 (2019). https://doi.org/10.1021/acsomega.8b03163
139. Chen, W., Liu, Y., Yang, L., Wu, J., Chen, Q., Zhao, Y., Wang, Y., Du, X.: Difference in anisotropic etching characteristics of alkaline and copper based acid solutions for single-crystalline Si. Sci. Rep. **8**, 3408 (2018). https://doi.org/10.1038/s41598-018-21877-x
140. Ji, J., Tay, F.E., Miao, J., Iliescu, C.: Microfabricated silicon microneedle array for transdermal drug delivery. J. Phys. Conf. Ser. 1127–1131 (2006). https://doi.org/10.1088/1742-6596/34/1/186
141. Zhang, J.X.J., Hoshino, K.: Molecular Sensors and Nanodevices: principles, Designs and Applications in Biomedical Engineering. Academic Press (2018)
142. Lu, H., Zhang, H., Jin, M., He, T., Zhou, G., Shui, L.: Two-layer microstructures fabricated by one-step anisotropic wet etching of Si in KOH solution. Micromachines (Basel) **7** (2016). https://doi.org/10.3390/mi7020019
143. Asoh, H., Fujihara, K., Ono, S.: Triangle pore arrays fabricated on Si (111) substrate by sphere lithography combined with metal-assisted chemical etching and anisotropic chemical etching. Nanoscale Res. Lett. **7**, 406 (2012). https://doi.org/10.1186/1556-276X-7-406
144. Ramakrishnaiah, R., Alkheraif, A.A., Divakar, D.D., Matinlinna, J.P., Vallittu, P.K.: The effect of hydrofluoric acid etching duration on the surface micromorphology, roughness, and wettability of dental ceramics. Int. J. Mol. Sci. **17** (2016). https://doi.org/10.3390/ijms17060822
145. Lim, K., Gupta, S., Ropp, C., Waks, E.: Development of metal etch mask by single layer lift-off for silicon nitride photonic crystals. Microelectron. Eng. **88**, 994–998 (2011). https://doi.org/10.1016/j.mee.2010.12.113
146. Miyake, S., Yamazaki, S.: Low-damage direct patterning of silicon oxide mask by mechanical processing. Nanoscale Res. Lett. **9**, 269 (2014). https://doi.org/10.1186/1556-276X-9-269
147. Lee, J.B., Choi, K.H., Yoo, K.: Innovative SU-8 lithography techniques and their applications. Micromachines (Basel) **6**, 1–18 (2015). https://doi.org/10.3390/mi6010001
148. Wang, P.-C., Wester, B.A., Rajaraman, S., Paik, S.-J., Kim, S.-H., Allen, M.G.: Hollow polymer microneedle array fabricated by photolithography process combined with micromolding technique. Annual International Conference of the IEEE Engineering in Medicine

and Biology Society 2009, pp. 7026–7029 (2009). https://doi.org/10.1109/IEMBS.2009.5333317
149. Acikgoz, C., Hempenius, M.A., Huskens, J., Vancso, G.J.: Polymers in conventional and alternative lithography for the fabrication of nanostructures. Eur. Polym. J. **47**, 2033–2052 (2011). https://doi.org/10.1016/j.eurpolymj.2011.07.025
150. Mishra, R., Maiti, T.K., Bhattacharyya, T.K.: Development of SU-8 hollow microneedles on a silicon substrate with microfluidic interconnects for transdermal drug delivery. J. Micromech. Microeng. **28**, 105017 (2018). https://doi.org/10.1088/1361-6439/aad301
151. Guillot, A.J., Martínez-Navarrete, M., Giner, R.M., Recio, M.C., Santos, H.A., Cordeiro, A.S., Melero, A.: Cyanocobalamin-loaded dissolving microneedles diminish skin inflammation in vivo. J. Control. Release **375**, 537–551 (2024). https://doi.org/10.1016/j.jconrel.2024.09.032
152. Boateng, J.S., Stevens, H.N.E., Eccleston, G.M., Auffret, A.D., Humphrey, M.J., Matthews, K.H.: Development and mechanical characterization of solvent-cast polymeric films as potential drug delivery systems to mucosal surfaces. Drug Dev. Ind. Pharm. **35**, 986–996 (2009). https://doi.org/10.1080/03639040902744704
153. Umar, Q.-U.-A., Khan, M.I., Ahmad, Z., Akhtar, M.F., Sohail, M.F., Madni, A., Erum, A., Ayesha, B., Ain, Q.U., Mushtaq, A.: Dissolving microneedles patch: a promising approach for advancing transdermal delivery of antischizophrenic drug. J. Pharm. Sci. **113**, 3078–3087 (2024). https://doi.org/10.1016/j.xphs.2024.08.011
154. Lobita, M.C., El-Sayed, N., Pinto, J.F., Santos, H.A.: Development of fast dissolving polymer-based microneedles for delivery of an antigenic melanoma cell membrane. Int. J. Pharm. **642**, 123143 (2023). https://doi.org/10.1016/j.ijpharm.2023.123143
155. Andranilla, R.K., Anjani, Q.K., Hartrianti, P., Donnelly, R.F., Ramadon, D.: Fabrication of dissolving microneedles for transdermal delivery of protein and peptide drugs: polymer materials and solvent casting micromoulding method. Pharm. Dev. Technol. **28**, 1016–1031 (2023). https://doi.org/10.1080/10837450.2023.2285498
156. Demartis, S., Anjani, Q.K., Volpe-Zanutto, F., Paredes, A.J., Jahan, S.A., Vora, L.K., Donnelly, R.F., Gavini, E.: Trilayer dissolving polymeric microneedle array loading Rose Bengal transfersomes as a novel adjuvant in early-stage cutaneous melanoma management. Int. J. Pharm. **627**, 122217 (2022). https://doi.org/10.1016/j.ijpharm.2022.122217
157. Anjani, Q.K., Pandya, A.K., Demartis, S., Domínguez-Robles, J., Moreno-Castellanos, N., Li, H., Gavini, E., Patravale, V.B., Donnelly, R.F.: Liposome-loaded polymeric microneedles for enhanced skin deposition of rifampicin. Int. J. Pharm. **646**, 123446 (2023). https://doi.org/10.1016/j.ijpharm.2023.123446
158. Nainggolan, A.D.C., Hartrianti, P., Anjani, Q.K., Donnelly, R.F., Putra, A.B.N., Kho, K., Kurniawan, A., Andranilla, R.K., Rattu, S.A., Ramadon, D.: Double-layer dissolving microneedles for delivery of mesenchymal stem cell Secretome: formulation, characterisation and skin irritation study. Eur. J. Pharm. Biopharm. **204**, 114495 (2024). https://doi.org/10.1016/j.ejpb.2024.114495
159. Ramadon, D., Karn, P.R., Anjani, Q.K., Kim, M.-H., Cho, D.Y., Hwang, H., Kim, D.H., Kim, D.H., Kim, G., Lee, K., Eum, J.H., Im, J.Y., Aileen, V., Hamda, O.T., Donnelly, R.F.: Development of ropivacaine hydrochloride-loaded dissolving microneedles as a local anesthetic agent: a proof-of-concept, Int. J. Pharm. **660**, 124347 (2024). https://doi.org/10.1016/j.ijpharm.2024.124347
160. Anjani, Q.K., Demartis, S., Moreno-Castellanos, N., Gavini, E., Donnelly, R.F.: Formulation and evaluation of ivermectin-loaded dissolving microarray patches for rosacea disease. J. Pharm. Investig. **54**, 683–698 (2024). https://doi.org/10.1007/s40005-024-00682-x
161. El-Domyati, M., Abdel-Wahab, H., Hossam, A., Elfar, N.N., Hasby, E.A., Vijaya Lakshmi, Y., Swetha Reddy, L., Naga Neelima Devi, K., Phani Kumar, K., Guru Karthik, G., Srinivas Chakravarthy, P., Nageswar Rao, K.: Combining microneedling with other minimally invasive procedures for facial rejuvenation: a split-face comparative study. Int. J. Dermatol. **31**, 620–626 (2020). https://doi.org/10.1111/ijd.14815

162. Chen, H., Wu, B., Zhang, M., Yang, P., Yang, B., Qin, W., Wang, Q., Wen, X., Chen, M., Quan, G., Pan, X., Wu, C.: A novel scalable fabrication process for the production of dissolving microneedle arrays. Drug Deliv. Transl. Res. **9**, 240–248 (2019). https://doi.org/10.1007/s13346-018-00593-z
163. Krieger, K.J., Bertollo, N., Dangol, M., Sheridan, J.T., Lowery, M.M., O'Cearbhaill, E.D.: Simple and customizable method for fabrication of high-aspect ratio microneedle molds using low-cost 3D printing. Microsyst. Nanoeng. **5**, 42 (2019). https://doi.org/10.1038/s41378-019-0088-8
164. Lee, J.W., Park, J.-H., Prausnitz, M.R.: Dissolving microneedles for transdermal drug delivery. Biomaterials **29**, 2113–2124 (2008). https://doi.org/10.1016/j.biomaterials.2007.12.048
165. Chu, L.Y., Choi, S.-O., Prausnitz, M.R.: Fabrication of dissolving polymer microneedles for controlled drug encapsulation and delivery: bubble and pedestal microneedle designs. J. Pharm. Sci. **99**, 4228–4238 (2010). https://doi.org/10.1002/jps.22140
166. Mistilis, M.J., Bommarius, A.S., Prausnitz, M.R.: Development of a thermostable microneedle patch for influenza vaccination. J. Pharm. Sci. **104**, 740–749 (2015). https://doi.org/10.1002/jps.24283
167. Wang, M., Hu, L., Xu, C.: Recent advances in the design of polymeric microneedles for transdermal drug delivery and biosensing. Lab Chip **17**, 1373–1387 (2017). https://doi.org/10.1039/C7LC00016B
168. Li, W., Li, S., Fan, X., Prausnitz, M.R.: Microneedle patch designs to increase dose administered to human subjects. J. Control. Release **339**, 350–360 (2021). https://doi.org/10.1016/j.jconrel.2021.09.036
169. Machekposhti, S.A., Soltani, M., Najafizadeh, P., Ebrahimi, S.A., Chen, P.: Biocompatible polymer microneedle for topical/dermal delivery of tranexamic acid. J. Control. Release **261**, 87–92 (2017). https://doi.org/10.1016/j.jconrel.2017.06.016
170. Champeau, M., Jary, D., Mortier, L., Mordon, S., Vignoud, S.: A facile fabrication of dissolving microneedles containing 5-aminolevulinic acid. Int. J. Pharm. **586**, 119554 (2020). https://doi.org/10.1016/j.ijpharm.2020.119554
171. Cordeiro, A.S., Tekko, I.A., Jomaa, M.H., Vora, L., McAlister, E., Volpe-Zanutto, F., Nethery, M., Baine, P.T., Mitchell, N., McNeill, D.W., Donnelly, R.F.: Two-photon polymerisation 3D printing of microneedle array templates with versatile designs: application in the development of polymeric drug delivery systems. Pharm. Res. **37**, 174 (2020). https://doi.org/10.1007/s11095-020-02887-9
172. Moga, K.A., Bickford, L.R., Geil, R.D., Dunn, S.S., Pandya, A.A., Wang, Y., Fain, J.H., Archuleta, C.F., O'Neill, A.T., Desimone, J.M.: Rapidly-dissolvable microneedle patches via a highly scalable and reproducible soft lithography approach. Adv. Mater. **25**, 5060–5066 (2013). https://doi.org/10.1002/adma.201300526
173. Dean, D.A.: Preparation (pulling) of needles for gene delivery by microinjection. CSH Protoc. **2006** (2006). https://doi.org/10.1101/pdb.prot4651
174. Dean, D.A., Gasiorowski, J.Z.: Preparing injection pipettes on a PUL-1 micropipette puller. Cold Spring Harb. Protoc. **2011**, prot5587 (2011). https://doi.org/10.1101/pdb.prot5587
175. Mahadevan, G., Sheardown, H., Selvaganapathy, P.: PDMS embedded microneedles as a controlled release system for the eye. J. Biomater. Appl. **28**, 20–27 (2013). https://doi.org/10.1177/0885328211433778
176. Bystrova, S., Luttge, R.: Micromolding for ceramic microneedle arrays, Microelectron. Eng. **88**, 1681–1684 (2011). https://doi.org/10.1016/j.mee.2010.12.067
177. Naglieri, V., Gutknecht, D., Garnier, V., Palmero, P., Chevalier, J., Montanaro, L.: Optimized slurries for spray drying: different approaches to obtain homogeneous and deformable alumina-zirconia granules. Materials (Basel) **6**, 5382–5397 (2013). https://doi.org/10.3390/ma6115382
178. Haj-Ahmad, R., Khan, H., Arshad, M.S., Rasekh, M., Hussain, A., Walsh, S., Li, X., Chang, M.-W., Ahmad, Z.: Microneedle coating techniques for transdermal drug delivery. Pharmaceutics **7**, 486–502 (2015). https://doi.org/10.3390/pharmaceutics7040486

179. Gill, H.S., Prausnitz, M.R.: Coated microneedles for transdermal delivery. J. Control. Release **117**, 227–237 (2007). https://doi.org/10.1016/j.jconrel.2006.10.017
180. DeMuth, P.C., Su, X., Samuel, R.E., Hammond, P.T., Irvine, D.J.: Nano-layered microneedles for transcutaneous delivery of polymer nanoparticles and plasmid DNA. Adv. Mater. **22**, 4851–4856 (2010). https://doi.org/10.1002/adma.201001525
181. Kim, Y.-C., Park, J.-H., Prausnitz, M.R.: Microneedles for drug and vaccine delivery. Adv. Drug Deliv. Rev. **64**, 1547–1568 (2012). https://doi.org/10.1016/j.addr.2012.04.005
182. Gill, H.S., Prausnitz, M.R.: Pocketed microneedles for drug delivery to the skin. J. Phys. Chem. Solids **69**, 1537–1541 (2008). https://doi.org/10.1016/j.jpcs.2007.10.059
183. Ono, A., Ito, S., Sakagami, S., Asada, H., Saito, M., Quan, Y.-S., Kamiyama, F., Hirobe, S., Okada, N.: Development of novel faster-dissolving microneedle patches for transcutaneous vaccine delivery. Pharmaceutics **9** (2017). https://doi.org/10.3390/pharmaceutics9030027
184. Lopez-Heredia, M.A., Weiss, P., Layrolle, P.: An electrodeposition method of calcium phosphate coatings on titanium alloy. J. Mater. Sci. Mater. Med. **18**, 381–390 (2007). https://doi.org/10.1007/s10856-006-0703-8
185. Mansoor, I., Liu, Y., Häfeli, U.O., Stoeber, B.: Fabrication of hollow microneedle arrays using electrodeposition of metal onto solvent cast conductive polymer structures. In: 2013 Transducers & Eurosensors XXVII: the 17th International Conference on Solid-State Sensors, Actuators and Microsystems (TRANSDUCERS & EUROSENSORS XXVII), pp. 373–376 (2013). https://doi.org/10.1109/Transducers.2013.6626780
186. FDA, Regulatory Considerations for Microneedling Products Guidance for Industry and Food and Drug Administration Staff Preface Public Comment, pp 1–12 (2020). https://www.regulations.gov
187. Zhao, C., Yang, Y., Cui, X., Shan, Y., Xue, J., Jiang, D., Sun, J., Li, N., Li, Z., Yang, A.: Self-powered electrical impulse chemotherapy for oral squamous cell carcinoma. Materials (Basel) **15** (2022). https://doi.org/10.3390/ma15062060
188. Rajput, A. Osmani, R.A.M., Khire, A., Jaiswal, S., Banerjee, R.: Levonorgestrel microneedle array patch for sustained release contraception: formulation, optimization and in vivo characterization. Molecules **27** (2022). https://doi.org/10.3390/molecules27072349
189. Baek, J.Y., Kang, K.M., Kim, H.J., Kim, J.H., Lee, J.H., Shin, G., Jeon, J.G., Lee, J., Han, Y., So, B.J., Kang, T.J.: Manufacturing process of polymeric microneedle sensors for mass production. Micromachines (Basel) **12** (2021). https://doi.org/10.3390/mi12111364
190. Li, Y., Hu, X., Dong, Z., Chen, Y., Zhao, W., Wang, Y., Zhang, L., Chen, M., Wu, C., Wang, Q.: Dissolving microneedle arrays with optimized needle geometry for transcutaneous immunization. Eur. J. Pharm. Sci. **151**, 105361 (2020). https://doi.org/10.1016/j.ejps.2020.105361
191. Sillankorva, S., Pires, L., Pastrana, L.M., Bañobre-López, M.: Antibiofilm efficacy of the *Pseudomonas aeruginosa Pbunavirus* vB_PaeM-SMS29 loaded onto dissolving polyvinyl alcohol microneedles. Viruses **14** (2022). https://doi.org/10.3390/v14050964
192. Min, H.S., Kim, Y., Nam, J., Ahn, H., Kim, M., Kang, G., Jang, M., Yang, H., Jung, H.: Shape of dissolving microneedles determines skin penetration ability and efficacy of drug delivery. Biomater. Adv. **145**, 213248 (2023). https://doi.org/10.1016/j.bioadv.2022.213248
193. Migdadi, E.M., Courtenay, A.J., Tekko, I.A., McCrudden, M.T.C., Kearney, M.-C., McAlister, E., McCarthy, H.O., Donnelly, R.F.: Hydrogel-forming microneedles enhance transdermal delivery of metformin hydrochloride. J. Control. Release **285**, 142–151 (2018). https://doi.org/10.1016/j.jconrel.2018.07.009
194. Permana, A.D., Tekko, I.A., McCrudden, M.T.C., Anjani, Q.K., Ramadon, D., McCarthy, H.O., Donnelly, R.F.: Solid lipid nanoparticle-based dissolving microneedles: a promising intradermal lymph targeting drug delivery system with potential for enhanced treatment of lymphatic filariasis. J. Control. Release **316**, 34–52 (2019). https://doi.org/10.1016/j.jconrel.2019.10.004
195. Teodorescu, M., Bercea, M., Morariu, S.: Biomaterials of PVA and PVP in medical and pharmaceutical applications: perspectives and challenges. Biotechnol. Adv. **37**, 109–131 (2019). https://doi.org/10.1016/j.biotechadv.2018.11.008

196. Larrañeta, E., Moore, J., Vicente-Pérez, E.M., González-Vázquez, P., Lutton, R., Woolfson, A.D., Donnelly, R.F.: A proposed model membrane and test method for microneedle insertion studies. Int. J. Pharm. **472**, 65–73 (2014). https://doi.org/10.1016/j.ijpharm.2014.05.042
197. Volpatti, L.R., Burns, D.M., Basu, A., Langer, R., Anderson, D.G.: Engineered insulin-polycation complexes for glucose-responsive delivery with high insulin loading. J. Control. Release **338**, 71–79 (2021). https://doi.org/10.1016/j.jconrel.2021.08.017
198. Faizi, H.S., Vora, L.K., Nasiri, M.I., Wu, Y., Mishra, D., Anjani, Q.K., Paredes, A.J., Thakur, R.R.S., Minhas, M.U., Donnelly, R.F.: Deferasirox nanosuspension loaded dissolving microneedles for intradermal delivery. Pharmaceutics **14** (2022). https://doi.org/10.3390/pharmaceutics14122817
199. Lhernould, M.S., Deleers, M., Delchambre, A.: Hollow polymer microneedles array resistance and insertion tests. Int. J. Pharm. **480**, 152–157 (2015). https://doi.org/10.1016/j.ijpharm.2015.01.019
200. Coulman, S.A., Birchall, J.C., Alex, A., Pearton, M., Hofer, B., O'Mahony, C., Drexler, W., Považay, B.: In vivo, in situ imaging of microneedle insertion into the skin of human volunteers using optical coherence tomography. Pharm. Res. **28**, 66–81 (2011). https://doi.org/10.1007/s11095-010-0167-x
201. Kight, E., Alfaro, R., Gadila, S.K.G., Chang, S., Evans, D., Embers, M., Haselton, F.: Direct capture and early detection of lyme disease spirochete in skin with a microneedle patch. Biosensors (Basel) **12** (2022). https://doi.org/10.3390/bios12100819
202. Permana, A.D., McCrudden, M.T.C., Donnelly, R.F.: Enhanced intradermal delivery of nanosuspensions of Antifilariasis drugs using dissolving microneedles: a proof of concept study. Pharmaceutics **11** (2019). https://doi.org/10.3390/pharmaceutics11070346
203. McCrudden, M.T.C., Larrañeta, E., Clark, A., Jarrahian, C., Rein-Weston, A., Lachau-Durand, S., Niemeijer, N., Williams, P., Haeck, C., McCarthy, H.O., Zehrung, D., Donnelly, R.F.: Design, formulation and evaluation of novel dissolving microarray patches containing a long-acting rilpivirine nanosuspension. J. Control. Release **292**, 119–129 (2018). https://doi.org/10.1016/j.jconrel.2018.11.002
204. Larrañeta, E., Stewart, S., Fallows, S.J., Birkhäuer, L.L., McCrudden, M.T.C., Woolfson, A.D., Donnelly, R.F.: A facile system to evaluate in vitro drug release from dissolving microneedle arrays. Int. J. Pharm. **497**, 62–69 (2016). https://doi.org/10.1016/j.ijpharm.2015.11.038
205. Ko, P.-T., Lee, I.-C., Chen, M.-C., Tsai, S.-W.: Polymer microneedles fabricated from PCL and PCL/PEG blends for transdermal delivery of hydrophilic compounds. J. Taiwan Inst. Chem. Eng. **51**, 1–8 (2015). https://doi.org/10.1016/j.jtice.2015.01.003
206. Ritger, P.L., Peppas, N.A.: A simple equation for description of solute release I. Fickian and non-fickian release from non-swellable devices in the form of slabs, spheres, cylinders or discs. J. Control. Release **5**, 23–36 (1987). https://doi.org/10.1016/0168-3659(87)90034-4
207. Ritger, P.L., Peppas, N.A.: A simple equation for description of solute release II. Fickian and anomalous release from swellable devices. J. Control. Release **5**, 37–42 (1987). https://doi.org/10.1016/0168-3659(87)90035-6
208. Kim, H., Fassihi, R.: Application of binary polymer system in drug release rate modulation. 2. Influence of formulation variables and hydrodynamic conditions on release kinetics. J. Pharm. Sci. **86**, 323–328 (1997). https://doi.org/10.1021/js960307p
209. Peppas, N.A., Sahlin, J.J.: A simple equation for the description of solute release. III. Coupling of diffusion and relaxation. Int. J. Pharm. **57**, 169–172 (1989). https://doi.org/10.1016/0378-5173(89)90306-2
210. Thakur, R.R.S., Tekko, I.A., Al-Shammari, F., Ali, A.A., McCarthy, H., Donnelly, R.F.: Rapidly dissolving polymeric microneedles for minimally invasive intraocular drug delivery. Drug Deliv. Transl. Res. **6**, 800–815 (2016). https://doi.org/10.1007/s13346-016-0332-9
211. Tian, Y., Li, Y., Xu, X., Jin, Z.: Starch retrogradation studied by thermogravimetric analysis (TGA). Carbohydr. Polym. **84**, 1165–1168 (2011). https://doi.org/10.1016/j.carbpol.2011.01.006
212. Kolluru, C., Gomaa, Y., Prausnitz, M.R.: Development of a thermostable microneedle patch for polio vaccination. Drug Deliv. Transl. Res. **9**, 192–203 (2019). https://doi.org/10.1007/s13346-018-00608-9

213. Govender, R., Kissi, E.O., Larsson, A., Tho, I.: Polymers in pharmaceutical additive manufacturing: a balancing act between printability and product performance. Adv. Drug Deliv. Rev. **177**, 113923 (2021). https://doi.org/10.1016/j.addr.2021.113923
214. Knopp, M.M., Löbmann, K., Elder, D.P., Rades, T., Holm, R.: Recent advances and potential applications of modulated differential scanning calorimetry (mDSC) in drug development. Eur. J. Pharm. Sci. **87**, 164–173 (2016). https://doi.org/10.1016/j.ejps.2015.12.024
215. Arshad, M.S., Hassan, S., Hussain, A., Abbas, N., Kucuk, I., Nazari, K., Ali, R., Ramzan, S., Alqahtani, A., Andriotis, E.G., Fatouros, D.G., Chang, M.-W., Ahmad, Z.: Improved transdermal delivery of cetirizine hydrochloride using polymeric microneedles. Daru **27**, 673–681 (2019). https://doi.org/10.1007/s40199-019-00301-3
216. Ramadon, D., Ulayya, F., Qur'ani, A.S., Iskandarsyah, I., Harahap, Y., Anjani, Q.K., Aileen, V., Hartrianti, P., Donnelly, R.F.: Combination of dissolving microneedles with nanosuspension and co-grinding for transdermal delivery of Ketoprofen. Pharmaceuticals **16**, 378 (2023). https://doi.org/10.3390/ph16030378
217. Ng, S.-F., Rouse, J.J., Sanderson, F.D., Meidan, V., Eccleston, G.M.: Validation of a static franz diffusion cell system for in vitro permeation studies. AAPS PharmSciTech **11**, 1432–1441 (2010). https://doi.org/10.1208/s12249-010-9522-9
218. Lahiji, S.F., Dangol, M., Jung, H.: A patchless dissolving microneedle delivery system enabling rapid and efficient transdermal drug delivery. Sci. Rep. **5**, 7914 (2015). https://doi.org/10.1038/srep07914
219. Guillot, A.J., Martínez-Navarrete, M., Garrigues, T.M., Melero, A.: Skin drug delivery using lipid vesicles: a starting guideline for their development. J. Control. Release **355**, 624–654 (2023). https://doi.org/10.1016/j.jconrel.2023.02.006
220. Pineau, A., Guillard, O., Fauconneau, B., Favreau, F., Marty, M.-H., Gaudin, A., Vincent, C.M., Marrauld, A., Marty, J.-P.: In vitro study of percutaneous absorption of aluminum from antiperspirants through human skin in the FranzTM diffusion cell. J. Inorg. Biochem. **110**, 21–26(2012). https://doi.org/10.1016/j.jinorgbio.2012.02.013
221. Ogunjimi, A.T., Carr, J., Lawson, C., Ferguson, N., Brogden, N.K.: Micropore closure time is longer following microneedle application to skin of color. Sci. Rep. **10**, 18963 (2020). https://doi.org/10.1038/s41598-020-75246-8
222. Donnelly, R.F., Singh, T.R.R., Tunney, M.M., Morrow, D.I.J., McCarron, P.A., O'Mahony, C., Woolfson, A.D.: Microneedle arrays allow lower microbial penetration than hypodermic needles in vitro. Pharm. Res. **26**, 2513–2522 (2009). https://doi.org/10.1007/s11095-009-9967-2
223. Guillot, A.J., Cordeiro, A.S., Donnelly, R.F., Montesinos, M.C., Garrigues, T.M., Melero, A.: Microneedle-based delivery: an overview of current applications and trends. Pharmaceutics **12**, 1–28 (2020). https://doi.org/10.3390/pharmaceutics12060569
224. Zhang, Q., Shi, L., He, H., Liu, X., Huang, Y., Xu, D., Yao, M., Zhang, N., Guo, Y., Lu, Y., Li, H., Zhou, J., Tan, J., Xing, M., Luo, G.: Down-regulating scar formation by microneedles directly via a mechanical communication pathway. ACS Nano **16**, 10163–10178 (2022). https://doi.org/10.1021/acsnano.1c11016
225. Tansathien, K., Suriyaamporn, P., Ngawhirunpat, T., Opanasopit, P., Rangsimawong, W.: A novel approach for skin regeneration by a potent bioactive placental-loaded microneedle patch: comparative study of deer, goat, and porcine placentas. Pharmaceutics **14** (2022). https://doi.org/10.3390/pharmaceutics14061221
226. Marshall, S., Sahm, L.J., Moore, A.C.: The success of microneedle-mediated vaccine delivery into skin. Hum. Vaccin. Immunother. **12**, 2975–2983 (2016). https://doi.org/10.1080/21645515.2016.1171440
227. Cheung, K., Das, D.B.: Microneedles for drug delivery: trends and progress. Drug Deliv. **23**, 2338–2354 (2016). https://doi.org/10.3109/10717544.2014.986309
228. Bhatnagar, S., Dave, K., Venuganti, V.V.K.: Microneedles in the clinic. J. Control. Release **260**, 164–182 (2017). https://doi.org/10.1016/j.jconrel.2017.05.029
229. Singh, T.R.R., Dunne, N.J., Cunningham, E., Donnelly, R.F.: Review of patents on microneedle applicators. Recent Pat. Drug Deliv. Formul. **5**, 11–23 (2011). https://doi.org/10.2174/187221111794109484

230. Lhernould, M.S., Tailler, S., Deleers, M., Delchambre, A.: Review of patents for microneedle application devices allowing fluid injections through the skin. Recent Pat. Drug Deliv. Formul. **9**, 146–157 (2015). https://doi.org/10.2174/1872211309666150313155503
231. Kirkby, M. Hutton, A.R.J., Donnelly, R.F.: Microneedle mediated transdermal delivery of protein, peptide and antibody based therapeutics: current status and future considerations. Pharm. Res. **37**, 117 (2020). https://doi.org/10.1007/s11095-020-02844-6
232. Donnelly, R.F., Prausnitz, M.R.: The promise of microneedle technologies for drug delivery. Drug Deliv. Transl. Res. (2023). https://doi.org/10.1007/s13346-023-01430-8
233. Smith, F., Sabri, A.H., Heppel, M., Fonseca, I., Chowdhury, F., Cheung, K., Willmor, S., Rawson, F., Marlow, M.: The clinical and translational prospects of microneedle devices, with a focus on insulin therapy for diabetes mellitus as a case study. Int. J. Pharm. **628**, 122234 (2022). https://doi.org/10.1016/j.ijpharm.2022.122234
234. Vicente-Pérez, E.M., Quinn, H.L., McAlister, E., O'Neill, S., Hanna, L.-A., Barry, J.G., Donnelly, R.F.: The use of a pressure-indicating sensor film to provide feedback upon hydrogel-forming microneedle array self-application in vivo. Pharm. Res. **33**, 3072–3080 (2016). https://doi.org/10.1007/s11095-016-2032-z
235. Norman, J.J., Arya, J.M., McClain, M.A., Frew, P.M., Meltzer, M.I., Prausnitz, M.R.: Microneedle patches: usability and acceptability for self-vaccination against influenza. Vaccine **32**, 1856–1862 (2014). https://doi.org/10.1016/j.vaccine.2014.01.076
236. Su, Y., Andrabi, S.M., Shahriar, S.M.S., Wong, S.L., Wang, G., Xie, J.: Triggered release of antimicrobial peptide from microneedle patches for treatment of wound biofilms. J. Control. Release **356**, 131–141 (2023). https://doi.org/10.1016/j.jconrel.2023.02.030

Chapter 2
Designing Microneedles for Intravaginal Delivery: Current Challenges and Prospects

Sopan Nangare, Swapnil Thakare, Lalitkumar Vora, Sai Hanuman Sagar Boddu, and Namdeo Jadhav

2.1 Introduction

The vagina is a tubular and fibromuscular female genital organ that extends from the cervix of the uterus to the vaginal vestibule. It has a length of about 9 cm [1] and a total surface area of 72 ± 21 cm^2 [2]. Its functions include menstruation discharge, sexual intercourse, conception [1], and immune defense [3]. The vagina is a collapsible S-shaped (longitudinal) fibromuscular tube [4] composed of various layers including the epithelial layer, muscular coat, and tunica adventitia [5]. Histopathologically, the vagina is divided into stratified squamous epithelium, lamina propria, muscularis layer, and adventitia (Fig. 2.1) [1]. The inner mucosal layer, or tunica mucosa, consists of non-keratinized stratified squamous epithelium and contains numerous vascular networks embedded in the lamina propria (connective tissue) [4]. The stratified squamous epithelium acts as a barrier against physical and biochemical invasion [3], while the rugae or folds on the vaginal surface provide support, expandability, and high surface area [5]. The entrance to the vagina is partially obscured by the labia

S. Nangare (✉) · N. Jadhav
Department of Pharmaceutics, Krishna Vishwa Vidyapeeth (Deemed to be University), Krishna Institute of Pharmacy, Karad, Maharashtra, India
e-mail: snangareopan@gmail.com

S. Thakare
Department of Pharmaceutics, Rajashri College of Pharmacy, Mehkar, Buldhana, Maharashtra, India

L. Vora
School of Pharmacy, Medical Biology Centre, Queen's University Belfast, Belfast, UK

S. H. S. Boddu
Department of Pharmaceutical Sciences, College of Pharmacy and Health Sciences, Ajman University, Ajman, United Arab Emirates

N. Talreja et al. (eds.), *Microneedles (MNs)-Based Technology*, Series in BioEngineering, https://doi.org/10.1007/978-981-96-3916-8_2

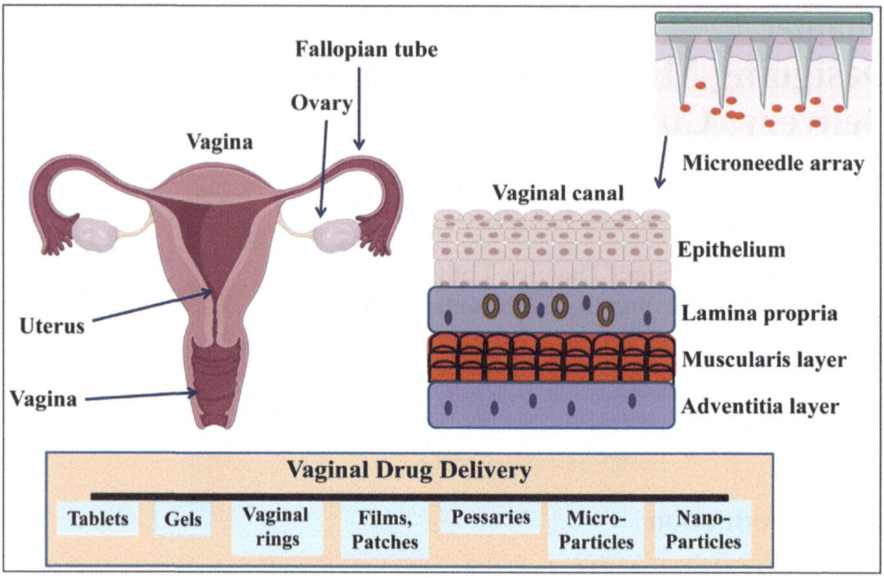

Fig. 2.1 Presentation of the vagina, layers of vaginal canal, and traditionally used intravaginal drug delivery systems

minora [4]. During menstruation, the thickness of the vaginal epithelial layer changes by approximately 200–300 μm. The vagina possesses good elastic properties due to the presence of smooth elastic fibers in the muscular coat, and the tunica adventitia containing loose connective tissue further enhances its elasticity [5]. According to the literature, although the vagina contains a mucosal layer, there are typically no glands present in a normal vagina [3]. Consequently, vaginal secretions consist of a blend of fluids from various sources [1] such as cervical mucus, tubal fluid, leukocytes, transudate from the vaginal wall, exfoliated vaginal epithelial cells, and others. The total amount of vaginal fluid produced daily is approximately 6 g, while at any given time; it typically ranges from 0.5 to 0.75 g within the vagina. It consists of ions, mucins, water (90–95%), proteins, and macromolecules [3]. A mucus layer in the vagina hinders the attachment of microorganisms to vaginal surfaces and facilitates their removal. Mucus layer functions as the body's primary defense mechanism, often referred to as the first line of defense [6]. The vascular network within the vagina comprises a plexus of arteries originating from arteries such as the internal iliac artery, middle rectal artery, uterine artery, and internal pudendal artery. In essence, the vaginal wall contains a substantial number of lymphatic vessels, blood vessels, and arteries [7]. In terms of the microbial environment of the vagina, lactobacilli is the most common bacteria present in women of reproductive age. Lactobacilli play a crucial role in regulating the balanced ecological environment of the vagina with pH typically ranging between 3.8 and 4.4 [8]. Furthermore, it has been reported that the composition of the vaginal microbiota is dynamic and influenced by various factors such as infections, age, pregnancy, blood glucose levels, sexual intercourse, and other factors [3].

2.2 Intravaginal Drug Delivery

The published literature reports that vaginal infections are extremely prevalent in women, specifically during their reproductive years. There are various types of common vaginal infections, including trichomoniasis, vulvovaginal candidiasis, vaginal bacteriosis, and aerobic vaginitis [9]. According to the literature, vulvovaginal candidiasis significantly impacts women's health worldwide. Recent statistics indicate that approximately 138 million women globally are affected by this condition, a number projected to rise to 158 million in the next few years. Fluconazole, a first-generation antifungal (triazole), has been widely used to treat Candida albicans infections [10]. In the United States, sexually transmitted diseases (STDs) affect more than 13 million men and women annually [11]. There is significant interest in STDs such as human immunodeficiency virus (HIV), human papillomavirus (HPV), and herpes simplex virus (HSV), as these diseases pose serious health risks and require attention and preventive measures [12]. Despite the low mortality rates associated with vaginal infections, their impact on women's quality of life and high levels of anxiety are major concerns. Published scientific reports indicate that the preference for topical treatment may be less significant than oral treatment. Therefore, there is a need for higher local drug concentrations, fewer drug interactions, and minimal side effects [9]. Extensive efforts have been made by the scientific community to explore alternative routes for delivering therapeutically active molecules that are poorly absorbed when administered orally. In this regard, the mucosal route has gained wide popularity in drug delivery [5]. To enhance the bioavailability and efficacy of drug molecules for treating vaginal infections, numerous formulations have been documented for poorly soluble molecules. These include solid dispersion, nanoparticles, nanocrystals, and nanosuspensions. However, oral administration of these dosage forms presents several limitations, such as rapid metabolism, the requirement for larger doses, which can lead to cell toxicity, and increased chances of drug resistance [10]. Since 1980 and 1990, the vaginal drug delivery systems have gained much attention [3]. Since its introduction, females have favored vaginal administration as a route for delivering a wide array of molecules for treating infections or contraception [13]. Nevertheless, vaginal dosage forms exhibit limited absorption due to the biological barriers present in the vagina, including the immune system, mucus, and the epithelium [12]. In this context, the vagina serves as a mucosal drug delivery route that can provide prolonged retention time [1]. Despite the advantages of this non-invasive route of administration, the existing literature underscores that the human vagina remains a relatively underexplored pathway for delivering therapeutically active molecules [5]. Nevertheless, the vagina holds significant importance as a site for the application of therapeutically active molecules, both for local and systemic delivery, to address female-associated health issues [14].

Historically, it has been utilized for the delivery of contraceptives, microbicides, labor inducers, and other medications [1]. According to published literature, vaginal drug delivery systems markets are broadly classified into various categories based on their therapeutic purposes. These categories encompass a wide range of applications, including contraception to reduce fertility hormones and sperm viability, promotion of cervical maturation, treatment of vaginal infections such as bacterial and fungal infections, and prevention of sexually transmitted diseases such as AIDS (acquired immune deficiency syndrome) [3]. More recently, intra-vaginal drug delivery has garnered increased attention due to its potential for more convenient drug release and enhanced therapeutic efficacy [15]. In this context, various forms of drug delivery are suitable for vaginal administration [5, 16]. It comprises solid dosage forms such as vaginal tablets, capsules, rings, and films, as well as semisolid dosage forms like vaginal creams, suppositories, pessaries, and gels. Additionally, liquid preparations such as emulsions (O/W, W/O, W/O/W), and self-emulsifying drug delivery systems are included [3]. Despite the advancements in pharmaceutical dosage forms for vaginal applications, there remain significant limitations. These include the short residence time of drugs at the vaginal site, poor drug absorption, limited permeability for hydrophilic molecules, and the need for frequent administration. These factors restrict their applications. Therefore, there is a pressing need to prioritize the design of advanced novel carriers and technologies for vaginal drug delivery systems [17]. Consequently, intra-vaginal drug delivery systems offer several advantages over other routes, such as the ability to bypass first-pass metabolism, achieve good systemic circulation [3], simplicity of administration [5], accessibility, and permeability for peptides and proteins (larger molecular weight therapeutic molecules) [18], and others. However, intravaginal drug delivery systems also present certain challenges [3] including potential for local irritation, gender specificity, issues related to personal hygiene, cultural sensitivities, and the influence of sexual intercourse [5]. Factors such as gravity and the effect of self-cleaning mechanisms may impact the absorption and retention of dosage forms administered vaginally [3], further complicating the process. Hence, there is a pressing need to develop advanced dosage forms to address the aforementioned challenges in intravaginal drug delivery systems [3, 5]. Additionally, it's important to note that the rate and extent of drug absorption through the vaginal route are linked to vaginal epithelium thickness, resulting in significant variability [5]. According to published literature, there are documented topical and systemic vaginal drug delivery systems. Topical systems are designed to deliver drugs directly to the vaginal surface. In contrast, systemic vaginal drug delivery systems are intended to deliver drug molecules through the blood vessels located near the vaginal site, allowing for systemic circulation and wider distribution of the drug [19]. The acceptability of vaginal dosage forms is a crucial factor in intra-vaginal drug delivery. It necessitates consideration of several aspects including patient compliance, ease of use, potential side effects, mode of application, effectiveness, and type of dosage forms [20]. Therefore, there is a demand to prioritize the quality, efficacy, and safety of vaginal drug delivery dosage forms [3]. Overall, it is imperative to design advanced dosage forms as the primary choice for women to address various health issues related to the vagina.

2.3 Microneedles for Drug Delivery

Microneedles is a dosage form based on micron-sized needles, with dimensions ranging from 100 to 1000 μm, cast on a suitable base plate with various shapes. It is considered a promising technology in biomedicine [21]. Moreover, it represents an efficient, unique, and minimally invasive innovation compared to traditional strategies [22]. The concept of miniature needles for delivering therapeutically active molecules was first documented in 1960, and patented by the Alza Corporation in 1971 [23]. The emergence of microneedles in the pharmaceutical field offers the potential to enhance patient access[24]. Patients do not require special medical assistance for applying microneedles, thereby increasing patient compliance due to this advanced pharmaceutical design [21]. Over the past two decades, microneedles have garnered significant attention from the scientific community in pharmaceutical platforms due to their ability to deliver therapeutically active molecules to the intended site [25]. Primarily, microneedle design has been favored for delivering drug molecules via the skin, piercing the stratum corneum, which is a major barrier to drug delivery. Microneedle design facilitates drug delivery to the epidermis layer without reaching the pain nerves and blood vessels [21]. Microneedles offer a viable alternative for delivering therapeutically active molecules to the target site, providing benefits such as painless and blood-free delivery [25]. Additionally, the use of microneedles for biosensing has been documented, where the interstitial fluid from the skin can be collected and analyzed for various biomarkers, analytes, or metabolites. This collected fluid can be utilized for disease diagnosis and drug concentration monitoring [26]. Easier penetration with microneedles can be achieved by ensuring that the tip diameter of the microneedle is less than 15 μM and there is sufficient spacing between microneedles on the base plate [27]. To date, various types of materials such as sugar, silicon, glass, polymer, metal, ceramic, and hydrogel have been utilized in the design of microneedles [22]. The polymers used for fabricating polymeric microneedles include sugars such as galactose, maltose, and sucrose, along with synthetic polymers like poly(vinyl alcohol), hyaluronic acid, and poly(vinyl pyrrolidone) [25]. Polymeric materials for microneedle design can be categorized into natural (chitosan, dextran, zein, galactose), biodegradable (polyglycolic acid, polycarbonate, polyvinylpyrrolidone, etc.), and non-biodegradable (alginic acid, polyvinyl acetate, carbopol 971 P-NF, polyetherimide). When designing microneedles, certain criteria must be considered for material selection, including good mechanical properties, adequate biocompatibility, stability, physical properties, and chemical resistance. Factors such as biodegradability, cost, and safety also play significant roles in material selection for microneedle fabrication [22]. According to the literature, five types of microneedles have been widely reported in pharmaceuticals: solid, hollow, coated, hydrogel-forming, and dissolving microneedles (Fig. 2.2) [21, 22, 28]. Microneedle shapes include pyramidal, conical, and multi-faceted piercing protrusions [28]. Due to their good mechanical strength, cone and pyramid-shaped microneedles have received more attention than other geometries. Cross-shaped microneedles pose challenges in the fabrication and removal from the application site, leading to its fewer applications in drug delivery systems [29]. Various fabrication methods such as wet etch technology, injection molding, hot embossing, isotropic etching, deep reactive ion etching techniques, micro-electromechanical systems, laser ablation, laser cutting, solvent casting, atomized spraying method, and stereolithography are crucial for microneedle design [23, 29]. Solid microneedles, composed of metal and silicon materials, are simpler types of microneedles that create micro-sized holes at the application site. Applying dosage forms to these micropore areas facilitates the diffusion of drug molecules from the application site to the skin layers [30]. Using solid microneedles, it is possible to deliver drugs from gels, solutions, creams, and transdermal patches to the target site. Cobalt alloy, nickel, silicon, palladium, and other metallic materials have been utilized for solid microneedle construction. Additionally, preferences for maltose, ceramic, sucrose, and various polymers have been documented in solid microneedle design [22]. Solid microneedles offer potential good mechanical properties even without incorporating therapeutically active molecules [31]. They operate as a poke and patch-based two-step system [28]. In the second type of microneedles, drugs are layered on the surface, termed coated microneedles. This method represents a single-step process compared to solid microneedles. However, the loading amount on the surface of the microneedle presents a significant challenge, particularly with larger drug contents [21]. In contrast to solid microneedles, coated microneedles offer a single-step process. However, the size of needles and the thickness of the coating layer significantly impact drug loading [28]. Limited loading capacity on the surface of the microneedle presents a major challenge, particularly for larger drug content applications[21]. Hollow microneedles feature a spacious cavity within the core of each microneedle and a bore on the tip, allowing for a drug solution to be filled into the empty cavity and injected into the target site based on the poke and flow principle. After complete drug delivery from the hollow microneedles, removing the microneedles patch results in the collapse of micro-sized pores formed on the skin at the application site [28]. Dissolving microneedles offer excellent drug-loading potential and possess several advantages that make them an ideal drug-delivery system. They feature a simplistic design method, easy disposal, and can be designed using a wide range of materials. Additionally, the use of dissolving microneedles enables direct drug entry based on the dissolution pattern of the microneedles, eliminating the need for additional excipients to facilitate drug release at the target site [28]. Here, the mixture of a drug and biodegradable matrix is used in the design of dissolving microneedles[31]. Hydrogel-forming microneedles are crafted using suitable types of crosslinker polymers with the support of a base plate [32]. As hydrogel-forming microneedles represent a new technology (first reported in 2012), they offer several advantages over conventional microneedles [29]. Hydrogel-forming microneedles can be drug-free, with the drug available in a reservoir affixed on the top of the base plate of the microneedles [32]. After application at the targeted site, hydrogel-forming microneedles may swell, potentially due to contact with dissolution media and the matrix used in their design [21]. In microneedles, polymeric materials require special attention with respect to non-toxicity, biodegradation, and biocompatibility [31]. Dissolvable microneedles have found diverse applications

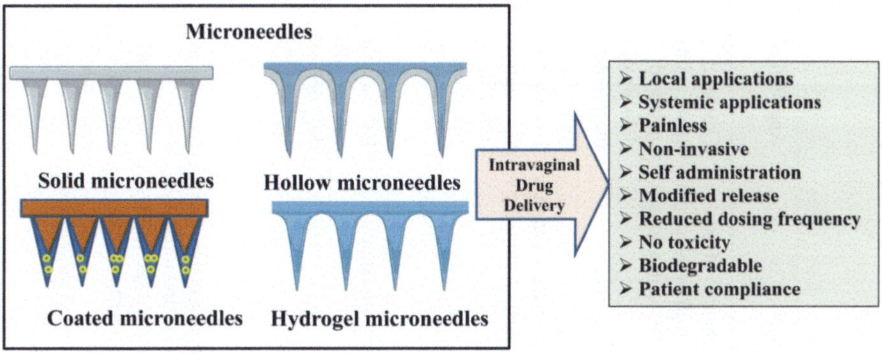

Fig. 2.2 Types of microneedles and their merits in intravaginal drug delivery systems

in cosmetics, vaccine delivery, diagnosis, drug monitoring, pain and inflammation management, treatment of diabetes, skin-related diseases, hair/scalp treatment, and autoimmune disorders management [28]. Hydrogel-forming microneedles have been utilized for skin cancer treatment, wound healing, and antibody delivery [29]. In addition, microneedles have been reported in the fields of contraception, rheumatoid arthritis, and osteoarthritis [33]. Microneedle design offers pharmaceutical advantages such as customized drug release, control over release kinetics, pulsatile drug delivery [33], good loading capacity, dosing accuracy [34], increased drug deposition, and improved bioavailability [35]. Table 2.1 overview the microneedles type.

2.4 Microneedles for Intravaginal Drug Delivery

Vaginal drug delivery has been recognized as a suitable alternative route for certain drug molecules in treating both local and systemic infections, overcoming the drawbacks associated with other routes, such as first-pass metabolism [36]. However, despite its merits, challenges persist, including the restriction of drug permeation to vaginal tissues by the outermost layer of vaginal cells [10]. In this context, microneedles represent an important and unique alternative for drug delivery across biological barriers, offering improved therapeutic potential and patient compliance [37]. Polymeric microneedles should ideally be painless and result in minimal biohazardous waste, necessitating their composition from biodegradable polymers such as polyvinylpyrrolidone (PVP), polyvinyl alcohol (PVA), among others. Additionally, the dissolution time of microneedles upon contact with vaginal mucosa is a crucial criterion, with reported times of up to 15 min in the literature [10]. Furthermore, consideration of the surface pH is essential to ensure the safety of microneedles in vaginal applications, preventing vaginal irritation. Therefore, the application of a microneedle array directly to the vaginal area within an appropriate pH range needs to be considered [10, 38]. The type of microneedles should also be carefully selected to

Table 2.1 Summary of different types of microneedles in biomedical applications

Type of microneedle	Material type	Manufacturing method	Characteristics	Advantages	Disadvantages	Applications
Solid	Silicon, polymer, metal	Lithography, etching, laser ablation, electroplating, micro molding	It provides suitable mechanical strength	It is simple to design and helps deliver the drug to the target site	There are chances of skin damage as well as infections due to the micro-incisions	Cosmetic applications include the delivery of therapeutic molecules
Hollow	Silicon, glass, stainless steel, titanium	Pulling pipettes, micro-molding, lithography, multiple pulse laser microhole drilling, etching	It can control the release of drugs over some time	There is the potential to deliver large quantities of therapeutic molecules	Weak microneedles, and may lead to chances of clogging and leakage	It is used for the diagnosis of health conditions
Coated	Silicon, mesoporous silicon, stainless steel, titanium	Post-synthetic grafting method, lithography, etching, electroplating, electropolishing, wet etching	It provides the ability to deliver proteins and peptides with a less invasive method	It enables the rapid delivery of therapeutic molecules to the skin	It is prone to infection	It is used for vaccine and drug delivery applications
Dissolving	Polymer	Photolithography, micro-molding, atomized spraying process, electro-discharge machining process, fused deposition modeling	It offers the rapid release of therapeutically active macromolecules	Easy to use/ administer and offers more patient compliance	It needs time to dissolve the microneedles, and it requires more technical expertise to design the microneedles	It is used for cosmetic, vaccine, and drug delivery applications

avoid causing injury in intravaginal drug delivery systems, necessitating the design of minimally invasive microneedles to mitigate the risk of injury. Moreover, the designed microneedles must exhibit good stability, undergo no structural damage during fabrication, possess favorable morphological properties, adequate sharpness, and flexibility [10]. The moisture adsorption ability of the designed microneedles is critical for their mechanical strength and insertion ability, thus requiring consideration of polymer properties and concentrations. Additionally, the cytotoxicity of designed microneedles to erythrocytes must be evaluated in intravaginal drug delivery to ensure non-toxicity to both erythrocytes and normal cells, ensuring their safety in treating vaginal infections. The use of microneedles for delivering active molecules should reduce dosing frequency and improve patient acceptance of intravaginal drug delivery systems [39]. The basic criteria for components used in microneedle design include an adequate manufacturing process that does not affect the stability of therapeutically active molecules, providing either immediate or controlled drug release patterns, and sufficient mechanical strength to penetrate the application site [23]. Drug molecules with low toxicity, high effectiveness, and broad applicability for fungal infections, such as fluconazole, are promising candidates for intravaginal drug delivery. These molecules are typically composed of various types of polymers, including biodegradable ones like PVA and PVP, which minimize biohazardous waste. Moreover, polymers with favorable chemical properties are preferred for microneedle design in vaginal drug delivery [10]. The use of hygroscopic polymers in microneedle design can offer a suitable water vapor transmission rate, crucial for maintaining flexibility during storage. Therefore, the selection of polymers based on their hydrophilicity and hydrophobicity is a critical consideration in microneedles for intravaginal drug delivery [10, 40]. Table 2.2 shows the suitable criteria for materials in design of microneedles for vaginal applications. Overall, the use of microneedles creates small channels in the epithelium, allowing drug molecules to penetrate to the underlying tissues. As a result of the microneedle array in intravaginal drug delivery systems, it offers improved drug absorption and enhanced bioavailability compared to traditionally used topical and oral dosage forms. Additionally, drug-loaded microneedles can provide targeted delivery of drug molecules to specific layers of the female vagina, directly benefiting the delivery of therapeutic molecules to the desired site of action. Moreover, microneedles can play a key role in reducing systemic exposure to active ingredients and help avoid potential side effects. The smaller size of microneedles reduces patient discomfort and pain compared to traditional needle injections, thus improving patient compliance. The controlled release provided by microneedles over time reduces dosing frequency, ensures sufficient therapeutic concentration of drugs, and thereby improves therapeutic outcomes. Therefore, microneedles can be considered a promising treatment for a range of therapeutic applications in intra-vaginal drug delivery systems.

Table 2.2 Criteria can help to identify the suitable materials for the design of vaginal microneedles

Sr. No	Criteria for material	Importance of criteria in the design of microneedles for vaginal applications
1	Biocompatibility	The materials used for designing microneedles must be biocompatible with the mucosa of the vagina. They should minimize the risk of adverse reactions and tissue damage upon insertion into the vaginal mucosa at the application site
2	Biodegradability	The materials used for the design of microneedles intended for dissolution or degradation after application must be biodegradable. As a result, they can provide safe removal from the body and reduce environmental impact
3	Drug compatibility	The materials selected for the design of microneedles for vaginal applications should be compatible with the drug molecules. They must ensure stable drug loading and release
4	Serializability	The materials selected for the fabrication of microneedles for vaginal applications must be compatible with widely used sterilization methods. As a result, they can provide safe and sterile microneedles for vaginal drug delivery
5	Stability	The materials intended for the design of microneedles for vaginal applications must be able to retain their structural and functional characteristics during storage, transportation, and use. They must demonstrate consistent performance over time
6	Flexibility	The materials selected for microneedles for vaginal applications must have the potential to adapt to the natural contours and movements of the vaginal canal. This can provide patient comfort and lessen the chances of tissue infection or damage
7	Mechanical properties	The materials used for microneedle fabrication for vaginal drug delivery must offer good mechanical properties to maintain structural integrity during application
8	Cost-effectiveness	During the selection of materials for the fabrication of microneedles, the cost factor should be considered. As a result, this can help prepare the device to be economically viable and accessible

2.5 Applications of Microneedles for Intravaginal Delivery

The microneedles for intravaginal delivery can help treat different vaginal infections, including bacterial vaginosis, fungal infections, and sexually transmitted infections. Here, the use of microneedles can help deliver antibiotics and antifungal, antiviral, and antibacterial agents. It offers targeted therapy and the potential to reduce systemic side effects. Moreover, the use of microneedles for intravaginal delivery offers targeted delivery for the treatment of other health ailments, such as hormonal therapies for menopausal treatment, endometriosis, and contraception. They can be used to deliver hormones such as estrogen and progesterone. Pain management (vaginal pain), vaccination, cancer (cervical cancer), and microbicide treatments are possible using microneedles via the vaginal route. Here, the use of microneedles for intravaginal delivery offers several merits to treat such infections, including localized delivery, minimized side effects, improved patient compliance, and enhanced absorption. Therefore, in this section, we have discussed the different applications of microneedles for intravaginal delivery. In concisely, the treatment of human immunodeficiency virus-1 (HIV-1) presents a significant challenge among adults. Various antiretroviral agents have been developed to address HIV-1 infection. In addition, there has been notable progress in the development of long-acting intramuscular formulations for clinical use. Examples such as novel long-acting antiretrovirals (implants and microneedle patches), novel long-acting approved oral antiretrovirals, and long-acting pro-drugs of oral antiretrovirals (tenofovir alafenamide vs tenofovir disoproxil fumarate). Despite advancements in injectable therapies, challenges such as patient compliance, self-administration, and discomfort associated with injectable treatments remain significant limitations requiring effective solutions. In 2019, McCrudden and his team designed rilpivirine-loaded long-acting nanosuspension incorporated into a dissolving microarray patch for intravaginal delivery (Fig. 2.3). The microneedle patch consisted of two layers: a drug-loaded layer of microneedle tips and a mucoadhesive baseplate. The mucoadhesive layer facilitates adhesion between the vaginal mucosa and the microneedle array. The formulation comprised 15% w/w poly (vinyl alcohol) [PVA, Mol. Wt: 9000–10000 Da] and 70% w/w long-acting drug-loaded suspension. The mucoadhesive baseplate was prepared using polyvinyl pyrrolidone (PVP, MW: 360000 Da), and Gantrez S97, with low molecular weight polyethylene glycol used as plasticizers. The interaction between Gantrez S97, containing acidic functional groups, and PVA, containing alcohol groups, resulted in a cross-linked hydrogel, making it suitable as the baseplate for drug-loaded microneedles. A dissolving drug-loaded long-acting microneedle array was fabricated using a needle-casting two-step method. When tested on synthetic tissue, the microneedle array demonstrated the creation of micropores under manual force with the drug-loaded suspension accumulating in the pores and on the tissue surface after patch removal. Similar results were observed under controlled forces of 10 N over 30 s. In both synthetic and excised bovine vaginal tissues, the designed microneedle array exhibited mechanical strength and retained its shape after the application of 10 N insertion force over 30 s. Preclinical studies showed a mean plasma concentration of rilpivirine of 116.5 ng/mL in rats at the study endpoint (56 days), comparable to the control group (intramuscular, 188.9 ng/mL). Furthermore, drug distribution studies in vaginal tissue and lymph nodes confirmed the systemic presence of rilpivirine at primary sites of viral challenge and replication. Overall, the designed microneedle array demonstrates the localized delivery capabilities, offering promising prospects for the treatment of HIV-1 infection [41].

Apart from HIV, vulvovaginal candidiasis is a major health concern in females, caused by Candida albicans in the mucosa of the vaginal cavity. Fluconazole has been reported effective in the treatment of vulvovaginal candidiasis, however, the poor aqueous solubility and first-pass metabolism limit its therapeutic outcomes against Candida albicans. In 2023, Aziz and co-workers developed fluconazole nanocrystals-based dissolving microneedles as a dual delivery system for enhanced local vaginal delivery. The fluconazole nanocrystals were designed using poly(vinyl

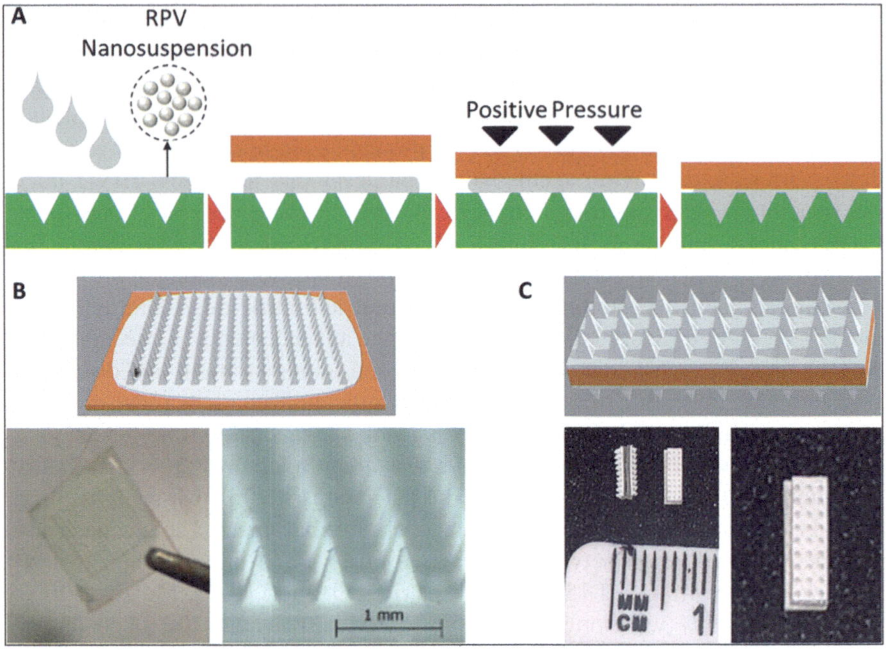

Fig. 2.3 **a** Presentation of drug-loaded long-acting microneedls. **b** Images of drug-loaded long-acting microneedles for mechanical studies. **c** Images of drug-loaded long-acting microneedles for preclinical studies. Reprinted with permission from Ref. [41]. Copyright 2019, Published by WILEY-VCH Verlag GmbH & Co. KGaA, Weinheim

alcohol) (PVA) as a stabilizer (1% w/v) with a top-down method, followed by 12 h of the modified media milling process. Subsequently, the microneedle array was designed using PVA (15% w/w) and PVP (25% w/w) via a molding method. The dissolving microneedles exhibited good recovery rates of 96.45 $\pm$ 2.38%, an average insertion percentage of 76.14 $\pm$ 2.28%, and a percentage of height decrease of 7.53 $\pm$ 0.56%. Moreover, they demonstrated rapid dissolution within 15 min. In an ex vivo vaginal infection model, the dissolving microneedles showed better anti-candidiasis efficacy compared to other treatment groups using film and hydrogel. Therefore, this strategy could be considered as a potential alternative in the treatment of vulvovaginal candidiasis [10]. Antimicrobial resistance to azoles due to increased efflux of the drug from the fungal cell remains an obstacle in the effective management of vaginal candidiasis. Probiotics have been used to overcome the antimicrobial resistance. For instance, Enggi et al. developed dissolving microneedles of PVP and PVA loaded with probiotic (*L. Plantarum*) multilayer microcapsules for treating vulvovaginal candidiasis. L. Plantarum-loaded multilayer microcapsules were designed using a multilayer double emulsion to improve the viability of probiotics and thus increase their therapeutic potential. The designed formulation exhibited good mechanical properties, sufficient insertion ability, and stability for up to

2 months. Additionally, the designed microneedle array showed rapid dissolution after insertion (in less than 15 min), offering the release of probiotics. Mucosal analysis has revealed that microneedles are non-toxic, safe, and non-irritating to the vaginal mucosa. The antimicrobial potential of microneedles against *Candida albicans* in an ex vivo infection model exhibited three times the potency compared to hydrogel and patches. In conclusion, the designed probiotic (*L. Plantarum*) multilayer microcapsule-loaded microneedle array provides a promising alternative for the treatment of vulvovaginal candidiasis [39]. Wand and co-authors developed a novel dual delivery system utilizing ammonium bicarbonate-loaded multifunctional liposomes incorporated into a microneedle array for vaginal mucosal vaccine adjuvant delivery (Fig. 2.4). The system consisted of stealth lipid A liposomes (SLLs) with a particle size of 50 nm and mannosylated lipid A liposomes (MLLs) with a particle size of 200 nm, both loaded with sodium bicarbonate as a pH-sensitive chemical and a model antigen. This combination forms the proSLL/MLL-based microneedle array. Upon administration to the vaginal mucosa, the microneedle array dissolved in tissue fluids, facilitating the release of MLLs and SLLs and enabling the delivery of active ingredients for optimal utilization. The delivery of antigen from prepared liposomes is presented by antigen presenting cells via major histocompatibility complex class I that leads to immune response. Here, it is possible due to the potential of liposomes to escape the lysosomes. It is because of the release of sodium bicarbonate that offers the rupture of lysosome and production of reactive oxygen species. Studies in mice revealed superior levels of antigen-specific cellular and humoral immunity from the microneedles compared to traditional intradermal administration. Additionally, MLLs were found to be taken up by dendritic cells within the vaginal mucosa, while SLLs were drained into lymph nodes. Here, the designed stealth nanocarriers move efficiently via the lymphatic system and body fluids eventually leading to the draining lymph nodes. Antigen release from the engineered liposomes was facilitated by lysosome escape and stimulation of reactive oxygen species, leading to a presentation of MHC-I by antigen-presenting cells. Consequently, the antigen-engineered liposomes protected mice against viral infection. Overall, this innovative vaginal mucosal vaccine adjuvant dual delivery system represents a promising approach to preventing sexually transmitted infections, opening new avenues in the field [42]. In conclusion, microneedle-based drug delivery systems offer a promising approach for administering antifungal drugs, probiotics, antiretroviral drugs, and other treatments for vaginal infections in females.

2.6 Current Challenges and Prospects

Recent reports on microneedle-based intravaginal drug delivery have demonstrated its potential in addressing infections prevalent among women, particularly antimicrobial infections. Despite significant progress in managing microbial infections, the exploration of microneedles for this purpose is still ongoing. A few studies have investigated the delivery of molecules such as azoles, retroviral, and probiotics via

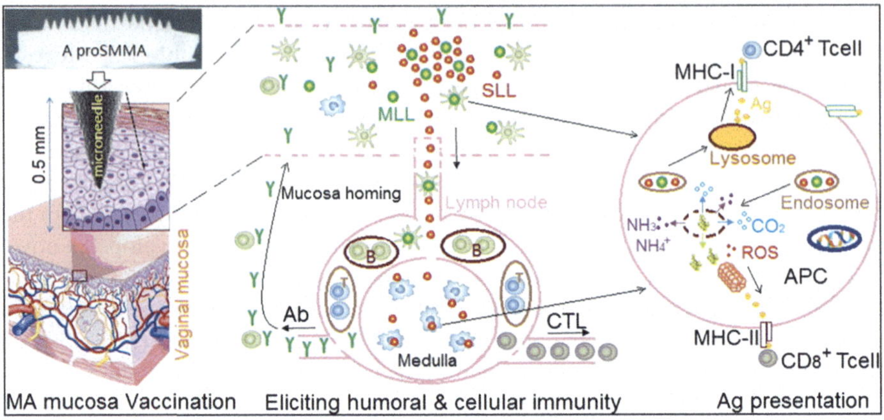

Fig. 2.4 Schematic presentation of proSMMA incorporated polymeric microneedles for vaginal mucosal vaccine adjuvant-dual delivery system. Reprinted with permission from Ref. [42]. Copyright 2016 Elsevier B.V. All rights reserved

microneedles to the vaginal site. In the future, designing microneedle arrays for the effective delivery of active pharmaceutical agents to treat vaginal infections remains a priority.

Several reports have discussed dissolvable microneedles capable of delivering therapeutically active molecules directly to the vaginal site. Moreover, advanced microneedles incorporating nanomaterials such as metal nanoparticles and carbon-based nanomaterials hold great promise. However, despite the advantages of microneedle-based intravaginal delivery, there are several critical factors to consider during development. These include the chemical and physical stability of the microneedles during fabrication, sterilization methods, process complexity, and cost.

For antimicrobial applications, both natural phytoconstituents and synthetic antimicrobial agents should be explored. Dual-delivery approaches, combining antibiotics with other therapeutic agents, could also enhance the effectiveness of infection management. While most studies have used artificial vaginal tissue models, future research should focus on animal models to improve real-time applicability. Toxicity assessments on vaginal cells are crucial, particularly regarding the safety of polymeric matrices used in microneedle fabrication. Only a few polymers have been deemed safe, so more testing of natural, synthetic, and semi-synthetic polymers is needed to identify suitable candidates for intravaginal drug delivery.

The compatibility of microneedle materials with blood cells should also be considered, as interactions could occur during application. Furthermore, microneedles could be used to deliver biomolecules aimed at boosting the immune system, thus enhancing women's reproductive health. When loading microneedles with active compounds, care must be taken to precisely load the molecules and maintain appropriate pH levels to avoid toxicity. Irritation at the application site is another important factor that needs to be addressed. Preclinical and clinical studies are essential to verify the safety and efficacy of microneedles in intravaginal applications. Regulatory guidelines for the good manufacturing practices (GMP) of microneedles, along with standardized pharmacopeial methods to assess their safety and performance, are still under development. Regulatory considerations must include evaluations of efficacy, safety, and quality standards for microneedle-based vaginal drug delivery systems. The potential for irritation, discomfort, or even microtears in the vaginal mucosa due to microneedle application is a critical concern. Proper design considerations, such as the number of applications per site and the long-term impact on reproductive health, require further investigation. To date, no comprehensive studies have assessed the long-term effects of microneedle use on fertility and the normal functioning of reproductive organs. In addition to tissue health and potential adverse reactions, the impact of microneedles on the natural vaginal microbiota should be carefully evaluated. Regulatory bodies must ensure rigorous testing and data collection to facilitate the practical use of microneedles for vaginal drug delivery. One key aspect is determining the optimal pressure required to puncture the vaginal mucosa effectively. Inadequate pressure may result in suboptimal drug delivery, compromising therapeutic outcomes.

The selection of drug molecules, microneedle types, and materials used in microneedle fabrication should be carefully considered. Additionally, the size and shape of microneedles must be optimized to minimize pain and invasiveness while ensuring efficient delivery of active compounds to the vaginal site. Psychological factors also play a significant role in the acceptance of microneedle-based drug delivery systems. Due to the invasive nature of the procedure, patients may experience discomfort or anxiety. Providing proper education and effective communication regarding the safety and benefits of microneedles can alleviate concerns and reduce anxiety. Cultural attitudes toward reproductive health and vaginal administration must also be considered, as these factors can affect patient acceptance. In conclusion, there is vast potential for microneedle-based intravaginal drug delivery systems, particularly for the treatment of vaginal infections. Addressing psychological, cultural, and educational factors is essential to increase patient acceptance and improve therapeutic outcomes. With further research and regulatory development, microneedles could revolutionize the treatment of vaginal infections and other reproductive health conditions.

2.7 Conclusions

The delivery of drugs to the vagina using suitable types of microneedles holds significant potential in the prevention and therapy of fungal, viral, and other infections in females. Importantly, the use of biodegradable polymers for the design of dissolvable microneedles offers rapid dissolution within 15 min, painlessness, and customized delivery. Additionally, microneedle arrays exhibit good chemical and mechanical properties that ensure safety, flexibility, and insertion ability. Moreover, the polymeric microneedle is biocompatibility with erythrocytes as well as vaginal tissues. To date, some antifungals, antiretrovirals, and biomolecules have been delivered using microneedle arrays. Drug delivery to the vagina can result in both local and systemic actions. Delivery of probiotics using microneedles was found to overcome antimicrobial resistance. In vivo and ex vivo studies showed the usefulness of microneedles for intra-vaginal drug delivery. In conclusion, the design of polymeric microneedles offers a new avenue for vaginal drug delivery over conventional dosage forms. Despite significant progress, standardized testing protocols for microneedle arrays in intravaginal delivery need to be developed.

Acknowledgements The Dr. Sopan Nangare is thankful to Krishna Vishwa Vidyapeeth (Deemed to be University), Krishna Institute of Pharmacy, Karad, for providing the facilities to complete this book chapter.

Conflict of Interest There is no conflict of interest.

References

1. Das Neves, J., Bahia, M.F.: Gels as vaginal drug delivery systems. Int. J. Pharm. **318**(1–2), 1–14 (2006)
2. Luo, J., Betschart, C., Ashton-Miller, J.A., et al.: Quantitative analyses of variability in normal vaginal shape and dimension on MR images. Int. Urogynecol. J. **27**, 1087–1095 (2016)
3. Yang, Z., Wu, X., Wang, H., et al.: Vagina, a promising route for drug delivery. J. Drug Deliv. Sci. Technol. 105397 (2024)
4. Graziottin, A., Gambini, D.: Anatomy and physiology of genital organs–women. Handb. Clin. Neurol. **130**, 39–60 (2015)
5. Hussain, A., Ahsan, F.: The vagina as a route for systemic drug delivery. J. Control. Release **103**(2), 301–313 (2005)
6. Adnane, M., Meade, K.G., O'Farrelly, C.: Cervico-vaginal mucus (CVM)—An accessible source of immunologically informative biomolecules. Vet. Res. Commun. **42**(4), 255–263 (2018)
7. Richardson, J., Illum, L.: VIII: the vaginal route of peptide and protein drug delivery. Adv. Drug Deliv. Rev. **8**(2–3), 341–366 (1992)
8. Ventolini, G.: Progresses in vaginal microflora physiology and implications for bacterial vaginosis and candidiasis. Women's Health **12**(3), 283–291 (2016)
9. Palmeira-de-Oliveira, R., Palmeira-de-Oliveira, A., Martinez-de-Oliveira, J.: New strategies for local treatment of vaginal infections. Adv. Drug Deliv. Rev. **92**, 105–122 (2015)
10. Aziz, A.Y.R., Mahfufah, U., Syahirah, N.A., et al.: Dual delivery systems combining nanocrystals and dissolving microneedles for improved local vaginal delivery of fluconazole. Drug Deliv. Trans. Res. 1–15 (2023)
11. Rohan, L.C., Sassi, A.B.: Vaginal drug delivery systems for HIV prevention. AAPS J. **11**, 78–87 (2009)
12. Osmałek, T., Froelich, A., Jadach, B., et al.: Recent advances in polymer-based vaginal drug delivery systems. Pharmaceutics **13**(6), 884 (2021)
13. Woolfson, A.D.: Intravaginal drug delivery technologies. Modified-release drug delivery technology, pp. 783–798. CRC Press (2002)
14. Baloglu, E., Bernkop-Schnürch, A., Karavana, S.Y., et al.: Strategies to prolong the intravaginal residence time of drug delivery systems. J. Pharm. Pharm. Sci. **12**(3), 312–336 (2009)
15. Kumar, A., Kumar, S.: Intra Vaginal drug delivery system (novel drug delivery system). Int. J. Res. Appl. Sci. Biotechnol. (IJRASB) **7**(6), 234–241 (2020)
16. Woolfson, A.D., Malcolm, R.K., Gallagher, R.: Drug delivery by the intravaginal route. Crit. Rev. Ther. Drug Carrier Syst. **17**(5), 509–555 (2000)

17. Li, W.-Z., Hao, X.-L., Zhao, N., et al.: Propylene glycol-embodying deformable liposomes as a novel drug delivery carrier for vaginal fibrauretine delivery applications. J. Control. Release **226**, 107–114 (2016)
18. Justin-Temu, M., Damian, F., Kinget, R., et al.: Intravaginal gels as drug delivery systems. J. Women's Health **13**(7), 834–844 (2004)
19. Ensign, L.M., Cone, R., Hanes, J.: Nanoparticle-based drug delivery to the vagina: a review. J. Control. Release **190**, 500–514 (2014)
20. Yu, T., Malcolm, K., Woolfson, D., et al.: Vaginal gel drug delivery systems: understanding rheological characteristics and performance. Expert Opin. Drug Deliv. **8**(10), 1309–1322 (2011)
21. Nainggolan, A.D.C., Anjani, Q.K., Hartrianti, P., et al.: Microneedle-mediated transdermal delivery of genetic materials, stem cells, and secretome: an update and progression. Pharmaceutics. **15**(12), 2767 (2023)
22. Dugam, S., Tade, R., Dhole, R., et al.: Emerging era of microneedle array for pharmaceutical and biomedical applications: recent advances and toxicological perspectives. Future J. Pharm. Sci. **7**, 1–26 (2021)
23. Tucak, A., Sirbubalo, M., Hindija, L., et al.: Microneedles: characteristics, materials, production methods and commercial development. Micromachines. **11**(11), 961 (2020)
24. Donnelly, R.F., Prausnitz, M.R.: The promise of microneedle technologies for drug delivery. Drug Deliv. Trans. Res. **14**(3), 573–580 (2024)
25. Roussel, S., Udabe, J., Sabri, A.B., et al.: Leveraging novel innovative thermoresponsive polymers in microneedles for targeted intradermal deposition. Int. J. Pharm. **652**, 123847 (2024)
26. Vora, L.K., Sabri, A.H., McKenna, P.E., et al.: Microneedle-based biosensing. Nat. Rev. Bioeng. **2**(1), 64–81 (2024)
27. Le, Z., Yu, J., Quek, Y.J., et al.: Design principles of microneedles for drug delivery and sampling applications. Mater. Today **63**, 137–169 (2023)
28. Sartawi, Z., Blackshields, C., Faisal, W.: Dissolving microneedles: applications and growing therapeutic potential. J. Control. Release **348**, 186–205 (2022)
29. Turner, J.G., White, L.R., Estrela, P., et al.: Hydrogel-forming microneedles: current advancements and future trends. Macromol. Biosci. **21**(2), 2000307 (2021)
30. Donnelly, R., Douroumis, D.: Microneedles for Drug and Vaccine Delivery and Patient Monitoring, pp. 311–12. Springer (2015)
31. Battah, B., Shbibe, L., Ahmad, O., et al.: *Juniperus oxycedrus* L. ssp. Essential oil microneedles: a promising antimicrobial and wound healing activity. Pharmaceuticals **17**(1), 40 (2023)
32. Abraham, A.M., Anjani, Q.K., Adhami, M., et al.: Novel SmartReservoirs for hydrogel-forming microneedles to improve the transdermal delivery of rifampicin. J. Mater. Chem. B. (2024)
33. Zhang, W., Zhang, W., Li, C., et al.: Recent advances of microneedles and their application in disease treatment. Int. J. Mol. Sci. **23**(5), 2401 (2022)
34. Avcil, M., Çelik, A.: Microneedles in drug delivery: progress and challenges. Micromachines **12**(11), 1321 (2021)
35. Cárcamo-Martínez, Á., Mallon, B., Domínguez-Robles, J., et al.: Hollow microneedles: a perspective in biomedical applications. Int. J. Pharm. **599**, 120455 (2021)
36. Permana, A.D., Paredes, A.J., Zanutto, F.V., et al.: Albendazole nanocrystal-based dissolving microneedles with improved pharmacokinetic performance for enhanced treatment of cystic echinococcosis. ACS Appl. Mater. Interfaces **13**(32), 38745–38760 (2021)
37. Wang, C., Jiang, X., Zeng, Y., et al.: Rapidly separable microneedle patches for controlled release of therapeutics for long-acting therapies. Med. Drug Discov. **13**, 100118 (2022)
38. Hayashida, S., Takada, K., Melnikov, V.G., et al.: How were Lactobacillus species selected as single dominant species in the human vaginal microbiota? Coevolution of humans and Lactobacillus. Med. Hypotheses **163**, 110858 (2022)
39. Enggi, C.K., Sulistiawati, S., Stephanie, S., et al.: Development of probiotic loaded multilayer microcapsules incorporated into dissolving microneedles for potential improvement treatment of vulvovaginal candidiasis: a proof of concept study. J. Colloid Interface Sci. **648**, 203–219 (2023)

40. Singh, A., Bali, A.: Formulation and characterization of transdermal patches for controlled delivery of duloxetine hydrochloride. J. Anal. Sci. Technol. **7**, 1–13 (2016)
41. Mc Crudden, M.T., Larrañeta, E., Clark, A., et al.: Design, formulation, and evaluation of novel dissolving microarray patches containing rilpivirine for intravaginal delivery. Adv. Healthc. Mater. **8**(9), 1801510 (2019)
42. Wang, N., Zhen, Y., Jin, Y., et al.: Combining different types of multifunctional liposomes loaded with ammonium bicarbonate to fabricate microneedle arrays as a vaginal mucosal vaccine adjuvant-dual delivery system (VADDS). J. Control. Release **246**, 12–29 (2017)

Chapter 3
Hollow-Microneedles-Based Injectable Delivery System

Mehdi Abedi, Younes Ghasemi, and Sara Ranjbar

3.1 Introduction

Injecting drugs subcutaneously with a hypodermic needle is more effective than oral administration, which often fails to yield the desired results. This occurs because the body might not effectively absorb the drug in the gastrointestinal tract, or it may be degraded and metabolized in the liver [1, 2]. Right now, using needles for under-the-skin injections is close to leading to new ways to deliver medicine. However, the administration of hypodermic needles needs specialized training for patients; otherwise, their application is restricted to clinical settings. Additionally, the pain, satisfaction and distress experienced by patients due to the size of the hypodermic needle are significant concerns [3, 4]. The practice of using needles multiple times contributes greatly to the spread of blood-transmitted infections [5, 6]. Different approaches to delivering medication have been investigated, but none match the efficacy of administering it directly into the body using a needle. MNs (MNs) are an excellent method for administering drug transdermal, as they minimize the risk of unintended needle injuries, are less intrusive for patients, and offer efficient drug delivery, potentially lowering the required dosage [7–11]. The majority of research conducted on MNs has mainly concentrated on transdermally administering drugs

M. Abedi (✉) · Y. Ghasemi · S. Ranjbar
Pharmaceutical Sciences Research Center, Shiraz University of Medical Sciences, Shiraz, Iran
e-mail: m_abedi@sums.ac.ir; mehdi.abedi.a@gmail.com

M. Abedi
Research and Development Department, Danesh Salamat Kowsar Co, Shiraz, Iran

Y. Ghasemi
Department of Pharmaceutical Biotechnology, School of Pharmacy, Shiraz University of Medical Sciences, Shiraz, Iran

S. Ranjbar
Computational Vaccine and Drug Design Research Center, Shiraz University of Medical Sciences, Shiraz, Iran

N. Talreja et al. (eds.), *Microneedles (MNs)-Based Technology*, Series in BioEngineering, https://doi.org/10.1007/978-981-96-3916-8_3

and vaccines. The effectiveness of conventional skin patches is hindered by the stratum corneum, which acts as a barrier on the skin's surface. Different methods using chemicals, biological processes, and physical techniques have been researched to enhance permeation, penetration and absorption through the skin [12–14]. Nevertheless, utilizing chemical and biochemical techniques has proven to be ineffective for administering drugs and vaccines through the skin. Although physical methods such as Iontophoresis, electroporation, and ultrasound show some success at delivering large molecules, they typically require sophisticated, costly equipment and specialized training for operation. MNs enable the development of a cost-effective patch that simplifies the process for patients to deliver vaccines and drugs [15, 16].

The design of hollow MNs enables a straightforward passage for the administration of drugs into the skin and other tissues. Similar to traditional needles, hollow MNs facilitate the flow of liquid when pressure is exerted, as shown in Fig. 3.1. Pressure and flow rates can be adjusted for quick injections, slow infusions, or changing delivery speeds over time. Liquid formulation makes it easier to use current injectable products with MNs. This approach may enhance the stability of the drugs and provide a more user-friendly method for patients to utilize a patch for administration. The application of hollow MNs has enabled drug delivery into the skin from a medication source that does not require any pressure [17]. Hollow MNs offer distinct advantages over solid MNs and other MN types. Here are some key benefits: (a) They enable the delivery variety volumes of liquid formulations, including vaccines, drug, insulin, and other biologic, directly into the skin through a 'poke and flow' mechanism, allowing for controlled drug administration [17–19], (b) While all MNs are designed to minimize pain compared to traditional hypodermic needles, hollow MNs can create a less traumatic experience, as they allow for a more uniform and controlled penetration of the skin, however due to infusion pressure but still usually caused little pain [20, 21], (c) Deliver drugs directly into the targeted tissue can enhance bioavailability and therapeutic efficacy, allowing for lower doses and potentially fewer systemic side effects [22], (d) Hollow MNs can accommodate a range of formulations, including liquid drugs, vaccines, and even gene therapies, making them versatile for different medical applications [23], (e) facilitate the delivery of vaccines with improved immune response, as they can be designed to preferentially target immune-rich areas of the skin [24], (f) Certain designs could enable controlled release of medications over time, providing sustained delivery of therapeutic agents [25], (g) Since hollow MNs can be designed for one-time use, there's a lower risk of contamination compared to multi-use solid MNs [25].

3.2 Hollow Microneedle Fabrication Methods

Hollow MNs, designed with a small hole in the center or on the side, have widths from 5 to 70 μm. They can be filled with a drug liquid or left empty and connected to a drug solution container [26]. These MNs establish a route for the medicine to access the outer skin layer immediately after insertion [27]. Different materials, such

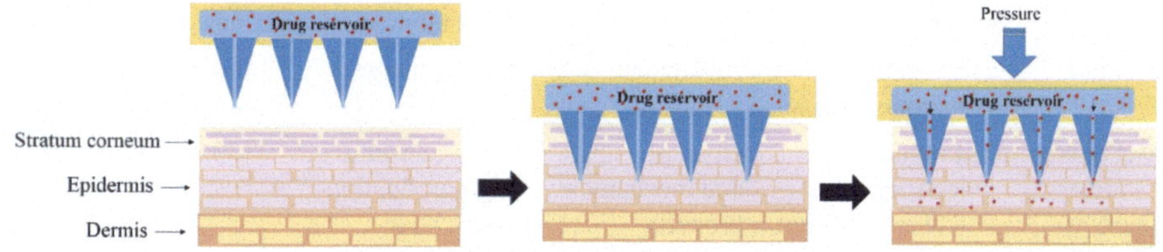

Fig. 3.1 Mechanism of drug release hollow MNs. Diffusion, capillary action, or application of pressure

as metal, silicon, Borosilicates and polymer, can be used to create hollow MNs [28]. Hollow MNs have been made from a material using advanced methods like microelectromechanical systems (MEMS) [29], laser cutting [30–33], metal electroplating [34], X-ray lithography [35–38], lithographic molding [19, 39–41], micro-molding [42], chemical etching [41, 43–45], reactive-ion etching [46–48], 3D print [49, 50], and other techniques to shape and create them. Figure 3.2 shows several MNs fabricated with different materials and techniques. The advantages and disadvantages of these material and methods are presented in Table 3.1.

However, for the fabrication of Hollow MNs, a combination of different techniques is usually used. Still, Different fabrication approaches are briefly described as

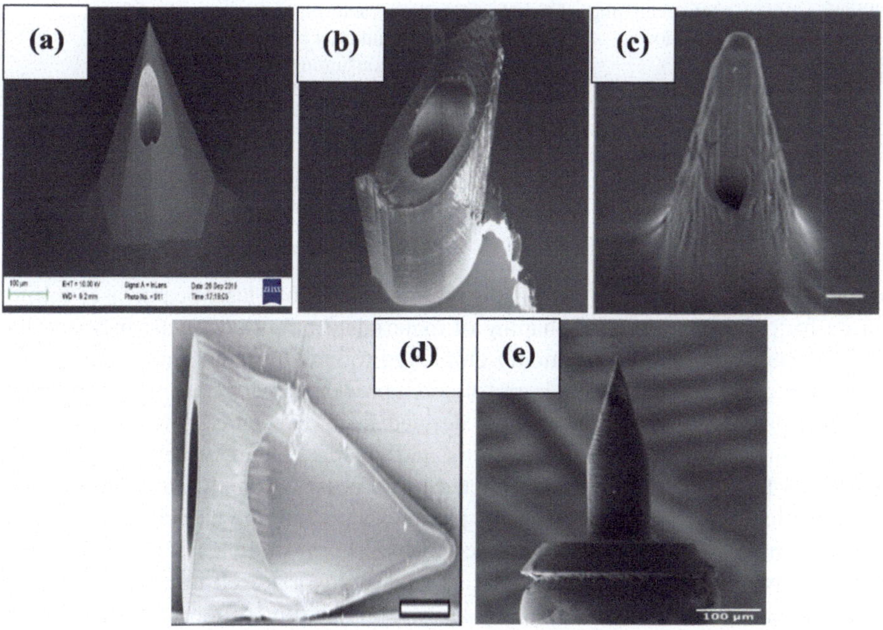

Fig. 3.2 MNs prepared by **a** Chemical etching [51], **b** reactive ion etching [47], **c** laser cutting molding [52], **d** photolithography [53], **e** 3D print [20]. The images were reproduced with permission [47, 51]

Table 3.1 Some advantages and drawbacks of hollow MNs fabrication technique

Method	Fabrication method	Advantage	Disadvantage	References
Metal (Stainless steel, Titanium, Nickel, Gold)	Molding, electroplating, electrodeposition, laser micromachined, laser cutting, 3D printing	Good physical and mechanical property	Time-consuming, high production costs, poor biocompatibility, expensive instrument, corrosion, poor hole quality, hole blockage	[30, 33, 34, 54–56]
Ceramic (Alumina, Zirconia, Calcium phosphate/ sulphate)	Lithography, micromoulding, ceramic sintering,	Good biocompatibility, good chemical and mechanical property	Fractures, the shape and size of the needles are not the same	[55–58]
Polymer	3D print, lithography, Injection micro-molding, Solvent casting, laser ablation	Low-cost material, good biocompatibility, degradable capability	Poor mechanical property, multi-step process, low resolution of needle tips	[19, 59, 60]
Silicon	Chemical etching, reactive ion etching, lithography, etching	Desirable physical and mechanical properties	Fractures, high material cost, time consuming fabrication	[46, 55, 61, 62]
Borosilicates	Pulling pipettes	Good biocompatibility	Fracture, time consuming fabrication	[63–65]

follows: laser cutting: By employing the thermal energy of a laser beam, laser cutting allows for precise cutting and shaping of materials. In the sheet metal industry, the laser cutting technique is commonly employed to slice through hard materials. In laser cutting, the material or the laser follows a predetermined route, which has been designed through a computer program derived from a specific drawing. The main parts are the laser, a control for opening and closing, a unit to deliver the beam, and lenses to focus the light. The directed beam passes through a nozzle, which is accompanied by a flow of gas. The gas jet plays a crucial role in facilitating cutting and ensuring the lenses remain unharmed from airborne particles. It also helps push materials away from the cutting area. Laser cutting works well in terms of making narrow cuts, smooth surfaces, shaping edges, and handling different types of materials compared to other cutting methods like plasma cutting [66]. In hollow MNs, holes are laser-drilled through a polymer or metal substrate.

3.2.1 X-Ray Lithography

The X-ray lithography (XRL) process utilizes unique substances referred to as resists. These resists react to X-rays by changing how easily they dissolve in a certain liquid after being exposed to the X-rays. This process allows them to be patterned through a mask placed in front of the X-ray source. The mask is made of heavy materials like gold and tungsten that help absorb things. The base layer is made of materials like silicon and silicon carbide that allow things to pass easily [36].

3.2.2 Micromolding

The process of micromolding involves crafting minuscule polymeric forms with the help of molds, often known as masters or tools, to achieve precise detailing. The three most common ways to make small shapes are injection molding, hot embossing, and casting. The process of micromolding consists of four primary stages, first construction of a mold that features an inverted version of the desired shape. Second, pour the plastic substance into the mold. Third, give the plastic time to solidify or decrease in temperature. Ultimately, the completed plastic will be extracted from the mold [66].

3.2.3 3D Printing

This process involves constructing solid items through the accumulation of material guided by a digitally designed template. The process of 3D printing involves accumulating layers of material to construct different designs. To print, the machine uses a computer program and adds layers of material, one on top of the other, to create the final shape. These layers autonomously join or fuse to establish the final configuration. This method lets you make almost any shape or design [67]. Currently, various mechanisms have been developed for 3D printing, such as Fused Deposition Modeling (FDM), Digital Light Processing (DLP), and Stereolithography Apparatus (SLA) [68]. FDM utilizes plastic materials that are formed into elongated filament strands. In FDM, an object is created by sequentially depositing melted material in layers, adhering to the design specified in the computer model. SLA is a 3D printing technique that involves a liquid resin that solidifies upon exposure to light. A tank is filled with resin, and UV light is directed at it to form solid shapes. The resin solidifies gradually under the UV light until the final product is completed. DLP is a three-dimensional printing technique that rapidly fabricates components using light-reactive plastic. The process is quite comparable to SLA, but the primary difference is that SLA printers draw layers individually with a laser, in contrast to DLP printers, which use a light source to solidify an entire layer at once [69, 70]. The procedures for MEMS techniques for hollow MNs involve a meticulously orchestrated three-step process, including Deposition, Patterning and Etching, which is detailed in the following sections [55].

3.2.4 Deposition

This stage involves the formation of a thin layer employing methods such as chemical or physical vapour deposition (CVD or PVD) or through the spinning of a liquid coating. In the process of PVD, a material is vaporized from either a solid or liquid form. The vapor moves to the surface in a low-pressure or air-depleted environment, where it cools and forms a delicate layer ranging from a few nanometers to several micrometres thick. Generally, deposition rates lie within the range of 1–10 nm/s. In the CVD process, gases are injected into a specific chamber, commonly operating in a vacuum, to fabricate new materials. As the chamber reaches the appropriate temperature, the gases undergo decomposition or react, forming a coating that adheres to the material's surface [71]. Spin coating is a traditional method employed to create a thin film of material on various surfaces. The technique relies on liquid solvents that evaporate during spinning, contributing to the formation of a solid layer that is thin as it dries [72].

3.2.5 Patterning

After laying down the necessary film layer, a flat design of the material is copied from the original photomask onto the light-sensitive film on the surface. Usually, a silicon wafer is used as a base, and the transfer process is done using photolithography, which uses light as the source. In this method, UV light traverses a designated pattern, transferring that pattern onto a silicon wafer. This method is employed to meticulously eliminate specific areas of a thin layer or the material beneath it by modifying the chemical properties or the dissolution rate of the photoresist. The photoresist is composed of a blend of polymers and a substance that reacts to light. A geometric design is shifted from a photomask to a light-sensitive layer on a surface, which may have either positive or negative characteristics. The interaction of UV light with a positive photoresist leads to the breakdown of polymer chains, enhancing its ability to dissolve in a chemical solvent. Conversely, exposure to UV light causes a negative photoresist to form polymers that are insoluble as they bond together [55].

3.2.6 *Etching*

There are two approaches to etching, one involves wet processes, while the other employs plasma-based dry methods. The initial method involves using etching to remove excess material by immersing the surface in a chemical solution. Conversely, a vapour or plasma etcher is used for dry etching. In plasma-based etching, which is also known as reactive ion etching (RIE), a gas is energized to become active. This promotes the reaction of the gas with the surface of the substance. Due to the slow etching speed in RIE processes, achieving a high width-to-height ratio when creating structures becomes challenging. Deep reactive ion etching (DRIE) is an effective technique for fabricating off-plane hollow microstructures featuring openings that measure several hundred μm in width. Although wet etching offers a cost-effective solution, the superior quality and sharpness of MN tips result from wet and dry etching [55]. Schematic overview fabrication of hollow MN by photolithography described in Fig. 3.3.

3.3 Skin Delivery of Drugs with Hollow MNs

A group of researchers examined the effectiveness of hollow MNs in administering tofacitinib citrate through the skin. The authors compared different types of m MNs, including dissolving MNs, hollow MN arrays filled with powder, and a standard Aqueous Cream BP. Hollow MNs were made and then filled with tofacitinib citrate using special. Insertion of hollow MN arrays was simple, and they demonstrated similar strength against pressure as dissolving MNs. MN arrays were slightly more effective in delivering tofacitinib into pig skin compared to Aqueous Cream BP. However, they did not perform as well as dissolving MN arrays for the delivery of tofacitinib [73]. Yadav created a hollow MNs using 3D printing and stereolithography. The hollow MNs were used to deliver rifampicin through the skin, which

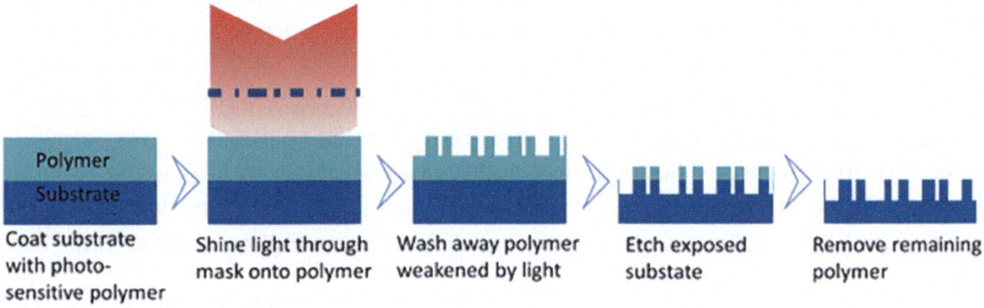

Fig. 3.3 Schematic illustration of hollow MNs fabrication via photolithography. Reproduced with permission from the cited Ref. [55]

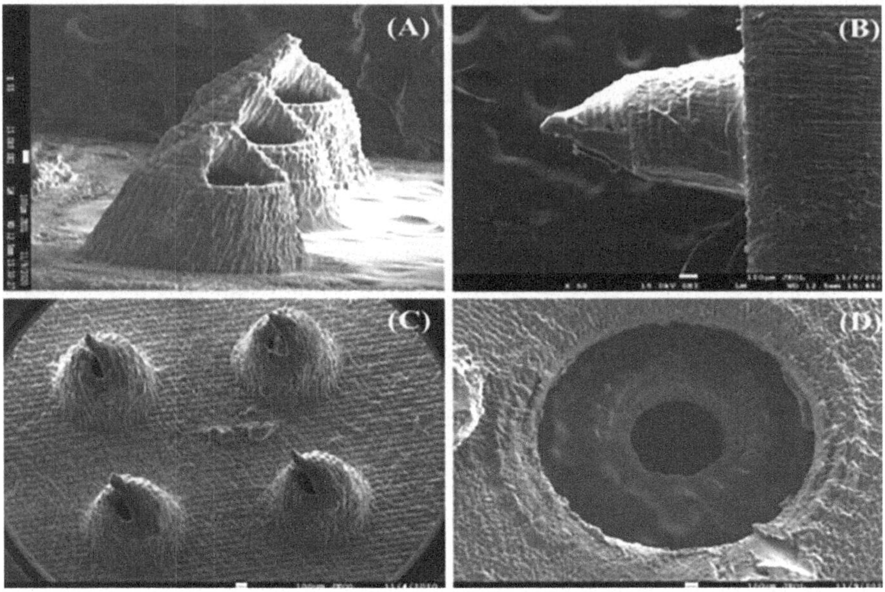

Fig. 3.4 FESEM images of the 3D printed HMNs array showing efficient printing of micron size intricate details of the HMNs and HMNs assembly. **a** and **b** Side view images of the HMNs; **c** top view: and **d** reservoir opening. Reproduced with permission from the cited Ref. [74]

suffers from many problems such as degradation due to gastric acid, is not effectively absorbed by the body, and poses significant risks to the liver. The design of the hollow MNs array incorporates small holes near the tips of one-fourth of the needles (Fig. 3.4) to enhance its strength and reliability. In experiments with Sprague-Dawley rats, it was found that using an MNs system for rifampicin delivery resulted in better drug absorption in the body [74].

Ryan's colleagues displayed the capability of bacteria to the bloodstream by MNs, and developed research to deliver *E. coli* bacteria through the skin to help treat infections. They used hollow MNs made from polycarbonate that were created using a Hollow MN, and these hollow MNs were attached to a syringe for administration. Research on skin pigs indicated that after injection of 1 ml *E. Coli* solution by hollow MNs, only a small quantity of *E. Coli* was detected remaining on the skin's. Experiments were conducted on rats, potentially related to the skin's resistance to the flow of the solution. They administered 250 μL of the solution at four separate locations. Thirty minutes after administration, bacteriophages were detected in the bloodstream. This study showed that we can deliver bacteriophages through the skin [75].

Wang and his team created Hollow MNs, which were made using 3D printing by using stereolithography. Then, they combined these structures with a MEMS. Using special imaging methods, they monitored how liquid moved through the skin tissue as the device was delivering it. Tests done on living animals showed that using the 3D

Table 3.2 Drugs delivered using hollow MNs

MNs material	Fabrication technique	Delivered drug	Target tissue	References
Methyl vinyl ether/maleic anhydride		Amphotericin B		[77]
Photosensitive resin (polymer)	3D printing/ lithography	Insulin delivery	Skin	[78]
Polyimide, a poly (vinylidene fluoride-co-trifluoroethylene) copolymer-single-walled carbon nanotube	Electroplating	Biosensor		[72]
Biocompatible/photo-curable resin	Casting	Sulforhodamine B	Skin	[79]
Borosilicate	Fire-polishing borosilicate glass pipettes	Sulforhodamine	Eye	[80]
Borosilicate	Fire-polishing borosilicate glass pipettes	Sulforhodamine	Eye	[81]
Glass	Fire-polishing borosilicate glass pipettes	Sulforhodamine	Skin	[82]
Borosilicate	Borosilicate micropipette tubes	Sulforhodamine B	Eye	[80]
Glass	Lithography/laser cutting	Phenylephrine	Eye	[83]
Polystyrene, polymethyl methacrylate, epoxy resin	Molding	FITC-BSA	Skin	[21]
PEEK	Laser micromachining	BSA-FITC	Skin	[84]
Resin	Stereolithographic 3D-print	Insulin	Skin	[85]

MNs-MEMS to deliver insulin worked better for managing blood sugar in diabetic animals than regular injections under the skin. These results showed that the 3D MNs-MEMS could be a flexible and effective way to deliver medication through the skin for individual care [76]. Some applications of hollow MNs for drug delivery are presented in Table 3.2.

3.3.1 Cell Delivery

Gualeni investigated the application of MN for cell therapy in the treatment of various dermatological issues, including burns, abrasions, scars, and discolouration. The main aim of their investigation was to employ hollow MNs to direct the application of cell therapy to the skin. This therapy involves using a person's epidermal cells, keratinocytes, and melanocytes. Compared to the traditional available methods of skin therapy, this new technology, which results in considerable pain and is associated with infection risk, is significantly more effective. To determine the viability of the cells after MNA treatment, they employed a blue dye test in conjunction with a counting device. They evaluated the adhesion of the cells and assessed the number of cells that had grown after 24 and 48 h. The type of cells was analyzed using Western blot tests on cell samples or immunofluorescence. According to the results, cells were able to survive passage through a 75 μm hole, whereas their mortality rate increased substantially with smaller holes. The cell quantity did not change following the MNs passage. After the process, the cell appearance remained unchanged, with a hole size of more than 75 μm. The cells maintained their activity after bieng inserted into the skin. Ex vivo human cells injected by hollow MNs revealed unique markers associated with keratinocytes and melanocytes, demonstrating that they preserved their identity following administration. The process of skin healing was also evident in organisms that are alive. By using skin scanner, at scheduled times, an image was taken from MN insertion of the skin of volunteer people to check how well the tiny holes were closing up. The paths made by MNA were blocked within 4–24 h [86].

A new technique for fabricating silicon based hollow MNs was introduced by Iliescu and his team. These MNs facilitate a more efficient delivery of cell solution to tissues during cell therapy. The integration of SiO_2 and Si_3N_4 provides MNs with a hydrophilic surface, a thin wall structure, and minimal pressure. Single-crystal silicon wafers were cleaned using piranha. A SiO_2 layer was deposited on the purified water, then a photoresist mask was employed to form the SiO_2 layer. A Reactive ion etching system using CHF_3 and O_2 was used to create a SiO_2 pattern. The photoresist and SiO_2 mask were used to create holes in the square pattern on the silicon wafer. For silicon etching, a traditional Bosch method was used on a deep Reactive Ion Etching (RIE) machine and then the photoresist mask has been removed. A layer of SiO_2 was created by heating it in a furnace, and a layer of Si_3N_4 was added on top of it. The SiO_2/Si_3N_4 layer was removed from the backside of the wafer. A fabricated silicon wafer was bonded to the side of the legitimate wafer, which still featured a SiO_2/Si_3N_4 layer, with the help of quartz wax. RIE machine and isotropic dry etching were utilized to process the bonded wafer assembly, employing a technique known as isotropic dry etching to remove approximately 200 μm of silicon. This helped to reveal the tips of the SiO_2/Si_3N_4 MNs. The MNs tips were smoothed out using both chemical methods and polishing. These hollow MNs were about 500 μm length, 30–50 μm wide, and had walls that were 3–5 μm thick. The hollow MNs were affixed to a plastic part made through 3D printing, and this assembly was then attached to a conventional syringe to develop a device for transplanting cells. Researchers have conducted experiments in a laboratory to utilize the cell transplant device for delivering luminescent MDCK cells into rat liver tissue. Images of the injected tissue indicate that the MDCK cells were effectively introduced into the tissue, with several of these cells persisting even after the MN was removed. The data suggests that MNs can facilitate the injection of donor cells into tissues at evenly distributed distances.

This suggests that they may play a crucial role in developing a novel approach to cell transplantation [87].

Antigen-specific immunotherapy is a treatment method for autoimmune diseases like type 1 diabetes. This therapy involves providing patients with specific substances sourced from their bodies, called autoantigens, to aid their immune system in avoiding attacks on their cells. The potential of gold particles to diminish inflammation makes them an appealing option for delivering pharmaceuticals. They anticipated that delivering C19–A3 GNP via Hollow MNs could optimize the medicine's performance and elicit a more accepting response from the immune system. They performed a trial with human participants to determine the safety and effectiveness of this method for the first time. Although the study involved only a limited group of individuals, the transdermal delivery of C19–A3 GNP seems to be both safe and positively received by those living with type 1 diabetes. The long-lasting presence of C19–A3 GNP in the skin after it's injected offers several chances to improve its ability to reduce immune reactions [88].

3.3.2 *Macromolecule and Insulin Delivery*

Given that MNs offer a relatively non-invasive approach, researchers are considering its application in insulin delivery. Most patients receive insulin through a subcutaneous injection with a needle, which can sometimes lead to hesitance in following their prescribed treatment. This may occur due to fear, discomfort, or difficulty in using and disposing of the needles properly [89]. Insulin is the most studied hormone for being given in ways other than through injection under the skin. About MNs technology, a researcher repotted a hollow silicone MNs to inject insulin in small amounts. Results from in vivo experiments showed that the fast-acting insulin was successful, which was confirmed by measuring the blood glucose levels. MNs delivery led to a smaller decrease in blood sugar levels but a big increase in insulin levels (40–50% increase). This is because it delivers insulin into the body more effectively [90].

Instead of using a catheter, Gupta and his team administer lispro insulin through a hollow MNs. Borosilicate glass pipettes were formed with a bevel angle of 30° and tip diameters ranging from 60 to 80 μM. Following the pulling process to create the glass, it was immersed in deionized water and cleaned with ultrasound for 2 min. After that, they were dried in an oven at 180 °C for 2 h and sterilized using steam in an autoclave. The MNs were placed at a 90° into the skin at depths between 1 and 5 mm using a special drilling tool. The MNs were connected to a 3 mL syringe that was part of a syringe pump controlling the rate at which insulin was administered. MNs penetrating 1 mm into the skin can aid in reducing post-meal blood sugar levels. A raised spot on the skin developed immediately after administering insulin, indicating that the medication was effectively injected into the skin tissue. Users reported only slight pain and favored MNs for delivering insulin rather than utilizing an under-the-skin catheter. This study suggests that using MNs might work faster than using subcutaneous catheters for delivering insulin. To validate this concept, further studies involving larger participant groups and consistent insulin dosages are necessary [91].

Resnik and his team conducted in vitro experiments to assess the effectiveness of drug delivery into human skin. He applied a methodology that evaluates the insulin levels against C-peptide levels, measured through the methylene blue. C-peptide is removed from the body by the kidneys, while most of the insulin made by the body is removed by the liver cells. So, the ratio between the two should be less than 1.0 when a person is fasting or eating, unless given insulin. They evaluated insulin administration via hollow MNs. Normal skin injection of insulin was done for comparison. We discovered that MN delivery led to a smaller reduction in blood sugar levels when compared to administering the same dose of medicine under the skin. However, it led to a larger rise in insulin levels (40–50%). This is mainly because MNs deliver a stronger form of insulin (200 IU/mL) more effectively. Long-lasting stable blood sugar levels were achieved by hollow MNs, which caused a smaller drop in glucose levels compared to subcutaneous delivery [90]. Davis and his team created an easy and strong Hollow MNs in an inexpensive way and large amounts. They used Lithographie/Galvanoformung/Abformung technique to shape molds from polyethylene terephthalate using a UV laser. Then, they covered those molds with nickel by depositing it onto a thin layer made earlier. Finally, they removed the metal MNs shapes from the molds by carefully dissolving the polymer. The experiments demonstrated that these MNs possessed sufficient strength to pass through living skin intact. Arrays with 16 MNs, each 500 μM long and 75 μM wide at the tip, were placed into the skin of diabetic rats. Administering Insulin with MNs reduced blood sugar levels to 47% of their pre-treatment values within four hours. Subsequently, the levels showed little variation and stayed consistent for an additional four hours. The direct measurement of insulin in the blood showed the highest level of 0.43 ng/ml [92].

Currently, utilizing hollow MNs in the healthcare sector is challenging due to the complexities involved in their production. Li and his team have created a simple 3D print lithography method for quickly producing hollow MNs. The process of generating printable images allows for easy customization of personalized Hollow MNs, which in turn supports scale-up production. The developed hollow MNs feature a smooth surface, lacking the typical layered appearance associated with 3D-printed materials. By this technique hollow MNs could be made in just 5 s using a single step of light treatment. The Hollow MNs were biocompatible, strong, and resistant to being pierced. After getting a small syringe, the hollow MNs can easily inject insulin into the skin, helping to control blood sugar levels in diabetic mice [78].

3.4 Hollow Microneedle for Nanoparticle (NPs) Delivery

Nanoparticles (NPs) are extensively employed for administering various drugs aimed at treating different diseases, such as diabetes, wound healing, and cancer, due to their unique advantages like resistance to degradation in the body, prolonged medication effects, precise targeting of specific areas, and gradual release over time [93]. By reaching beyond the outer layer of the skin, hollow MNs allow small nanoparticles to more easily access the body [94]. Despite the availability of multiple administration routes for drug-loaded NPs, including oral, intravenous, and ocular methods, there is growing interest in using MNs to penetrate the skin for delivery. This method helps avoid the degradation of drugs in the gastrointestinal tract and makes it easier for patients because it hurts less than regular needle injections [95]. Hollow MNs are effective tools for the transdermal delivery of NPs. The hollow MNs inserted into the skin help steadily inject liquid mixtures of this NPs. Compared to solid MNs, these types of MNs are more capable of precisely targeting the delivery of NPs in both location and timing [95, 96]. As an example, Mir constructed a liquid injection system known as AdminPen® by blending specialized NPs that are sensitive to bacteria with hollow MNs. Skin tests indicated that the application of NPs resulted in approximately eight times more delivery of the drug carvacrol (CAR) compared to a standard gel containing pure CAR. This indicates that hollow MNs might significantly enhance the ability of NPs to penetrate the skin [97].

Bolton applied deep reactive-ions etching technology to produce silicon MNs with hollow interiors. The researchers were able to create straight up-and-down edges and a sloped tip for the MNs. The ability of the hollow MNs to deliver the drug was tested by using isothiocyanate-labeled insulin (FITC-insulin). FITC-insulin passed smoothly through the tiny openings of three out of five MNs. The authors reported that an obstruction inside them caused the malfunction of the two MNs. To illustrate that the needles can penetrate pig skin, a syringe with a specific fluorescent dye is linked to a pump. A constant rate of 10 mL per minute was used to deliver fluorescent dye into the pig skin. The dye penetrated the pig skin through the minuscule needles without leaking out through the small openings [47]. A group of researchers developed a hollow MNs array containing noisome to deliver plasmid DNA loaded with ovalbumin for skin immunization. The findings indicated that the administration of cationic niosomes via hollow MNs to deliver ovalbumin triggered immune responses in cells that are regulated by interferon-gamma, immunoglobulin G antibodies, and interleukin-4. According to the authors, employing hollow MNs in conjunction with cationic niosome/ovalbumin complexes resulted in a stronger immune response than simply using naked ovalbumin. This was attributed to an increase in the release of cytokines and a greater production of IgG antibodies. It was shown that using hollow MNs with special niosome/ovalbumin mixtures caused more cytokine release and a stronger IgG immune response than regular under-the-skin injections [98].

The use of photodynamic therapy (PDT) for the treatment of non-melanoma skin cancers has become more popular in recent years. Hypericin is a potent compound known for its ability to absorb light, which may aid in cancer treatment. Due to its poor solubility in water, propensity to clump in biological systems, and ineffective skin absorption, its applications remain constrained. Heba Abd-El-Azim utilized a specialized needle device known as AdminPen along with a custom-made apparatus to effectively inject Hypericin into lipid nanocapsules into the skin. The results showed that Hypericin-Lipid nano carriers (LNCs) were created successfully with very small particles. They were good at holding onto drugs, worked well when exposed to light, placed more medicine on the skin, were taken in better by cells, and

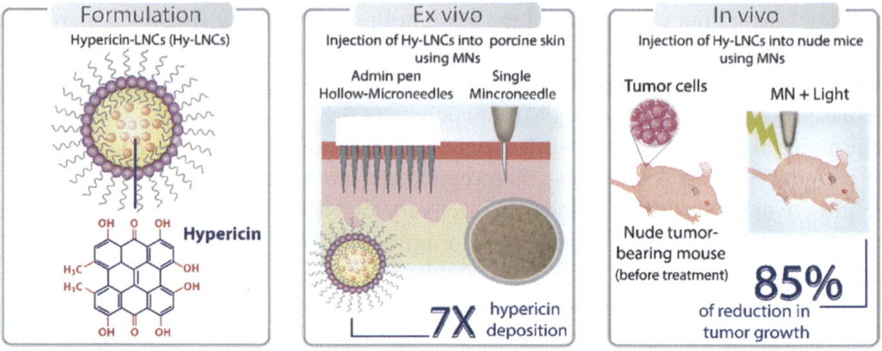

Fig. 3.5 Schematic representation Hypericin loaded Lipid nanocarrier with hollow MNs for melanoma photodynamic therapy. Reproduced with permission from the cited Ref. [99]

were more harmful to cells when exposed to light compared to regular Hypericin. The therapeutic impact of Hypericin-LNCs was assessed in live mice with implanted tumors. Interestingly, Hypericin-LNCs given by Ho-MN showed great ability to fight tumors, destroying 85.84% of them after being exposed to 595 nm light. This research revealed that the combination of hollow MNs for delivering hypericin-lipid nanocarriers and radiation therapy could provide an effective and targeted method for addressing non-melanoma skin cancers [99]. Schematic representation Hypericin loaded Lipid nanocarrier with hollow MNs for melanoma photodynamic therapy presented in Fig. 3.5.

The team led by Patel used hollow MNs to penetrate the sclera of the eye and introduced NPs into the space behind it. Research was carried out on the eye parts of rabbits, pigs, and humans, removed from their bodies for the tests. They utilized brightfield and fluorescence microscopy along with microcomputed tomography to visualize the Mechanism of particle delivery. MNs were used to inject sulforhodamine B and NPs into suprachoroidal space in rabbits, pigs, and humans. Uniform quantities of up to 35 μl were delivered. Enhancing the delivery device's parameters demonstrated that the MNs length, applied pressure, and particle size played a crucial role in ensuring effective delivery within the suprachoroidal space. Using MNs that are 800–1000 μm long and a pressure of 250–300 kPa gave the best and most consistent results. This points to the possibility that MNs might serve as a gentle means of delivering treatment to the back of the eye [80].

3.5 Hollow Microneedle for Vaccine Delivery

Maaden and his team designed a machine that utilizes computer control to manage a miniature needle for automatic skin injections (Fig. 3.6). Their injection system contained an applicator, micrometre actuator, capillary, hollow MN, connectors,

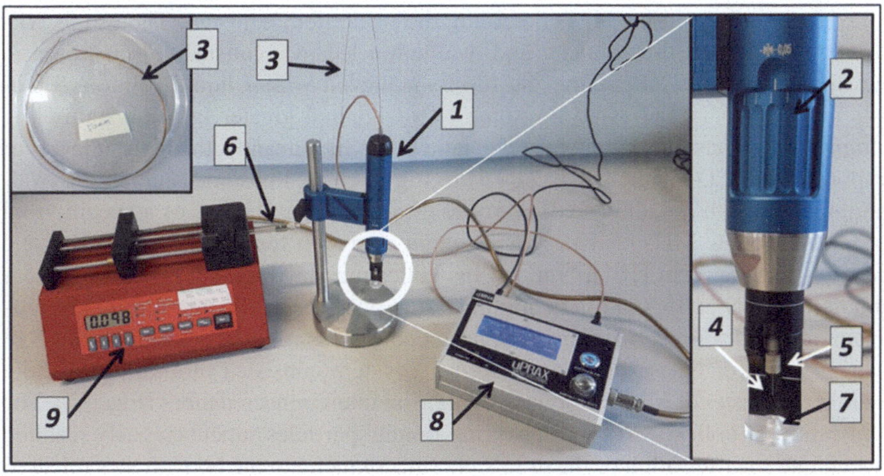

Fig. 3.6 Digitally controlled hollow MNs injection system. MN applicator (1); micrometre actuator (2); capillary (3); hollow MN (4); connectors (5, 6); supporting plateau (7); applicator controller unit (8); syringe pump. The image was reproduced with the permission [100]

supporting plateau, applicator controller unit, syringe pump. The system is composed of an MN applicator (1) equipped with a small mechanism that allows precise adjustments to the protrusion length of the MNs (2). A slender tube (3) is attached to hollow MNs (4) at one end within the applicator, secured by a connector (5). The smooth side of the capillary is linked to the syringe with connectors (6). When the applicator's solenoid is turned on, the MN tip (4) will stick out of the support platform (7). In addition, the system has a controller unit (8) that can manage both the MN applicator and the syringe pump at the same time. The syringe pump was changed so that it could be controlled digitally (9). This allows for precise small injections through the hollow MN using the controller unit. Using hydrofluoric acid, fused silica tubes were molded into the form of hollow MNs. The longest edge of the made hollow MNs was 97 mm, and to ensure there was no leakage when injecting, they needed to be inserted at least 100 mm deep. The MN system, operated by a computer, delivered small and exact doses of 1 mL with high accuracy. The designed system was subsequently utilized to deliver small doses of a liposomal cancer vaccine. An engineered protein from the human papillomavirus was incorporated into liposomal carriers. Prepared liposomal vaccine formulations were injected into the skin using a designed MN system. The vaccine induced a significant activation of CD4 and CD8 cells in mice [100].

Some researchers, based on this hypothesis, used hollow MNs for subcutaneous vaccine delivery. Du and his team studied hollow MNs for nano-formulated vaccine delivery into the skin. MN applicator was used to Inject mixtures into the skin of mice at around 120 mm deep. They used liposomes, PLGA (poly (lactic-co-glycolic)), gelatin and mesoporous silica NPs as a carrier for vaccine delivery. The result shows that similar quantities of IgG1 and IgG antibodies were produced by the ovalbumin/

adjuvant solution and the NPs containing the ovalbumin/adjuvant. The use of NPs to incorporate both the adjuvant and ovalbumin led to a considerable increase in the IgG2a response, surpassing the results achieved with a liquid solution. Findings revealed that PLGA and liposome NPs triggered greater IgG2a responses in comparison to mesoporous silica and gelatin NPs. Significantly, the highest levels of response from CD4 and CD8 T cells were triggered by liposomes containing ovalbumin along with an adjuvant. The study illustrated that the use of a specific tool to operate hollow MNs provides an excellent approach for delivering nanoparticle vaccines into the skin [101]. Niu and his colleagues investigated the effectiveness of a unique MN configuration in administering polymeric NPs into the skin of rats. A polymeric device containing 18 small needles was used to administer the NPs through the skin. The experimental delivery system consisted of a miniature fluid chamber, a cluster of MNs, an adhesive ring, and interconnected tubes (Fig. 3.7). The application of hollow MNs for delivering minute particles subcutaneously resulted in a unique manifestation of the drug's pharmacokinetic in the body. It was released quickly at first through the lymph nodes but didn't spread widely throughout the rest of the body. Administering medicine in this manner led to improved absorption and effectiveness compared to traditional methods such as subcutaneous or intravenous delivery. By this assembly, a considerable amount of nanoparticles were achieved in the lymphatic system, and it can be used in a different way for subcutaneous vaccine administration. The researchers employed Poly (D, L-lactide-coglycolide)-NPs to load ovalbumin, monophosphoryl Lipid A and imiquimod. The ovalbumin -encapsulated NPs elicited a quicker enhancement in antibody levels.

In contrast to a vaccine formulated with soluble ovalbumin, the newly developed formulation shows faster antibody affinity maturation kinetics. Additionally, employing a hollow MNs array to administer NPs containing antigens resulted in a significantly greater rise in the production of interferon (IFN)-g by lymphocytes, as well as a more robust IgG2a antibody response compared to delivering the antigens in liquid form or injecting the NPs directly into muscle tissue. The findings indicated that employing hollow MNs to administer small polymer particles into the skin significantly enhances vaccine efficacy [102].

The team led by De Groot examined the influence of hollow MNs on the vaccine delivery method. A Schematic MNs design are presented on Fig. 3.8. By employing wet etching, the authors produced hollow MNs from fused silica. The inside of capillaries with a 20 μm diameter, which is coated with polyimide, was filled with silicone oil overnight in a vacuum oven heated to 100 °C. To create hollow MNs, the ends of the silicone oil-filled capillaries were treated by etching them in a hydrofluoric acid solution. The researchers employed a device to insert hollow MNs into the abdomens of mice. Using an applicator makes it easier to control how deep, how much, and how fast the injections are given. Researchers found that delivering PLGA-encapsulated ovalbumin and poly(I:C) via hollow MNs stimulated immune cells and prepared the CD8+ T cells to respond to vaccination. They show that a single dose of ovalbumin -loaded PLGA NPs was significantly more effective than standard ovalbumin in

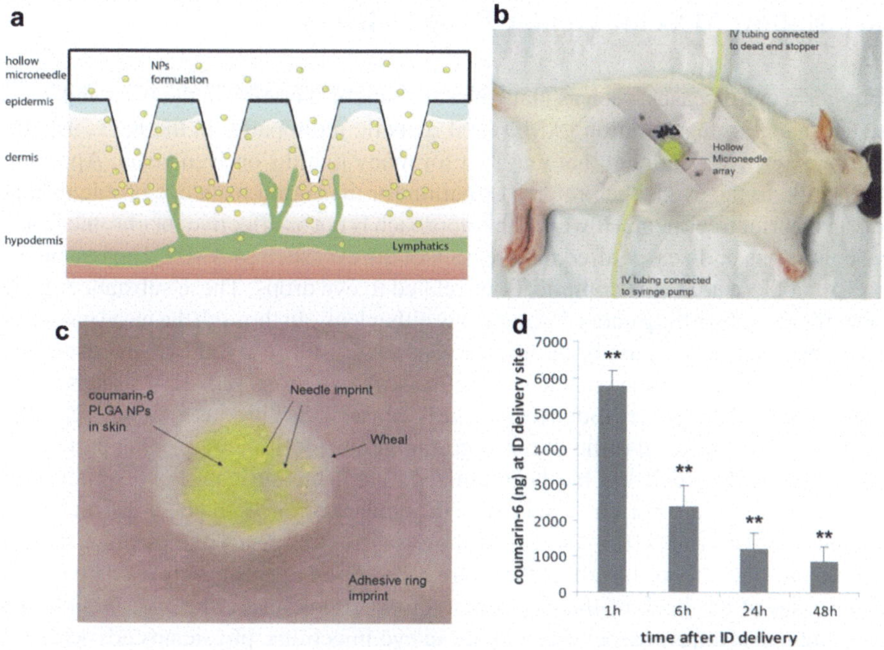

Fig. 3.7 **a** A diagram showing how PLGA NPs are delivered under the skin using hollow MN; **b** A setup with a hollow MNs array used to deliver PLGA NPs under the skin; **c** A photo of rat skin right after giving 5 mg of fluorescent PLGA NPs using hollow MNs system. The hollow MNs system device was taken out before the photo was taken; **d** Retention profile of NPs at the injection site under the skin. Reproduced with permission from the cited Ref. [102]

priming specialized CD8+ and CD4+ T cells that target specific antigens. By implementing a three-phase vaccination strategy, PLGA NPs facilitated the activation of the body's OVA-specific CD8+ T cells [103].

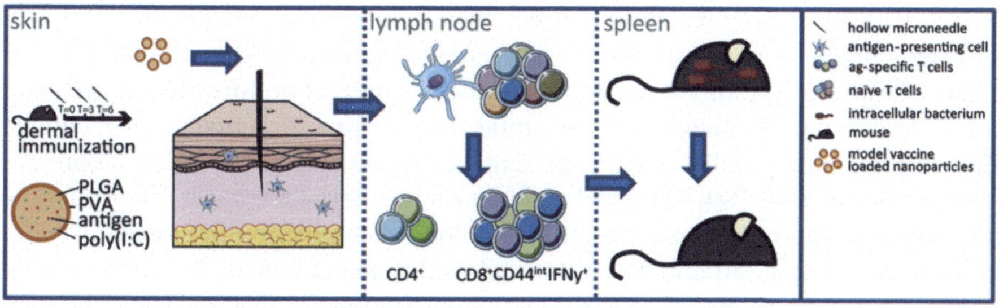

Fig. 3.8 Schematic represented CD4 and CD8+ activation by Hollow MNs evaluated by De Groot. Reproduced with permission from the cited RRef. [103]

3.6 Hollow MNs for Ocular Drug Delivery

Long-term eye diseases, such as glaucoma and amebic keratitis in the front and Age-related macular degeneration (AMD) and diabetic retinopathy in the back, usually require extensive treatment that may last for many months or even years. Applying medication to the eye is the preferred solution for managing front eye problems that only need brief treatment. However, this approach is not as effective for chronic front-eye conditions or diseases affecting the back of the eye, such as AMD. The reason for this is that there are some complications related to eye drops. These substances have poor efficacy (absorbing under 5%) and may either leak out through the nasal passages or be absorbed into the body in other ways. As a result, it's difficult to ensure the correct dosage of medicine reaches the retina with eye drops [22, 104]. Issues in the posterior part of the eye are tough to manage because of the vitreous humour, a gel-like substance. This makes it difficult for important molecules to move to that part of the eye. Systemic drug delivery has been utilized to manage intricate eye disorders, but these medications are not very effective in the ocular region, with under 2% making it past the retinal barriers [105]. Increasing the dosage or frequency of these medications may lead to side effects that affect the entire body. Another commonly used treatment for diseases at the back of the eye is getting injections inside the eye. Despite the risks and invasiveness associated with these eye injections, physicians advocate for them due to the significant advantages they might provide [106]. These regular eye injections can cause some short-term problems, like serious eye infections, retinal detachment, bleeding inside the eye, and an increased chance of cataracts. Recently, researchers have looked into using injections around the eyes as a simpler way to treat problems in the eye's inner space. The subventricular, subconjunctival, and peribulbar methods include injections around the eyes. Although these methods of administration are less intrusive, they must overcome the eye's defences to access the required area for the medication [107]. Scientists are currently investigating hollow MNs as a method for administering medication to the eyes, to improve treatments for eye disorders. These tiny needles allow drugs to be delivered accurately and directly to certain parts of the eye, like the clear front part or the space in the front of the eye. By overcoming the eye's barriers, they optimize the performance of drugs, allowing a higher concentration to be delivered to the target area and reducing side effects elsewhere. The tiny size and less invasive nature of hollow MNs make them more comfortable for patients. Their ability to give precise amounts of medicine allows for the correct delivery of drugs, and they might also be useful for taking eye fluid samples or tests [80, 81]. Patel and his team explored the capability of hollow MNs for suprachoroidal delivering triamcinolone acetonide in rabbit eyes. A 33 G needle attached to a 750 mm MN was employed to inject a medicine solution with a concentration of 40 mg/ml. After drug administration via Hollow MNs, the total drug quantity present in the eye was 98.8%, indicating that there was no leakage into the other tissues. Most of the drug concentration was observed in the sclera, choroid, and retina, with just a trace amount found in the vitreous humour [108]. Patel and team investigated hollow MNs to deliver sulforhodamine B and nanoparticle and microparticle suspensions into the suprachoroidal space of rabbits, pigs, and humans. Consistent doses of up to 35 μl were administered. Improving the delivery device showed that the length of the MNs, the pressure used, and the size of the particles were key factors for successfully delivering into the suprachoroidal space. The optimal results for delivery were achieved using needle lengths ranging from 800 to 1,000 μm and pressures set between 250 and 300 kPa. They show that hollow MNs could provide a non-invasive and successful route for administering drugs and another drug-containing suspensions to the eye [80]. Jiang and his team inserted hollow MNs into the outer layer of human eyes and subsequently filled them with solutions containing special dyes or luminescent particles. They achieved this through the application of constant pressure. Fluid flow was very slow after the microneedle was put into the eye's sclera without pulling it back. By pulling the microneedle back around 200–300 μm, 10–35 μl of fluid was able to penetrate into the tissue. Tiny NPs were injected into the sclera (the white part of the eye), but larger microparticles could only be delivered when special enzymes called hyaluronidase and collagenase were used. This shows that some substances in the sclera, like glycosaminoglycans and collagen fibers, make it harder for the larger particles to enter. This study shows that hollow MNs can inject liquids into the sclera for a quick and easy way to deliver tiny molecules, NPs, and micro particles [81]. In another experimental design, Verdugo and colleagues investigated introducing cells beneath the retina of dogs, utilizing Hollow MNs angled at 30°. Research has demonstrated that injecting around 100–150 μL of liquid under the retina can be done without risking retinal detachment. Injections were done using a high-quality microscope, and special needles with a 30° point and rounded tips were used to prevent any unintentional harm to the retina [109]. Hollow MNs are a new and helpful technology for delivering medicine to the eyes. They could solve some problems caused by traditional methods. Recently, their use has grown in beauty products and vaccines, but they are becoming very important for delivering treatments to the eyes. Studies on using hollow MNs for delivering treatments to the eyes have been successful in similar uses. It is clear that using hollow MNs is important for creating new ways to deliver medicine to the eyes. The key aim is to guarantee patient comfort, facilitate the administration of treatments, and ensure that they are effective. This will help us manage eye diseases better in the future.

3.7 Conclusion

Numerous studies conducted in recent years indicate that hollow MNs enhance drug delivery and can be utilized for monitoring purposes. Various studies have explored the potential benefits of hollow MNs in administering drugs and vaccines due to their ability to improve the body's reaction to immunizations. Many companies are also interested in using hollow MN in real medical treatment. Unlike dissolving MNs, hollow MNs require additional assembly and parts, such as a pump, chamber, etc., for drug delivery, which requires additional consideration. In addition, it is essential to assess the stability and sterility of liquid molecules in prospective designs, as well as their microbial presence and the requirement for hygienic manufacturing practices. Studies and experiments carried out in laboratories may pave the way for hollow MNs to become a valuable resource for administering medicine without pain or difficulty. These needles are capable of transporting a variety of substances, ranging from tiny to large particles, including water-loving and water-repelling liquids. They can also be used for checking health conditions.

References

1. Singh, R., Singh, S., Lillard, J.W., Jr.: Past, present, and future technologies for oral delivery of therapeutic proteins. J. Pharm. Sci. **97**(7), 2497–2523 (2008)
2. Wang, L., Coppel, R.L.: Oral vaccine delivery: can it protect against non-mucosal pathogens? Expert Rev. Vaccines **7**(6), 729–738 (2008)
3. Nir, Y., Paz, A., Sabo, E., Potasman, I.: Fear of injections in young adults: prevalence and associations. Am. J. Trop. Med. Hyg. **68**(3), 341–344 (2003)
4. Hamilton, J.G.: Needle phobia: a neglected diagnosis. J. Fam. Pract. **41**(2), 169–175 (1995)
5. Kermode, M.: Unsafe injections in low-income country health settings: need for injection safety promotion to prevent the spread of blood-borne viruses. Health Promot. Int. **19**(1), 95–103 (2004)
6. Drucker, E., Alcabes, P.G., Marx, P.A.: The injection century: massive unsterile injections and the emergence of human pathogens. The Lancet **358**(9297), 1989–1992 (2001)
7. Zhang, X., Sun, L., Wang, Y., Bian, F., Wang, Y., Zhao, Y.: Multibioinspired slippery surfaces with wettable bump arrays for droplets pumping. Proc Natl Acad Sci U S A **116**(42), 20863–20868 (2019)
8. Zhang, X., Chen, G., Bian, F., Cai, L., Zhao, Y.: Encoded microneedle arrays for detection of skin interstitial fluid biomarkers. Adv. Mater. **31**(37), e1902825 (2019)
9. Zhang, X., Wang, F., Yu, Y., Chen, G., Shang, L., Sun, L., Zhao, Y.: Bio-inspired clamping microneedle arrays from flexible ferrofluid-configured moldings. Sci. Bull. **64**(15), 1110–1117 (2019)
10. Moon, S., Wang, Y., Edens, C., Gentsch, J.R., Prausnitz, M.R., Jiang, B.: Dose sparing and enhanced immunogenicity of inactivated rotavirus vaccine administered by skin vaccination using a microneedle patch. Vaccine **31**(34), 3396–3402 (2013)
11. Unver, N., Odabas, S., Demirel, G.B., Gul, O.T.: Hollow microneedle array fabrication using a rational design to prevent skin clogging in transdermal drug delivery. J. Mater. Chem. B **10**(41), 8419–8431 (2022)
12. Yu, Y.Q., Yang, X., Wu, X.F., Fan, Y.B.: Enhancing permeation of drug molecules across the skin via delivery in nanocarriers: novel strategies for effective transdermal applications. Front Bioeng Biotechnol **9**, 646554 (2021)
13. Schafer, N., Balwierz, R., Biernat, P., Ochędzan-Siodłak, W., Lipok, J.: Natural ingredients of transdermal drug delivery systems as permeation enhancers of active substances through the stratum Corneum. Mol. Pharm. **20**(7), 3278–3297 (2023)
14. Barry, B.W.: Mode of action of penetration enhancers in human skin. J. Control. Release **6**(1), 85–97 (1987)
15. Prausnitz, M.R., Mikszta, J.A., Cormier, M., Andrianov, A.K.: Microneedle-based vaccines. Vaccines Pandemic Influenza 369–393 (2009)
16. Nikolić, I., Simić, M., Pantelić, I., Stojanović, G., Antić Stanković, J., Marković, B., Savić, S.: Chemical versus physical methods to improve dermal drug delivery: a case study with nanoemulsions and iontophoresis. Pharmaceutics **14**(6) (2022)

17. Kim, Y.C., Park, J.H., Prausnitz, M.R.: Microneedles for drug and vaccine delivery. Adv. Drug Deliv. Rev. **64**(14), 1547–1568 (2012)
18. Nguyen, H.X., Nguyen, C.N.: Microneedle-mediated transdermal delivery of biopharmaceuticals. Pharmaceutics **15**(1) (2023)
19. Wang, P.C., Wester, B.A., Rajaraman, S., Paik, S.J., Kim, S.H., Allen, M.G.: Hollow polymer microneedle array fabricated by photolithography process combined with micromolding technique. In: Annual International Conference of the IEEE Engineering in Medicine and Biology Society 2009, pp. 7026–7029 (2009)
20. Szeto, B., Aksit, A., Valentini, C., Yu, M., Werth, E.G., Goeta, S., Tang, C., Brown, L.M., Olson, E.S., Kysar, J.W., Lalwani, A.K.: Novel 3D-printed hollow microneedles facilitate safe, reliable, and informative sampling of perilymph from guinea pigs. Hear. Res. **400**, 108141 (2021)
21. Zhou, C., Tang, H., Zhang, L., An, H., Wu, Y., Tang, D., Zhou, J.: Hollow microneedle arrays produced by low-cost, high-fidelity replication of hypodermic needle tips for high-dose transdermal drug delivery. Adv. Eng. Mater. **23**(6), 2001355 (2021)
22. Gade, S., Glover, K., Mishra, D., Sharma, S., Guy, O., Donnelly, R.F., Vora, L.K., Thakur, R.R.S.: Hollow microneedles for ocular drug delivery. J. Control. Release **371**, 43–66 (2024)
23. Luo, X., Yang, L., Cui, Y.: Microneedles: materials, fabrication, and biomedical applications. Biomed. Microdevices **25**(3), 20 (2023)
24. Edwards, C., Shah, S.A., Gebhardt, T., Jewell, C.M.: Exploiting unique features of microneedles to modulate immunity. Adv. Mater. **35**(52), 2302410 (2023)
25. Wonglertnirant, N., Todo, H., Opanasopit, P., Ngawhirunpat, T., Sugibayashi, K.: Macromolecular delivery into skin using a hollow microneedle. Biol. Pharm. Bull. **33**(12), 1988–1993 (2010)
26. Parhi, R.: Recent advances in 3D printed microneedles and their skin delivery application in the treatment of various diseases. J. Drug Deliv. Sci. Technol. **84**, 104395 (2023)
27. Sonetha, V., Majumdar, S., Shah, S.: Step-wise micro-fabrication techniques of microneedle arrays with applications in transdermal drug delivery—A review. J. Drug Deliv. Sci. Technol. **68**, 103119 (2022)
28. Zhang, R., Miao, Q., Deng, D., Wu, J., Miao, Y., Li, Y.: Research progress of advanced microneedle drug delivery system and its application in biomedicine. Colloids Surf. B **226**, 113302 (2023)
29. Economidou, S.N., Uddin, M.J., Marques, M.J., Douroumis, D., Sow, W.T., Li, H., Reid, A., Windmill, J.F., Podoleanu, A.: A novel 3D printed hollow microneedle microelectromechanical system for controlled, personalized transdermal drug delivery. Addit. Manuf. **38**, 101815 (2021)
30. Dong, C.-W., Jeon, J.-Y., Kang, H.-M., Park, W.-T.: Tip fabrication methods of hollow metal microneedles. J. Mech. Sci. Technol. **37**(1), 261–269 (2023)
31. Nejad, H.R., Sadeqi, A., Kiaee, G., Sonkusale, S.: Low-cost and cleanroom-free fabrication of microneedles. Microsyst. Nanoeng. **4**(1), 17073 (2018)
32. Van Hileghem, L., Kushwaha, S., Piovesan, A., Verboven, P., Nicolaï, B., Reynaerts, D., Dal Dosso, F., Lammertyn, J.: Innovative fabrication of hollow microneedle arrays enabling blood sampling with a self-powered microfluidic patch. Micromachines **14**(3), 615 (2023)
33. Davis, S.P., Prausnitz, M.R., Allen, M.G.: Fabrication and characterization of laser micromachined hollow microneedles. In: TRANSDUCERS '03. 12th International Conference on Solid-State Sensors, Actuators and Microsystems. Digest of Technical Papers (Cat. No.03TH8664), vol. 2, pp. 1435–1438 (2003)
34. Miller, P.R., Moorman, M., Boehm, R.D., Wolfley, S., Chavez, V., Baca, J.T., Ashley, C., Brener, I., Narayan, R.J., Polsky, R.: Fabrication of hollow metal microneedle arrays using a molding and electroplating method. Mrs. Adv. **4**(24), 1417–1426 (2019)
35. Fabrication of a polymer micro needle array by mask-dragging x-ray lithography and alignment x-ray lithography. Chinese Phys. Lett. **28**(3), 038101 (2011)
36. Bharti, A., Turchet, A., Marmiroli, B.: X-ray lithography for nanofabrication: is there a future? Front. Nanotechnol. **4** (2022)

37. Lee, K., Jung, H.: Drawing lithography for microneedles: a review of fundamentals and biomedical applications. Biomaterials **33**(30), 7309–7326 (2012)
38. Sang Jun, M., Chun Yan, J., Seung, S.L.: A novel method of microneedle array fabrication using inclined deep x-ray exposure. J. Phys.: Conf. Ser. **34**(1), 180 (2006)
39. Wang, J., Wang, H., Lai, L., Li, Y.: Preparation of microneedle array mold based on MEMS lithography technology. Micromachines (Basel) **12**(1) (2020)
40. Luttge, R., Berenschot, E.J.W., Boer, M.J.d., Altpeter, D.M., Vrouwe, E.X., Berg, A.V.D., Elwenspoek, M.: Integrated lithographic molding for microneedle-based devices. J. Microelectromech. Syst. **16**(4), 872–884 (2007)
41. Chen, Z., Ren, L., Li, J., Yao, L., Chen, Y., Liu, B., Jiang, L.: Rapid fabrication of microneedles using magnetorheological drawing lithography. Acta Biomater. **65**, 283–291 (2018)
42. Khumpuang, S., Ruther, P., Paul, O.: Micromolding methods for hollow microneedle arrays using megasonically enhanced casting. In: TRANSDUCERS 2009–2009 International Solid-State Sensors, Actuators and Microsystems Conference, pp. 220–223 (2009)
43. Faraji Rad, Z., Prewett, P.D., Davies, G.J.: An overview of microneedle applications, materials, and fabrication methods.Beilstein. J. Nanotechnol. **12**, 1034–1046 (2021)
44. Lyubarskaya, A.V., Chaplygin, Y.A., Golishnikov, A.A., Pankratov, O.V.: Study of anisotropic wet chemical etching for silicon microneedles fabrication. In: 2021 IEEE Conference of Russian Young Researchers in Electrical and Electronic Engineering (ElConRus), pp. 2579–2582 (2021)
45. Deng, Y.L., Juang, Y.J.: Polydimethyl siloxane wet etching for three dimensional fabrication of microneedle array and high-aspect-ratio micropillars. Biomicrofluidics **8**(2), 026502 (2014)
46. Li, Y., Zhang, H., Yang, R., Laffitte, Y., Schmill, U., Hu, W., Kaddoura, M., Blondeel, E.J.M., Cui, B.: Fabrication of sharp silicon hollow microneedles by deep-reactive ion etching towards minimally invasive diagnostics. Microsyst. Nanoeng. **5**(1), 41 (2019)
47. Bolton, C.J.W., Howells, O., Blayney, G.J., Eng, P.F., Birchall, J.C., Gualeni, B., Roberts, K., Ashraf, H., Guy, O.J.: Hollow silicon microneedle fabrication using advanced plasma etch technologies for applications in transdermal drug delivery. Lab Chip **20**(15), 2788–2795 (2020)
48. Liu, Y., Eng, P.F., Guy, O.J., Roberts, K., Ashraf, H., Knight, N.: Advanced deep reactive-ion etching technology for hollow microneedles for transdermal blood sampling and drug delivery. IET Nanobiotechnol. **7**(2), 59–62 (2013)
49. Islam, H., Poly, T.S., Tisha, Z.T., Rahman, S., Naveed, A.I., Ahmed, A., Ahmed, S.N., Hassan, J., Uddin, M.J., Das, D.B.: 3D printed hollow microneedles for treating skin wrinkles using different anti-wrinkle agents: a possible futuristic approach. Cosmetics (2023)
50. Ghaznavi, A., Xu, J., Lee, C.U., Hara, S.A.: 3D-printed hollow microneedles array with Luer lock connection for facile and painless intradermal injection: a proof of concept. Adv. Mater. Technol. **9**(18), 2400286 (2024)
51. O'Mahony, C., Sebastian, R., Tjulkins, F., Whelan, D., Bocchino, A., Hu, Y., O'Brien, J., Scully, J., Hegarty, M., Blake, A., Slimi, I., Clover, A.J.P., Lyness, A., Kelleher, A.-M.: Hollow silicon microneedles, fabricated using combined wet and dry etching techniques, for transdermal delivery and diagnostics. Int. J. Pharm. **637**, 122888 (2023)
52. Jiang, X., Lillehoj, P.B.: Microneedle-based skin patch for blood-free rapid diagnostic testing. Microsyst. Nanoeng. **6**(1), 96 (2020)
53. Dardano, P., De Martino, S., Battisti, M., Miranda, B., Rea, I., De Stefano, L.: One-shot fabrication of polymeric hollow microneedles by standard photolithography. Polymers **13**(4), 520 (2021)
54. Reeves, N.B.: Fabrication of flexible metal hollow microneedles using microsecond pulsing of single mode fiber laser (2012)
55. Cárcamo-Martínez, Á., Mallon, B., Domínguez-Robles, J., Vora, L.K., Anjani, Q.K., Donnelly, R.F.: Hollow microneedles: a perspective in biomedical applications. Int. J. Pharm. **599**, 120455 (2021)
56. Guillot, A.J., Cordeiro, A.S., Donnelly, R.F., Montesinos, M.C., Garrigues, T.M., Melero, A.: Microneedle-based delivery: an overview of current applications and trends. Pharmaceutics **12**(6), 569 (2020)

57. Ita, K.: Ceramic microneedles and hollow microneedles for transdermal drug delivery: two decades of research. J. Drug Deliv. Sci. Technol. **44**, 314–322 (2018)
58. van der Maaden, K., Luttge, R., Vos, P.J., Bouwstra, J., Kersten, G., Ploemen, I.: Microneedle-based drug and vaccine delivery via nanoporous microneedle arrays. Drug Deliv. Transl. Res. **5**, 397–406 (2015)
59. Evens, T., Van Hileghem, L., Dal Dosso, F., Lammertyn, J., Malek, O., Castagne, S., Seveno, D., Van Bael, A.: Producing hollow polymer microneedles using laser ablated molds in an injection molding process. J. Micro- Nano-Manuf. **9**(3), 030902 (2021)
60. Mansoor, I., Häfeli, U.O., Stoeber, B.: Hollow out-of-plane polymer microneedles made by solvent casting for transdermal drug delivery. J. Microelectromech. Syst. **21**, 44–52 (2012)
61. Howells, O., Blayney, G.J., Gualeni, B., Birchall, J.C., Eng, P.F., Ashraf, H., Sharma, S., Guy, O.J.: Design, fabrication, and characterization of a silicon microneedle array for transdermal therapeutic delivery using a single step wet etch process. Eur. J. Pharm. Biopharm. **171**, 19–28 (2022)
62. Aldawood, F.K., Andar, A., Desai, S.: Investigating laser ablation process parameters for the fabrication of customized microneedle arrays for therapeutic applications. Pharmaceutics **16**(7) (2024)
63. Sivamani, R.K., Liepmann, D., Maibach, H.I.: Microneedles and transdermal applications. Expert Opin. Drug Deliv. **4**(1), 19–25 (2007)
64. Gupta, J., Park, S.S., Bondy, B., Felner, E.I., Prausnitz, M.R.: Infusion pressure and pain during microneedle injection into skin of human subjects. Biomaterials **32**(28), 6823–6831 (2011)
65. McAllister, D.V., Wang, P.M., Davis, S.P., Park, J.-H., Canatella, P.J., Allen, M.G., Prausnitz, M.R.: Microfabricated needles for transdermal delivery of macromolecules and nanoparticles: fabrication methods and transport studies. Proc. Natl. Acad. Sci. **100**(24), 13755–13760 (2003)
66. Bhattacharyya, B., Doloi, B.: Chapter Four—Machining processes utilizing thermal energy. In: Bhattacharyya, B., Doloi, B. (eds.), Modern Machining Technology, pp. 161–363. Academic Press
67. Niaounakis, M.: Chapter 8—Surface Treatment. In: Niaounakis, M. (ed.) Biopolymers: Processing and Products, pp. 303–326. William Andrew Publishing, Oxford (2015)
68. Jeong, J., Park, J., Lee, S.: 3D printing fabrication process for fine control of microneedle shape. Micro Nano Syst. Lett. **11**(1), 1 (2023)
69. Li, W., Wang, M., Ma, H., Chapa-Villarreal, F.A., Lobo, A.O., Zhang, Y.S.: Stereolithography apparatus and digital light processing-based 3D bioprinting for tissue fabrication. iScience **26**(2), 106039 (2023)
70. Lakkala, P., Munnangi, S.R., Bandari, S., Repka, M.: Additive manufacturing technologies with emphasis on stereolithography 3D printing in pharmaceutical and medical applications: a review. Int. J. Pharm.: X **5**, 100159 (2023)
71. Abedi, M., Rahiminezhad, Z., Mehrzadeh, M., Alimardani, V., Ghasemi, Y., Chauhan, D., Ashfaq, M., Viswanathan, M.R., Talreja, N.: Sulfide-based 2D nanomaterials for photocatalytic activity. IN: Materials Technology for the Energy and Environmental Nexus, Volume 1, pp. 7-1–7-28, IOP Publishing (2023)
72. Chinnamani, M.V., Hanif, A., Kannan, P.K., Kaushal, S., Sultan, M.J., Lee, N.-E.: Soft microfiber-based hollow microneedle array for stretchable microfluidic biosensing patch with negative pressure-driven sampling. Biosens. Bioelectron. **237**, 115468 (2023)
73. Cárcamo-Martínez, Á., Mallon, B., Anjani, Q.K., Domínguez-Robles, J., Utomo, E., Vora, L.K., Tekko, I.A., Larrañeta, E., Donnelly, R.F.: Enhancing intradermal delivery of tofacitinib citrate: comparison between powder-loaded hollow microneedle arrays and dissolving microneedle arrays. Int. J. Pharm. **593**, 120152 (2021)
74. Yadav, V., Sharma, P.K., Murty, U.S., Mohan, N.H., Thomas, R., Dwivedy, S.K., Banerjee, S.: 3D printed hollow microneedles array using stereolithography for efficient transdermal delivery of rifampicin. Int. J. Pharm. **605**, 120815 (2021)
75. Ryan, E., Garland, M.J., Singh, T.R., Bambury, E., O'Dea, J., Migalska, K., Gorman, S.P., McCarthy, H.O., Gilmore, B.F., Donnelly, R.F.: Microneedle-mediated transdermal bacteriophage delivery. Eur. J. Pharm. Sci. **47**(2), 297–304 (2012)

76. Wang, P.M., Cornwell, M., Hill, J., Prausnitz, M.R.: Precise microinjection into skin using hollow microneedles. J. Investig. Dermatol. **126**(5), 1080–1087 (2006)
77. Machekposhti, S.A., Kadian, S., Vanderwal, L., Stafslien, S., Narayan, R.J.: Novel hollow biodegradable microneedle for amphotericin B delivery. MedComm **4**(4), e321 (2023)
78. Li, R., Liu, X., Yuan, X., Wu, S., Li, L., Jiang, X., Li, B., Jiang, X., Gou, M.: Fast Customization of hollow microneedle patches for insulin delivery. Int. J. Bioprint **8**(2), 553 (2022)
79. Sadeqi, A., Kiaee, G., Zeng, W., Rezaei Nejad, H., Sonkusale, S.: Hard polymeric porous microneedles on stretchable substrate for transdermal drug delivery. Sci. Rep. **12**(1), 1853 (2022)
80. Patel, S.R., Lin, A.S.P., Edelhauser, H.F., Prausnitz, M.R.: Suprachoroidal drug delivery to the back of the eye using hollow microneedles. Pharm. Res. **28**(1), 166–176 (2011)
81. Jiang, J., Moore, J.S., Edelhauser, H.F., Prausnitz, M.R.: Intrascleral drug delivery to the eye using hollow microneedles. Pharm. Res. **26**(2), 395–403 (2009)
82. Martanto, W., Moore, J.S., Kashlan, O., Kamath, R., Wang, P.M., O'Neal, J.M., Prausnitz, M.R.: Microinfusion using hollow microneedles. Pharm. Res. **23**(1), 104–113 (2006)
83. Lee, C.Y., Lee, K., You, Y.S., Lee, S.H., Jung, H.: Tower microneedle via reverse drawing lithography for innocuous intravitreal drug delivery. Adv Healthc Mater **2**(6), 812–816 (2013)
84. Detamornrat, U., Parrilla, M., Domínguez-Robles, J., Anjani, Q.K., Larrañeta, E., De Wael, K., Donnelly, R.F.: Transdermal on-demand drug delivery based on an iontophoretic hollow microneedle array system. Lab Chip **23**(9), 2304–2315 (2023)
85. Economidou, S.N., Pere, C.P.P., Reid, A., Uddin, M.J., Windmill, J.F.C., Lamprou, D.A., Douroumis, D.: 3D printed microneedle patches using stereolithography (SLA) for intradermal insulin delivery. Mater Sci Eng C Mater Biol Appl **102**, 743–755 (2019)
86. Gualeni, B., Coulman, S., Shah, D., Eng, P., Ashraf, H., Vescovo, P., Blayney, G., Piveteau, L.D., Guy, O., Birchall, J.: Minimally invasive and targeted therapeutic cell delivery to the skin using microneedle devices. Br. J. Dermatol. **178**(3), 731–739 (2018)
87. Iliescu, F.S., Teo, J.C.M., Vrtacnik, D., Taylor, H., Iliescu, C.: Cell therapy using an array of ultrathin hollow microneedles. Microsyst. Technol. **24**(7), 2905–2912 (2018)
88. Tatovic, D., McAteer, M.A., Barry, J., Barrientos, A., Rodríguez Terradillos, K., Perera, I., Kochba, E., Levin, Y., Dul, M., Coulman, S.A., Birchall, J.C., von Ruhland, C., Howell, A., Stenson, R., Alhadj Ali, M., Luzio, S.D., Dunseath, G., Cheung, W.Y., Holland, G., May, K., Ingram, J.R., Chowdhury, M.M.U., Wong, F.S., Casas, R., Dayan, C., Ludvigsson, J.: Safety of the use of gold nanoparticles conjugated with proinsulin peptide and administered by hollow microneedles as an immunotherapy in type 1 diabetes. Immunotherapy Adv. **2**(1) (2022)
89. Zhang, Y., Yu, J., Kahkoska, A.R., Wang, J., Buse, J.B., Gu, Z.: Advances in transdermal insulin delivery. Adv. Drug Deliv. Rev. **139**, 51–70 (2019)
90. Resnik, D., Možek, M., Pečar, B., Janež, A., Urbančič, V., Iliescu, C., Vrtačnik, D.: In vivo experimental study of non-invasive insulin microinjection through hollow Si microneedle array. Micromachines **9**(1), 40 (2018)
91. Gupta, J., Felner, E.I., Prausnitz, M.R.: Minimally invasive insulin delivery in subjects with type 1 diabetes using hollow microneedles. Diabetes Technol. Ther. **11**(6), 329–337 (2009)
92. Davis, S.P., Martanto, W., Allen, M.G., Prausnitz, M.R.: Hollow metal microneedles for insulin delivery to diabetic rats. IEEE Trans. Biomed. Eng. **52**, 909–915 (2005)
93. Abedi, M., Abolmaali, S.S., Heidari, R., Mohammadi Samani, S., Tamaddon, A.M.: Hierarchical mesoporous zinc-imidazole dicarboxylic acid MOFs: surfactant-directed synthesis, pH-responsive degradation, and drug delivery. Int. J. Pharm. **602**, 120685 (2021)
94. Alimardani, V., Abolmaali, S.S., Yousefi, G., Rahiminezhad, Z., Abedi, M., Tamaddon, A., Ahadian, S.: Microneedle arrays combined with nanomedicine approaches for transdermal delivery of therapeutics. J. Clin. Med. **10**(2), 181 (2021)
95. Jiang, X., Zhao, H., Li, W.: Microneedle-mediated transdermal delivery of drug-carrying nanoparticles. Front. Bioeng. Biotechnol. **10** (2022)
96. Roxhed, N., Griss, P., Stemme, G.R.: Membrane-sealed hollow microneedles and related administration schemes for transdermal drug delivery. Biomed. Microdevice **10**, 271–279 (2008)

97. Mir, M., Permana, A.D., Tekko, I.A., McCarthy, H.O., Ahmed, N., Donnelly, R.F.: Microneedle liquid injection system assisted delivery of infection responsive nanoparticles: a promising approach for enhanced site-specific delivery of carvacrol against polymicrobial biofilms-infected wounds. Int. J. Pharm. **587**, 119643 (2020)
98. Pamornpathomkul, B., Niyomtham, N., Yingyongnarongkul, B.E., Prasitpuriprecha, C., Rojanarata, T., Ngawhirunpat, T., Opanasopit, P.: Cationic Niosomes for enhanced skin immunization of plasmid DNA-encoding ovalbumin via hollow microneedles. AAPS PharmSciTech **19**(1), 481–488 (2018)
99. Abd-El-Azim, H., Tekko, I.A., Ali, A., Ramadan, A., Nafee, N., Khalafallah, N., Rahman, T., McDaid, W., Aly, R.G., Vora, L.K., Bell, S.J., Furlong, F., McCarthy, H.O., Donnelly, R.F.: Hollow microneedle assisted intradermal delivery of hypericin lipid nanocapsules with light enabled photodynamic therapy against skin cancer. J. Control. Release **348**, 849–869 (2022)
100. van der Maaden, K., Heuts, J., Camps, M., Pontier, M., Terwisscha van Scheltinga, A., Jiskoot, W., Ossendorp, F., Bouwstra, J.: Hollow microneedle-mediated micro-injections of a liposomal HPV E7(43–63) synthetic long peptide vaccine for efficient induction of cytotoxic and T-helper responses. J. Control. Release **269**, 347–354 (2018)
101. Du, G., Hathout, R.M., Nasr, M., Nejadnik, M.R., Tu, J., Koning, R.I., Koster, A.J., Slütter, B., Kros, A., Jiskoot, W., Bouwstra, J.A., Mönkäre, J.: Intradermal vaccination with hollow microneedles: a comparative study of various protein antigen and adjuvant encapsulated nanoparticles. J. Control. Release **266**, 109–118 (2017)
102. Niu, L., Chu, L.Y., Burton, S.A., Hansen, K.J., Panyam, J.: Intradermal delivery of vaccine nanoparticles using hollow microneedle array generates enhanced and balanced immune response. J. Control. Release **294**, 268–278 (2019)
103. de Groot, A.M., Du, G., Mönkäre, J., Platteel, A.C.M., Broere, F., Bouwstra, J.A., Sijts, A.: Hollow microneedle-mediated intradermal delivery of model vaccine antigen-loaded PLGA nanoparticles elicits protective T cell-mediated immunity to an intracellular bacterium. J. Control. Release **266**, 27–35 (2017)
104. Glover, K., Mishra, D., Gade, S., Vora, L.K., Wu, Y., Paredes, A.J., Donnelly, R.F., Singh, T.R.R.: Microneedles for advanced ocular drug delivery. Adv. Drug Deliv. Rev. **201**, 115082 (2023)
105. Yasukawa, T., Ogura, Y., Tabata, Y., Kimura, H., Wiedemann, P., Honda, Y.: Drug delivery systems for vitreoretinal diseases. Prog. Retin. Eye Res. **23**(3), 253–281 (2004)
106. Nagarwal, R.C., Kant, S., Singh, P., Maiti, P., Pandit, J.: Polymeric nanoparticulate system: a potential approach for ocular drug delivery. J. Control. Release **136**(1), 2–13 (2009)
107. Ranta, V.-P., Mannermaa, E., Lummepuro, K., Subrizi, A., Laukkanen, A., Antopolsky, M., Murtomäki, L., Hornof, M., Urtti, A.: Barrier analysis of periocular drug delivery to the posterior segment. J. Control. Release **148**(1), 42–48 (2010)
108. Edelhauser, H.F., Verhoeven, R.S., Burke, B., Struble, C.B., Patel, S.R.: Intraocular distribution and targeting of triamcinolone acetonide suspension administered into the suprachoroidal space. Invest. Ophthalmol. Vis. Sci. **55**(13), 5259–5259 (2014)
109. Verdugo, M.E., Alling, J., Lazar, E.S., Del Cerro, M., Ray, J., Aguirre, G.: Posterior segment approach for subretinal transplantation or injection in the canine model. Cell Transplant. **10**(3), 317–327 (2001)

Chapter 4
Hydrogel-Based Microneedles Via 3D DLP Printing: A New Frontier in Controlled Drug Delivery

Esra Pilavci, Eray Altan, Yeliz Guldorum, Aslihan Calhan, Ali Alpaslan Celik, Cem Bulent Ustundag, Canan Dogan, Nazmi Ekren, and Oguzhan Gunduz

4.1 Introduction

Hydrogel-based microneedles (MNs) have emerged as a transformative technology in the field of drug delivery and biomedical applications. These MNs, fabricated using hydrophilic polymer matrices, offer unique advantages such as minimally invasive administration, biocompatibility, and the ability to encapsulate and release therapeutic agents in a controlled manner. One key advancement that enables the practical use of hydrogel MNs is the integration of three-dimensional (3D) printing technologies, particularly Digital Light Processing (DLP). DLP-based 3D printing allows for

E. Pilavci · E. Altan · Y. Guldorum · A. Calhan · A. A. Celik · C. Dogan · N. Ekren · O. Gunduz (✉)
Center for Nanotechnology and Biomaterials Application and Research (NBUAM), Marmara University, Istanbul, Turkey
e-mail: ucemogu@gmail.com

E. Pilavci · E. Altan · A. A. Celik · C. Dogan · O. Gunduz
Department of Metallurgical and Materials Engineering, Faculty of Technology, Marmara University, Istanbul, Turkey

Y. Guldorum · C. B. Ustundag
Department of Bioengineering, Faculty of Chemical and Metallurgical Engineering, Yildiz Technical University, Istanbul, Turkey

A. Calhan
Chemical Engineering Department, Yildiz Technical University, Istanbul, Turkey

C. B. Ustundag
Health Biotechnology Joint Research and Application Center of Excellence, Istanbul, Turkey

N. Ekren
Electrical Engineering, Faculty of Technology, Marmara University, Istanbul, Turkey

N. Talreja et al. (eds.), *Microneedles (MNs)-Based Technology*, Series in BioEngineering, https://doi.org/10.1007/978-981-96-3916-8_4

high-resolution, customisable designs of MNs that can be tailored to specific clinical requirements [1].

Recent research has explored various materials and fabrication techniques to optimise the performance of hydrogel MNs. For example, Yao et al. demonstrated the fabrication of multifunctional hydrogel MNs via high-precision DLP printing, highlighting their potential for drug delivery and biosensing applications [1]. Similarly, silk fibroin-based MNs, produced through a one-step DLP printing process, have shown promise for their biocompatibility and capacity to deliver molecules transdermally [2]. Furthermore, the development of intelligent microneedle arrays (iμNA) using PEGDA hydrogels has enabled stimuli-responsive drug release, adding another dimension of functionality to this technology [3].

Hydrogel-based MNs have garnered significant attention in drug delivery due to their minimally invasive nature, biocompatibility, and the ability to achieve controlled and sustained release of therapeutic agents. These MNs are made from hydrophilic polymers, which can absorb body fluids, swell, and enable the release of drugs into the skin. Combining hydrogels with MNs has opened new avenues for transdermal drug delivery, especially with the integration of advanced fabrication techniques like DLP-based 3D printing. One notable development is using hydrogel MNs for transdermal delivery, which bypasses the limitations of the stratum corneum, the skin's natural barrier, and enhances drug permeability. For example, hydrogel-forming MNs have been shown to deliver macromolecules by forming a continuous hydrogel conduit between the drug reservoir and the dermal microcirculation, improving transdermal drug delivery efficiency. These MNs can be produced in various geometries and sizes, making them highly customisable for different clinical applications [4].

Hydrogel-based MNs offer several advantages in drug delivery systems, particularly in enhancing patient compliance and minimising skin damage due to their biocompatibility and flexibility. Their ability to swell upon contact with interstitial fluid allows for the sustained and controlled release of therapeutic agents, which is particularly beneficial for macromolecules that are traditionally challenging to administer transdermally. Furthermore, the tunability of the hydrogel's mechanical properties allows for the customisation of penetration depth and drug release kinetics. These MNs can also bypass the stratum corneum, the primary barrier of the skin, enabling more efficient and targeted drug delivery to the dermal microcirculation. Additionally, using hydrogels in microneedle fabrication offers the advantage of incorporating stimuli-responsive properties, such as pH or temperature sensitivity, which can be tailored for site-specific and on-demand drug release. This versatility positions hydrogel-forming MNs as a promising platform for applications ranging from vaccines to biologics, potentially transforming the landscape of transdermal drug delivery systems. Applications of MNs are given in Fig. 4.1.

In this chapter, the focus is on drug delivery systems utilising MNs, particularly hydrogel-based MNs. The use of MNs for enhancing transdermal drug delivery emphasises their potential to improve the efficiency of delivering a wide range of therapeutic agents. A detailed classification of different types of MNs is provided, including their design, material composition, and functionality, with particular attention to the advantages of hydrogel-based MNs. Key aspects such as forming a

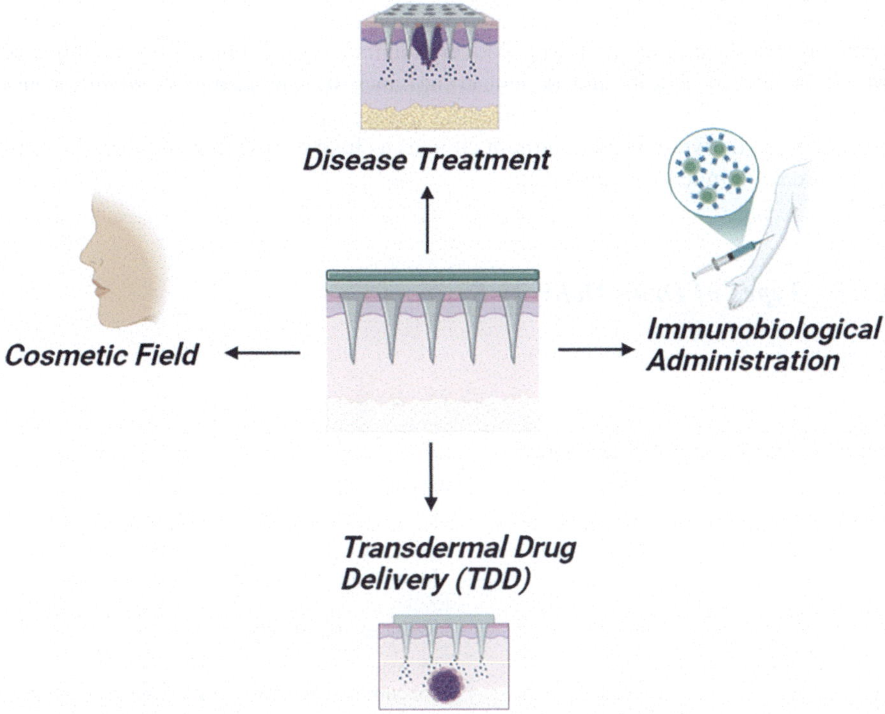

Fig. 4.1 Applications of MNs

hydrogel conduit between the drug reservoir and dermal microcirculation, biocompatibility, and the capacity for controlled and sustained drug release are examined. Additionally, recent advancements in stimuli-responsive hydrogel MNs and their potential for site-specific and on-demand drug delivery are discussed. Challenges related to the mechanical properties, fabrication techniques, and clinical translation of hydrogel MNs are also addressed, providing a comprehensive overview of their role in the future of transdermal drug delivery systems.

4.2 Drug Delivery Systems

Drug delivery systems are cutting-edge techniques created to move medicinal chemicals throughout the body for the best possible therapeutic results. These systems are essential to modern medicine because they guarantee that medications are administered precisely and under control, increasing their effectiveness while lowering their adverse effects [1]. They fall into several categories, including targeted delivery systems, controlled-release systems, and sophisticated drug delivery tools like liposomes, nanoparticles, and MNs. Treatment methods for various illnesses, from acute infections to chronic ailments like diabetes and cancer, have been entirely transformed by the development of drug-delivery technology [1]. These systems improve overall healthcare results and patient compliance by increasing the stability and bioavailability of medications. Future drug delivery provides the prospect of more advanced and individualised treatment choices as long as research and development in this area continue to advance [2].

4.2.1 Types of Drug Delivery Systems

See Table 4.1.

Table 4.1 Types of drug delivery systems

	Description	Advantages	Examples
Controlled-release systems [3]	Release drugs over a specific period of time	–Reduced dosing frequency –Maintained therapeutic levels –Reduced side effects	Oral tablets, transdermal patches, implantable devices
Targeted delivery systems [4]	Deliver drugs to specific sites in the body	–Increased drug concentration at target site –Reduced systemic toxicity –Enhanced efficacy	Monoclonal antibodies, targeted nanoparticles, conjugated drugs
Nanoparticles [5]	Tiny particles that can carry drugs to specific cells or tissues	–High surface area –Enhanced solubility –Targeted delivery	Liposomes, polymeric nanoparticles, dendrimers
Liposomes [6]	Spherical vesicles with a lipid bilayer that encapsulate drugs	–Biocompatible –Protect drugs from degradation –Can target specific tissues	Liposomal doxorubicin, liposomal amphotericin B
Transdermal patches [7]	Patches applied to the skin that release drugs over time	–Non-invasive –Long-term release –Bypasses gastrointestinal tract	Nicotine patches, fentanyl patches
Implantable devices [8]	Devices placed under the skin to release drugs over a prolonged period	–Continuous drug release –Patient compliance –Long-term treatment	Hormone implants, cancer treatment implants
MNs [9]	Tiny needles that deliver drugs through the skin without pain	–Painless –Easy to use –Suitable for vaccines	Microneedle patches for insulin, vaccines

4.2.1.1 Controlled Release Systems

Controlled release systems are an innovative approach to drug delivery that allows for the gradual release of a therapeutic agent over an extended period. They are intended to maintain steady drug levels in the bloodstream, improving therapeutic efficacy and patient compliance, in contrast to conventional drug delivery techniques, which frequently call for frequent dosing [10]. These systems use a variety of processes, including osmosis, diffusion, and degradation, to regulate the drug's release rate. They come in various forms, each specifically designed to address a particular medical need, such as oral tablets, transdermal patches, and implanted devices. Controlled release systems decrease drug concentration changes by delivering a steady drug release, which lowers the risk of side effects and enhances treatment results overall [11].

4.2.1.2 Targeted Delivery Systems

Targeted delivery systems are a cutting-edge approach in drug delivery designed to transport therapeutic agents directly to specific cells, tissues, or organs in the body. To minimise side effects and enhance patient outcomes, this precision optimises the drug's therapeutic effects while minimising its impact on non-targeted areas [12]. Molecular targeting with ligands or antibodies, using nanoparticles, and adding stimuli-responsive materials—which release the medicine in response to particular environmental triggers like pH, temperature, or enzymes—are some of the tactics used in targeted delivery systems [13]. This technology is beneficial for treating conditions like cancer when targeting malignant cells while preserving good tissue is critical. Future treatment methods could be considerably more complex and successful thanks to the ongoing innovations in this field, which incorporate developments in materials science, biotechnology, and nanotechnology.

4.2.1.3 Nanoparticles

These systems encapsulate drugs and deliver them to particular cells or tissues using nanoparticles, which are minuscule particles that usually range in size from 1 to 100 nm. Nanoparticles can interact with biological systems at the molecular level because of their small size and huge surface area, which enables regulated and targeted drug release [14]. Nanoparticles can enhance the solubility and stability of drugs, protect therapeutic agents from degradation, and improve bioavailability. Additionally, by modifying the surface properties of nanoparticles with antibodies or other targeting moieties, these systems can achieve precise delivery to specific disease sites, such as tumours or inflamed tissues, minimising the impact on healthy cells and reducing side effects [15]. Drug delivery uses a variety of nanoparticle kinds, each with special qualities suited to various therapeutic purposes, such as liposomes, polymeric nanoparticles, dendrimers, and metallic nanoparticles. These adaptable carriers are undergoing intensive study and development to treat various illnesses, including cancer, heart disease, infectious diseases, and neurological disorders [16].

As nanotechnology develops, nanoparticle drug delivery technologies, which provide safer, more effective, and more patient-friendly therapy alternatives, have enormous potential for the future of customised medicine.

4.2.1.4 Transdermal Patches

Transdermal patches are an innovative drug delivery system that allows medications to be absorbed directly into the bloodstream through the skin. This method provides a convenient, non-invasive alternative to oral or injectable routes, enhancing patient compliance and comfort [17]. Transdermal patches consist of an adhesive layer containing the drug, which is released at a controlled rate when applied to the skin. Transdermal patches are used for many therapeutic purposes, such as hormone replacement therapy (e.g., estrogen patches), nicotine replacement therapy (e.g., nicotine patches for quitting smoking), pain management (e.g., fentanyl patches), and the treatment of ailments like motion sickness and hypertension [18]. Depending on the therapeutic need, their architecture can be modified to release medications over various periods, from hours to days [19]. As research in materials science and pharmaceutical technology advances, the potential applications of transdermal patches continue to expand. Innovations such as ***MNs patches*** and iontophoresis enhance the efficiency and versatility of transdermal delivery, paving the way for new treatments and improving patient outcomes.

4.3 MNs for Drug Delivery

The most common drug delivery methods have drawbacks, such as limited bioavailability, toxicity, and drug degradation [5]. The right amount of the prescribed medication must be delivered at the right time to the correct location. Despite being the quickest, most direct, and most dependable method of drug delivery, the parenteral route is limited in its use in several situations due to its associated characteristics, which include lower bioavailability at the targeted site, drug accumulation in unwanted surrounding organs, and rapid blood clearance. These factors collectively result in reduced efficacy overall. Similarly, the mucosal enzyme of the intestinal environment will inevitably break down vaccines or drugs administered orally or by buccal or oral route. Scientists developed the transdermal method to get around the problems with parenteral or oral drug delivery [6]. The extreme pH of stomach acid, the degradation of digestion enzymes, and the possible risks of inhaled drugs and their auxiliary particles to the lungs can all be avoided with the MN transdermal drug delivery system [7]. When viewed, the protective, inflammatory, and immunological characteristics of the skin make the MN delivery system an attractive alternative system to drug delivery to describe the constraints associated with traditional methods [8].

MN, a revolutionary transdermal delivery technique, has been studied over the past 30 years to increase the transferability of molecules into the skin. Since the first report on the use of MNs for drug delivery was published in the 1990s, MN technology has been acknowledged as a successful means of delivering genes or drugs, significantly improving the passage of molecules through skin tissue [9].

In addition, MNs are a self-administering tool that can painlessly and minimally invasively administer medications, vaccinations, and other treatments by transdermally absorbing interstitial fluid. Furthermore, MNs can be made and altered into any shape, size, geometry, or substance depending on the application and release mechanism. They can also be used in a combinatory manner with two distinct therapies at the same time [10].

These minuscule needle patches can load genes, antibodies, vaccines, heavy molecular weighted proteins, small or large medication molecules, nanoparticles, and many other substances [6, 11]. These nanoparticle-loaded MNs stimulate the necessary immune response while delivering medications to deep dermal areas close to underlying dendritic cells, langerhans, and neutrophils [6].

4.3.1 Disease Treatment with Microneedle Drug Delivery

Many medications, including vaccinations, nanoparticles, and small and macromolecular drugs, have been inserted into MNs to treat various illnesses. Drug-loaded MNs have the ability to momentarily rupture the stratum corneum, increasing skin permeability and, as a result, enhancing the effectiveness of drug penetration for improved therapeutic results. Over the past ten years, there has been a notable surge in microneedle research [12]. The MN drug delivery technology has been used and made popular in increasing fields, including family planning, dermatocosmetology, chronic illnesses like diabetes and arthritis, cancer, and epidemic disease prevention. It has achieved significant success with the advancement of materials science and fabrication technology [13]. Due to the disadvantages of traditional cancer treatment techniques, attention has switched to using MNs for early cancer diagnostics and the delivery of anti-cancer vaccines and drugs. Additionally, synergistic techniques like gene therapy or chemotherapy in conjunction with photothermal or photodynamic therapy are appropriate for this delivery system. The noninvasive self-administration tool provides an option to conventional drug delivery methods, such as systemic administration through hypodermic needle insertion [6].

Since patients can insert the MNs themselves, this could provide cancer patients with a more accurate and effective method of personalised treatment. The capacity to affix MNs directly or very close to the targeted tumour site improves the accuracy of drug delivery to tumour cells, hence boosting therapeutic efficacy. MNs are more accurate in targeting tumours and delivering drugs than traditional therapy methods, such as oral or hypodermic injections.

Treatment of the most prevalent degenerative joint condition, osteoarthritis and its symptoms are desperately needed, as osteoarthritis is the most frequent form of arthritis and a significant source of disability [13, 14]. There are still insufficient efficient treatment options for OA despite the limitations of the ones available today. The avoidance of liver metabolism, the decrease in severe adverse reactions of oral drugs to the gastrointestinal tract, the ensuring prolonged and sustained drug release, and the ease of stopping drug delivery in the event of toxicity are just a few potential benefits of using MNs in the treatment of OA [13].

In addition to offering improved immunogenicity, thermostability, simplicity, safety, and compliance, MN patches can provide a viable alternative vaccine modality that reduces sharp waste, increases cost-effectiveness, and allows for self-administration, which could enhance vaccine distribution [15]. The superiority of MNs in vaccine and insulin administration tests was examined. MNs are reported to have benefits over conventional treatments, including the ability to avoid needle phobia, lower costs due to fewer vaccines, and the absence of specialist training requirements for administration [16].

Viral antigens or antibodies that may not be present in blood can sometimes be found locally at the skin lesion level by MN extraction for cutaneous viral infections. To aid in the diagnosis and distinction of varicella/herpes zoster or herpes simplex, MN-based bioassays for virus antigen capture may be employed. According to the literature, MN application is available to combat common viral infectious disorders that affect the skin or the body, such as COVID-19, herpes simplex, varicella/herpes zoster, warts, influenza, and measles [15].

Additionally, there is a promising area for using MNs in cosmetic medicine. MNs work by puncturing the epidermal skin, which facilitates the release of potent active substances and can reduce wrinkles, discolouration, and scars [12].

4.3.2 Types of MNs/Classification of MNs

MNs, characterised as micron-sized needles ranging in length from 25 to 2000 μm and located on a supporting base or patch [17], can be classified according to their structure, materials, and drug transportation method [18]. Each category offers its advantages and challenges, making the selection of the appropriate microneedle type for specific therapeutic targets critical. Initially, the delivery potential of MNs for administering therapeutic molecules that cannot cross the skin barrier across biological membranes such as skin, mucosal tissue and sclera made them an attractive option for drug delivery. However, their applications have swiftly broadened to include various regions, such as ocular, gastrointestinal, and buccal areas [19]. Furthermore, the manipulation of microneedle geometry and composition is crucial for facilitating controlled drug delivery to target organs. MNs designed for drug delivery are classified into five main categories: solid, coated, dissolvable, hydrogel-forming, and hollow [20]. This classification provides a framework for understanding the role of MNs in modern medical practice. As shown in Table 4.2, different MNs

Table 4.2 Overview of microneedle types, delivery strategies, materials, advantages, and limitations

Type of Microneedles	Drug Delivery Strategies	Schematic Illustrations	Examples of Material	Advantages	Limitations
Solid	Poke and Patch	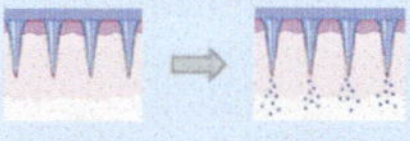	Metals (stainless steel, titanium), ceramics (alumina, zirconia), polymers (PLA, PLGA), silicone	-Durable with high mechanical stability -Effective in creating microchannels for enhanced transdermal delivery	-Challenges in delivering precise doses -Requires post-treatment for wound recovery -Increased risk of contamination if reused
Coated	Coat and Poke		Metals (stainless steel, titanium), coating materials (PVP, sucrose, glucose)	-Simplifies direct drug application without additional steps - Maintains structural integrity during insertion.	-Limited drug capacity due to coating thickness - Coating might detach during application, reducing efficiency
Dissolvable	Poke and Release		Water-soluble polymers (PVP, PVA), biodegradable polymers (PLGA, poly(acrylic acid), silk, chitosan)	-Cost-effective and straightforward production process - Supports controlled and sustained drug release in a single step	-Reduced drug penetration due to limited structural robustness -Weaker mechanical properties compared to other types
Hollow	Poke and Flow		Metals (stainless steel, titanium), silicone, polymers (PMMA, polycarbonate)	- High accuracy in delivering liquid formulation -Enables delivery of large molecules and biologics	-Blockage of the needle tip in tissues is a common challenge -Reuse can raise safety and hygiene concerns
Hydrogel-Forming	Poke and Release	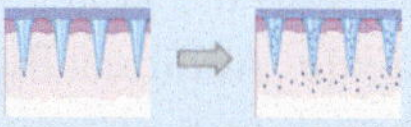	Synthetic polymers (poly(acrylic acid), PEG, poly(acrylamide)), natural polymers (gelatin, alginate, chitosan)	- Leaves no harmful residues on the skin after use -Straightforward to fabricate with potential for controlled drug release	-Mechanically weaker compared to other microneedle types

offer unique advantages and limitations depending on their materials and mechanisms of action. Each category will be detailed in this section, and the basic properties of MNs, their advantages and disadvantages, areas of use, mechanisms of action, and their place in healthcare applications will be explained.

4.3.2.1 Solid MNs

Solid MNs, the first type of microneedle technology created, penetrate or scrape the skin surface to create microchannels through which active ingredients can be transported, thus acting as a skin pretreatment. Solid MNs themselves do not contain an active ingredient; instead, drugs are applied to the microporous skin in topical pharmaceutical forms such as gel, ointment, spray, lotion, cream, foam or patch. This method facilitates the diffusion of the drug through the created microchannels, allowing molecules to reach the target tissue. To improve penetration, microneedle therapy can also be used in conjunction with methods like iontophoresis and near-infrared technologies [21].

Solid MNs systems are typically fabricated from metals such as palladium, nickel, gold, stainless steel, and titanium, as well as silicon material and polymers. These systems can be directly produced from these materials using techniques such as laser cutting, mechanical or chemical etching or manufacturing through mould casting or electroplating. The MNs' force must be kept low while sufficiently penetrating the skin to ensure a painless and least invasive application. In addition, MNs must have high mechanical strength and adaptability to prevent breakage or kinking. Solid MNs can increase skin permeability up to four times by creating fine incisions in the skin [22]. However, they also present some challenges. One of the main disadvantages is that the micropores are reported to remain open for a limited time, such as only 2 h. This affects the adequate delivery of the active ingredient. Although micropores can be extended for up to three days under occlusive conditions using formulations such as tapes or patches, the risk of infection is significantly increased under these circumstances [23].

4.3.2.2 Coated MNs

Coated MNs are obtained by coating the surfaces of solid MNs with a film layer of diverse hydrophilic or hydrophobic active compounds, including nucleic acids, proteins, small molecules, vaccines, and peptides, to the surfaces of solid MNs. In these types of MN, therapeutic agents are delivered to the skin and dissolved through the created microchannels [24]. Once placed on the skin, the coated drug formulations gradually dissolve and permeate into the systemic circulation, while the MNs can be removed from the insertion site. Additionally, the micron size of the MNs significantly reduces medical risks such as nerve stimulation, acute pain or skin injury during application. Compared to oral and parenteral administration, transdermal drug delivery with coated MNs offers significant advantages in terms of both drug efficacy and patient comfort [25]. Although the amount of drug that can be loaded in coated microneedle (MN) systems is limited, they offer the advantage of providing long-term efficacy [26]. To optimise the drug delivery efficacy of coated MNs (MN), it is essential to evaluate parameters like the simplicity of coating solution preparation, coating efficiency, disintegration characteristics, and physicochemical features. Consequently, the correlation between various compositions of the coating solution and drug delivery efficacy warrants investigation [27].

4.3.2.3 Dissolvable/Biodegradable MNs

Dissolvable MNs, first introduced in 2005, utilise a "poke and release" delivery strategy. These polymer-based systems incorporate drugs that are dissolved or suspended within the microneedle formulation. The needle tips dissolve upon application to the skin, releasing the therapeutic payload and facilitating its distribution to the target tissue [28]. These MN arrays are composed of water-soluble or biodegradable polymers that dissolve or disintegrate upon contact with skin interstitial fluid, enabling drug release. Commonly used materials include water-soluble polymers such as poly(methylvinylether/maelic acid), poly(vinyl pyrrolidone) (PVP), and poly(vinyl alcohol) (PVA), as well as biodegradable polymers like silk, poly(acrylic acid), poly(lactide-co-glycolide), and chitosan. The primary advantage of biodegradable and dissolvable MNs is their ability to ensure that the entire drug load is administered without generating biohazardous waste. Furthermore, biodegradable needles can enable long-term therapeutic effects through the controlled release of specific molecules [29]. Dissolvable MNs also have several other benefits: the polymers are low-cost, do not leave biohazardous or sharp waste, and can be safely disposed of due to their self-deactivating properties, eliminating the risk of reuse [30]. The dissolution process of dissolvable MNs can be controlled by external triggers such as electric current, ultrasound, near-infrared light, or other specially designed conditions, allowing for controlled drug release. However, due to the polymer matrix, their skin penetration capacity is lower than solid and hollow MNs. On the other hand, the needle bodies of dissolvable MNs dissolve entirely in the body, offering high biocompatibility and reducing the risks of inflammation and infection. Nonetheless, the selection and use of dissolvable MNs materials should be carefully evaluated, particularly regarding biological toxicity and material properties [31]. Dissolvable MNs, made from polymeric materials with superior biodegradability and biocompatibility compared to other types of MNs, are ideal for drug delivery and other biomedical applications.

4.3.2.4 Hollow MNs

Hollow MNs, commonly used to deliver insulin, vaccines, and biomarkers, feature a lumen or internal channel with a width of 5–70 μm. These hollow structures, acting as channels, allow active ingredients to be transported by passive diffusion or by applying pressure through a pump, syringe, or gas. Liquid formulations typically flow from a reservoir through hollow MNs at a rate of 10–100 μL/min, continuously delivering their content to specific layers of the skin via propelling flow mechanisms [32]. Hollow MNs are innovative structures made from various materials, including polymers, metals, and silicone. They are produced using advanced techniques such as three-dimensional (3D) printing, drawing lithography, and etching. However, the complexity and high cost of these manufacturing methods limit their accessibility to a broader patient population [33].

4.3.2.5 Hydrogel-Forming MNs

The term hydrogel, first used by Van Bemmelen in 1894 to describe a colloid composed of inorganic salts, has since moved away from the original definition. Hydrogels (HGs) are a collection of classes of soft substances, including various polymeric substances of varying origins that can create swollen form three-dimensional networks in water through chemical or physical cross-links, where the type of bond that forms dictates the reversibility of the hydrogel [34]. The most common microneedle designs are solid, drug-coated, dissolved and hollow. Hydrogel-forming MNs (HFMs), the first time was developed by Leese's research group in 2012, are derived from swellable polymers [35]. HFMNs have attracted considerable interest from researchers owing to their superior water retention capabilities and biocompatibility [36]. Also, HFMs are safe, non-toxic, without any residual matrix material, and appropriate for elevated drug loading capacity and long-term, multi-drug administration [37]. The drug release mechanism of HFM differs from other forms of MN. The release of high concentration drugs from hydrogel-forming MNs is particularly noteworthy due to the swelling behaviour observed after insertion into the site, attributed to the superior hydrophilicity and water absorption capabilities of hydrogel substrates [35]. HFMs also have a higher drug loading capacity and a tunable drug release rate by directly manipulating the polymer crosslinking rate, overcoming the limitations of traditional MNs [38].

Structure of Hydrogel

Hydrogels (HGs) are three-dimensional networks of hydrophilic cross-linked polymers capable of absorbing significant amounts of water and reacting to stimuli. Hydrogels can be synthesised by cross-linking monomers within the polymeric chain through chemical interactions (covalent bonds) and physical interactions with cross-linking agents (hydrogen bonds, electrostatic interactions), which subsequently influence their mechanical and physicochemical properties [39].

Classification of Hydrogels

Hydrogels are versatile materials that can be categorised based on various characteristics, as shown in Fig. 4.2, including preparation methods, degradation rates, crosslinking approaches, ionic charges, origins, and physical properties [40]. Based on their biodegradability, hydrogels are classified as biodegradable or non-biodegradable. Their preparation methods further divide them into homopolymeric, copolymeric, and interpenetrating polymeric hydrogels. Depending on their ionic change status, hydrogels can be categorised as nonionic, anionic, cationic, or ampholytic structures [41]. Considering their origins, hydrogels are classified as natural, semi-synthetic, and synthetic materials, and based on their formation mechanisms, they are categorised as physically or chemically crosslinked hydrogels [42]. From a physical properties perspective, hydrogels are divided into traditional and "smart" gels that respond to remote or endogenous triggers to control substance distribution. Additionally, hydrogels can be classified based on their responsiveness to stimuli, such as temperature, light, magnetic fields, pH, and sound [43]. These multifaceted classifications highlight the complex structure and extensive application potential of hydrogels (Fig. 4.2).

4.4 Remote or Endogenous Triggers to Control Delivery: "Smart" Gels

A unique class of hydrogels, known as smart or stimuli-responsive hydrogels, has become an area of great interest for researchers in pharmaceutical, biotechnological and biomedical applications. These hydrogels enable the controlled and regulated release of encapsulated drugs or biomolecules by responding to environmental factors such as temperature, pH, light and magnetic fields [44]. This class of hydrogels can be

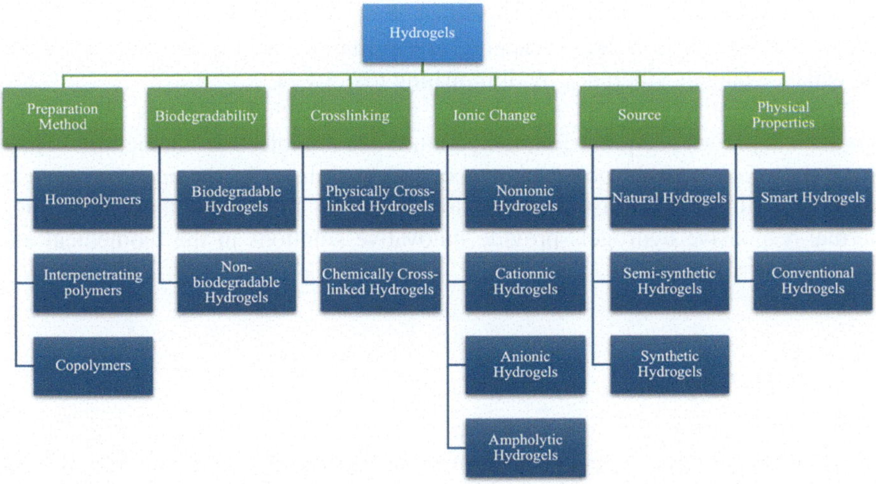

Fig. 4.2 Classification of hydrogels by key properties

developed based on physical stimuli, chemical stimuli and dual-responsive systems. Physical stimuli include pressure, ultrasound, magnetism, temperature, light and electric fields, while chemical stimuli include pH, ionic factors and biochemical agents such as glucose [45]. Dual-responsive hydrogels attract attention with their ability to respond to more than one type of stimuli simultaneously in a system [46].

4.4.1 Thermal-Responsive Hydrogels

Thermal-responsive hydrogels are defined as polymeric structures that reversibly change their physical phase and morphological properties in response to temperature changes [47]. The most prevalent thermo-responsive polymers are poly(N-isopropylacrylamide) (PNIPAM) and its derivatives; however, other polymers such as poly(2-methoxyethyl vinyl ether) (MOVE), poly 3-[N-(3-methacrylamidopropyl)-N, N-dimethyl] ammoniopropane sulfonate (SPP), poly(2-ethoxyethyl vinyl ether) (EOVE), poly oligo(ethylene glycol) methyl ether methacrylate (OEGMA), and poly di(ethylene glycol) methyl ether methacrylate (MEO2MA) are also utilised in research. Thermal-responsive hydrogels comprise thermo-responsive polymers demonstrating low critical solution temperature (LCST) or high critical solution temperature (UCST) characteristics. In LCST polymers, phase separation transpires when the temperature exceeds the LCST, whereas UCST polymers exhibit phase separation at lower temperatures [48]. Thermal-responsive polymers are significant in biological applications, particularly for their ability to demonstrate sol–gel transition within a temperature range near body temperature (25–37 °C). Due to their injectable and biocompatible characteristics, these polymers are utilised in applications including drug delivery, therapeutic encapsulation, and tissue filling. They also can enhance patient adherence by diminishing toxicity [49].

Phase separation in UCST polymers is driven by enthalpic effects, which form two distinct phases at equilibrium temperature due to molecular interactions. These phase transitions enable polymers to be widely utilised in therapeutic encapsulation and minimally invasive application processes [50]. With these properties, thermal-responsive hydrogels provide innovative solutions in the biomedical and pharmaceutical fields.

4.4.2 PH-Responsive Hydrogels

pH-sensitive hydrogels are generally composed of polymers containing weakly basic or acidic groups in the polymer backbone, such as poly(diethyl-aminoethyl methacrylate) (PDEAEMA), polyacrylic acid (PAA), and polymethacrylic acid (PMAA). These polymers tend to ionise more at certain pH conditions [51]. Ionic groups in the polymer backbone (e.g., amines, sulfonic acids, carboxylic acids, and imines) ionise under appropriate pH conditions, creating electrostatic repulsive forces. These forces lead to the expansion of the polymeric network and consequent swelling of the hydrogel [46].

Cationic hydrogels tend to swell at low pH levels ($pH < pKa$) due to the protonation of amino or imine groups. This is due to the repulsive interaction of positive charges in the polymer chains with each other. In contrast, anionic hydrogels swell at high pH levels ($pH > pKa$) due to interchain repulsive forces resulting from the ionisation of acidic groups [52]. Conformational changes in the pH-sensitive polymer can occur through destabilisation (collapse or swelling), dissociation, and drug/carrier partition coefficient changes. The release of drugs can be affected by conformational changes or solubility changes, and therefore, the release kinetics can be controlled by manipulating the degree and rate of conformational changes [42].

4.4.3 Photo-Responsive Hydrogels

Photo-responsive hydrogels are materials that undergo changes in their physical and chemical properties under light exposure (near-infrared (NIR), UV, or visible). These hydrogels, prepared by incorporating photosensitiser groups into the polymer backbone, respond to specific wavelengths of light and initiate photoreactions such as isomerisation, cleavage, or dimerisation [53]. These photoreactions induce swelling or shrinking in the hydrogel through mechanisms like free radical polymerisation or breaking chemical bonds. Additionally, macroscopic changes such as gelation or degradation can occur under light exposure, enabling their application in biomedical fields, particularly in drug delivery systems [54]. Photoisomerisation typically results in reversible hydrogels, whereas photorefractive processes produce irreversible structures. Molecules like azobenzene and spiropyran facilitate isomerisation, while o-nitrobenzyl derivatives are commonly used for photorefractive structures. Furthermore, hydrogels enriched with photothermal agents can induce phase transitions in thermoresponsive hydrogels through light-triggered heat generation. The NIR sensitivity of these materials offers significant advantages for deep tissue applications [55].

4.4.3.1 Healing Sound: Ultrasounds as a Trigger

Ultrasound waves, as high-frequency mechanical waves with a frequency of $\geq$20 kHz, are used as a non-invasive and cost-effective method in many clinical applications due to their ability to focus and transmit in specific environments [56]. Ultrasound offers high penetration capacity into soft tissues compared to light and can trigger drug release by thermal or mechanical means through cavitation phenomena and acoustic convection [57]. These features allow us to solve the problem of site specificity in drug delivery systems, overcoming one of the most significant challenges encountered in developing these systems.

Ultrasound-responsive systems enable spatially and temporally targeted drug release by non-toxic acoustic mechanisms. These mechanisms transfer energy to initiate a specific biological response. Acoustic waves are used for energy transfer in a sensitive hydrogel by thermal and non-thermal (mechanical) mechanisms. In the thermal mechanism, increasing the temperature increases the diffusion rate and provides an effective release; in the non-thermal mechanism, the strong mechanical effects of acoustic cavitation support targeted release in drug delivery systems [58]. The efficacy of these mechanisms facilitates novel uses of ultrasound in the biomedical field, increasing its potential for drug delivery and delivery systems.

4.4.3.2 Magnetically Responsive Hydrogels

Magnetic hydrogels are composite materials formed by embedding magnetic particles such as Fe_3O_4 and γ-Fe_2O_3 into the hydrogel matrix. Their working principle is based on properties such as remote control by magnetic field, noninvasiveness, high biocompatibility, and low toxicity [59]. Thanks to these properties, magnetic hydrogels offer great potential in biomedical engineering applications such as muscle regeneration, long-term magnetic resonance imaging, and drug delivery. However, their ion sensitivity to physiological fluids and inadequate mechanical properties constitute significant limitations in their practical use [60].

4.5 Fabrication Techniques of MNs/The Manufacture of MNs/Fabrication of MNs

MNs are a drug delivery technology that provides a minimally invasive technique of giving treatments via the skin. Their production includes several specialised processes [61], each designed to create MNs with unique properties suited for various applications. Production methods include; lithography [62], molding [63], electroforming [64], etching [65] and 3D printing [66]. This section will examine 3D printing technologies, especially **DLP**, in detail (Fig. 4.3).

4.5.1 Digital Light Processing (DLP)

DLP 3D printing is an advanced additive manufacturing technique that utilises a digital light projector to cure photopolymer resin layer by layer, creating highly detailed and precise objects [67]. Unlike other 3D printing methods that use a point-by-point process, DLP uses a digital micromirror device (DMD) to project an entire image of a layer simultaneously onto the resin. This results in faster print times and high-resolution prints with fine details and smooth surfaces (Table 4.3).

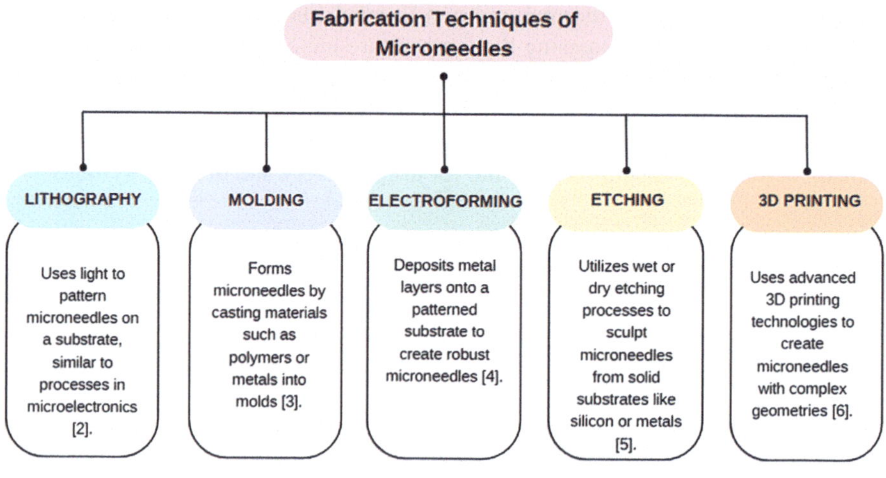

Fig. 4.3 Fabrication techniques of MNs

Table 4.3 Advantages of DLP printing [68–71]

Advantage	Description
High resolution and detail	Produces highly detailed parts with smooth surfaces, ideal for intricate designs
First print times	Prints faster by curing an entire layer at once, compared to point-by-point methods
Excellent surface finish	Results in smooth, high-quality finishes, reducing the need for post-processing
Material versatility	Compatible with a variety of resins, including flexible, rigid, and biocompatible materials
Cost-effective for high-resolution	Provides cost-effective high-resolution printing for small to medium-sized objects
Consistency and reliability	Ensures uniform curing and consistent quality across prints
Minimal material waste	Uses photopolymer resin, reducing waste compared to subtractive manufacturing

4.5.2 Microneedle Design

Microneedle design for DLP (Digital Light Processing) 3D printing has various advantages. It is an excellent approach for producing precise, detailed, and functional MNs for medicine delivery, vaccinations, and diagnostic devices [72]. Figure 4.4 shows the design and production step of MNs.

4.5.2.1 High Precision and Detail

DLP printing provides high resolution, allowing for the production of MNs with extremely fine tips and sharp geometries. MNs require this level of precision because they must be tiny enough to penetrate the skin without causing discomfort while still being durable enough to store and distribute therapeutic ingredients [73].

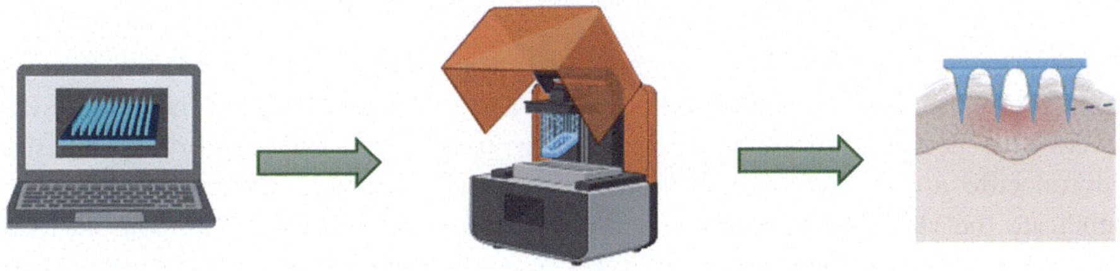

Fig. 4.4 Design and production steps of a microneedle

4.5.2.2 Customisation and Complexity

DLP 3D printing also allows for customising microneedle designs to suit specific applications. Designers can modify the height, shape, and array of the MNs to optimise them for targeted drug delivery or specific skin penetration capabilities. DLP 3D printing also enables the modification of microneedle patterns for particular applications. Designers can change the height, shape, and array of the MNs to improve targeted drug administration or skin penetration capabilities [74].

4.5.2.3 Material Versatility

DLP technology is compatible with various photopolymer resins, including flexible, stiff, and biocompatible materials. This allows for the development of MNs with various mechanical qualities customised to specific applications, such as flexible MNs that can bend to fit the skin or rigid ones that maintain sharpness. Biocompatible resins are very valuable in medical applications, as they ensure that MNs are safe to use in the human body [75]. Using all these design parameters, different types of MNs can be produced. Standard microneedle designs for DLP Printing are [76]:

- Solid MNs: Simple, solid MNs are used for skin penetration.
- Hollow MNs: MNs with hollow cores designed for the delivery of drugs directly into the skin or bloodstream.
- Coated MNs: MNs coated with drugs or vaccines, designed to dissolve in the skin for sustained release.
- Arrays: DLP printing allows the creation of microneedle arrays, which consist of multiple MNs arranged in a grid or pattern to cover larger surface areas for more efficient drug delivery.

4.5.3 Physical Characterization of MNs

4.5.3.1 Morphology and Surface Structure

Microscopic techniques like scanning electron microscopy (SEM) or optical microscopy are used to examine the surface and structure of MNs. These methods reveal the needle's tip sharpness, surface roughness, and uniformity of the microneedle array. The texture of the microneedle surface affects its ability to penetrate the skin and interact with drugs or biological tissues [77]. The surface morphology, such as the presence of pores or grooves, can influence the drug release rate and skin penetration efficiency. The height and length of MNs are crucial for reaching the desired skin depth (typically the stratum corneum or epidermis) for effective drug delivery. MNs generally are between 100 μm to 1 mm in height. The number of MNs per unit area in an array determines how effectively the MNs cover the skin and deliver the therapeutic agent. Arrays are usually designed with a spacing that balances skin coverage and ease of insertion [78].

4.5.3.2 Mechanical Properties

The mechanical strength of MNs is crucial for preventing them from breaking during insertion into the skin. Testing the microneedle's strength and stiffness guarantees it will remain intact during application stresses. Compression and bending tests are used to determine these qualities [79].

4.5.3.3 Drug Release Properties

Testing how they break down in the skin (e.g., polymeric MNs) is essential if they are designed to dissolve or degrade after insertion. Dissolution tests in simulated skin environments or buffers can evaluate the release profile of drugs encapsulated in the MNs. For MNs designed for drug delivery, the rate and pattern of drug release must be evaluated. This can involve testing how the encapsulated drug is released after insertion into the skin, which is crucial for optimising therapeutic outcomes [80].

4.6 Biomedical Applications of 3D-Printed MNs

MNs fabricated using 3D printing technologies have been shown to improve the transdermal delivery of therapeutics by creating minimally invasive channels for drug administration. For example, hydrogel MNs created via DLP 3D printing demonstrated their ability to deliver drugs like diclofenac sodium through controlled and stimuli-responsive release mechanisms [3]. Additionally, silk fibroin MNs, which are biocompatible and can be used for molecular delivery, have been successfully fabricated using DLP 3D printing [2].

4.6.1 Disease Treatment

In the medical field, MNs intended for therapeutic applications are subject to more stringent regulatory requirements than those used for cosmetic purposes. The passive nature of drug delivery through MNs necessitates rigorous experimental validation and certification to ensure stable and precise delivery. As a result, there is an increasing focus within the transdermal microneedle patch domain on developing detachable MNs. These MNs are engineered to effectively administer drugs by separating from their substrate upon application, addressing the need for enhanced precision and reliability in drug administration.

Detachable MN technologies can be categorised into several types. The first type, dissolvable MNs or substrate-separable MNs (DeMN), involves loading the entire microneedle structure with the drug. Upon skin application, the MNs dissolve in the interstitial fluid, detaching from the substrate and facilitating controlled drug release into the skin. This method enables a high drug load but carries the risk of potential drug loss at the skin's surface and exposed areas post-insertion. The second type, multi-layered DeMN, incorporates MNs with drug-loaded tips connected to the substrate. Similar to the dissolvable MNs, the drug is released as the MNs dissolve after skin penetration. This approach allows for deeper skin insertion while minimising drug exposure on the surface, enhancing precision in drug delivery. Lastly, a third method introduces an additional layer between the drug-loaded portion of the microneedle and the substrate, further improving delivery accuracy and minimising drug loss.

MNs have emerged as a promising technology for treating various diseases due to their ability to deliver therapeutic agents in a minimally invasive manner, overcoming the limitations of conventional drug delivery methods. In cancer treatment, MNs offer precise, localised drug delivery, allowing for targeted chemotherapy and immunotherapy, reducing systemic toxicity, and enhancing therapeutic efficacy. Studies have demonstrated their effectiveness in delivering large molecules, including vaccines and genes, to tumour sites, potentially combining chemotherapy with photothermal therapy for improved outcomes [6]. Similarly, in cardiovascular diseases, MNs provide controlled and sustained release of medications such as antihypertensive drugs, offering an alternative to traditional invasive techniques like intramyocardial injections. This technology can improve drug retention at disease sites, thereby enhancing therapeutic effects in conditions like atherosclerosis and myocardial diseases [81]. Moreover, in neurodegenerative disorders such as Alzheimer's disease, MNs have been developed to transdermally deliver drugs like Donepezil hydrochloride, which bypasses gastrointestinal degradation and first-pass metabolism, achieving higher drug bioavailability and improved patient outcomes compared to oral administration [82].

4.6.2 Immunobiological Administration

MNs have emerged as a promising tool for administering immunobiological due to their minimally invasive nature and ability to enhance immune responses. These small needle-like structures can penetrate the skin's outer layer to deliver vaccines, drugs, and other therapeutic agents directly into the dermis, where immune cells like dendritic cells are abundant, making them an effective vaccine delivery and immunotherapy platform. MNs (MNs) can effectively penetrate the skin's stratum corneum (SC), which is the primary obstacle for transdermal drug delivery, allowing them to deliver therapeutic agents into deeper skin layers. This enhances drug effectiveness for long-term disease treatment and facilitates the administration of immunobiologicals. Malek-Khatabi et al. [83] conducted a comprehensive study on applying PLGA-based MNs (MNs) for transdermal drug and vaccine delivery. Their work demonstrated that polymeric MNs can effectively bypass the skin's stratum corneum, delivering therapeutic agents directly to the dermal layer. This approach enhances the bioavailability of drugs and vaccines while minimising invasiveness. The study further highlighted the biocompatibility and biodegradability of PLGA, making it an ideal material for long-term drug delivery systems. Additionally, the research emphasised the versatility of PLGA-based MNs in delivering a range of therapeutic molecules, including proteins, vaccines, and small-molecule drugs [83]. Currently, numerous microneedle-based immunotherapies are in preclinical or clinical stages, including vaccines for influenza, measles-rubella, rabies, and spinal cord grey matter. Compared to conventional subcutaneous injections, microneedle-based immunotherapies offer several advantages: they are minimally invasive and painless; they can carry immunotherapeutic agents in a dry form, enhancing thermal stability and reducing costs related to storage and transportation; and they allow for self-administration without requiring special equipment. The skin, which is rich in immune cells, allows for stronger immune responses and a reduced dose of the immunotherapeutic agent. Clinical trials have demonstrated that microneedle-based immunotherapies can produce immune responses that are comparable to or better than traditional subcutaneous injections. Early developments in this field have primarily utilised hollow MNs.

4.6.3 Cosmetic Field

MNs have gained significant attention in the cosmetic field due to their ability to enhance the delivery of active ingredients through the skin's outermost barrier, the stratum corneum. Traditionally, cosmetic treatments face challenges in effectively penetrating ingredients like collagen, hyaluronic acid, and vitamins, as the skin's barrier function prevents deep absorption. MNs provide a minimally invasive solution by creating microchannels that allow these ingredients to bypass the stratum corneum, reaching the deeper layers of the skin where they can be more effective. This enhances the overall efficacy of cosmetic products aimed at anti-ageing, skin rejuvenation, and acne treatment. Moreover, MNs offer several advantages in cosmetic applications, including pain-free delivery, minimal recovery time, and improved patient compliance. Unlike more invasive procedures, MNs can be applied safely at home or clinics with minimal infection or scarring risk. These features make MNs an attractive and innovative approach in the cosmetic industry for enhancing skin appearance and addressing dermatological concerns. Serrano-Castañeda et al. highlighted using MNs to treat various skin conditions such as acne, scars, depigmentation, and alopecia [84]. They demonstrated how MNs improve drug absorption without significant side effects, making them suitable for cosmetic applications. Similarly, McCrudden et al. explored how MNs induce elastin and collagen production, which are critical for skin rejuvenation, while also serving as efficient vehicles for delivering cosmetic molecules to deeper skin layers [85]. This method enhances the effectiveness of skincare products by overcoming the barrier function of the skin, making treatments more efficient and less invasive. Fonseca et al. further developed microneedle patches combining hyaluronic acid with bacterial nanocellulose, demonstrating their potential to deliver bioactive molecules while maintaining skin safety and promoting hydration [86]. These studies underscore the growing role of MNs in cosmetics, providing innovative solutions for skin care and improving the overall effectiveness of treatments.

4.6.4 Examples of Various MN Arrays for 3D Printing-Based TDD Systems

Transdermal drug delivery (TDD) systems using MNs (MNs) are designed to deliver drugs and vaccines through the skin minimally invasively. MNs are tiny structures that penetrate the skin's outer barrier, allowing drugs to be delivered directly to the skin layers where therapeutic action is needed. MNs can be made of various materials such as metals, polymers, and ceramics, and they offer several advantages for transdermal drug delivery. In the context of TDD systems, MNs can deliver a range of drugs and vaccines, bypassing the gastrointestinal system and enhancing bioavailability. Microneedle arrays allow for controlled and targeted drug release, with some systems designed to dissolve after delivering their payload. This method increases the efficacy of treatments by ensuring more consistent delivery of active pharmaceutical ingredients (APIs). One notable benefit of MNs in TDD systems is that they can be self-administered by patients, reducing the need for healthcare visits and offering a more convenient treatment option, especially for vaccines or chronic disease management. Furthermore, they can improve the thermal stability of vaccines and reduce the need for refrigeration, which can be a significant advantage in areas with limited healthcare infrastructure. 3D printing technology has been increasingly used to manufacture MNs, providing precise control over their size, shape, and material composition. This method allows for the customisation of MNs to meet specific drug delivery needs, such as controlling the release rate of therapeutic agents, improving patient comfort, and optimising therapeutic outcomes [24]. Chen et al. [87] developed an innovative approach using near-infrared (NIR) light-responsive composite MNs designed for on-demand drug delivery. Their study demonstrated the incorporation of silica-coated lanthanum hexaboride nanostructures into polycaprolactone MNs, enabling controlled light-to-heat transduction. The MNs melted at 50 °C upon NIR light exposure, initiating drug release from the matrix stepwise. This system provided exact control over drug administration without significant leakage during off periods, making it a promising advancement in transdermal drug delivery technologies that respond to external stimuli [87]. Xenikakis et al. [88] conducted a comprehensive evaluation of 3D-printed MNs through both experimental testing and finite element analysis (FEA). In their study, polymer-based microneedle arrays were fabricated using stereolithography, and their mechanical properties were assessed by applying axial load. The researchers verified the structural integrity of the MNs, ensuring they could penetrate human skin without fracturing.

The study further included in vitro permeation tests using two model dyes to demonstrate that MNs could significantly enhance the transport of molecules across the skin barrier, highlighting the potential of 3D-printed MNs for effective drug delivery [88]. Loizidou et al. explored using sugar-based dissolvable MNs for transdermal drug delivery, mainly focusing on their structural characterisation and mechanical properties [89]. Using a combination of experimental tests and finite element modelling, the researchers compared MNs made from various sugar compositions. The study found that MNs composed of carboxymethylcellulose/maltose exhibited superior mechanical strength and drug delivery capabilities compared to MNs made from trehalose or sucrose. This work highlights the role of material composition in optimising MNs for effective transdermal administration, particularly for delivering biomolecules like proteins and peptides [88]. Liu et al. [90] focused on the transdermal delivery of drugs with high molecular weight using novel dissolving MNs fabricated from hyaluronic acid (HA). The study demonstrated that HA MNs could significantly enhance the skin's permeability to large molecules, using fluorescein isothiocyanate-labeled dextran as a model drug. The researchers reported that MNs improved the accumulation of the drug in the skin while minimising any adverse effects, indicating the MNs' safety and efficacy for delivering more prominent therapeutic agents transdermally [90]. Harvey et al. conducted an experimental study comparing the pharmacokinetics of microneedle-based intradermal (ID) delivery with traditional subcutaneous (SC) injections for protein drugs [91]. The results revealed that MNs facilitated faster systemic drug absorption, as evidenced by a shorter time to maximum concentration (tmax) and higher peak concentration (Cmax) compared to SC delivery. Additionally, the study demonstrated rapid lymphatic uptake of protein drugs using MNs, showcasing their potential for delivering therapeutic dosages of biologics more effectively than conventional methods [91].

4.7 Conclusions and Future Perspectives

The development of MNs from solid shapes to intricate designs has allowed potential applications in drug delivery. Although initially created for transdermal drug delivery, MNs now hold promise for bridging biological barriers in the delivery of drugs to the oral, buccal, intramyocardial, and ophthalmic regions [92]. The stages at which microneedle research has advanced are preclinical investigations to clinical trials, small-scale lab production to commercial manufacturing, and academic research to industrial collaborations [93]. Even while the existing methods for manufacturing MNs can produce acceptable resolution, they still have drawbacks such as labor-intensiveness, manual stages, and expensive costs. By enabling direct design and manufacture, 3D printing provides a solution and reduces the time required for prototyping. However, slow production, low resolution, material selection, and biocompatibility are disadvantages. Although 3D printing makes it possible to create a variety of patient-specific MNs at a reduced cost, the technology still needs to be faster than more conventional techniques like injection moulding, and it is challenging to improve resolution. Research is still being done to improve 3D printing's speed, resolution, and material compatibility [24]. MNs have been categorised based on their purpose, including dissolving, coated, solid, hollow, and hydrogel-forming MNs. The performance of MNs is mainly dependent on the mechanical characteristics and biocompatibility of the material used in their production. High mechanical strength and low production costs are common factors to consider when selecting materials for fabrication. In this sense, polymers are the favoured materials for making MNs since they can be made to perform with various strengths and functions, do not cause an immune reaction in the body, and disintegrate naturally [94]. Using DLP 3D printing for microneedle design offers several benefits, such as high resolution for creating precise, crisp, and detailed structures. This accuracy is crucial for creating MNs with tiny tips that can penetrate the skin painlessly while being robust enough to deliver therapeutics. Hydrogel MNs must have sufficient mechanical strength when dried to pierce the epidermis. Their three-dimensional hydrogel network allows them to absorb interstitial fluid (ISF) and swell, turning them into a gel. ISF contains biomarkers; by extracting it, hydrogel MNs can work as a minimally invasive diagnostic tool. Drugs can also permeate into deeper layers of the skin thanks to the channels formed by the hydrogel structure. The hydrogel network's cross-linking density affects the swelling capacity, drug release rate and loading capacity. The back of the HMNs can have patches with drug reservoirs, microneedle arrays, or needle tips for storing drugs. By swapping out the patch, the medication can be altered as it diffuses via the channels the absorbed ISF generated [95].

Polymers, including synthetic and natural polymers, are typically used to produce hydrogel MN. Despite having less mechanical strength than metal and silicon, natural and manufactured polymeric materials have excellent biocompatibility and minimal toxicity. The main components of natural polymers are collagen, silk fibroin protein, gelatin, etc.. Natural polymer-based hydrogel MNs can be biocompatible and biodegraded with minimal toxicity in the body since they contain natural components and share similarities with extracellular matrix components. This makes them suitable for a variety of biomedical applications. In comparison to natural polymers, synthetic polymers like polyvinyl alcohol (PVA), polylactic acid (PLA), polyglycolic acid (PGA), and polyvinyl pyrrolidone (PVP) are more designable [36]. Compared to other 3D printing technologies, DLP's ultra-fast speed, excellent precision, and affordability make it the best option for fabricating hydrogel MN [96]. 3D DLP bioprinting addresses the main drawbacks of its extrusion counterpart, forming structures from photo-curable polymers utilising regulated light patterns. Its capacity to quickly construct complex geometries at high resolutions (1–50 μm) and speeds has made significant progress in regenerating various tissues possible [97].

MNs have emerged as a transformative technology for precise, minimally invasive drug delivery. They benefit areas like transdermal peptide administration and cancer treatments, including breast cancer and skin carcinoma. The rise of 3D-printed MNs marks a new phase in personalised cancer therapy, while dissolvable MNs show promise for targeted delivery of active ingredients in cosmetics. Integrating MNs with immunotherapy also offers significant potential to enhance cancer treatment effectiveness [28]. This chapter focuses on hydrogel-based MNs in drug delivery systems and highlights their ability to enhance delivery and therapeutic efficacy. MNs are classified according to their design and function, emphasising hydrogel-based types due to their biocompatibility, controlled drug release, and ability to form a hydrogel for efficient delivery. Advances in stimuli-responsive MNs for site-specific and on-demand delivery and challenges in mechanical properties and clinical translation are reviewed, providing a comprehensive perspective on their potential in transdermal drug delivery.

References

1. Yao, W., et al.: 3D Printed multi-functional hydrogel microneedles based on high-precision digital light processing. Micromachines (Basel)**11**(1) (2020). https://doi.org/10.3390/mi11010017
2. Shin, D., Hyun, J.: Silk fibroin microneedles fabricated by digital light processing 3D printing. J. Ind. Eng. Chem. **95** (2021). https://doi.org/10.1016/j.jiec.2020.12.011
3. Kundu, A., et al.: DLP 3D printed 'Intelligent' microneedle array (iμNA) for stimuli responsive release of drugs and its in vitro and ex vivo characterization. J. Microelectromech. Syst. **29**(5), 685–691 (2020). https://doi.org/10.1109/JMEMS.2020.3003628
4. Donnelly, R.F., et al.: Hydrogel-forming microneedle arrays for enhanced transdermal drug delivery. Adv. Funct. Mater. **22**(23), 4879–4890 (2012). https://doi.org/10.1002/adfm.201200864
5. Queiroz, M.L.B., et al.: Microneedles as an alternative technology for transdermal drug delivery systems: a patent review (2020). https://doi.org/10.1080/13543776.2020.1742324
6. Singh, V., Kesharwani, P.: Recent advances in microneedles-based drug delivery device in the diagnosis and treatment of cancer. J. Control. Release**338** (2021). https://doi.org/10.1016/j.jconrel.2021.08.054
7. Xie, Y., et al.: Improved therapeutic efficiency against obesity through transdermal drug delivery using microneedle arrays. Pharmaceutics**13**(6) (2021). https://doi.org/10.3390/pharmaceutics13060827
8. Avcil, M., Çelik, A.: Microneedles in drug delivery: progress and challenges (2021). https://doi.org/10.3390/mi12111321
9. Zhi, D., Yang, T., Zhang, T., Yang, M., Zhang, S., Donnelly, R.F.: Microneedles for gene and drug delivery in skin cancer therapy (2021). https://doi.org/10.1016/j.jconrel.2021.05.009
10. Ganeson, K., Alias, A.H., Murugaiyah, V., Amirul, A.A.A., Ramakrishna, S., Vigneswari, S.: Microneedles for efficient and precise drug delivery in cancer therapy (2023). https://doi.org/10.3390/pharmaceutics15030744
11. Nagarkar, R., Singh, M., Nguyen, H.X., Jonnalagadda, S.: A review of recent advances in microneedle technology for transdermal drug delivery (2020). https://doi.org/10.1016/j.jddst.2020.101923
12. Chen, J., et al.: Microneedle-mediated drug delivery for cutaneous diseases (2022). https://doi.org/10.3389/fbioe.2022.1032041
13. Zhang, W., Zhang, W., Li, C., Zhang, J., Qin, L., Lai, Y.: Recent advances of microneedles and their application in disease treatment (2022). https://doi.org/10.3390/ijms23052401
14. Liang, R., et al.: Implantable and degradable antioxidant poly(ε-caprolactone)-lignin nanofiber membrane for effective osteoarthritis treatment. Biomaterials**230** (2020). https://doi.org/10.1016/j.biomaterials.2019.119601

15. Wang, F.Y., Chen, Y., Huang, Y.Y., Cheng, C.M.: Transdermal drug delivery systems for fighting common viral infectious diseases. Drug Deliv. Transl. Res. **11**(4) (2021). https://doi.org/10.1007/s13346-021-01004-6
16. Jeong, S.Y., Park, J.H., Lee, Y.S., Kim, Y.S., Park, J.Y., Kim, S.Y.: The current status of clinical research involving microneedles: a systematic review (2020). https://doi.org/10.3390/pharmaceutics12111113
17. Duarah, S., Sharma, M., Wen, J.: Recent advances in microneedle-based drug delivery: special emphasis on its use in paediatric population. Elsevier B.V. (2019).https://doi.org/10.1016/j.ejpb.2019.01.005
18. Ma, G., Wu, C.: Microneedle, bio-microneedle and bio-inspired microneedle: a review. Elsevier B.V. (2017) https://doi.org/10.1016/j.jconrel.2017.02.011
19. Moore, L.E., Vucen, S., Moore, A.C.: Trends in drug- and vaccine-based dissolvable microneedle materials and methods of fabrication. Elsevier B.V. (2022). https://doi.org/10.1016/j.ejpb.2022.02.013
20. Xu, P., et al.: Potential strategy of microneedle-based transdermal drug delivery system for effective management of skin-related immune disorders. Elsevier B.V. (2024) https://doi.org/10.1016/j.ejpb.2023.11.013
21. Han, X., et al.: 3D micro-nano printing technology as a transformative tool apply for microneedle drug delivery. Editions de Sante (2024). https://doi.org/10.1016/j.jddst.2024.105709
22. Oliveira, C., Teixeira, J.A., Oliveira, N., Ferreira, S., Botelho, C.M.: Microneedles' device: design, fabrication, and applications. Multidisciplinary Digital Publishing Institute (MDPI) (2024). https://doi.org/10.3390/macromol4020019
23. Guillot, A.J., Cordeiro, A.S., Donnelly, R.F., Montesinos, M.C., Garrigues, T.M., Melero, A.: Microneedle-based delivery: An overview of current applications and trends. MDPI AG (2020). https://doi.org/10.3390/pharmaceutics12060569
24. Al-Nimry, S.S., Daghmash, R.M.: Three Dimensional printing and its applications focusing on microneedles for drug delivery. Multidisciplinary Digital Publishing Institute (MDPI) (2023). https://doi.org/10.3390/pharmaceutics15061597
25. Chen, B.Z., He, M.C., Zhang, X.P., Fei, W.M., Cui, Y., Guo, X.D.: A novel method for fabrication of coated microneedles with homogeneous and controllable drug dosage for transdermal drug delivery. Drug Deliv. Transl. Res. **12**(11), 2730–2739 (2022). https://doi.org/10.1007/s13346-022-01123-8
26. Xu, J., Xu, D., Xuan, X., He, H.: Advances of microneedles in biomedical applications. MDPI (2021). https://doi.org/10.3390/molecules26195912
27. Jin, M., et al.: Preparation and evaluation of rapid disintegrating formulation from coated microneedle. Drug Deliv. Transl. Res. **12**(2), 415–425 (2022). https://doi.org/10.1007/s13346-021-01046-w
28. Aldawood, F.K., Andar, A., Desai, S.: A comprehensive review of microneedles: types, materials, processes, characterizations and applications. MDPI AG (2021). https://doi.org/10.3390/polym13162815
29. Pires, L.R., Amado, I.R., Gaspar, J.: Dissolving microneedles for the delivery of peptides—Towards tolerance-inducing vaccines. Int. J. Pharm. **586** (2020). https://doi.org/10.1016/j.ijpharm.2020.119590
30. Hamdan, I.: Microneedle and drug delivery across the skin: an overview. Pharmacia **71**, 1–12 (2024). https://doi.org/10.3897/pharmacia.71.e112503
31. Tan, S., Wang, Y., Wei, X., Xiao, X., Gao, L.: Microneedle-mediated drug delivery for neurological diseases. Elsevier B.V. (2024).https://doi.org/10.1016/j.ijpharm.2024.124400.
32. Tucak, A., et al.: Microneedles: characteristics, materials, production methods and commercial development. MDPI AG (2020). https://doi.org/10.3390/mi11110961
33. Wu, C., Yu, Q., Huang, C., Li, F., Zhang, L., Zhu, D.: Microneedles as transdermal drug delivery system for enhancing skin disease treatment. Acta Pharm. Sin B (2024). https://doi.org/10.1016/j.apsb.2024.08.013

34. Shriky, B., Babenko, M., Whiteside, B.R.: Dissolving and swelling hydrogel-based microneedles: an overview of their materials, fabrication, characterization methods, and challenges. Multidisciplinary Digital Publishing Institute (MDPI) (2023). https://doi.org/10.3390/gels9100806
35. Wei, H., et al.: Hydrogel-based microneedles of chitosan derivatives for drug delivery. React. Funct. Polym. **172** (2022). https://doi.org/10.1016/j.reactfunctpolym.2022.105200
36. Yao, S., Zhang, C., Ping, J., Ying, Y.: Recent advances in hydrogel microneedle-based biofluid extraction and detection in food and agriculture. Elsevier Ltd. (2024). https://doi.org/10.1016/j.bios.2024.116066
37. Hou, X., et al.: Advances and prospects for hydrogel-forming microneedles in transdermal drug delivery. Multidisciplinary Digital Publishing Institute (MDPI) (2023). https://doi.org/10.3390/biomedicines11082119
38. Turner, J.G., White, L.R., Estrela, P., Leese, H.S.: Hydrogel-forming microneedles: current advancements and future trends. Wiley-VCH Verlag (2021). https://doi.org/10.1002/mabi.202000307
39. Filho, D., Guerrero, M., Pariguana, M., Marican, A., Durán-Lara, E.F.: Hydrogel-based microneedle as a drug delivery system. Multidisciplinary Digital Publishing Institute (MDPI) (2023). https://doi.org/10.3390/pharmaceutics15102444
40. Chauhan, N., Saxena, K., Jain, U.: Hydrogel based materials: a progressive approach towards advancement in biomedical applications. Elsevier Ltd. (2022). https://doi.org/10.1016/j.mtcomm.2022.104369
41. Kesharwani, P., Bisht, A., Alexander, A., Dave, V., Sharma, S.: Biomedical applications of hydrogels in drug delivery system: an update. Editions de Sante (2021). https://doi.org/10.1016/j.jddst.2021.102914
42. Yadav, P., Singh, S., Jaiswal, S., Kumar, R.: Synthetic and natural polymer hydrogels: a review of 3D spheroids and drug delivery. Elsevier B.V. (2024).https://doi.org/10.1016/j.ijbiomac.2024.136126
43. Ahmad, Z., et al.: Versatility of hydrogels: from synthetic strategies, classification, and properties to biomedical applications. MDPI (2022). https://doi.org/10.3390/gels8030167
44. Hameed, H., Faheem, S., Paiva-Santos, A.C., Sarwar, H.S., Jamshaid, M.: A comprehensive review of hydrogel-based drug delivery systems: classification, properties, recent trends, and applications. Springer Science and Business Media Deutschland GmbH (2024). https://doi.org/10.1208/s12249-024-02786-x
45. Xiao, Z., Li, Q., Liu, H., Zhao, Q., Niu, Y., Zhao, D.: Adhesion mechanism and application progress of hydrogels. Elsevier Ltd. (2022). https://doi.org/10.1016/j.eurpolymj.2022.111277
46. Ullah, A., Lim, S.I.: Bioinspired tunable hydrogels: an update on methods of preparation, classification, and biomedical and therapeutic applications. Elsevier B.V. (2022).https://doi.org/10.1016/j.ijpharm.2021.121368
47. Shi, Q., Liu, H., Tang, D., Li, Y., Li, X.J., Xu, F.: Bioactuators based on stimulus-responsive hydrogels and their emerging biomedical applications. Nat. Res. (2019). https://doi.org/10.1038/s41427-019-0165-3
48. Li, Z., Zhou, Y., Li, T., Zhang, J., Tian, H.: Stimuli-responsive hydrogels: Fabrication and biomedical applications. View**3**(2) (2022). https://doi.org/10.1002/VIW.20200112
49. Zhang, K., Xue, K., Loh, X.J.: Thermo-responsive hydrogels: from recent progress to biomedical applications. MDPI AG (2021). https://doi.org/10.3390/gels7030077
50. Chatterjee, S., Hui, P.C.L.: Review of applications and future prospects of stimuli-responsive hydrogel based on thermo-responsive biopolymers in drug delivery systems. MDPI AG (2021). https://doi.org/10.3390/polym13132086
51. Hendi, A., et al.: Healthcare applications of pH-sensitive hydrogel-based devices: a review. Dove Medical Press Ltd. (2020).https://doi.org/10.2147/IJN.S245743
52. Singh, J., Nayak, P.: pH-responsive polymers for drug delivery: trends and opportunities. Wiley (2023).https://doi.org/10.1002/pol.20230403
53. Mehta, P., Sharma, Devi, M.: Hydrogels: an overview of its classifications, properties, and applications. Elsevier Ltd. (2023). https://doi.org/10.1016/j.jmbbm.2023.106145

54. Khan, F., et al.: Synthesis, classification and properties of hydrogels: Their applications in drug delivery and agriculture. R. Soc. Chem. (2022) https://doi.org/10.1039/d1tb01345a
55. Andrade, F., Roca-Melendres, M.M., Durán-Lara, E.F., Rafael, D., Schwartz, S.: Review stimuli-responsive hydrogels for cancer treatment: the role of pH, light, ionic strength and magnetic field. MDPI (2021). https://doi.org/10.3390/cancers13051164
56. Wei, P., Cornel, E.J., Du, J.: Ultrasound-responsive polymer-based drug delivery systems. Drug Deliv. Transl. Res. **11**(4), 1323–1339 (2021). https://doi.org/10.1007/s13346-021-00963-0
57. Dreiss, C.A.: Hydrogel design strategies for drug delivery. Elsevier Ltd. (2020). https://doi.org/10.1016/j.cocis.2020.02.001
58. Yeingst, T.J., Arrizabalaga, J.H., Hayes, D.J.: Ultrasound-induced drug release from stimuli-responsive hydrogels. MDPI (2022). https://doi.org/10.3390/gels8090554
59. Tang, J., et al.: Super tough magnetic hydrogels for remotely triggered shape morphing. J. Mater. Chem. B **6**(18), 2713–2722 (2018). https://doi.org/10.1039/c8tb00568k
60. Tang, J., et al.: Magnetic double-network hydrogels for tissue hyperthermia and drug release. J. Mater. Chem. B **7**(8), 1311–1321 (2019). https://doi.org/10.1039/C8TB03301C
61. Rad, Z.F., Prewett, P.D., Davies, G.J.: An overview of microneedle applications, materials, and fabrication methods. Beilstein J. Nanotechnol. **12**, 1034–1046 (2021). https://doi.org/10.3762/BJNANO.12.77
62. Tan, J.Y., et al.: Experimental validation of diffraction lithography for fabrication of solid microneedles. Materials**15**(24) (2022). https://doi.org/10.3390/ma15248934
63. Badnikar, K., et al.: Generic molding platform for simple, low-cost fabrication of polymeric microneedles. Macromol. Mater. Eng. **305**(5) (2020). https://doi.org/10.1002/mame.202000072
64. Zhang, H., Zhang, N., et al.: Advances in precision micro/nano-electroforming: a state-of-the-art review. J. Micromech. Microeng. (2020)
65. Bolton, C.J.W., et al.: Hollow silicon microneedle fabrication using advanced plasma etch technologies for applications in transdermal drug delivery. Lab Chip **20**(15), 2788–2795 (2020). https://doi.org/10.1039/d0lc00567c
66. Li, R., et al.: 3D-printed microneedle arrays for drug delivery (2022). https://doi.org/10.1016/j.jconrel.2022.08.022
67. Li, H., et al.: Digital light processing (DLP)-based (bio)printing strategies for tissue modeling and regeneration. Wiley (2023).https://doi.org/10.1002/agt2.270
68. Carbajo, J., Nam, S.H., Fang, N.X.: Fabrication of micro-perforated panel (MPP) sound absorbers using digital light processing (DLP) 3D printing technology. Appl. Acoust. **216** (2024). https://doi.org/10.1016/j.apacoust.2023.109788
69. Gong, J., et al.: Digital light processing (DLP) in tissue engineering: from promise to reality, and perspectives (2022). https://doi.org/10.1088/1748-605X/ac96ba
70. Wang, H., Xia, Y., Zhang, Z., Xie, Z.: 3D gradient printing based on digital light processing (2023). https://doi.org/10.1039/d3tb00763d
71. Zhu, G., Hou, Y., Xu, J., Zhao, N.: Reprintable polymers for digital light processing 3D printing. Adv. Funct. Mater. **31**(9) (2021). https://doi.org/10.1002/adfm.202007173
72. Parhi, R.: Recent advances in microneedle designs and their applications in drug and cosmeceutical delivery (2022). https://doi.org/10.1016/j.jddst.2022.103639
73. Ozyilmaz, E.D., Turan, A., Comoglu, T.: An overview on the advantages and limitations of 3D printing of microneedles. Taylor and Francis Ltd. (2021). https://doi.org/10.1080/10837450.2021.1965163
74. Reinke, A., Whiteside, E.J., Windus, L., Desai, D., Stehr, E., Rad, Z.F.: The advantages of microneedle patches compared to conventional needle-based drug delivery and biopsy devices in medicine. Biomed. Eng. Adv. 100127 (2024). https://doi.org/10.1016/j.bea.2024.100127
75. Jung, J.H., Jin, S.G.: Microneedle for transdermal drug delivery: current trends and fabrication. Springer (2021). https://doi.org/10.1007/s40005-021-00512-4
76. Henriquez, F., Morales-Ferreiro, J.O., Celentano, D.: Structural evaluation by the finite-element method of hollow microneedle geometries for drug delivery. Adv. Eng. Mater. **24**(10) (2022). https://doi.org/10.1002/adem.202200049

77. Pineda-Álvarez, R.A., Bernad-Bernad, M.J., Rodríguez-Cruz, I.M., Escobar-Chávez, J.J.: Development and characterization of starch/gelatin microneedle arrays loaded with lecithin-gelatin nanoparticles of losartan for transdermal delivery. J. Pharm. Innov. **17**(1), 71–84 (2022). https://doi.org/10.1007/s12247-020-09494-6
78. Puri, A., Nguyen, H.X., Tijani, A.O., Banga, A.K.: Characterization of microneedles and microchannels for enhanced transdermal drug delivery. Taylor & Francis (2021). https://doi.org/10.4155/tde-2020-0096
79. Ando, D., et al.: Mechanical characterization of dissolving microneedles: factors affecting physical strength of needles. Pharmaceutics**16**(2) (2024). https://doi.org/10.3390/pharmaceutics16020200
80. Sabbagh, F., Saha, P., Kim, B.S.: Transdermal delivery of catechin using dissolving Poly(vinyl alcohol)-based microneedles: effect of microneedle composition on drug release. ACS Appl. Polym. Mater. **5**(11), 8919–8928 (2023). https://doi.org/10.1021/acsapm.3c01293
81. Zhou, R., Yu, J., Gu, Z., Zhang, Y.: Microneedle-mediated therapy for cardiovascular diseases. Drug Deliv. Transl. Res. **12**(2) (2022). https://doi.org/10.1007/s13346-021-01073-7
82. Kim, J.Y., Han, M.R., Kim, Y.H., Shin, S.W., Nam, S.Y., Park, J.H.: Tip-loaded dissolving microneedles for transdermal delivery of donepezil hydrochloride for treatment of Alzheimer's disease. Eur. J. Pharm. Biopharm. **105** (2016). https://doi.org/10.1016/j.ejpb.2016.06.006
83. Malek-Khatabi, A., et al.: Recent progress in PLGA-based microneedle-mediated transdermal drug and vaccine delivery. Royal Society of Chemistry (2023). https://doi.org/10.1039/d3bm00795b
84. Serrano-Castañeda, P., Escobar-Chávez, J.J., Rodríguez-Cruz, I.M., Melgoza-Contreras, L.M., Martínez-Hernández, J.: Microneedles as enhancer of drug absorption through the skin and applications in medicine and cosmetology. J. Pharm. Pharm. Sci. **21**, 73–93 (2018). https://doi.org/10.18433/jpps29610
85. McCrudden, M.T.C., McAlister, E., Courtenay, A.J., González-Vázquez, P., Raj Singh, T.R., Donnelly, R.F.: Microneedle applications in improving skin appearance. Exp. Dermatol. **24**, 561–566 (2015). https://doi.org/10.1111/exd.12723
86. Fonseca, D.F., Vilela, C., Pinto, R.J., Bastos, V., Oliveira, H., Catarino, J., Faísca, P., Rosado, C., Silvestre, A. J., Freire, C.S.: Bacterial nanocellulose-hyaluronic acid microneedle patches for skin applications: in vitro and in vivo evaluation. Mater. Sci. Eng. C **118**, 111350 (2020). https://doi.org/10.1016/j.msec.2020.111350
87. Chen, M.C., Ling, M.H., Wang, K.W., Lin, Z.W., Lai, B.H., Chen, D.H.: Near-infrared light-responsive composite microneedles for on-demand transdermal drug delivery. Biomacromolecules**16**(5) (2015). https://doi.org/10.1021/acs.biomac.5b00185
88. Xenikakis, I., et al.: Fabrication and finite element analysis of stereolithographic 3D printed microneedles for transdermal delivery of model dyes across human skin in vitro. Eur. J. Pharm. Sci. **137** (2019). https://doi.org/10.1016/j.ejps.2019.104976
89. Loizidou, E.Z., Williams, N.A., Barrow, D.A., Eaton, M.J., McCrory, J., Evans, S.L., Allender, C.J.: Structural characterisation and transdermal delivery studies on sugar microneedles: experimental and finite element modelling analyses. Eur. J. Pharm. Biopharm. **89**, 224–231 (2015). https://doi.org/10.1016/j.ejpb.2014.11.023
90. Liu, S., et al.: Transdermal delivery of relatively high molecular weight drugs using novel self-dissolving microneedle arrays fabricated from hyaluronic acid and their characteristics and safety after application to the skin. Eur. J. Pharm. Biopharm. **86**(2) (2014). https://doi.org/10.1016/j.ejpb.2013.10.001
91. Harvey, A.J., Kaestner, S.A., Sutter, D.E., Harvey, N.G., Mikszta, J.A., Pettis, R.J.: Microneedle-based intradermal delivery enables rapid lymphatic uptake and distribution of protein drugs. Pharm. Res. **28**(1), 107–116 (2011). https://doi.org/10.1007/s11095-010-0123-9
92. Le, Z., et al.: Design principles of microneedles for drug delivery and sampling applications (2023). https://doi.org/10.1016/j.mattod.2022.10.025
93. Donnelly, R.F., Prausnitz, M.R.: The promise of microneedle technologies for drug delivery. Drug Deliv. Transl. Res. **14**(3) (2024). https://doi.org/10.1007/s13346-023-01430-8.

94. Al-Japairai, K.A.S., et al.: Current trends in polymer microneedle for transdermal drug delivery (2020). https://doi.org/10.1016/j.ijpharm.2020.119673
95. Ji, M., et al.: Research progress of hydrogel microneedles in wound management. American Chemical Society (2024). https://doi.org/10.1021/acsbiomaterials.4c00972
96. Zhou, X., Liu, H., Li, H.: DLP 3D printing of high-resolution, temperature-responsive hydrogel microneedles with rapid customization capability. J. Phys.: Conf. Ser. Inst. Phys. (2024). https://doi.org/10.1088/1742-6596/2808/1/012041
97. Yang, M., et al.: Multi-material digital light processing (DLP) bioprinting of heterogeneous hydrogel constructs with perfusable networks. Adv. Funct. Mater. **34**(32) (2024). https://doi.org/10.1002/adfm.202316456
98. Baykara, D., et al.: Fabrication and optimization of 3D printed gelatin methacryloyl microneedle arrays based on vat photopolymerization. Front. Bioeng. Biotechnol. **11** (2023). https://doi.org/10.3389/fbioe.2023.1157541
99. Erkus, H., et al.: Innovative transdermal drug delivery system based on amoxicillin-loaded gelatin methacryloyl microneedles obtained by 3D printing. Materialia (Oxf)**27**, 101700 (2023). https://doi.org/10.1016/j.mtla.2023.101700

Chapter 5
Combining the Electrospun Nanofibers with Microneedle Arrays Enhance Their Efficacy

Maliheh Gharibshahian, Amir Abbas Seraji, and Nima Beheshtizadeh

5.1 Introduction

Injuries and tissue disorders caused by trauma, disease, and aging are major challenges in different societies [1]. Today's clinical treatments including surgery, autografts and allografts, artificial implants, and mechanical devices have helped patients to some extent in restoring and restoring lost function. However, problems such as surgical risks, limited access to donors, stimulation of the immune system, and insufficient regeneration have limited the efficacy of these methods [2, 3].

Tissue engineering and drug release systems are two vital fields in medical and therapeutic science that play a significant role in tissue damage repair and disease management [4]. These technologies are critical in treating chronic diseases, injuries, and tissue disorders. Tissue engineering refers to the design and construction of biological structures to repair or replace damaged tissues, and helps to improve tissue function and reduce complications caused by treatments [5]. Drug delivery systems

M. Gharibshahian
Department of Tissue Engineering and Applied Cell Sciences, School of Medicine, Semnan University of Medical Sciences, Semnan, Iran

A. A. Seraji
Department of Mechanical and Industrial Engineering, University of Toronto, Toronto, Canada

Department of Polymer Engineering and Color Technology, Amirkabir University of Technology, Tehran, Iran

N. Beheshtizadeh (✉)
Department of Tissue Engineering, Faculty of Advanced Medical Sciences, Tabriz University of Medical Sciences, Tabriz, Iran
e-mail: n.beheshtizadeh@tbzmed.ac.ir

M. Gharibshahian · N. Beheshtizadeh
Regenerative Medicine Group (REMED), Universal Scientific Education and Research Network (USERN), Tehran, Iran

N. Talreja et al. (eds.), *Microneedles (MNs)-Based Technology*, Series in BioEngineering, https://doi.org/10.1007/978-981-96-3916-8_5

are a set of techniques and methods designed to deliver drugs to target areas in the body [6]. These systems are essential for the management of chronic and complex diseases, including diabetes, cancer, and autoimmune disorders. The primary objective is to enhance the efficacy of pharmaceuticals across different organs and tissues while extending their longevity within the body [7, 8].

Tissue engineering scaffolds are made by various methods such as electrospinning [9], freeze-drying [10], 3D printing [11], and foam casting [12], and are suitable options for drug release systems. Nanofibers produced by the electrospinning technique have a structure with high surface area, high porosity, and significant absorption and imitation of the extracellular matrix [13]. These nanofibers can lock various drugs in themselves and manage their release process by controlling conditions, such as temperature, humidity, and time [14]. Nanofibers are used in various applications such as wound dressings [15], targeted drug delivery [14], and tissue repair [16]. However, their poor mechanical properties have somewhat limited the use of these nanofibers.

Microneedle arrays, tiny rods designed to penetrate tissue, help reduce pain from traditional injections, and can effectively deliver drugs and vaccines [17]. The small size of these microneedles allows them to cause minimal damage to the skin while creating tiny channels in the layers of the skin, allowing drugs that normally cannot be absorbed through the skin to easily enter the bloodstream [17, 18]. Integration of electrospun nanofibers with microneedle arrays can lead to a revolution in drug delivery science [19]. By creating systems that combine the features of both technologies, it is possible to deliver more complex drugs with tunable properties. This combination will help reduce side effects caused by drugs and increase the effectiveness of treatments. In addition, the design of combined systems can lead to combined treatments with greater effects and faster recovery [19, 20]. Nanofiber as a drug reservoir and microneedle as a tissue-penetrating component allow accurate administration of drugs and their gradual release over time [21]. This chapter aims to investigate the synergistic potential of electrospun nanofibers and microneedles in drug delivery. In this chapter, the introduction of electrospinning technology and its types, microneedle technology and its different types, and the need to integrate these two technologies to achieve higher efficiency will be discussed. Also, case studies, advantages, and challenges in these technologies are reviewed and the future path for these technologies' development will be considered.

5.2 Electrospinning Technology

Electrospinning is one of the cheap, simple, versatile, and repeatable methods for creating scalable and continuous 3D structures in the form of nano and microfibers [19]. This method can control the percentage of porosity, flexibility, mechanical strength, and high surface-to-volume ratio, and is used in various applications such as filtration membranes [22], energy storage and conversion [23], biosensors [24], tissue engineering [25], and the release of drugs/ biomolecules [26]. The fibers generated through this process enhance the attachment of cells, their proliferation, and migration by mimicking the extracellular matrix (ECM) [14].

The electrospinning process typically employs a high-voltage power source, either direct current or alternating current, to facilitate the assembly of polymer solution fibers at room temperature. Applying an electrical charge to the exterior of the melt polymer allows the power supply to overcome the surface tension force, facilitating the movement of the charged polymer toward the oppositely charged collector. This motion results in the creation of a Taylor-shaped conical jet. The charged jet is extended linearly and thinned under the influence of an electric field and solvent evaporation, resulting in the collection of solid fibers on the collector [27, 28]. The rise in bending instabilities during this process results in a decrease in the jet diameter to the nanometer scale [28].

Solution electrospinning, melt spinning, and coaxial spinning are the three main subcategories of this method, which are classified based on the molten or soluble state of the rotating material and the type of needle [29]. Coaxial electrospinning was first used by Sun and his colleagues to produce structural core-sheath fibers (in the submicron range) and combine different properties of polymers in one fiber [30]. Similar to solution electrospinning, applying an electric field on polymer solutions causes polymer jets to stretch. The nozzle comprises a smaller inner capillary positioned concentrically within a larger outer capillary, with two distinct polymer solutions for the shell and core materials stored separately in two reservoirs. The reservoir with the sheath solution is linked to the outer needle, while the reservoir with the core solution is linked to the inner needle, allowing for the simultaneous removal of both solutions. The electric field induces charge accumulation on the surface of the sheath liquid exiting the outer coaxial capillary, resulting in a conical shape. Upon reaching the threshold voltage, a fine jet is emitted from the core material contained within the sheath material. As the solvents evaporate, the core-sheath fibers solidify [31, 32]. Applying an electrical field to the polymer melt is used to produce nanofibers in melt electrospinning, and unlike the previous methods, no solvent is needed to dissolve the polymer. Melt electrospinning uses the same equipment as solution electrospinning, except that a heating assembly and a temperature controller (including a heating gun, laser beam, and heating element) are included to melt the polymer. Elongated jets are formed by applying electrical potential and melting the polymer until the appropriate viscosity is reached. This method faces limitations in terms of the molecular weight of the polymer, and only polymers with a molecular weight between 40,000 and 80,000 g/mol can be electrospun [33, 34]. Also, the weakness or absence of bending instability (due to the high viscosity of the polymer melt and the electrical insulation of most polymers) in this process leads to the production of fibers with a larger diameter and in the range of micrometers [35]. Advantages of melt electrospinning include the ability to customize mechanical properties, drug release profiles, and fiber degradation rates for various tissues. Modifying parameters associated with the solution (including electrical conductivity, tension on the surface, and viscosity), the process (encompassing electric field, tip-to-collector distance, flow rate, and needle diameter), and the environment (considering temperature and humidity) facilitates the production of fibers with specific properties [36, 37]. As the electric potential rises, leading to an increase in charge within the jet, both the electrostatic force and the Coulombic repulsion, which acts in opposition, also intensify. An increase in the Coulombic force results in heightened tension of the Taylor cone, reduced fiber diameter, and accelerated solvent evaporation. Conversely, an increase in the electrostatic force correlates with greater fluid removal and an increase in fiber diameter [38]. The solution's flow rate influences the production of uniform beadless fibers with the specified diameter. An elevated flow rate may result in larger fiber diameters, the creation of beads (attributable to heightened flow and insufficient solvent evaporation), and unspun droplets (due to gravitational effects), thus indicating a preference for a minimum flow rate [39]. The distance between the tip and the collector must be selected to enable solvent evaporation from the jet while ensuring an adequate electrostatic field for fiber formation. The properties of the fibers are influenced by the concentration and viscosity of the polymer solution. Insufficient concentration may result in inadequate entanglement among polymer chains, leading to fiber breakage prior to reaching the collector, influenced by applied voltage and surface tension. A high concentration may obstruct the needle tip, hindering the solution's flow. An increase in surface tension beyond the optimal limit may lead to jet instability and droplet formation [40].

Solution conductivity is a critical factor; the lack of surface charge may restrict the electrostatic force necessary to form a Taylor cone. Moreover, enhancing the conductivity of the solution may result in a reduction of fiber quality [41]. The speed and material of the drum significantly influence the diameter and orientation of the nanofibers. Increasing the drum speed results in a decrease in the diameter of the nanofibers and an enhancement in the alignment of the fibers along the voltage direction [42]. High temperature can decrease the viscosity of the solution and increase the fiber diameter, while elevated humidity in the environment facilitates fiber discharge [43].

5.2.1 Microneedle Arrays

As a new technology in various fields of pharmaceutical and medicine, micro-needles have great potential for anti-tumor treatments [44], tissue repair [45], drug delivery [17], hypoglycemia [46], and vaccination [47]. Microneedles were first patented in 1976 by Gressel and Pleiss, but the practical use of this technology for drug delivery began in earnest in 1998 with the work of Henry et al. This system consists of an array of one or more microneedles with a length of 100–1000 μm and a tip diameter of 1–10 μm [48]. These needles effectively pass through the stratum corneum of the skin and create microscopic pores on the surface of the skin. This capability increases the delivery efficiency of drugs, tissue fluids, nanoparticles, proteins, lipids, and nucleic acids, without damaging blood vessels and deeper nerves (1–2 mm under the skin) [49, 50].

The geometry, core-shell structure, and material type of the needles influence their mechanical performance, penetration dimension, and drug absorption features. With these techniques, micro-needles are customized specifically to the patient's needs [51]. Common materials utilized in microneedle fabrication encompass metals, ceramics, hydrogels, as well as biodegradable and non-biodegradable polymers. Various methods are used to construct these needles, including micro molding, lithography, thermal drawing, electric drawing, magnetic traction lithography, and 3D printing [52].

Microneedles come in a variety of forms, such as solid, coated, hollow, soluble, hydrogel-forming, and cryogenic (frozen) microneedles. This diversity in design allows researchers to have more options for drug delivery according to the patient's condition and the type of treatment [53]. As innovative tools in the delivery of drugs and vaccines, micro-needles have different designs and materials, each of which has its own characteristics and applications. In the following, different types of microneedles and the characteristics of each are described:

- **Solid micro-needles**: These needles are usually made of non-biodegradable materials, such as stainless steel [54], titanium [55], and silicone [56]. The laser-cutting process is used for their fabrication. By piercing the skin, these needles create microscopic pores and enable passive drug release. Despite its strong mechanical properties, breaking the needle during treatment and leaving its tip in the skin can be dangerous [57, 58].
- **Coated micro-needles**: These needles are produced through dipping, spraying, and layer coating. The drug is loaded on the surface of the needle tip and released under certain conditions. However, the limited surface area and easy removal of the coating can reduce the load capacity. These needles are used to administer macromolecules, proteins, and vaccines [59].
- **Hollow micro-needles**: These types of needles are manufactured by digital micromachining systems and have an inner diameter of 50–70 μm. The drug is loaded into the pores of the tip or base of the needle. These needles are capable of injecting controlled drug doses and are used to deliver liquid formulations into the dermis. The risk of clogging and the complexity of the preparation process are common problems of these microneedles [60, 61].
- **Dissolvable microneedles**: These needles are constructed of biodegradable polymer materials such as polyvinyl alcohol [62] and dextran [63] and do not need to be removed. The drug is contained at the needle's tip and is released upon application of the stimulus. This design enhances safety and minimizes side effects.
- **Hydrogel-forming microneedles**: made of polymer matrix and multi-step micromolding. The needles penetrate the skin, absorbing tissue fluid and causing the hydrogel to swell, thereby facilitating the controlled release of the drug. The rate of drug release from these needles is inversely correlated with the concentration and cross-linking duration of the substance [64].
- **Cryomicro-needle**: These needles are used to transport and store live cells and bacteria. By penetrating the skin, the loaded substances are released. These needles

can preserve large numbers of cells and bacteria in liquid nitrogen for long periods, facilitating complex vaccination processes [65, 66].

In general, microneedles have great potential in the field of medicine and drug delivery, and with the development of new technologies, they continue to be an efficient tool for various treatments. Various kinds of microneedles can penetrate the skin [67], cornea [68], and other tissues with minimal invasion and deliver the drug to the intended site with minimal inflammation and tissue damage (compared to the usual injections of clinical implants). Currently, microneedles serve not only as an efficient means for localized drug delivery but also contribute to tissue repair by offering mechanical support and physical stimuli that influence cellular behavior.

Zhao et al. [69] employed methacrylated HA hydrogel microneedles loaded with human growth factor D (PDGF-D) and human adipose stem cells (ADSCs) to facilitate healing of diabetic ulcers. The results showed that these hydrogel microneedles kept 90% of ADSC activity over 24 h and helped diabetic wounds heal by improving blood flow and the production of collagen. The shape of the microneedles and the area or the material of the needle tip are the factors that determine their mechanical properties, and the available microneedles with a power of more than 0.1N per needle easily penetrate the skin, endometrium, cornea, and heart and overcome the complex barriers of different tissues [66, 70]. Microneedles with their unique structures cause the integration of bioactive substances and the host tissue improve the contact surface of the biological substance with the tissue and provide the possibility of better regulation of cell behavior to promote the repair of damaged tissue. However, the type, number, mechanical strength, and height of the needles must be fully considered when designing for the repair of different tissues [71].

In addition, microneedles should have the ability to load several drugs at the same time. For example, to repair an infected wound with microneedles, antimicrobial, pro-angiogenic, and tissue-regenerating drugs should be used at the same time. This problem can be easily overcome by combining several active substances in polymer solutions using the micro-molding method and producing microneedles. Wang et al. [72] used a combination of HA nanoparticles with recombinant human collagen and Poly(lactic-co-glycolic acid) (PLGA) nanoparticles encapsulated with naproxen in microneedles and placed in the wound bed. The microneedles were rapidly dissolved, resulting in the release of recombinant human collagen and PLGA nanoparticles and slowly naproxen released in derm for 14 days. Moreover, to enhance the loading of drugs and active substances, simultaneous loading in both the tip and support bed of the microneedle can improve the location allocation of pharmaceuticals within the tissue of interest. A study demonstrated the encapsulation of prussian blue nanoenzyme and vascular endothelial growth factor (VEGF) at the tip of the microneedle, with polymyxin loaded at its base [73]. The utilization of materials exhibiting varying dissolution or decomposition rates facilitates the regulation of drug release rates. The interface between the needles and the base, such as the introduction of air bubbles, can be utilized to regulate the release rate and efficiency of drug delivery [74].

5.3 Combining Nanofibers with Microneedles

The integration of electrospun nanofibers with microneedle arrays presents a promising strategy for enhancing drug delivery via the skin and for applications in tissue engineering. This approach leads to an increased surface area, sustained drug release at the injury site, accelerated tissue repair, and improved patient compliance. This integration may occur as direct spin on the outermost layer of microneedles or through the inclusion of microneedles within a polymer matrix. The former enhances drug loading capacity and enables controlled release of therapeutic compounds, while the latter improves mechanical properties and drug delivery due to the incorporation of microneedles [75]. In fact, the combination of these two technologies overcomes the problems faced by microneedles (such as not fully connecting to the skin and reducing the efficiency of drug delivery, hypersensitivity reactions, inflammation, and Irritation) and electrospinning (insufficient mechanical strength and inability to penetrate epithelial barriers) [76, 77].

A study developed dissolving microneedles on a pillar array that was covered with a PLGA nanofiber membrane, referred to as DEPA, to establish a connection between the microneedle and the skin's surface, replacing the need for an adhesive patch (Fig. 5.1) [78]. The microneedle was implanted in porcine skin, demonstrating its ability to address issues related to flat patch-based dissolving microneedles, including restricted penetration and reduced drug efficacy [78].

This integration has caused the resulting microneedles to become stiffer and increase their breaking force, so that they have been able to pierce the tissue with more compressive force than needles without fibers and penetrate deeper into the buccal

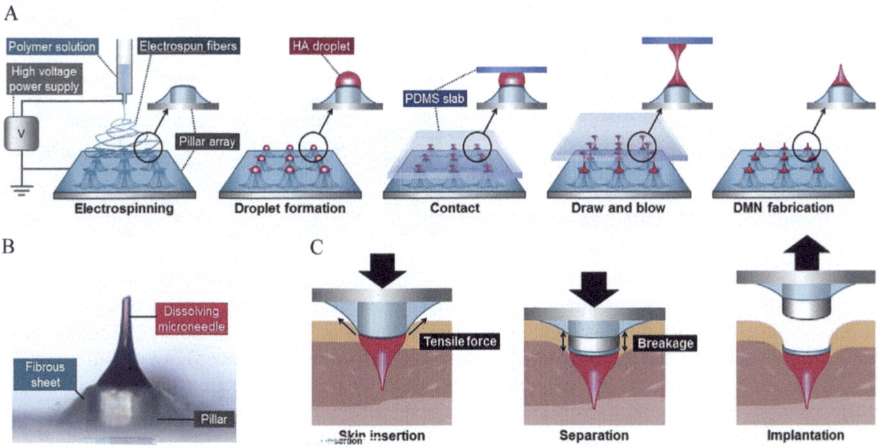

Fig. 5.1 Development and production of dissolving microneedles. **a** Fabrication of a dissolving microneedle on a pillar array coated with electrospun nanofibers; **b** product designing incorporating a dissolving microneedle, pillar array, and electrospun nanofibers; **c** phases of developed product implantation onto the skin surface. Reprinted with permission from [78]

tissue (penetration depth of about 240 μm) [79, 80]. A Janus-type wound dressing incorporating a dissolving microneedle with W379 peptide was developed for biofilm penetration, alongside an antimicrobial encapsulated W379 nanofiber membrane aimed at eliminating biofilm in chronic wound beds (Fig. 5.2a) [81]. In the ex vivo test using a human skin model, microneedles containing W379 combined with nanofiber exhibited a more pronounced effect on biofilm destruction compared to microneedles with W379 alone. Additionally, in the in vivo study involving diabetic rats, the application of microneedles loaded with W379 and nanofiber resulted in significantly improved wound healing outcomes. Figure 5.2b–e illustrate the developed microneedles integrated with electrospun nanofibers in this study [81].

In addition, the simultaneous delivery of antimicrobial drugs using nanofiber mats and dissolvable microneedles as a new approach can help improve the treatment of bacterial biofilms. These compounds are continuously supplied to bacteria in order to achieve better antimicrobial activity [78, 81]. The implementation of electronic nanofibers through biocompatible and degradable materials (such as proteins and natural polymers) can help reduce unwanted side effects such as irritation and inflammation associated with microneedles and improve the strength and life of microneedles. Also, the electrospinning technique helps in the production of nano fibers with various characteristics that allow the design of micro needles suitable for different types and stages of tissue damages [78, 82].

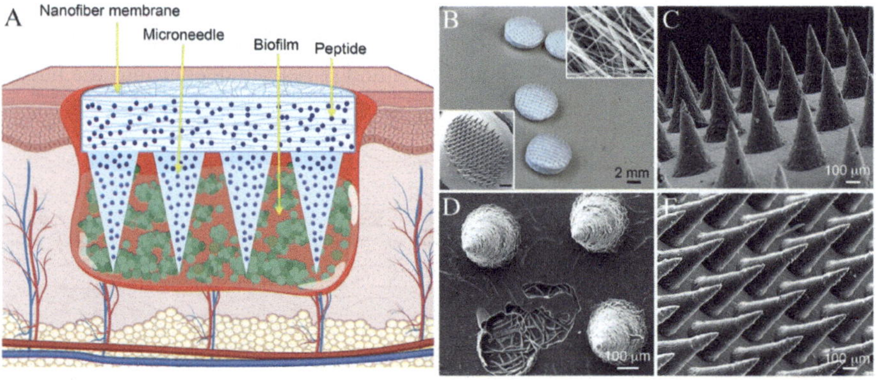

Fig. 5.2 Janus-type wound dressing and corresponding micrographs. **a** The commercial Janus-type wound dressing comprises integrated microneedle and nanofiber membranes that incorporate antimicrobial peptides for the treatment of biofilms. **b** Image depicting a Janus-type dressing. Inset (left bottom): scanning electron microscopy image depicting the complete view of a microneedle array fixed on the surface of the nanofiber membrane. Scale bar measures 1 mm. Inset (upper right): scanning electron microscopy (SEM) image depicting the electrospun nanofiber substrate. Scale bar measures 5 μm. **c** Scanning electron microscopy image depicting peptides integrated into dissolvable microneedle arrays on nanofiber membranes, with each membrane containing 100 microneedles. **d** SEM image depicting a magnified view of (**c**), with one microneedle intentionally removed to expose the underlying nanofiber matrices. SEM image depicting peptides integrated within dissolvable microneedle arrays on nanofiber membranes, comprising 150 microneedles per membrane. Reprinted with permission from [81]

5.4 Future Perspectives

The integration of microneedles and electrospun nanofibers is one of the new and turbulent fields in science and technology, which has great potential in medicine, pharmaceuticals, and tissue engineering. This combination can lead to performance enhancement and optimization of new product features. The primary focus of integrating these technologies is the enhancement and innovation of drug delivery systems. Nanofibers, characterized by their porous structure and high surface area, serve as effective carriers for biological and anticancer drugs. The combination of these technologies can make microneedle delivery methods more efficient and contribute to higher absorption of drugs. On the other hand, the integration of microneedles and nanofibers allows the development of personalized medicines. Using these technologies, it is possible to produce special medicines for individuals with different genetic characteristics. This can help optimize the toxicity and efficacy of treatments. In addition, the integration of microneedles and nanofibers can revolutionize tissue engineering and wound healing. Electrospun nanofibers can act as a foundation for cell growth, and microneedles can be used to inject nutrients and growth factors into damaged tissues. This could provide more effective treatments for tissue damage and wound repair. Research on the use of microneedles and electrospun nanofibers in the identification and diagnosis of diseases is expanding. These materials can act as biosensors, especially for detecting skin and infectious diseases that require rapid diagnosis. Hybrid systems can simultaneously sample and analyze body fluids. Although the integration of these technologies offers many opportunities, there are also challenges related to the optimization of the production process and production of nanofibers and microneedles simultaneously and on a large scale, safety evaluation of these compounds and their effects on living tissues, precise control of drug release from nanofibers and How it affects microneedles, and the costs of manufacturing and developing these technologies, may be a barrier to their commercialization and clinical use. Consequently, future research is recommended to ensure the safety and sterilization of products, scale up production, optimize storage conditions, refine production methods and materials, and assess the incorporation of various therapeutic and antimicrobial agents to enhance their synergistic effects in the healing process.

References

1. de Kanter, A.-F.J., Jongsma, K.R., Verhaar, M.C., Bredenoord, A.L.: The ethical implications of tissue engineering for regenerative purposes: a systematic review. Tissue Eng. Part B: Rev. **29**(2), 167–187 (2023)
2. Varshan, M.G., Josephson, P.J., Nayak, B.B., Hariram, V., Balachandar, K.: History, challenges, and opportunities in tissue engineering. In: Handbook of Research on Advanced Functional Materials for Orthopedic Application, pp. 148–167 (2023)
3. Younes, H.M., Kadavil, H., Ismail, H.M., Adib, S.A., Zamani, S., Alany, R.G., Al-Kinani, A.A.: Overview of tissue engineering and drug delivery applications of reactive electrospinning

and crosslinking techniques of polymeric nanofibers with highlights on their biocompatibility testing and regulatory aspects. Pharmaceutics **16**(1), 32 (2023)
4. Ding, Y., Li, W., Zhang, F., Liu, Z., Zanjanizadeh Ezazi, N., Liu, D., Santos, H.A.: Electrospun fibrous architectures for drug delivery, tissue engineering and cancer therapy. Adv. Funct. Mater. **29**(2), 1802852 (2019)
5. Echeverria Molina, M.I., Malollari, K.G., Komvopoulos, K.: Design challenges in polymeric scaffolds for tissue engineering. Front. Bioeng. Biotechnol. **9**, 617141 (2021)
6. Jain, K.K.: An overview of drug delivery systems. Drug Deliv. Syst. 1–54
7. Zielińska, A., Karczewski, J., Eder, P., Kolanowski, T., Szalata, M., Wielgus, K., Szalata, M., Kim, D., Shin, S.R., Słomski, R.: Scaffolds for drug delivery and tissue engineering: the role of genetics. J. Ophthalmol. Clinics Res. **359**, 207–223 (2023)
8. Wang, Z., Wang, Y., Yan, J., Zhang, K., Lin, F., Xiang, L., Deng, L., Guan, Z., Cui, W., Zhang, H.: Pharmaceutical electrospinning and 3D printing scaffold design for bone regeneration. Adv. Drug Deliv. Rev. **174**, 504–534 (2021)
9. Han, S., Nie, K., Li, J., Sun, Q., Wang, X., Li, X., Li, Q.: 3D electrospun nanofiber-based scaffolds: from preparations and properties to tissue regeneration applications. Stem Cells Int. **2021**(1), 8790143 (2021)
10. Aghmiuni, A.I., Keshel, S.H., Sefat, F., AkbarzadehKhiyavi, A.: Fabrication of 3D hybrid scaffold by combination technique of electrospinning-like and freeze-drying to create mechanotransduction signals and mimic extracellular matrix function of skin. Mater. Sci. Eng.: C **120**, 111752 (2021)
11. Zhou, X., Zhou, G., Junka, R., Chang, N., Anwar, A., Wang, H., Yu, X.: Fabrication of polylactic acid (PLA)-based porous scaffold through the combination of traditional bio-fabrication and 3D printing technology for bone regeneration. Colloids Surf. B: Biointerfaces **197**, 111420 (2021)
12. Kapłon, H., Blawert, C., Chęcmanowski, J., Naplocha, K.: Development of open-porosity magnesium foam produced by investment casting. J. Magn. Alloys **10**(7), 1941–1956 (2022)
13. Owida, H.A., Al-Nabulsi, J.I., Alnaimat, F., Al-Ayyad, M., Turab, N.M., Al Sharah, A., Shakur, M.: Recent applications of electrospun nanofibrous scaffold in tissue engineering. Appl. Bionics Biomech. **2022**(1), 1953861 (2022)
14. Li, L., Hao, R., Qin, J., Song, J., Chen, X., Rao, F., Zhai, J., Zhao, Y., Zhang, L., Xue, J.: Electrospun fibers control drug delivery for tissue regeneration and cancer therapy. Adv. Fiber Mater. **4**(6), 1375–1413 (2022)
15. Alven, S., Buyana, B., Feketshane, Z., Aderibigbe, B.A.: Electrospun nanofibers/nanofibrous scaffolds loaded with silver nanoparticles as effective antibacterial wound dressing materials. Pharmaceutics **13**(7), 964 (2021)
16. Lyu, S., Huang, C., Yang, H., Zhang, X.: Electrospun fibers as a scaffolding platform for bone tissue repair. J. Orthop. Res. **31**(9), 1382–1389 (2013)
17. Donnelly, R.F., Singh, T.R.R., Woolfson, A.D.: Microneedle-based drug delivery systems: microfabrication, drug delivery, and safety. Drug Deliv. **17**(4), 187–207 (2010)
18. Al-Japairai, K.A.S., Mahmood, S., Almurisi, S.H., Venugopal, J.R., Hilles, A.R., Azmana, M., Raman, S.: Current trends in polymer microneedle for transdermal drug delivery. Int. J. Pharm. **587**, 119673 (2020)
19. Wang, C., Su, Y., Xie, J.: Advances in electrospun nanofibers: versatile materials and diverse biomedical applications. Accounts Mater. Res. **5**(8), 987–999 (2024)
20. Tiwari, R., Pathak, K.: Local drug delivery strategies towards wound healing. Pharmaceutics **15**(2), 634 (2023)
21. Azar, H.K., Monfared, M.H., Seraji, A.A., Nazarnezhad, S., Nasiri, E., Zeinanloo, N., Sherafati, M., Sharifianjazi, F., Rostami, M., Beheshtizadeh, N.: Integration of polysaccharide electrospun nanofibers with microneedle arrays promotes wound regeneration: a review. Int. J. Biol. Macromol. 128482 (2023)
22. Zhu, M., Han, J., Wang, F., Shao, W., Xiong, R., Zhang, Q., Pan, H., Yang, Y., Samal, S.K., Zhang, F.: Electrospun nanofibers membranes for effective air filtration. Macromol. Mater. Eng. **302**(1), 1600353 (2017)

23. Zhu, J., Yan, C., Li, G., Cheng, H., Li, Y., Liu, T., Mao, Q., Cho, H., Gao, Q., Gao, C.: Recent developments of electrospun nanofibers for electrochemical energy storage and conversion. Energy Storage Mater. 103111 (2023)
24. Ji, G., Chen, Z., Li, H., Awuye, D.E., Guan, M., Zhu, Y.: Electrospinning-based biosensors for health monitoring. Biosensors **12**(10), 876 (2022)
25. Ghosal, K., Agatemor, C., Špitálsky, Z., Thomas, S., Kny, E.: Electrospinning tissue engineering and wound dressing scaffolds from polymer-titanium dioxide nanocomposites. Chem. Eng. J. **358**, 1262–1278 (2019)
26. Wang, C., Tong, S.N., Tse, Y.H., Wang, M.: Conventional electrospinning vs. emulsion electrospinning: a comparative study on the development of nanofibrous drug/biomolecule delivery vehicles. Adv. Mater. Res. **410**, 118–121 (2012)
27. Abdul Hameed, M.M., Mohamed Khan, S.A.P., Thamer, B.M., Rajkumar, N., El-Hamshary, H., El-Newehy, M.: Electrospun nanofibers for drug delivery applications: methods and mechanism. Polym. Adv. Technol. **34**(1), 6–23 (2023)
28. Stanger, J., Tucker, N., Kirwan, K., Staiger, M.P.: Effect of charge density on the Taylor cone in electrospinning. Int. J. Modern Phys. B **23**(06n07), 1956–1961 (2009)
29. Zhou, H.: Electrospun fibers from both solution and melt: processing, structure and property. The American Chemical Society (2006)
30. Sun, Z., Zussman, E., Yarin, A.L., Wendorff, J.H., Greiner, A.: Compound core–shell polymer nanofibers by co-electrospinning. Adv. Mater. **15**(22), 1929–1932 (2003)
31. Greiner, A., Wendorff, J.H., Yarin, A.L., Zussman, E.: Biohybrid nanosystems with polymer nanofibers and nanotubes. Appl. Microbiol. Biotechnol. **71**, 387–393 (2006)
32. Yoon, J., Yang, H.S., Lee, B.S., Yu, W.R.: Recent progress in coaxial electrospinning: new parameters, various structures, and wide applications. Adv. Mater. **30**(42), 1704765 (2018)
33. Brown, T.D., Dalton, P.D., Hutmacher, D.W.: Direct writing by way of melt electrospinning. Adv. Mater. (Deerfield Beach, Fla.) **23**(47), 5651–5657 (2011)
34. Brown, T.D., Dalton, P.D., Hutmacher, D.W.: Melt electrospinning today: an opportune time for an emerging polymer process. Progr. Polym. Sci. **56**, 116–166 (2016)
35. Zhou, H., Green, T.B., Joo, Y.L.: The thermal effects on electrospinning of polylactic acid melts. Polymer **47**(21), 7497–7505 (2006)
36. Beachley, V., Wen, X.: Effect of electrospinning parameters on the nanofiber diameter and length. Mater. Sci. Eng.: C **29**(3), 663–668 (2009)
37. Jacobs, V., Anandjiwala, R.D., Maaza, M.: The influence of electrospinning parameters on the structural morphology and diameter of electrospun nanofibers. J. Appl. Polym. Sci. **115**(5), 3130–3136 (2010)
38. Purwar, R., Sai Goutham, K., Srivastava, C.M.: Electrospun Sericin/PVA/Clay nanofibrous mats for antimicrobial air filtration mask. Fibers Polym. **17**, 1206–1216 (2016)
39. Zargham, S., Bazgir, S., Tavakoli, A., Rashidi, A.S., Damerchely, R.: The effect of flow rate on morphology and deposition area of electrospun nylon 6 nanofiber. J. Eng. Fibers Fabrics **7**(4), 155892501200700400 (2012)
40. Haider, S., Al-Zeghayer, Y., Ahmed Ali, F.A., Haider, A., Mahmood, A., Al-Masry, W.A., Imran, M., Aijaz, M.O.: Highly aligned narrow diameter chitosan electrospun nanofibers. J. Polym. Res. **20**, 1–11 (2013)
41. Hayati, I., Bailey, A., Tadros, T.F.: Investigations into the mechanisms of electrohydrodynamic spraying of liquids: I. Effect of electric field and the environment on pendant drops and factors affecting the formation of stable jets and atomization. J. Colloid Interface Sci. **117**(1), 205–221 (1987)
42. Wannatong, L., Sirivat, A., Supaphol, P.: Effects of solvents on electrospun polymeric fibers: preliminary study on polystyrene. Polym. Int. **53**(11), 1851–1859 (2004)
43. Hardick, O., Stevens, B., Bracewell, D.G.: Nanofibre fabrication in a temperature and humidity controlled environment for improved fibre consistency. J. Mater. Sci. **46**, 3890–3898 (2011)
44. Li, D., Hu, D., Xu, H., Patra, H.K., Liu, X., Zhou, Z., Tang, J., Slater, N., Shen, Y.: Progress and perspective of microneedle system for anti-cancer drug delivery. Biomaterials **264**, 120410 (2021)

45. Zhang, X., Hasani-Sadrabadi, M.M., Zarubova, J., Dashtimighadam, E., Haghniaz, R., Khademhosseini, A., Butte, M.J., Moshaverinia, A., Aghaloo, T., Li, S.: Immunomodulatory microneedle patch for periodontal tissue regeneration. Matter **5**(2), 666–682 (2022)
46. GhavamiNejad, A., Li, J., Lu, B., Zhou, L., Lam, L., Giacca, A., Wu, X.Y.: Glucose-responsive composite microneedle patch for hypoglycemia-triggered delivery of native glucagon. Adv. Mater. **31**(30), 1901051 (2019)
47. Suh, H., Shin, J., Kim, Y.-C.: Microneedle patches for vaccine delivery. Clin. Exp. Vacc. Res. **3**(1), 42–49 (2014)
48. Prausnitz, M.R.: Engineering microneedle patches for vaccination and drug delivery to skin. Ann. Rev. Chem. Biomol. Eng. **8**(1), 177–200 (2017)
49. Jin, X., Zhu, D.D., Chen, B.Z., Ashfaq, M., Guo, X.D.: Insulin delivery systems combined with microneedle technology. Adv. Drug Deliv. Rev. **127**, 119–137 (2018)
50. Blicharz, T.M., Gong, P., Bunner, B.M., Chu, L.L., Leonard, K.M., Wakefield, J.A., Williams, R.E., Dadgar, M., Tagliabue, C.A., El Khaja, R.: Microneedle-based device for the one-step painless collection of capillary blood samples. Nat. Biomed. Eng. **2**(3), 151–157 (2018)
51. Jung, J.H., Jin, S.G.: Microneedle for transdermal drug delivery: current trends and fabrication. J. Pharm. Investigat 1–15 (2021)
52. Rad, Z.F., Prewett, P.D., Davies, G.J.: An overview of microneedle applications, materials, and fabrication methods. Beilstein J. Nanotechnol. **12**(1), 1034–1046 (2021)
53. Aldawood, F.K., Andar, A., Desai, S.: A comprehensive review of microneedles: types, materials, processes, characterizations and applications. Polymers **13**(16), 2815 (2021)
54. Rajabi, M., Roxhed, N., Shafagh, R.Z., Haraldson, T., Fischer, A.C., Wijngaart, W.V.D., Stemme, G., Niklaus, F.: Flexible and stretchable microneedle patches with integrated rigid stainless steel microneedles for transdermal biointerfacing. PloS One **11**(12), e0166330 (2016)
55. Li, J., Liu, B., Zhou, Y., Chen, Z., Jiang, L., Yuan, W., Liang, L.: Fabrication of a Ti porous microneedle array by metal injection molding for transdermal drug delivery. PLoS One **12**(2), e0172043 (2017)
56. Pradeep Narayanan, S., Raghavan, S.: Solid silicon microneedles for drug delivery applications. Int. J. Adv. Manuf. Technol. **93**, 407–422 (2017)
57. Zhu, Q., Zarnitsyn, V.G., Ye, L., Wen, Z., Gao, Y., Pan, L., Skountzou, I., Gill, H.S., Prausnitz, M.R., Yang, C.: Immunization by vaccine-coated microneedle arrays protects against lethal influenza virus challenge. Proc. Natl. Acad. Sci. **106**(19), 7968–7973 (2009)
58. Bernadete Riemma Pierre, M., Cristina Rossetti, F.: Microneedle-based drug delivery systems for transdermal route. Curr. Drug Targets **15**(3), 281–291 (2014)
59. Gill, H.S., Prausnitz, M.R.: Coated microneedles for transdermal delivery. J. Control. Release **117**(2), 227–237 (2007)
60. Liu, T., Chen, M., Fu, J., Sun, Y., Lu, C., Quan, G., Pan, X., Wu, C.: Recent advances in microneedles-mediated transdermal delivery of protein and peptide drugs. Acta Pharmaceutica Sinica B **11**(8), 2326–2343 (2021)
61. Cárcamo-Martínez, Á., Mallon, B., Domínguez-Robles, J., Vora, L.K., Anjani, Q.K., Donnelly, R.F.: Hollow microneedles: a perspective in biomedical applications. Int. J. Pharm. **599**, 120455 (2021)
62. Li, Q., Yu, X., Zheng, X., Yang, J., Hui, J., Fan, D.: Rapid dissolution microneedle based on polyvinyl alcohol/chitosan for local oral anesthesia. Int. J. Biol. Macromol. **257**, 128629 (2024)
63. Huang, S., Liu, H., Huang, S., Fu, T., Xue, W., Guo, R.: Dextran methacrylate hydrogel microneedles loaded with doxorubicin and trametinib for continuous transdermal administration of melanoma. Carbohydr. Polym. **246**, 116650 (2020)
64. Li, J.Y., Feng, Y.H., He, Y.T., Hu, L.F., Liang, L., Zhao, Z.Q., Chen, B.Z., Guo, X.D.: Thermosensitive hydrogel microneedles for controlled transdermal drug delivery. Acta Biomater. **153**, 308–319 (2022)
65. Zheng, M., Hu, T., Yang, Y., Qie, X., Yang, H., Zhang, Y., Zhang, Q., Yong, K.-T., Liu, W., Xu, C.: In situ-formed cryomicroneedles for intradermal cell delivery. NPG Asia Mater. **16**(1), 11 (2024)

66. Cui, M., Zheng, M., Wiraja, C., Chew, S.W.T., Mishra, A., Mayandi, V., Lakshminarayanan, R., Xu, C.: Ocular delivery of predatory bacteria with cryomicroneedles against eye infection. Adv. Sci. **8**(21), 2102327 (2021)
67. Kochhar, J.S., Soon, W.J., Choi, J., Zou, S., Kang, L.: Effect of microneedle geometry and supporting substrate on microneedle array penetration into skin. J. Pharm. Sci. **102**(11), 4100–4108 (2013)
68. Shi, H., Zhou, J., Wang, Y., Zhu, Y., Lin, D., Lei, L., Vakal, S., Wang, J., Li, X.: A rapid corneal healing microneedle for efficient ocular drug delivery. Small **18**(4), 2104657 (2022)
69. Xu, Y., Wu, X., Zhang, X., Zu, Y., Tan, Q., Zhao, Y.: Living microneedle patch with adipose-derived stem cells embedding for diabetic ulcer healing. Adv. Funct. Mater. **33**(1), 2209986 (2023)
70. Lim, S., Park, T.Y., Jeon, E.Y., Joo, K.I., Cha, H.J.: Double-layered adhesive microneedle bandage based on biofunctionalized mussel protein for cardiac tissue regeneration. Biomaterials **278**, 121171 (2021)
71. Makvandi, P., Kirkby, M., Hutton, A.R., Shabani, M., Yiu, C.K., Baghbantaraghdari, Z., Jamaledin, R., Carlotti, M., Mazzolai, B., Mattoli, V.: Engineering microneedle patches for improved penetration: analysis, skin models and factors affecting needle insertion. Nano-Micro Lett. **13**, 1–41 (2021)
72. Long, L.-Y., Liu, W., Li, L., Hu, C., He, S., Lu, L., Wang, J., Yang, L., Wang, Y.-B.: Dissolving microneedle-encapsulated drug-loaded nanoparticles and recombinant humanized collagen type III for the treatment of chronic wound via anti-inflammation and enhanced cell proliferation and angiogenesis. Nanoscale **14**(4), 1285–1295 (2022)
73. Guan, G., Zhang, Q., Jiang, Z., Liu, J., Wan, J., Jin, P., Lv, Q.: Multifunctional silk fibroin methacryloyl microneedle for diabetic wound healing. Small **18**(51), 2203064 (2022)
74. Ning, T., Yang, F., Chen, D., Jia, Z., Yuan, R., Du, Z., Liu, S., Yu, Y., Dai, X., Niu, X.: Synergistically detachable microneedle dressing for programmed treatment of chronic wounds. Adv. Healthc. Mater. **11**(11), 2102180 (2022)
75. Ali, R., Mehta, P., Monou, P.K., Arshad, M.S., Panteris, E., Rasekh, M., Singh, N., Qutachi, O., Wilson, P., Tzetzis, D.: Electrospinning/electrospraying coatings for metal microneedles: a design of experiments (DOE) and quality by design (QbD) approach. Eur. J. Pharm. Biopharm. **156**, 20–39 (2020)
76. Fukushima, K., Ise, A., Morita, H., Hasegawa, R., Ito, Y., Sugioka, N., Takada, K.: Two-layered dissolving microneedles for percutaneous delivery of peptide/protein drugs in rats. Pharm. Res. **28**, 7–21 (2011)
77. Malone, M., Bjarnsholt, T., McBain, A.J., James, G.A., Stoodley, P., Leaper, D., Tachi, M., Schultz, G., Swanson, T., Wolcott, R.D.: The prevalence of biofilms in chronic wounds: a systematic review and meta-analysis of published data. J. Wound Care **26**(1), 20–25 (2017)
78. Yang, H., Kim, S., Huh, I., Kim, S., Lahiji, S.F., Kim, M., Jung, H.: Rapid implantation of dissolving microneedles on an electrospun pillar array. Biomaterials **64**, 70–77 (2015)
79. Lee, J.W., Park, J.-H., Prausnitz, M.R.: Dissolving microneedles for transdermal drug delivery. Biomaterials **29**(13), 2113–2124 (2008)
80. McNeilly, C.L., Crichton, M.L., Primiero, C.A., Frazer, I.H., Roberts, M.S., Kendall, M.A.: Microprojection arrays to immunise at mucosal surfaces. J. Control. Release **196**, 252–260 (2014)
81. Su, Y., Mainardi, V.L., Wang, H., McCarthy, A., Zhang, Y.S., Chen, S., John, J.V., Wong, S.L., Hollins, R.R., Wang, G., Xie, J.: Dissolvable microneedles coupled with nanofiber dressings eradicate biofilms via effectively delivering a database-designed antimicrobial peptide. ACS Nano **14**(9), 11775–11786 (2020)
82. Tort, S., Mutlu Agardan, N.B., Han, D., Steckl, A.J.: In vitro and in vivo evaluation of microneedles coated with electrosprayed micro/nanoparticles for medical skin treatments. J. Microencapsul. **37**(7), 517–527 (2020)

Chapter 6
Nanocellulose-Based Microneedles for Enhanced Topical Drug Delivery

Mohd Jahir Khan, Abdullah Ramzan, and Chularat Sakdaronnarong

6.1 Introduction

Drugs are administered to the body through many routes, including enteral (oral, sublingual, and rectal), parenteral (intravascular, intramuscular, subcutaneous, and inhalation), or topical (skin and mucosal membranes). Each route possesses distinct functions, benefits, and drawbacks. The accessibility of drug molecules to their intended site and the efficacy of treatment are significantly influenced by the route of administration used [1]. The method of drug administration significantly influences the onset and efficacy of the drug's action. The rate with which drugs attain their site of action is reliant upon their absorption and distribution characteristics. Furthermore, individuals exhibit varied responses to drugs, influenced by numerous factors. No singular method of drug delivery is optimal for all drugs in every circumstance. In recent years, the advancement of controlled oral drug delivery and release systems has significantly increased [2].

Among different modes of drug administration, oral method is the predominant and most convenient technique; yet, it is constrained by challenges such as first-pass metabolism and delayed beginning of action. Drug delivery through injection is highly efficient; however, it poses many drawbacks, including discomfort, bleeding, the possibility of infection, and requirements for trained personnel [3]. Tropical drug administration (TDA) is significant as it offers a non-invasive, convenient and controlled method for administering medicines through the skin, presenting various benefits compared to conventional oral and injectable routes. TDA is suitable for localized treatments, limited by the skin's barrier characteristics, which restrict

M. J. Khan (✉) · C. Sakdaronnarong
Department of Chemical Engineering, Faculty of Engineering, Mahidol University, 25/25 Putthamonthon 4 Road, Salaya, Putthamonthon, Nakhon Pathom 73170, Thailand
e-mail: khanmohdjahir@gmail.com

A. Ramzan
University of New Brunswick, Fredericton, Canada

N. Talreja et al. (eds.), *Microneedles (MNs)-Based Technology*, Series in BioEngineering, https://doi.org/10.1007/978-981-96-3916-8_6

absorption for systemic effects. Human skin is an organ capable of delivering drugs systemically and/or locally by diffusion [4, 5]. Human skin consists of multiple layers, arranged from the innermost to the outermost, including the hypodermis, dermis, and epidermis. The dermis, approximately 3–5 mm in thickness, comprises elastin and collagen, contains fibres, nerve endings, sweat glands, lymphatic veins, and blood [6]. It has been established that TDA can be utilized for the delivery of small drug molecules, proteins and vaccines through the dermis [7, 8]. This system minimizes substantial drug degradation and facilitates reduced doses while enhancing bioavailability.

Transdermal systems, including patches and MNs, can be developed to deliver drugs continuously for hours or even days as the release of drugs into the bloodstream is regulated and uniform. TDDS minimizes dose frequency and improves the efficacy of drugs, particularly in chronic diseases. It offers ease of application and generally necessitates less frequent dosing in comparison to oral or injection methods. TDDS maintains consistent drug levels and minimizes the peaks, reducing the fluctuations commonly seen with oral or injectable medications. Moreover, this stable dug concentration diminishes side effects often induced by variations in drug levels, thereby improving both safety and therapeutic efficacy. In conditions requiring restricted treatment, including pain management or dermatological conditions, TDDS are capable of administering medicines directly to a specific area. This local delivery method facilitates elevated drug concentrations at the target site while minimizing systemic exposure and thereby decreasing the likelihood of side effects. TDDS is capable of accommodating a range of pharmaceuticals, including small molecules such as analgesics, as well as large entities like peptides, hormones, and vaccines. Technological advancements, particularly in microneedle development, are broadening the range of drugs suitable for effective transdermal delivery. MNs provide opportunities for biodegradable and environmentally sustainable drug delivery solutions, minimizing the waste linked to syringes, needles, and oral packaging. It improves the quality of care through precise, sustained, and patient-friendly drug administration, proving to be beneficial for both localized and systemic therapies across various medical conditions. Although, TDDS is a significant and non-invasive approach to administering medicines, it encounters many obstacles that restrict its efficacy. Significant challenges are the efficacy of traditional topical drug formulations, including the skin's protective barrier, restricted permeability, variable drug release, and potential skin irritation. Drug distribution, instability, and variability are the other obstacles which limit the development of reliable and successful topical therapies. Additionally, transdermal delivery requires penetration of the outer barrier of the stratum corneum, which is not easily achieved with drugs with high molecular weights and, therefore, requires a method for enhancing the efficiency of delivery. Recent advancements in the field of nanotechnology, MNs technology, and sophisticated drug delivery systems promise to tackle these difficulties by boosting skin penetration, optimizing drug release profiles, and facilitating precise dose control. However, understanding and conquering these obstacles is essential for the future of transdermal drug delivery.

6.2 MNs for Transdermal Drug Delivery

Transdermal drug delivery has currently been constrained by the absence of molecules with the necessary physicochemical qualities to passively penetrate the skin's outermost layer, known as the stratum corneum. Numerous chemical and physical techniques have been explored to diminish the barrier properties of the stratum corneum to harness the advantages of transdermal drug administration. MNs technology for improved transdermal drug delivery has garnered significant attention and is rapidly advancing towards commercialization [9]. MNs have emerged as a viable platform for minimally invasive drug delivery, facilitating the accurate administration of drugs, vaccines, and other biomolecules subcutaneously. MNs are fine needle-shaped entities that penetrate the skin with minimal discomfort and pain, circumventing the outermost barrier, the stratum corneum, to administer substances directly into the dermis. MNs are developed in the 1970s, and ever since then, they have been a subject of attention.

6.2.1 Classification of MNs

MNs can be classified into five different categories: solid, drug-coated, dissolvable, hollow, and hydrogel-based MNs (Fig. 6.1) [7].

6.2.1.1 Solid MNs

Solid MNs consist of a number of microscales tapered pointed tips composed of a single substance free of medicines or excipients. They generally vary between 150 and 300 μm in length, with a tapered tip angle of 15–20°. Their frequency of use varies from 30 s to 10 min [10]. They penetrate the skin, forming micron-sized pores on the surface. When applied to the affected area, medicine penetrates the stratum corneum, the principal barrier of the skin, via these pores. It is readily absorbed by the capillaries in the superficial dermis, enhancing the drug's bioavailability [11, 12]. Solid MNs can be made from metal, silicon, or polymer to facilitate precise piercings and enhance medicine permeability. The efficacy of drug administration using solid MNs is affected by several factors, namely insertion force, tip sharpness, and MNs density [13]. The most significant advantage of employing solid MNs is their simple and easy fabrication, inexpensive cost, coupled with various mass production technologies, and their adequate mechanical strength, which prevents breakage within the skin [14]. Sabri et al. reported that solid MNs enhanced the ex-vivo transdermal absorption of imiquimod (a medicine for the treatment of basal cell carcinoma), with the intradermal depot remaining effective for up to 24 h [15]. The prevalent drawbacks of solid MNs include the production of biohazardous sharp waste, which poses a potential risk to public health if inappropriately disposed of,

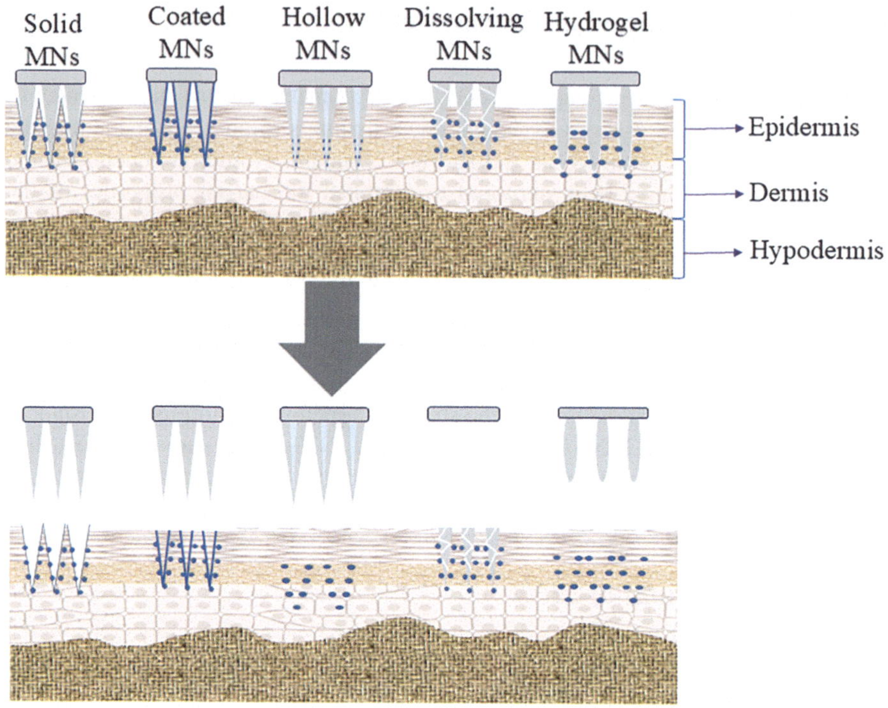

Fig. 6.1 Schematic illustration of the types of MNs. Solid MNs, coated MNs, hollow MNs, dissolving MNs and hydrogel MNs

and the prolonged recovery of the produced pores, normally recovered in 2 h but potentially extending up to 7 days [16, 17].

6.2.1.2 Coated MNs

Coated MNs are first coated with the drug prior to its delivery. Therefore, a specific quantity of the drugs can be administered upon the introduction of MNs into the dermis [18]. This methodology suggests a one-step application in contrast to the two-step method with uncoated solid MNs [19]. Drug-coated MNs facilitate the delivery of active compounds, including small drug molecules, DNA, proteins, and vaccines. The integration of siRNA via a minimally invasive approach has also been demonstrated in this method of drug administration [20, 21]. The MNs can be coated using either a dipping or spraying method with a water-soluble drug formulation. The coating formulation must form a film on the MNs surface and preserve adherence throughout storage and skin penetration. To accomplish this goal, the coating composition must possess appropriate viscosity. When pierced to the dermis with the coated MNs, the drug dissolves into the adjacent tissue and diffuses into microcirculation. Thereafter, the MNs may be extracted from the application site [21, 22]. Coated MNs are one-step drug delivery, allowing for rapid dissolution of the drugs in the skin and, consequently, a quick commencement of pharmacological action. A constraint of this method is that the requisite dosage of a medicine can solely be applied to the tip and shaft of MNs. This may limit the application of coated MNs for the delivery of powerful compounds [23, 24].

6.2.1.3 Dissolving MNs

Dissolving MNs can be produced from water-soluble or biodegradable polymers that enclose the medicines and have adequate mechanical strength for skin penetration. They rapidly dissolve after introduction into the skin and exhibit good biocompatibility due to the utilization of safe polymers as raw materials, which also provide a substantial loading capacity [25, 26]. Dissolving MNs are typically composed of biodegradable polymers that release the enclosed drugs within minutes. They are constructed to form pathways for medicines, penetrate the skin, and subsequently disintegrate and release more substances for transdermal absorption. These MNs are favoured over alternative varieties because of their advantageous properties of self-dissolution and lower risk of toxicity [27]. The initial report on the dissolving of MNs was published in 2005 by Miyano et al. [28]. The choice of polymers in the production of dissolving MNs is a critical factor since it influences the drug release kinetics. Dissolving MNs systems have been exploited to administer a diverse array of substances, ranging from hydrophilic low molecular weight pharmaceuticals to large biopharmaceutical entities, illustrating the capacity of this technology to augment TDDS. The literature demonstrated many examples involving dissolving MNs used for synergistic drug delivery, wherein additional approaches are utilized alongside MNs administration. The utilization of dissolving MNs extends beyond drug inclusion as it has also improved the formulation of MNs patches for vaccine delivery against influenza and adenovirus vectors [29, 30]. The primary benefit of these MNs is their biodegradable material characteristic when interacting with the skin's interstitial fluid. It encompasses a simplified one-step procedure that is advantageous for patients. The main concern with these MNs involves the compatibility of polymers with active substances, as manufacturing variables, including severe pH, elevated temperatures, and solvents, can affect the stability of integrated proteins, vaccines, and other pharmaceuticals. Another concern is the drug-to-polymer, as it substantially affects the mechanical strength of MNs [21].

6.2.1.4 Hollow MNs

Hollow MNs are different compared to other MNs in both length and thickness. These are made employing a 30-gauge hypodermic needle, with the unique MNs measuring 300 μm in length [31]. Such MNs comprise a drug reservoir containing approximately 200 μL of drug formulation, with a hollow bore within the needle's centre. It was primarily designed to deliver a substantial dosage of large molecular weight compounds such as proteins, vaccines and oligonucleotides to circumvent the constraints of coated MNs. This technique facilitates the quick delivery of medicine by passive diffusion, contrary to other MNs technology. Moreover, additional stimuli such as pressure and electrically induced transport are equally viable within this category of MNs [32]. A variety of substances are used in the manufacture of these MNs, including silicone, glass, ceramic, and polymers. While hollow MNs effectively regulate drug dosage and release timing, their fabrication methods such as laser micromachining, deep reactive etching, integrated lithographic molding, and X-ray photolithography—are costly and involve sophisticated techniques. It also poses risks of needle fracture and lumen obstruction [23, 24].

6.2.1.5 Hydrogel-Based MNs

In hydrogel, medicines are distributed throughout the microneedle tip, core substrate, and patch backing and are released gradually while the patch is adhered to the skin [33]. A significant quantity of the drugs in the hydrogel is administered to the skin via diffusion [34]. This MNs technology permits the inclusion of diverse drug formulations without restriction. This method was initially developed by Donnelly et al., who included an extremely swellable polymer in the microneedle framework. The MNs comprise a hydrophilic hydrogel matrix that absorbs neighbouring tissue fluid and expands to create microchannels or passageways, facilitating the diffusion of beneficial drugs into the microcirculation [35]. The hydrophilic properties enable the hydrogel-based MNs to absorb interstitial fluid and expand upon insertion into the skin. Materials exhibiting biocompatibility and favourable swelling characteristics should be chosen for the fabrication [36, 37]. The constraints of traditional microarray methods can be mitigated by hydrogel MNs, which enhance drug loading capacity, regulate release extent, and ensure exact drug coating. It offers potential tuning advantages, allowing for the fabrication of hydrogel-loaded MNs in the required form and size, which can be easily sterilized. Hydrogel-based MNs is a multifunctional device capable of incorporating several drugs with distinct therapeutic windows for individualized treatment alternatives [21, 38, 39].

6.3 Nanocellulose: Structure and Properties

Cellulose is one of the most abundant natural polymers on earth, and it is present in a variety of materials, such as wood, cotton, and bacteria. It is a complex carbohydrate composed of oxygen, carbon, and hydrogen [40, 41]. Nonetheless, it is produced extensively from wood and cotton. Cellulose is an unbranched, linear polymer comprised of glucose monomers interconnected solely by β-(1,4) glycosidic bonds [42]. Diverse approaches are employed in the production of nanocellulose. The low molecular weight of nanocellulose facilitates its successful synthesis with other substances during application, presenting notable advantages and promising development potential. The prevalent procedures for preparing nanocellulose encompass acid hydrolysis [43], TEMPO (2,2,6,6-tetramethylpiperidin-1-oxyl) oxidation [44], mechanical treatment [45], high-pressure homogenization [46], ultrasonic treatment [47] and enzyme hydrolysis [48]. Different preparation methods are applied based on the individual's application during the preparation process.

Nanocellulose possesses a distinctive characteristic of dispersing in water, resulting in a high-viscosity dispersion. The linear, unbranched configuration of nanocellulose facilitates the tight clustering of its molecules, leading to a compact network of hydrogen bonds among them. It, therefore, provides rigidity and significant tensile strength [49]. It is biodegradable and sustainable, and is characterized by its high efficiency, low density, high porosity, and significant specific surface area [50]. Consequently, it possesses diverse applications in oil/water separation, thermal insulation, adsorption, and medical science. The demand for eco-friendly products has heightened awareness of nanocellulose's applications in biomedical [51, 52], cosmetic [53], and electrochemical science [54] in recent years.

6.3.1 Nanocellulose Types

Nanocellulose denotes nanoscale cellulose molecules obtained from plant fibres and bacteria. Because of its exceptional properties, nanocellulose is extensively applied in areas including drug delivery and tissue engineering. Nanocellulose is categorized into three main types according to their structure, shape, and origin: Nanocrystalline cellulose (NCC), Nanofibrillated cellulose (NFC), and Bacterial nanocellulose (BNC). Each kind possesses distinctive features that render it appropriate for particular uses.

6.3.1.1 Nanocrystalline Cellulose (NCC)

Nanocrystalline cellulose, referred to as cellulose nanocrystals (CNC), has a high degree of crystallinity and is generally characterized by rod or needle morphology. The technique involves deconstructing the amorphous parts inside cellulose fibres, yielding small, highly organized crystalline particles. NCC particles generally range from 100 to 200 nm in length and possess diameters of approximately 5–20 nm. Increased crystallinity imparts robust mechanical characteristics, significant stiffness, and thermal stability to NCC. NCC demonstrates distinctive optical characteristics, including birefringence, owing to its nanoscale dimensions and elevated surface area. NCC is extensively applied as a reinforcing ingredient in composites and pharmaceuticals, where its superior strength and crystallinity improve the durability and efficacy of drug compositions [55].

6.3.1.2 Nanofibrillated Cellulose (NFC)

Nanofibrillated cellulose, or cellulose nanofibers (CNF), consists of elongated and flexible fibrils exhibiting both crystalline and amorphous domains. It is formed by physically or chemically fibrillating cellulose fibres, resulting in entangled, web-like formations. NFC fibres typically measure several micrometres in length and possess diameters ranging from 5 to 100 nm. The high aspect ratio of NFC (length-to-diameter) imparts superior flexibility, mechanical strength, and viscosity-enhancing characteristics. The entangled configuration enables NFC to make stable gels in water, rendering it very absorbent and appropriate for hydrogel synthesis. NFC is frequently utilized in applications demanding flexible, absorbent materials, such as wound dressings, drug delivery systems, and emulsion stabilizers, owing to its gel-forming properties and extensive surface area [56].

6.3.1.3 Bacterial Nanocellulose (BNC)

BNC is synthesized by specific bacterial species, including *Acetobacter xylinum*, via a fermentation process. In contrast to NCC and NFC, BNC is exclusively nanofibrillar, characterized by a network of extremely crystalline nanofibers. BNC generally possesses nanofibers with widths ranging from 20 to 100 nm, yielding a highly porous, interlinked structure. BNC is extremely pure (devoid of lignin and hemicellulose), biocompatible, and possesses superior water retention abilities. It is very flexible during manufacture, aiding the formation of membranes, sheets, or other configurations suitable for medical purposes. BNC's excellent purity, moldability, and biocompatibility make it especially useful for biomedical applications, including wound dressings, scaffolds for tissue engineering, and drug delivery systems [57].

6.3.2 Physicochemical Properties of Nanocellulose

Nanocellulose is a multifunctional biomaterial made from cellulose, receiving considerable interest because of its distinctive physicochemical attributes. The physicochemical characteristics of nanocellulose are largely influenced by its molecular configuration, crystalline characteristics and nanoscale measurements. Nanocellulose exhibits exceptional mechanical strength, extensive surface area, adjustable surface chemistry, biocompatibility, and distinctive rheological and optical properties, rendering it a versatile material applicable in medicines and engineering. These features not only augment its functionality but also offer new prospects for advancements in materials science and healthcare [58].

6.3.3 Functionalization Potential of Nanocellulose

Nanocellulose is an emerging substance in drug delivery owing to its distinct physicochemical features. However, surface alteration may be important to improve its drug-binding capacity and facilitate regulated release. Surface modification approaches provide the bonding of functional groups to the surface of nanocellulose, enhancing its capacity for drug binding, improving stability, and enabling adjustable release profiles. Nanocellulose is exceptionally versatile for use in tailored and persistent drug delivery systems [59].

6.3.3.1 Increased Binding Capacity via Functional Groups

Nanocellulose intrinsically possesses reactive surface hydroxyl (−OH) groups that can be altered with diverse functional groups to enhance drug-binding affinity. Carboxylation, sulfonation, and amination are prevalent changes that insert carboxyl (−COOH), sulfonic ($-SO_3H$), or amino ($-NH_2$) groups, respectively. Such functional groups can engage with various drugs via ionic or hydrogen bonding. By optimizing these interactions, tailored nanocellulose could be customized to bind individual drugs with varying affinities, hence improving loading efficiency and stability [59].

6.3.3.2 Controlled Release via PH-Sensitive and Stimuli-Responsive Modifications

Through certain modifications, nanocellulose can respond to changes in temperatures or pH, enabling controlled drug release in certain circumstances. pH-sensitive alterations incorporate groups (e.g., carboxyl groups) that respond to pH fluctuations. In acidic conditions, such as those found in specific tumour microenvironments, these groups may facilitate drug release by weakening the nanocellulose-drug combination. Thermo-responsive and photo-responsive modifications entail the grafting of thermally sensitive or light-sensitive polymers onto nanocellulose, enabling external activation for on-demand drug release and facilitating fine temporal control over the release process [60, 61].

6.3.3.3 Prolonged Drug Release Through Hydrophobic Modifications

Hydrophobic changes can be implemented to regulate the release rate of hydrophobic medicines. For instance, including hydrophobic groups (e.g., alkyl chains) in nanocellulose facilitates a more regulated release profile by impeding drug diffusion. This hydrophobic interaction improves the binding of particular drugs that have low solubility in water, providing nanocellulose as a versatile transporter for many hydrophobic medicines [62, 63].

6.3.3.4 Polymer Grafting for Enhanced Stability and Sustained Drug Release

Attaching polymers to the surface of nanocellulose, like poly(ethylene glycol) (PEG), chitosan, or poly(lactic-co-glycolic acid) (PLGA), can enhance the stability of the nanocellulose-drug complex and facilitate sustained release. Polymer-coated nanocellulose particles can produce hydrogels or nanocomposites that serve as a matrix for controlled release. This is especially beneficial for sensitive drugs that require protection and progressive release over time. This technique enables adjustable medicine release rates by modifying the thickness or composition of the grafted polymer layer [63, 64].

6.3.3.5 Electrostatic and Covalent Bonding for Improved Drug Stability

Specific drugs, including proteins and peptides, are susceptible to degradation and impaired bioactivity. Surface modification allows nanocellulose to attach drugs via electrostatic or covalent interactions, so improving their stability and shielding them from environmental influences. The covalent bonding of pharmaceuticals to modified nanocellulose facilitates the formation of a very stable complex, enabling progressive release via hydrolysis or enzyme-mediated cleavage, ensuring prolonged therapeutic benefits [65, 66].

6.3.4 Fabrication of Polymer-Based MNs

Different types of materials, such as silicon, metals and polymers, are commonly used in MNs fabrication. Silicon is flexible yet susceptible to breaking, requiring a sterile environment and incurring substantial manufacturing costs. Metals are more economical; nonetheless, they generate waste and biohazardous substances. Polymers have high viscosity, resistance to fracture, and are predominantly biocompatible, making them ideal for economical mass production, particularly in the fabrication of MNs for biomedical applications. Numerous biocompatible materials have been engineered for MNs manufacturing, including carboxymethyl cellulose [67, 68]. However, carboxymethyl cellulose lacks special structural features; it simply operates as a polymer [69]. In addition to cellulose, several cellulose derivatives are used in the manufacturing of MNs. Due to its nanoscale properties, nanocellulose may effortlessly amalgamate with other biocompatible matrices, producing a robust composite ideal for MNs applications. Nanocellulose exhibits a distinctive array of properties that render it very appropriate for drug delivery purposes. The amalgamation characteristics renders nanocellulose an optimal material for the fabrication of dissolvable or biodegradable MNs that facilitate efficient skin penetration, improved drug loading, and regulated release. Nanocellulose-based MNs possess the capability to overcome the constraints of conventional topical delivery systems, hence broadening the spectrum of drugs that can be delivered trans-dermally and enhancing patient outcomes.

Drawing lithography and micro-molding are the most common techniques applied to polymer-based MN production. A schematic diagram showing the micro-molding and drawing lithography methods of MNs fabrication in Fig. 6.2. In micro-molding, material initially shaped with or without a mold and subsequently solidified through external stimuli. Drawing lithography is an easier technique for fabricating MNs without requiring the use of molds. The polymer was initially placed on a temperature-regulated lower substrate. The upper substrate, including a series of cylinders, is in contact with the polymer-infused lower substrate. The upper component is elevated to elongate the polymer, resulting in the formation of MNs structures that can be cemented using heat [71], electric field [72], magnetic field [73], or centrifugal force [74]. The shape of MNs can be solidified through crosslinking via heat or ultraviolet light when a mold is employed [75]. Presently, the three-dimensional (3D) printing method is applied for the fabrication of MNs. In this technology, computer-aided design software facilitated virtual design, after which the virtual model was transferred to a 3D printer, which ultimately enabled the high-throughput production of physical items with exceptional precision and reproducibility [76]. The solvent casting process is another predominant technique for fabricating polymer MNs. This technique acquired an inverse mold from the MNs structure, applying a polymer formulation onto it, allowing it to cure, and subsequently detaching it from the inverse mold. Dissolving or hydrogel MNs are produced via the solvent casting technique with several polymers, including hydroxypropyl methylcellulose [77].

6.4 Application of Nanocellulose MNs in Transdermal Drug Delivery

The US Food and Drug Administration (FDA) approved the use of cellulose in the drugs and several other products [78]. The unique properties of cellulose make it an important material among researchers for drug delivery applications in recent years [79]. The potential for chemical modifications of nanocellulose renders it a unique material for binding and releasing diverse drug substances. Cellulose nanofibers (CNFs), cellulose nanocrystals (CNCs), and bacterial cellulose (BC) exhibit notable similarities, yet distinct properties render each kind of nanocellulose more appropriate for specific drug delivery systems. Furthermore, the drug release duration of nanocellulose-based systems ranges from a few minutes to several days or months. In

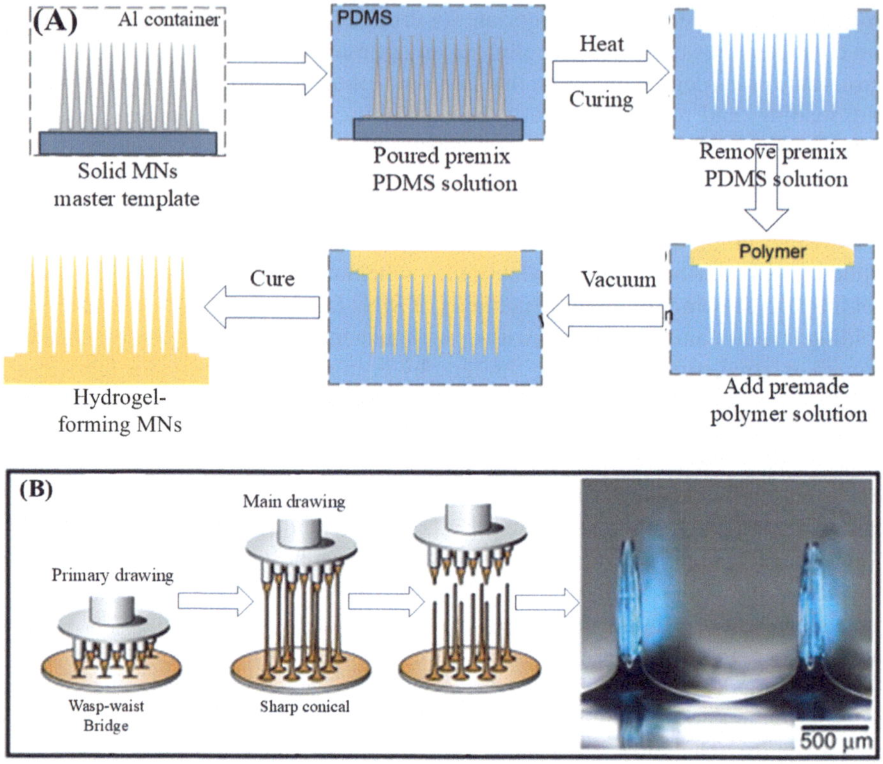

Fig. 6.2 **a** Schematic diagram showing the micro-molding method for the production of hydrogel-forming MNs [70]. Copyright 2021, Elsevier and **b** Drawing lithography for the fabrication of sharp-conical cone shapes dissolved MNs [36]. Copyright 2019 Wiley and Son

the last decade, nanocellulose has gained significant popularity and has been utilized in numerous drug delivery studies [80, 81]. A MNs array composed of hyaluronic acid (HA) utilized a bacterial nanocellulose membrane loaded with the bioactive molecule rutin as its backing layer (Fig. 6.3a). The MNs were designed to efficiently permeate the skin and decompose the HA matrix, thus releasing the active molecules that supported the HA MNs from the BNC membrane (Fig. 6.3b). This work illustrates the effective development of a multifunctional, dual-action MNs patch utilizing two biopolymers to incorporate active compounds [82]. Cellulose and various polymer nanocomposite materials, as well as methods of MN fabrication and drug loading, are given in Table 6.1.

In a study, Kim et al. exploited donepezil hydrochloride (DPH) enclosed in dissolving MNs for the treatment of Alzheimer's disease. They encapsulated a high amount of DPH in the tips of MNs for accelerated transdermal administration. Micro-molding technique was utilized for making MNs. They employed hydroxypropyl methylcellulose (HPMC)-ethanol/water mixture for the tips and carboxymethyl cellulose (CMC)-water for the MNs bases. Because of the distribution of DPH in

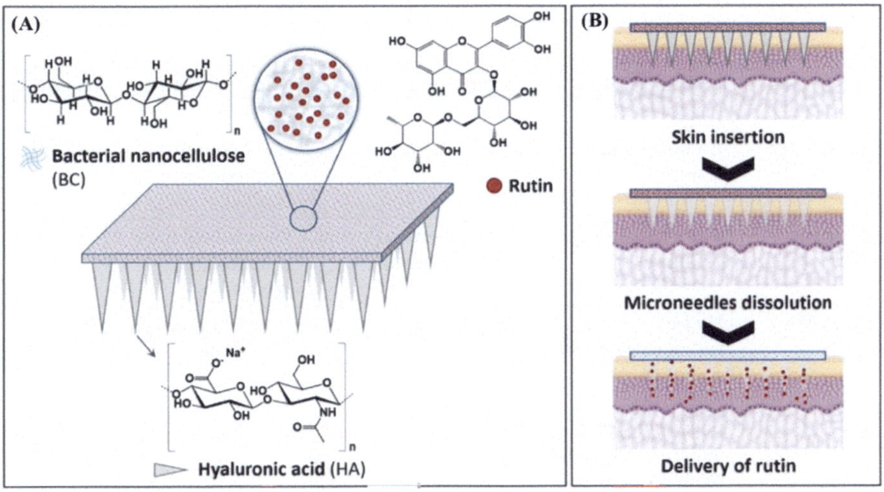

Fig. 6.3 **a** Schematic diagram showing hyaluronic acid-(Rutin loaded bacterial nanocellulose) MNs structure and **b** of the process of hyaluronic acid-(Rutin loaded bacterial nanocellulose) MNs insertion and rutin release [82]. Copyright 2021, Elsevier

the tips, 95% of the DPH was administered into pig skin after 5 min of insertion [83]. Another study demonstrated a significant enhancement in drug delivery with dissolvable CMC-based MNs. Laser ablation for molding markedly enhanced efficiency and decreased the cost of MNs synthesis. The mechanical characteristics and regulated drug release were enhanced by combining amylopectin with CMC. They used fluorescent molecules as model drugs to apply on pig skin. They exhibited a threefold increase in permeability, while the antioxidant properties of ascorbic acid encapsulated in the MNs demonstrated approximately a sixfold enhancement compared to direct application [84]. Applying hydrogel-formed MNs was shown to improve transdermal permeability and manage drug release. No barrier existed to deliver the hydrogel-MNs for the administration of medicine into the dermis. Lidocaine hydrochloride (LidH) was formed in a sodium carboxymethyl cellulose/gelatin hydrogel. The 'poke and patch' MNs administration technique was employed to augment the permeation flow of LidH. The microparticles were generated through electrostatic interactions between NaCMC and GEL macromolecules in a water/oil emulsion within paraffin oil, with covalent crosslinking facilitated by glutaraldehyde. The LidH was encapsulated, followed by an in vitro skin permeation assessment. The NaCMC/GEL drug formulation surpassed the minimal therapeutic drug threshold through MNs skin permeation in under 70 min [85].

Among different kinds of MNs, dissolving MNs received significant interest in transdermal drug delivery techniques. However, inadequate drug loading poses difficulty due to their small size. Song et al. used a single step to synthesize semi-dissolving MNs (SDMN) patches, which enable effective transdermal delivery without loading drugs directly [86]. The semi-dissolving microneedle patch was

Table 6.1 Summary of the cellulose based polymeric materials for drugs/molecules delivery and techniques for MNs fabrications

Types of MNs	MNs fabrication materials	Drugs/molecules loaded	MNs fabrication technology	References
Coated MNs	Carboxymethyl cellulose	VLP vaccine	Laser cutting and electropolishing method	[95]
Dissolving MNs	Carboxymethyl cellulose	Amylopectin, Bovine serum albumin	Photolithography and molding process	[93]
Coated MNs	Hydroxyethyl cellulose, Sodium carboxymethyl cellulose, polyvinylpyrrolidone and hydroxypropyl methylcellulose	Riboflavin	Isotropic-reactive ion etching process	[96]
Dissolving MNs	Carboxymethyl cellulose	Recombinant coronavirus vaccines	Casting-moulding	[97]
Dissolving MNs	Carboxymethyl cellulose, polyvinylpyrrolidone and hyaluronic acid	Lysozyme	Droplet-born air blowing method	[98]
Dissolving/ biodegradable MNs	Carboxymethyl cellulose	Calcitonin gene related peptide antagonist	Two-step centrifugation casting method	[99]
Dissolving MNs	Hydroxypropyl methylcellulose/ carboxymethyl cellulose	Donepezil hydrochloride	Micro-molding	[83]
Dissolving MNs	Carboxymethyl cellulose/ Amylopectin	Fluorescent molecules	Laser ablation	[84]
Hydrogel-MNs	Sodium carboxymethyl cellulose/gelatin hydrogel	Lidocaine hydrochloride	Electrostatic interactions between NaCMC and GEL	[85]
Semi-dissolving MNs	TEMPO-oxidized bacterial cellulose	Rhodamine B and fluorescein	Micro-molding	[86]
Dissolving MNs	Carboxymethyl cellulose/polystyrene	Rhodamine B/Insulin	–	[87]
Dissolving MNs	Carboxymethyl cellulose	Rhodamine/ascorbic acid	Laser writing and replica molding	[84]
Dissolving MNs	Carboxymethyl cellulose/hyaluronic acid	Rutin	Molding and solvent casting	[82]

(continued)

Table 6.1 (continued)

Types of MNs	MNs fabrication materials	Drugs/molecules loaded	MNs fabrication technology	References
Dissolving MNs	Polyvinylpyrrolidone (PVP)/cellulose	–	Curing and photolithography	[89]
Dissolving MNs	Gelatin/CMC	–	Centrifuge-casting method	[90]
Dissolving MNs	Dextran/oxidized cellulose	Timolol maleate/doxorubicin	Micro-molding	[91]
Dissolving MNs	Gelatin and sodium carboxymethyl cellulose	Rhodamine 6G and Fluorescein isothiocyanate-labeled insulin	Casting and centrifugation technique	[92]
Dissolving MNs	Carboxymethyl cellulose or amylopectin	Sulfo-rhodamine B, bovine serum albumin, and lysozyme	Casting centrifugation	[93]
Dissolving MNs	Carboxymethyl cellulose/polyvinylpyrrolidone	Amphotericin	Two-step casting process	[94]

produced through the micro-molding of TEMPO-oxidized bacterial cellulose and hydrophilic biopolymer mixture in a two-layer structure that includes a water-soluble needle layer and an insoluble supporting layer composed of TEMPO-oxidized bacterial cellulose. Furthermore, the SDMN patch, with an aperture in the backing layer for the homogeneous distribution of TEMPO-oxidized bacterial cellulose MNs provides significant benefits for the administration of substantial amounts of active substances. The needle layer concurrently dissolved in body fluids and the drug solution, forming micro-channels that facilitated the delivery of high amount of drugs to the skin than solution application alone. A different technique called a film-trigger applicator system for dissolved MNs was applied and compared to a dissolved MNs patch, indicating that the film can function as an applicator. The film-trigger applicator system comprises dissolved MNs produced on a CMC film, a holder for the dissolved MNs array, including aligned holes, and micropillars constructed from polystyrene. The film-trigger applicator method, including a basic polymer film and micropillars, demonstrated a superior distribution of dissolved MNs compared to the current dissolved MNs patch system. The film-trigger applicator operates solely with finger pressure, eliminating the need for adhesive patches or other devices. The developed film-trigger applicator system addresses the shortcomings of traditional microneedle patch delivery systems [87].

The transdermal route for drug administration offers several benefits, including a substantial absorption area, reduced dose frequency, non-invasiveness, and the circumvention of first-pass metabolism. The outer layer of the skin serves as a physical barrier, allowing only specific molecules with appropriate physicochemical properties to permeate. A study demonstrated that the skin permeability to the molecules can be regulated by modifying the composition of MNs materials. Sodium hyaluronate and CMC were employed as structural components, while amylopectin was used to enhance the mechanical strength of the MNs. High CMC fraction in MNs fabrication improved the mechanical strength and skin permeability of rhodamine B, a model compound exploited in the study [88]. Yong-Hun Park and colleagues reported increased drug permeability in a separate study [84]. They developed MNs for transdermal administration using laser writing and replica molding techniques. A threefold increase in drug permeability was noticed. They reported that this fabrication technique is most appropriate for cosmetic items made with CMC-based MNs. Daniela F.S. Fonseca et al. constructed dissolved hyaluronic acid (HA) MNs integrated with bacterial nanocellulose (BC) MNs patches for dermo-cosmetic purposes. The HA- and BC-blended MNs displayed adequate mechanical strength, facilitating a regulated release of the drug molecule. The in vivo toxicity studies revealed that the fabricated MNs was safe and biocompatible [82].

Besides increasing the mechanical properties of MNs, nanocellulose plays a significant role in the controlled release of the active molecules at the site of applications. Dissolvable MNs patches based on polyvinylpyrrolidone (PVP) exhibit diverse dissolution characteristics when PVP is co-polymerized with cellulose materials. The MNs patches were produced using a combination of thermal curing and photolithography. Multiple types of pharmaceutical cellulose, including hydroxypropyl methylcellulose and methyl cellulose, have been examined as dissolving modifiers integrated in the MNs patches. The resulting MNs patches exhibited disintegration patterns varying from 45 min to 48 h. The dissolving rates differed according to the levels of cellulose components. In addition to dissolving analysis, MNs patches were evaluated for their mechanical strength, moisture absorption, and skin penetration efficacy. All the MNs patches successfully penetrated human skin in-vitro. The PVP MNs patches exhibit significant potential for dermatological applications as drug carriers with adjustable dissolving characteristics [89]. Another study revealed a significant mechanical strength with the incorporation of CMC. Kamal et al. showed that the characteristics of gelatin-based dissolving MNs have markedly enhanced with CMC, obtained from kenaf bast fiber via alkaline treatment and esterification. MNs were fabricated utilizing the centrifuge-casting process. Gelatin/CMC MNs had a mechanical strength of 11.2 N. The blend of gelatin CMC holds significant potential for drug delivery via MNs, owing to the biodegradability and enhancement of the mechanical qualities of gelatin-based drug delivery systems [90].

A diverse array of drugs is delivered by dissolving MNs and the spectrum of transported molecules is extensive, encompassing both low molecular weight medicines and high molecular proteins and vaccines. Delivery of small to large molecular weight molecules using MNs through the skin can be employed in the treatment of numerous diseases. Jiang et al. produced dissolved MNs from dextran to treat hemangiomas, a superficial tumor marked by thick vascular structures that frequently impact the patient's aesthetic appearance due to a red discolouration on the skin. Timolol maleate is a medicine used in the treatment of hemangiomas, administered either as eye drops or in hydrogel form. Nonetheless, in both instances, there was a reduced drug delivery rate and an extended treatment duration. Dissolving dextran MNs was used for the administration of timolol maleate, while oxidized cellulose was integrated into the dextran MNs to promote thrombosis in the vascular-dense regions of hemangiomas. This dissolving MNs patch, composed of dextran and oxidized cellulose, can pierce the skin to a depth of 430 μm and dissolves swiftly within 3 min, facilitating direct drug delivery to the subcutaneous layer. The treated skin region restored its original appearance within two hours post-treatment. Besides superior skin permeability and swift breakdown, these patches markedly enhanced apoptosis and suppressed cell migration in mouse hemangioendothelioma EOMA cells [91]. Furthermore, Lee and colleagues employed a two-step casting and centrifugation technique to produce two-layer dissolving polymeric microneedle patches employing gelatin and sodium CMC. They applied confocal microscopy and demonstrated that rhodamine 6G and fluorescein isothiocyanate-labeled insulin (insulin-FITC) were progressively translocated from the puncture sites to deeper tissues. At the specified time, 50% of the insulin was released and permeated through the skin after 1 h, with cumulative penetration reaching 80% after 5 h. The cumulative area under the glucose level versus time curve, indicative of the hypoglycemic effect, was 128.4 ± 28.3 (% h) at a dosage of 0.25 IU/kg, while the relative pharmacologic availability and relative bioavailability (RBA) of insulin from MNs patches were 95.6% and 85.7%, respectively [92]. Dissolving MNs is designed to encapsulate large molecules, such as proteins and vaccines, enabling bolus or sustained delivery to the skin without biohazardous sharp medical waste. Microfabrication technology was applied to produce MNs. The process involved the casting of a viscous aqueous solution amid centrifugation to fill a micro-fabricated mold with biocompatible formulations of CMC or amylopectin. The authors analyzed encapsulated sulfo-rhodamine B, bovine serum albumin, and lysozyme and showed that the lysozyme retains complete enzymatic activity post-encapsulation and maintains 96% activity after 2 months of storage at room temperature. MNs demonstrated sufficient strength to penetrate cadaver skin and thereafter dissolve within minutes [93]. Zare et al. exploited dissolving CMC/polyvinylpyrrolidone MNs arrays for the transdermal administration of the antibiotic amphotericin B for the treatment of cutaneous leishmaniasis. This research showed that the MNs, upon insertion into the skin, rapidly disintegrated to release the encapsulated drugs, and the resultant micropores in the skin were rapidly resealed within 30 min. MNs patches demonstrated non-toxicity when exposed to the HT-29 cell line [94]. They exhibited significant *in-vitro* leishmanicidal activity of AMB-loaded MN patches against Leishmania parasites, resulting in up to 86% parasite mortality.

6.5 Conclusions

The number of research studies on nanocellulose in drug delivery and methods for nanocellulose modifications has increased over the past decade. This chapter is intended to outline the use of polymer-based MNs in transdermal drug delivery applications. Numerous formulations of nanocellulose-based medicines can be administered via multiple routes and have displayed controlled and often sustained drug release ascribed to the incorporation of nanocellulose. Furthermore, nanocellulose exhibits significant promise for additional research and enhanced application in controlled drug delivery systems.

Cellulose-based composite hydrogels provide significant benefits for the transdermal delivery of various drugs. Some of these involve rapid and straightforward drug administration, together with prolonged therapy via specific engineered controlled release of the drug, which is seen as a considerable benefit in topical administration. Consequently, cellulose-based hydrogels serve as an eco-friendly substitute for synthetic materials, exhibiting advantageous features that provide them a compelling choice for bioengineering applications and drug delivery systems. Despite the fact that most of the MN-based drug delivery studies remain in preclinical phases and safety concerns persist, it is expected that advancements in the research domain will foster belief for the imminent clinical deployment of these devices.

References

1. Alternative routes of drug administration—Advantages and disadvantages (subject review). American Academy of Pediatrics. Committee on Drugs, Pediatrics **100**(1), 143–52 (1997)
2. Ratnaparkhi, M., Gupta Jyoti, P.: Sustained release oral drug delivery system-an overview. Terminology **3**(4), 10–22270 (2013)
3. Kermode, M.: Unsafe injections in low-income country health settings: need for injection safety promotion to prevent the spread of blood-borne viruses. Health Promot. Int. **19**(1), 95–103 (2004)
4. Dhiman, S., Singh, T.G., Rehni, A.K.: Transdermal patches: a recent approach to new drug delivery system. Int. J. Pharm. Pharm. Sci. **3**(5), 26–34 (2011)
5. Sabbagh, F., Kim, B.S.: Recent advances in polymeric transdermal drug delivery systems. J. Control. Release **341**, 132–146 (2022)
6. Prausnitz, M.R., Elias, P.M., Franz, T.J., Schmuth, M., Tsai, J.-C., Menon, G.K., Holleran, W.M., Feingold, K.R.: Skin barrier and transdermal drug delivery. Dermatology **3**(18), 2065–2073 (2012)
7. Prausnitz, M.R.: Microneedles for transdermal drug delivery. Adv. Drug Deliv. Rev. **56**(5), 581–587 (2004)
8. Prausnitz, M.R., Mitragotri, S., Langer, R.: Current status and future potential of transdermal drug delivery. Nat. Rev. Drug Discov. **3**(2), 115–124 (2004)
9. Donnelly, R., Majithiya, R., Singh, T., Morrow, D., Garland, M., Demir, Y.K., Migalska, K., Ryan, E., Gillen, D., Scott, C., Woolfson, A.: Design, optimization and characterisation of polymeric microneedle arrays prepared by a novel laser-based micromoulding technique. Pharm. Res. **28**, 41–57 (2011)
10. More, S., Ghadge, T., Dhole, S.: Microneedle: an advanced technique in transdermal drug delivery system. Asian J. Res. Pharm. Sci. **3**(3), 141–148 (2013)
11. Henry, S., McAllister, D.V., Allen, M.G., Prausnitz, M.R.: Microfabricated microneedles: a novel approach to transdermal drug delivery. J. Pharm. Sci. **87**(8), 922–925 (1998)
12. Hoang, M.T., Ita, K.B., Bair, D.A.: Solid microneedles for transdermal delivery of amantadine hydrochloride and pramipexole dihydrochloride. Pharmaceutics **7**(4), 379–396 (2015)
13. Bhatnagar, S., Gadeela, P.R., Thathireddy, P., Venuganti, V.V.K.: Microneedle-based drug delivery: materials of construction. J. Chem. Sci. **131**(9), 90 (2019)
14. Sivamani, R.K., Liepmann, D., Maibach, H.I.: Microneedles and transdermal applications. Expert Opin. Drug Deliv. **4**(1), 19–25 (2007)

15. Sabri, A., Ogilvie, J., McKenna, J., Segal, J., Scurr, D., Marlow, M.: Intradermal delivery of an immunomodulator for basal cell carcinoma; expanding the mechanistic insight into solid microneedle-enhanced delivery of hydrophobic molecules. Mol. Pharm. **17**(8), 2925–2937 (2020)
16. Donnelly, R.F., Moffatt, K., Alkilani, A.Z., Vicente-Pérez, E.M., Barry, J., McCrudden, M.T., Woolfson, A.D.: Hydrogel-forming microneedle arrays can be effectively inserted in skin by self-application: a pilot study centred on pharmacist intervention and a patient information leaflet. Pharm. Res. **31**, 1989–1999 (2014)
17. Ghosh, P., Brogden, N.K., Stinchcomb, A.L.: Fluvastatin as a micropore lifetime enhancer for sustained delivery across microneedle-treated skin. J. Pharm. Sci. **103**(2), 652–660 (2014)
18. Caudill, C.L., Perry, J.L., Tian, S., Luft, J.C., DeSimone, J.M.: Spatially controlled coating of continuous liquid interface production microneedles for transdermal protein delivery. J. Control. Release **284**, 122–132 (2018)
19. Tuan-Mahmood, T.-M., McCrudden, M.T., Torrisi, B.M., McAlister, E., Garland, M.J., Singh, T.R.R., Donnelly, R.F.: Microneedles for intradermal and transdermal drug delivery. Eur. J. Pharm. Sci. **50**(5), 623–637 (2013)
20. Chong, R.H., Gonzalez-Gonzalez, E., Lara, M.F., Speaker, T.J., Contag, C.H., Kaspar, R.L., Coulman, S.A., Hargest, R., Birchall, J.C.: Gene silencing following siRNA delivery to skin via coated steel microneedles: in vitro and in vivo proof-of-concept. J. Control. Release **166**(3), 211–219 (2013)
21. Chege, M., McConville, A., Davis, J.: Microneedle drug delivery systems: appraising opportunities for improving safety and assessing areas of concern. J. Chem. Health Saf. **24**(2), 6–14 (2017)
22. Shakya, A.K., Ingrole, R.S., Joshi, G., Uddin, M.J., Anvari, S., Davis, C.M., Gill, H.S.: Microneedles coated with peanut allergen enable desensitization of peanut sensitized mice. J. Control. Release **314**, 38–47 (2019)
23. Waghule, T., Singhvi, G., Dubey, S.K., Pandey, M.M., Gupta, G., Singh, M., Dua, K.: Microneedles: a smart approach and increasing potential for transdermal drug delivery system. Biomed. Pharmacother. **109**, 1249–1258 (2019)
24. Kim, Y.-C., Park, J.-H., Prausnitz, M.R.: Microneedles for drug and vaccine delivery. Adv. Drug Deliv. Rev. **64**(14), 1547–1568 (2012)
25. Sullivan, S.P., Koutsonanos, D.G., del Pilar Martin, M., Lee, J.W., Zarnitsyn, V., Choi, S.-O., Murthy, N., Compans, R.W., Skountzou, I., Prausnitz, M.R.: Dissolving polymer microneedle patches for influenza vaccination. Nat. Med. **16**(8), 915–920 (2010)
26. Arya, J., Henry, S., Kalluri, H., McAllister, D.V., Pewin, W.P., Prausnitz, M.R.: Tolerability, usability and acceptability of dissolving microneedle patch administration in human subjects. Biomaterials **128**, 1–7 (2017)
27. Rejinold, N.S., Shin, J.-H., Seok, H.Y., Kim, Y.-C.: Biomedical applications of microneedles in therapeutics: recent advancements and implications in drug delivery. Expert Opin. Drug Deliv. **13**(1), 109–131 (2016)
28. Miyano, T., Tobinaga, Y., Kanno, T., Matsuzaki, Y., Takeda, H., Wakui, M., Hanada, K.: Sugar micro needles as transdermic drug delivery system. Biomed. Microdev. **7**(3), 185–188 (2005)
29. Garland, M.J., Caffarel-Salvador, E., Migalska, K., Woolfson, A.D., Donnelly, R.F.: Dissolving polymeric microneedle arrays for electrically assisted transdermal drug delivery. J. Control. Release **159**(1), 52–59 (2012)
30. Prausnitz, M.R.: Engineering microneedle patches for vaccination and drug delivery to skin. Annu. Rev. Chem. Biomol. Eng. **8**(1), 177–200 (2017)
31. Verbaan, F., Bal, S., Van den Berg, D., Dijksman, J., Van Hecke, M., Verpoorten, H., Van Den Berg, A., Luttge, R., Bouwstra, J.: Improved piercing of microneedle arrays in dermatomed human skin by an impact insertion method. J. Control. Release **128**(1), 80–88 (2008)
32. Larrañeta, E., Lutton, R.E.M., Woolfson, A.D., Donnelly, R.F.: Microneedle arrays as transdermal and intradermal drug delivery systems: materials science, manufacture and commercial development. Mater. Sci. Eng. R. Rep. **104**, 1–32 (2016)

33. He, R., Niu, Y., Li, Z., Li, A., Yang, H., Xu, F., Li, F.: A hydrogel microneedle patch for point-of-care testing based on skin interstitial fluid. Adv. Healthc. Mater. **9**(4), 1901201 (2020)
34. Courtenay, A.J., McAlister, E., McCrudden, M.T., Vora, L., Steiner, L., Levin, G., Levy-Nissenbaum, E., Shterman, N., Kearney, M.-C., McCarthy, H.O.: Hydrogel-forming microneedle arrays as a therapeutic option for transdermal esketamine delivery. J. Control. Release **322**, 177–186 (2020)
35. Donnelly, R.F., McCrudden, M.T.C., Zaid Alkilani, A., Larrañeta, E., McAlister, E., Courtenay, A.J., Kearney, M.-C., Singh, T.R.R., McCarthy, H.O., Kett, V.L., Caffarel-Salvador, E., Al-Zahrani, S., Woolfson, A.D.: Hydrogel-forming microneedles prepared from "Super Swelling" polymers combined with lyophilised wafers for transdermal drug delivery, Plos One **9**(10), e111547 (2014)
36. Lee, K., Lee, C.Y., Jung, H.: Dissolving microneedles for transdermal drug administration prepared by stepwise controlled drawing of maltose. Biomaterials 32(11), 3134–3140 (2011)
37. Aung, N.N., Ngawhirunpat, T., Rojanarata, T., Patrojanasophon, P., Pamornpathomkul, B., Opanasopit, P.: Fabrication, characterization and comparison of α-arbutin loaded dissolving and hydrogel forming microneedles. Int. J. Pharm. **586**, 119508 (2020)
38. Jamaledin, R., Yiu, C.K.Y., Zare, E.N., Niu, L.-N., Vecchione, R., Chen, G., Gu, Z., Tay, F.R., Makvandi, P.: Advances in antimicrobial microneedle patches for combating infections. Adv. Mater. **32**(33), 2002129 (2020)
39. Migdadi, E.M., Courtenay, A.J., Tekko, I.A., McCrudden, M.T.C., Kearney, M.-C., McAlister, E., McCarthy, H.O., Donnelly, R.F.: Hydrogel-forming microneedles enhance transdermal delivery of metformin hydrochloride. J. Control. Release **285**, 142–151 (2018)
40. Seddiqi, H., Oliaei, E., Honarkar, H., Jin, J., Geonzon, L.C., Bacabac, R.G., Klein-Nulend, J.: Cellulose and its derivatives: towards biomedical applications. Cellulose **28**(4), 1893–1931 (2021)
41. Ragauskas, A.J., Williams, C.K., Davison, B.H., Britovsek, G., Cairney, J., Eckert, C.A., Frederick Jr, W.J., Hallett, J.P., Leak, D.J., Liotta, C.L.: The path forward for biofuels and biomaterials. Science **311**(5760), 484–489 (2006)
42. Heinze, T.: Cellulose: structure and Properties, Cellulose Chemistry and Properties: fibers, Nanocelluloses and Advanced Materials, pp. 1–52 (2016)
43. Khan, M.J., Karim, Z., Charnnok, B., Poonsawat, T., Posoknistakul, P., Laosiripojana, N., Wu, K.C.-W., Sakdaronnarong, C.: Fabrication and characterization of functional biobased membranes from postconsumer cotton fabrics and palm waste for the removal of dyes. Int. J. Mol. Sci. **24**(7), 6030 (2023)
44. Barbash, V.A., Yashchenko, O.V., Gondovska, A.S., Deykun, I.M.: Preparation and characterization of nanocellulose obtained by TEMPO-mediated oxidation of organosolv pulp from reed stalks. Appl. Nanosci. **12**(4), 835–848 (2022)
45. Kargupta, W., Seifert, R., Martinez, M., Olson, J., Tanner, J., Batchelor, W.: Sustainable production process of mechanically prepared nanocellulose from hardwood and softwood: a comparative investigation of refining energy consumption at laboratory and pilot scale. Ind. Crops Prod. **171**, 113868 (2021)
46. Sonyeam, J., Chaipanya, R., Suksomboon, S., Khan, M.J., Amatariyakul, K., Wibowo, A., Posoknistakul, P., Charnnok, B., Liu, C.G., Laosiripojana, N., Sakdaronnarong, C.: Process design for acidic and alcohol based deep eutectic solvent pretreatment and high pressure homogenization of palm bunches for nanocellulose production. Sci. Rep. **14**(1), 7550 (2024)
47. Ma, Y., Xia, Q., Liu, Y., Chen, W., Liu, S., Wang, Q., Liu, Y., Li, J., Yu, H.: Production of nanocellulose using hydrated deep eutectic solvent combined with ultrasonic treatment. ACS Omega **4**(5), 8539–8547 (2019)
48. Ribeiro, R.S.A., Pohlmann, B.C., Calado, V., Bojorge, N., Pereira, N., Jr.: Production of nanocellulose by enzymatic hydrolysis: trends and challenges. Eng. Life Sci. **19**(4), 279–291 (2019)
49. Wohlert, M., Benselfelt, T., Wågberg, L., Furó, I., Berglund, L.A., Wohlert, J.: Cellulose and the role of hydrogen bonds: not in charge of everything. Cellulose **29**(1), 1–23 (2022)

50. Randhawa, A., Dutta, S.D., Ganguly, K., Patil, T.V., Patel, D.K., Lim, K.-T.: A review of properties of nanocellulose, its synthesis, and potential in biomedical applications. Appl. Sci. (2022)
51. Abeer, M.M., Mohd Amin, M.C.I., Martin, C.: A review of bacterial cellulose-based drug delivery systems: their biochemistry, current approaches and future prospects. J. Pharm. Pharmacol. **66**(8), 1047–1061 (2014)
52. Abdul Khalil, H.P.S., Adnan, A.S., Yahya, E.B., Olaiya, N.G., Safrida, S., Hossain, M.S., Balakrishnan, V., Gopakumar, D.A., Abdullah, C.K., Oyekanmi, A.A., Pasquini, D.: A review on plant cellulose nanofibre-based aerogels for biomedical applications. Polymers **12**(8), 1759 (2020)
53. Pacheco, G., de Mello, C.V., Chiari-Andréo, B.G., Isaac, V.L.B., Ribeiro, S.J.L., Pecoraro, É., Trovatti, E.: Bacterial cellulose skin masks—Properties and sensory tests. J. Cosmet. Dermatol. **17**(5), 840–847 (2018)
54. Chen, W., Yu, H., Lee, S.-Y., Wei, T., Li, J., Fan, Z.: Nanocellulose: a promising nanomaterial for advanced electrochemical energy storage. Chem. Soc. Rev. **47**(8), 2837–2872 (2018)
55. Karimian, A., Parsian, H., Majidinia, M., Rahimi, M., Mir, S.M., Samadi Kafil, H., Shafiei-Irannejad, V., Kheyrollah, M., Ostadi, H., Yousefi, B.: Nanocrystalline cellulose: preparation, physicochemical properties, and applications in drug delivery systems. Int. J. Biol. Macromol. **133**, 850–859 (2019)
56. Osong, S.H., Norgren, S., Engstrand, P.: Processing of wood-based microfibrillated cellulose and nanofibrillated cellulose, and applications relating to papermaking: a review. Cellulose **23**(1), 93–123 (2016)
57. Jozala, A.F., de Lencastre-Novaes, L.C., Lopes, A.M., de Carvalho Santos-Ebinuma, V., Mazzola, P.G., Pessoa Jr, A., Grotto, D., Gerenutti, M., Chaud, M.V.: Bacterial nanocellulose production and application: a 10-year overview. Appl. Microbiol. Biotechnol. **100**(5), 2063–2072 (2016)
58. Li, M., He, B., Chen, Y., Zhao, L.: Physicochemical properties of nanocellulose isolated from cotton stalk waste. ACS Omega **6**(39), 25162–25169 (2021)
59. Thakur, V., Guleria, A., Kumar, S., Sharma, S., Singh, K.: Recent advances in nanocellulose processing, functionalization and applications: a review. Mater. Adv. **2**(6), 1872–1895 (2021)
60. Sampath Udeni Gunathilake, T.M., Ching, Y.C., Chuah, C.H., Rahman, N.A., Liou, N.-S.: Recent advances in celluloses and their hybrids for stimuli-responsive drug delivery. Int. J. Biol. Macromol. **158**, 670–688 (2020)
61. Gong, J., Hou, L., Ching, Y.C., Ching, K.Y., Hai, N.D., Chuah, C.H.: A review of recent advances of cellulose-based intelligent-responsive hydrogels as vehicles for controllable drug delivery system. Int. J. Biol. Macromol. **264**, 130525 (2024)
62. Qing, W., Wang, Y., Wang, Y., Zhao, D., Liu, X., Zhu, J.: The modified nanocrystalline cellulose for hydrophobic drug delivery. Appl. Surf. Sci. **366**, 404–409 (2016)
63. Ning, L., Jia, Y., Zhao, X., Tang, R., Wang, F., You, C.: Nanocellulose-based drug carriers: functional design, controllable synthesis, and therapeutic applications. Int. J. Biol. Macromol. **222**, 1500–1510 (2022)
64. Numata, Y., Kono, H., Tsuji, M., Tajima, K.: Structural and mechanical characterization of bacterial cellulose–polyethylene glycol diacrylate composite gels. Carbohyd. Polym. **173**, 67–76 (2017)
65. Coughlan, D.C., Corrigan, O.I.: Drug–polymer interactions and their effect on thermoresponsive poly(N-isopropylacrylamide) drug delivery systems. Int. J. Pharm. **313**(1), 163–174 (2006)
66. Srivastava, A., Yadav, T., Sharma, S., Nayak, A., Kumari, A.A., Mishra, N.: Polymers in drug delivery. J. Biosci. Med. **4**(1), 69–84 (2015)
67. Banga, A.K.: Microporation applications for enhancing drug delivery. Expert Opin. Drug Deliv. **6**(4), 343–354 (2009)
68. Ma, G., Wu, C.: Microneedle, bio-microneedle and bio-inspired microneedle: a review. J. Control. Release **251**, 11–23 (2017)

69. Lee, B.-M., Lee, C., Lahiji, S.F., Jung, U.-W., Chung, G., Jung, H.: Dissolving microneedles for rapid and painless local anesthesia. Pharmaceutics **12**(4), 366 (2020)
70. Turner, J.G., White, L.R., Estrela, P., Leese, H.S.: Hydrogel-forming microneedles: current advancements and future trends. Macromol. Biosci. **21** (2), 2000307 (2021)
71. Choi, C.K., Lee, K.J., Youn, Y.N., Jang, E.H., Kim, W., Min, B.-K., Ryu, W.: Spatially discrete thermal drawing of biodegradable microneedles for vascular drug delivery. Eur. J. Pharm. Biopharm. **83**(2), 224–233 (2013)
72. Vecchione, R., Coppola, S., Esposito, E., Casale, C., Vespini, V., Grilli, S., Ferraro, P., Netti, P.A.: Electro-drawn drug-loaded biodegradable polymer microneedles as a viable route to hypodermic injection. Adv. Func. Mater. **24**(23), 3515–3523 (2014)
73. Chen, Z., Ye, R., Yang, J., Lin, Y., Lee, W., Li, J., Ren, L., Liu, B., Jiang, L.: Rapidly fabricated microneedle arrays using magnetorheological drawing lithography for transdermal drug delivery. ACS Biomater. Sci. Eng. **5**(10), 5506–5513 (2019)
74. Yang, H., Kim, S., Kang, G., Lahiji, S.F., Jang, M., Kim, Y.M., Kim, J.-M., Cho, S.-N., Jung, H.: Centrifugal lithography: self-shaping of polymer microstructures encapsulating biopharmaceutics by centrifuging polymer drops. Adv. Healthc. Mater. **6**(19), 1700326 (2017)
75. Lin, Y.-H., Lee, I.C., Hsu, W.-C., Hsu, C.-H., Chang, K.-P., Gao, S.-S.: Rapid fabrication method of a microneedle mold with controllable needle height and width. Biomed. Microdev. **18**(5), 85 (2016)
76. Yadav, V., Sharma, P.K., Murty, U.S., Mohan, N.H., Thomas, R., Dwivedy, S.K., Banerjee, S.: 3D printed hollow microneedles array using stereolithography for efficient transdermal delivery of rifampicin. Int. J. Pharm. **605**, 120815 (2021)
77. Hwa, K.-Y., Chang, V.H.S., Cheng, Y.-Y., Wang, Y.-D., Jan, P.-S., Subramani, B., Wu, M.-J., Wang, B.-K.: Analyzing polymeric matrix for fabrication of a biodegradable microneedle array to enhance transdermal delivery. Biomed. Microdev. **19**(4), 84 (2017)
78. Tran, M., Wang, C.: Semi-solid materials for controlled release drug formulation: current status and future prospects. Front. Chem. Sci. Eng. **8**(2), 225–232 (2014)
79. Abitbol, T., Rivkin, A., Cao, Y., Nevo, Y., Abraham, E., Ben-Shalom, T., Lapidot, S., Shoseyov, O.: Nanocellulose, a tiny fiber with huge applications. Curr. Opin. Biotechnol. **39**, 76–88 (2016)
80. Islam, M.T., Alam, M.M., Zoccola, M.: Review on modification of nanocellulose for application in composites. Int. J. Innov. Res. Sci. Eng. Technol. **2**(10), 5444–5451 (2013)
81. Tan, T.H., Lee, H.V., Dabdawb, W.A.Y., Abd Hamid, S.B.B.O.: A review of nanocellulose in the drug-delivery system. Mater. Biomed. Eng. 131–164 (2019)
82. Fonseca, D.F.S., Vilela, C., Pinto, R.J.B., Bastos, V., Oliveira, H., Catarino, J., Faísca, P., Rosado, C., Silvestre, A.J.D., Freire, C.S.R.: Bacterial nanocellulose-hyaluronic acid microneedle patches for skin applications: in vitro and in vivo evaluation. Mater. Sci. Eng. C **118**, 111350 (2021)
83. Kim, J.-Y., Han, M.-R., Kim, Y.-H., Shin, S.-W., Nam, S.-Y., Park, J.-H.: Tip-loaded dissolving microneedles for transdermal delivery of donepezil hydrochloride for treatment of Alzheimer's disease. Eur. J. Pharm. Biopharm. **105**, 148–155 (2016)
84. Park, Y.-H., Ha, S.K., Choi, I., Kim, K.S., Park, J., Choi, N., Kim, B., Sung, J.H.: Fabrication of degradable carboxymethyl cellulose (CMC) microneedle with laser writing and replica molding process for enhancement of transdermal drug delivery. Biotechnol. Bioprocess Eng. **21**(1), 110–118 (2016)
85. Nayak, A., Das, D.B., Vladisavljević, G.T.: Microneedle-assisted permeation of lidocaine carboxymethylcellulose with gelatine co-polymer hydrogel. Pharm. Res. **31**(5), 1170–1184 (2014)
86. Song, J.E., Jun, S.-H., Park, S.-G., Kang, N.-G.: A semi-dissolving microneedle patch incorporating TEMPO-oxidized bacterial cellulose nanofibers for enhanced transdermal delivery. Polymers (2020)
87. Kim, Y., Min, H.S., Shin, J., Nam, J., Kang, G., Sim, J., Yang, H., Jung, H.: Film-trigger applicator (FTA) for improved skin penetration of microneedle using punching force of carboxymethyl cellulose film acting as a microneedle applicator. Biomater. Res. **26**(1), 53 (2022)

88. Park, Y., Kim, B.: Skin permeability of compounds loaded within dissolving microneedles dependent on composition of sodium hyaluronate and carboxymethyl cellulose. Korean J. Chem. Eng. **34**(1), 133–138 (2017)
89. Kathuria, H., Lim, D., Cai, J., Chung, B.G., Kang, L.: Microneedles with tunable dissolution rate. ACS Biomater. Sci. Eng. **6**(9), 5061–5068 (2020)
90. Kamal, N.A.M., Mahmood, T.M.T., Ahmad, I., Ramli, S.: Improving rate of gelatin/carboxymethylcellulose dissolving microneedle for transdermal drug delivery. Sains Malays **49**(9), 2269–2279 (2020)
91. Jiang, H., Guo, Y., Tan, X., Jiang, Y., Pang, N., Niu, C., Liu, L., Zhou, Z., Liu, L., Li, H.: Oxidized cellulose microneedle patch combined with vascular embolization and local delivery of timolol maleate for hemangiomas. Colloids Surf. B **244**, 114174 (2024)
92. Lee, I.C., Lin, W.-M., Shu, J.-C., Tsai, S.-W., Chen, C.-H., Tsai, M.-T.: Formulation of two-layer dissolving polymeric microneedle patches for insulin transdermal delivery in diabetic mice. J. Biomed. Mater. Res. Part A **105**(1), 84–93 (2017)
93. Lee, J.W., Park, J.-H., Prausnitz, M.R.: Dissolving microneedles for transdermal drug delivery. Biomaterials **29**(13), 2113–2124 (2008)
94. Zare, M.R., Khorram, M., Barzegar, S., Sarkari, B., Asgari, Q., Ahadian, S., Zomorodian, K.: Dissolvable carboxymethyl cellulose/polyvinylpyrrolidone microneedle arrays for transdermal delivery of Amphotericin B to treat cutaneous leishmaniasis. Int. J. Biol. Macromol. **182**, 1310–1321 (2021)
95. Kim, Y.C., Quan, F.S., Compans, R.W., Kang, S.M., Prausnitz, M.R.: Formulation of microneedles coated with influenza virus-like particle vaccine. AAPS PharmSciTech **11**(3), 1193–1201 (2010)
96. Chen, J., Qiu, Y., Zhang, S., Yang, G., Gao, Y.: Controllable coating of microneedles for transdermal drug delivery. Drug Dev. Ind. Pharm. **41**(3), 415–422 (2015)
97. Kim, E., Erdos, G., Huang, S., Kenniston, T.W., Balmert, S.C., Carey, C.D., Raj, V.S., Epperly, M.W., Klimstra, W.B., Haagmans, B.L., Korkmaz, E., Falo, L.D., Gambotto, A.: Microneedle array delivered recombinant coronavirus vaccines: immunogenicity and rapid translational development, eBioMedicine **55**, 102743 (2020)
98. Lahiji, S.F., Jang, Y., Ma, Y., Dangol, M., Yang, H., Jang, M., Jung, H.: Effects of dissolving microneedle fabrication parameters on the activity of encapsulated lysozyme. Eur. J. Pharm. Sci. **117**, 290–296 (2018)
99. Xie, X., Pascual, C., Lieu, C., Oh, S., Wang, J., Zou, B., Xie, J., Li, Z., Xie, J., Yeomans, D.C., Wu, M.X., Xie, X.S.: Analgesic microneedle patch for neuropathic pain therapy. ACS Nano **11**(1), 395–406 (2017)

Chapter 7
Polylactic Acid-Based Microneedles in Dermatology

Yu Ruixing

7.1 Introduction

Polylactic acid (PLA) is a biodegradable and biocompatible polymer extensively utilized in medical applications. Microneedles, which are minuscule needles capable of penetrating the skin's outermost layer without reaching deeper tissues, provide a minimally invasive method for the delivery of drugs and other therapeutic agents. PLA microneedles synergize the advantages of PLA's biodegradability with the precision afforded by microneedle technology, rendering them highly suitable for a wide range of dermatological applications.

7.2 Development of PLA Microneedles

The development of PLA microneedles is a multi-stage process encompassing material selection, fabrication, and optimization. PLA is chosen for its biocompatibility, mechanical strength, and ability to degrade into non-toxic lactic acid, making it a safe and effective material for medical applications [1]. The fabrication of PLA microneedles involves several advanced techniques to ensure precision and efficacy. Commonly used methods include molding, etching, and 3D printing, which allow for the production of microneedles with precise dimensions and intricate shapes [2, 3].

Y. Ruixing (✉)
Department of Dermatology, China-Japan Friendship Hospital, Beijing, China
e-mail: yuruixing521@163.com

N. Talreja et al. (eds.), *Microneedles (MNs)-Based Technology*, Series in BioEngineering, https://doi.org/10.1007/978-981-96-3916-8_7

7.2.1 Material Selection

Material selection is a crucial aspect in the development of PLA microneedles. PLA is favored due to its FDA approval, well-documented safety profile, and its degradation into lactic acid, a naturally occurring substance in the human body [4]. The safety and biocompatibility of PLA make it an ideal material for medical applications, reducing the risk of adverse reactions and complications. Moreover, PLA can be synthesized from renewable resources, enhancing its appeal as an environmentally friendly option. Utilizing renewable resources for PLA production supports sustainability and mitigates the environmental impact of medical waste.

7.2.2 Fabrication Techniques

The fabrication of PLA microneedles can be achieved through several methods, each offering distinct advantages:

Molding: This technique involves casting PLA into micro-molds to create the desired microneedle shape [5]. Molding is cost-effective and well-suited for mass production, allowing for the rapid and consistent manufacture of large quantities of microneedles. The molds can be designed to produce microneedles with specific dimensions and features, ensuring uniformity and reliability in the final product. Molding is particularly advantageous for applications requiring high-volume production and consistent quality.

Etching: Techniques such as laser etching allow for precise control over microneedle dimensions and are ideal for creating complex geometries [6]. Laser etching can produce microneedles with sharp tips and intricate designs that enhance penetration ability and drug delivery efficiency. This method is highly precise and can generate microneedles with fine details that are difficult to achieve with other fabrication techniques. Etching is suitable for applications that demand high precision and customized microneedle designs.

3D Printing: Advanced 3D printing technologies enable the fabrication of microneedles with high precision and customization capabilities [7]. 3D printing allows for the creation of microneedles with complex internal structures and varying material compositions, providing unparalleled design flexibility. This method is ideal for developing microneedles for specialized applications requiring unique features and functionalities. Additionally, 3D printing facilitates rapid prototyping and iterative design, making it a valuable tool for research and development.

7.2.3 Optimization

Optimization of PLA microneedles involves fine-tuning their mechanical properties to ensure adequate skin penetration without breaking. This process includes adjusting parameters such as microneedle height, base diameter, and tip sharpness [8, 9]. The mechanical properties can be tailored to match the specific requirements of the application, ensuring effective drug delivery and patient comfort. The biodegradation rates of PLA can be controlled by modifying its molecular weight or blending it with other biocompatible materials [10, 11]. This allows for precise regulation of the therapeutic agent's release profile, ensuring sustained and controlled delivery. Drug loading capacities are optimized by incorporating the drug into the PLA matrix or coating the microneedle surface with the drug [12], ensuring the microneedles deliver the desired amount of the therapeutic agent for effective treatment outcomes.

7.3 Mechanisms of Action

PLA microneedles function by creating microchannels in the skin, which facilitate the targeted delivery of therapeutic agents. These microchannels bypass the stratum corneum, the skin's outermost barrier, enabling efficient transport of drugs, vaccines, and other compounds directly into the deeper layers of the skin [13]. The biodegradability of PLA ensures that the microneedles gradually dissolve, releasing their payload over time and eliminating the need for manual removal, thereby minimizing the risk of infection and discomfort [14].

7.3.1 Microchannel Formation

The primary mechanism of action for microneedles is the formation of microchannels in the skin. When applied, PLA microneedles penetrate the stratum corneum, creating pathways that enable drug delivery. These microchannels, which can be as small as a few micrometers in diameter, are sufficiently large to permit the passage of therapeutic agents while being small enough to minimize discomfort and reduce the risk of infection [15]. This minimally invasive approach preserves the integrity of the deeper skin tissues while providing an effective route for drug administration.

The process begins with the application of a microneedle patch to the skin. The microneedles penetrate the stratum corneum and epidermis, reaching the dermal layer where capillaries and immune cells are abundant. This penetration creates microchannels through which drugs can be delivered directly into the bloodstream or to targeted cells. The depth of penetration and the size of the microchannels can be precisely controlled by adjusting the length and diameter of the microneedles, ensuring that therapeutic agents are delivered to the intended depth without causing significant damage to the surrounding tissues.

The creation of microchannels not only facilitates drug delivery but also stimulates a mild inflammatory response, which can enhance the uptake of therapeutic agents. These microchannels act as conduits, allowing drugs to bypass the dense network of cells and lipids that constitute the stratum corneum. This bypassing significantly improves the bioavailability of drugs that would otherwise struggle to penetrate the skin barrier. Additionally, the microchannels close naturally within a short period, further reducing the risk of infection and enhancing patient comfort.

7.3.2 Drug Release

PLA microneedles can be engineered for controlled drug release, ensuring sustained therapeutic effects. This controlled release can be achieved by encapsulating the drug within the PLA matrix or by coating the microneedles with a drug-containing layer [16]. As the PLA microneedles degrade, the encapsulated drug is gradually released, providing sustained delivery that can extend over several hours to days, depending on the formulation and the degradation rate of the PLA [17].

Encapsulation involves integrating the therapeutic agent directly into the PLA during the fabrication process, allowing for uniform distribution of the drug throughout the microneedle structure and ensuring a consistent release profile. The rate of drug release can be modulated by adjusting the molecular weight of the PLA, which influences the degradation rate. Higher molecular weight PLA degrades more slowly, thereby providing a prolonged release of the encapsulated drug. Alternatively, drugs can be coated onto the surface of PLA microneedles. This approach is beneficial for delivering higher doses of drugs in a shorter timeframe. The coating can be designed to dissolve rapidly upon insertion into the skin, releasing the drug in a burst-like manner, which is particularly useful for vaccines and other applications requiring a quick onset of action. The degradation of PLA is influenced by various factors, including environmental pH, temperature, and the presence of enzymes. PLA is hydrolyzed into lactic acid, a naturally occurring metabolite in the body. This hydrolysis process is biocompatible and results in the gradual dissolution of the microneedles, ensuring that no remnants are left in the skin. The kinetics of drug release can be finely tuned by altering the composition and structure of the PLA microneedles, providing flexibility in designing treatments for different medical conditions. Controlled drug release from PLA microneedles offers several advantages over traditional drug delivery methods. It reduces the frequency of drug administration, thereby improving patient compliance and convenience. Additionally, localized delivery minimizes systemic side effects, as the drug is concentrated at the site of action. This precision in drug delivery enhances the therapeutic efficacy and safety of treatments, making PLA microneedles a versatile tool in dermatology and other medical fields.

7.4 Applications in Dermatology

7.4.1 Drug Delivery

PLA microneedles are primarily used for transdermal drug delivery, offering a minimally invasive and efficient means of delivering therapeutic agents through the skin. They have been shown to enhance the delivery of various drugs, including small molecules, proteins, and vaccines [18]. This technology leverages the ability of microneedles to penetrate the stratum corneum, the outermost layer of the skin, which typically serves as a significant barrier to drug delivery. Studies have demonstrated the effectiveness of PLA microneedles in delivering local anesthetics for pain relief and vaccines for immunization [7, 19].

7.4.1.1 Local Anesthetics

PLA microneedles have demonstrated efficacy in delivering local anesthetics for pain management. This approach offers several advantages over traditional methods, including reduced pain at the application site and a faster onset of analgesia. The microneedles create microchannels in the skin that facilitate the rapid absorption of the anesthetic, providing quick relief. Research has shown the successful delivery of lidocaine and other local anesthetics using PLA microneedles, underscoring their potential for clinical use in procedures requiring local anesthesia [20, 21] (Fig. 7.1).

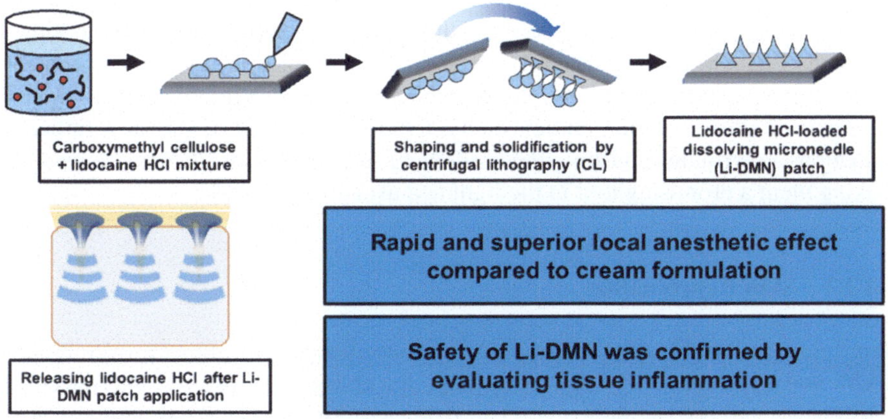

Fig. 7.1 Microneedle patch with carboxymethyl cellulose is a promising candidate method for the painless delivery of lidocaine-HCl. Image was reproduced with permission under common creative license [22]

7.4.1.2 Vaccines

The delivery of vaccines through PLA microneedles is an emerging area with significant potential. Microneedles can enhance the immune response by targeting antigen-presenting cells in the skin, leading to improved vaccine efficacy. This delivery method not only increases the immunogenicity of vaccines but also simplifies the administration process, making it less painful and more acceptable to patients, particularly children and those with needle phobia. Studies have demonstrated that PLA microneedles can effectively deliver various vaccines, including those for influenza and human papillomavirus (HPV), showcasing their versatility and effectiveness [23, 24].

7.4.2 Cosmetic Treatments

In the field of cosmetics, PLA microneedles are used for anti-aging treatments, skin rejuvenation, and scar reduction. These treatments benefit from the ability of microneedles to enhance the delivery of active ingredients such as hyaluronic acid, peptides, and growth factors, thereby improving skin texture and elasticity. The minimally invasive nature of microneedles makes them an attractive option for cosmetic applications, as they cause minimal discomfort and downtime.

7.4.2.1 Anti-Aging Treatments

Anti-aging treatments with PLA microneedles involve the delivery of compounds that stimulate collagen production and skin renewal. Hyaluronic acid, a popular ingredient in anti-aging products, can be effectively delivered using PLA microneedles, leading to improved skin hydration and elasticity. Clinical studies have shown significant reductions in wrinkles and fine lines following treatment with PLA microneedles, making them a promising tool in the fight against aging [25, 26].

7.4.2.2 Skin Rejuvenation

Skin rejuvenation therapies using PLA microneedles aim to improve overall skin health and appearance. By delivering peptides and growth factors, microneedles can enhance cell regeneration and repair damaged skin. Research has demonstrated that PLA microneedles can improve skin tone, texture, and firmness, providing a non-invasive alternative to more aggressive cosmetic procedures [27, 28].

7.4.2.3 Scar Reduction

Scar reduction treatments with PLA microneedles involve the delivery of agents that promote tissue remodeling and reduce scar formation. Studies have shown that microneedle-based delivery of transforming growth factor-beta (TGF-β) and other regenerative compounds can effectively reduce the appearance of scars. PLA microneedles provide a minimally invasive alternative to traditional scar treatments such as laser therapy and surgery, offering patients a less painful and more convenient option [29, 30] (Fig. 7.2).

7.4.3 Therapeutic Interventions

Beyond drug delivery and cosmetics, PLA microneedles have therapeutic applications in treating various skin conditions such as acne, hyperpigmentation, vitiligo, psoriasis, dermatitis and skin cancer. They enable the localized delivery of therapeutic agents, reducing systemic side effects and improving treatment efficacy. Additionally, PLA microneedles can be used in photodynamic therapy, where they deliver photosensitizing agents to target and destroy abnormal skin cells.

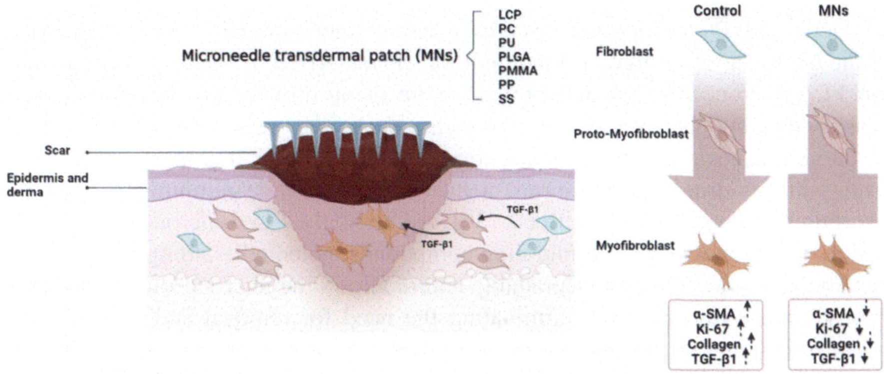

Fig. 7.2 Schematic illustration of MN treatment of a hypertrophic scar. The key scar biomarkers including collagen deposition, α-SMA and Ki-67 protein expression, and TGF-β1 mRNA expression were used to evaluate the efficacy of seven types of MNs for hypertrophic scars. Note: LCP: liquid-crystal polymers, PC: polycarbonates, PU: polyurethane, PLGA: polylactic-co-glycolic acid, PMMA: polymethylmethacrylate, PP: polypropylene, SS: stainless steel. The figure was reproduced with permission [31]

7.4.3.1 Acne

PLA microneedles can be used to deliver anti-acne medications directly to the affected areas, providing targeted treatment and reducing systemic side effects. Studies have shown that microneedle-assisted delivery of antibiotics and retinoids can effectively reduce acne lesions and inflammation. This approach offers a promising alternative to traditional topical and oral treatments, with the added benefit of improved patient compliance due to the reduced risk of systemic side effects [32, 33].

7.4.3.2 Hyperpigmentation

Hyperpigmentation treatments with PLA microneedles involve the delivery of skin-lightening agents such as hydroquinone and vitamin C. Microneedles enhance the penetration of these agents, leading to more effective reduction of dark spots and uneven skin tone. Clinical studies have demonstrated significant improvements in hyperpigmentation following treatment with PLA microneedles, making them a valuable tool in dermatological practice [34, 35].

7.4.3.3 Vitiligo

Polylactic acid (PLA) microneedles present a novel and minimally invasive approach for the treatment of vitiligo, a skin condition characterized by the loss of pigmentation. PLA microneedles can deliver therapeutic agents directly into the depigmented areas, enhancing the efficacy of treatments such as corticosteroids, calcineurin inhibitors, and novel agents like Janus kinase (JAK) inhibitors. The localized delivery ensures higher drug concentrations at the target site while minimizing systemic absorption and potential side effects [36]. Additionally, PLA microneedles can be used to deliver melanocyte-stimulating compounds, potentially aiding in the repigmentation process. Their biodegradable nature allows the microneedles to dissolve after administering the drug, eliminating the need for removal and reducing the risk of infection. Preliminary studies suggest that PLA microneedles could significantly improve the management of vitiligo, offering targeted treatment and improving patient outcomes [37, 38].

7.4.3.4 Psoriasis

Psoriasis treatment with PLA microneedles focuses on delivering anti-inflammatory and immunomodulatory agents directly to the affected skin. This localized delivery can reduce the severity of psoriatic plaques and improve overall skin condition. Research has shown that PLA microneedles can effectively deliver agents such as corticosteroids and biologics for psoriasis management, offering a less invasive

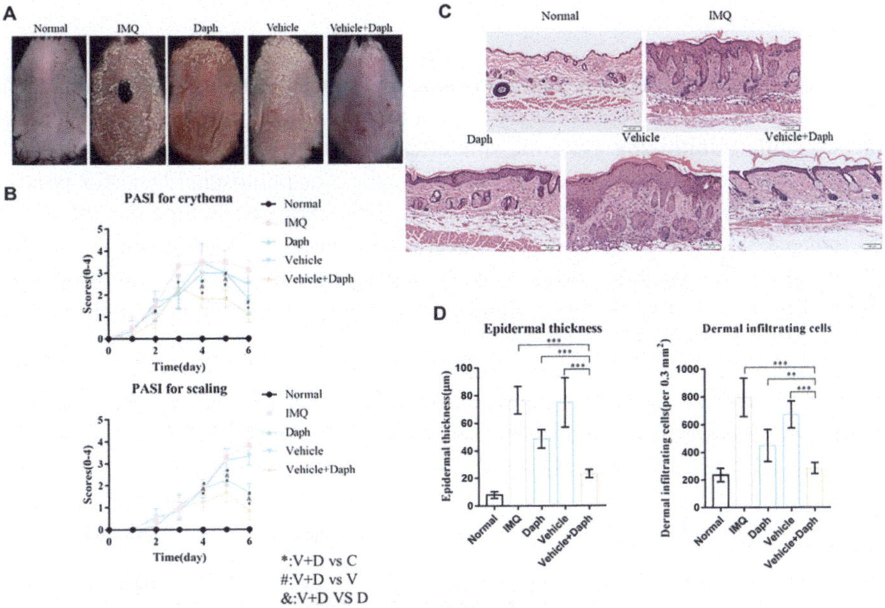

Fig. 7.3 Symptoms and histological results of psoriasis-like mice in different groups. **a** Photos of psoriasis modeling sites in different groups of mice. **b** H&E staining of psoriasis models in different groups of mice. **c** PASI scores of psoriatic mice in different groups. **d** Epidermal thickness and the number of inflammatory cells in each group. Data are shown as mean ± SE ($*p < 0.05$; $**p < 0.01$). The figure was reproduced with permission [39]

alternative to systemic treatments and reducing the risk of adverse effects [5, 10] (Fig. 7.3).

7.4.3.5 Dermatitis

PLA microneedles have shown promise in the treatment of dermatitis, a common inflammatory skin condition characterized by itching, redness, and swelling. PLA microneedles offer a minimally invasive method for delivering therapeutic agents directly to the affected skin layers, enhancing drug efficacy and reducing systemic side effects. These microneedles can be loaded with anti-inflammatory agents, such as corticosteroids and immunomodulators, to provide localized treatment and rapid relief from dermatitis symptoms. The biodegradable nature of PLA ensures that the microneedles dissolve after delivering the medication, eliminating the need for removal and reducing the risk of infection. Studies have demonstrated the potential of PLA microneedles in improving the management of dermatitis by achieving targeted drug delivery and minimizing the frequency of application compared to traditional topical therapies [2, 5, 10, 40].

7.4.3.6 Skin Cancer

PLA microneedles offer a minimally invasive and efficient method for the localized delivery of anticancer agents directly to skin tumors. This approach enhances drug efficacy while minimizing systemic side effects. PLA microneedles can be loaded with chemotherapeutic agents, immunotherapeutics, or photosensitizers for photodynamic therapy, providing targeted treatment that maximizes therapeutic outcomes and reduces systemic toxicity. The biodegradable nature of PLA ensures that the microneedles dissolve after drug delivery, eliminating the need for removal and minimizing the risk of infection. Studies have demonstrated the potential of PLA microneedles in effectively reducing tumor size and inhibiting cancer cell proliferation, highlighting their promise in skin cancer treatment. Ongoing research continues to explore their use in delivering innovative therapies, such as gene therapy, to further enhance treatment efficacy and patient outcomes [2, 5, 10] (Fig. 7.4).

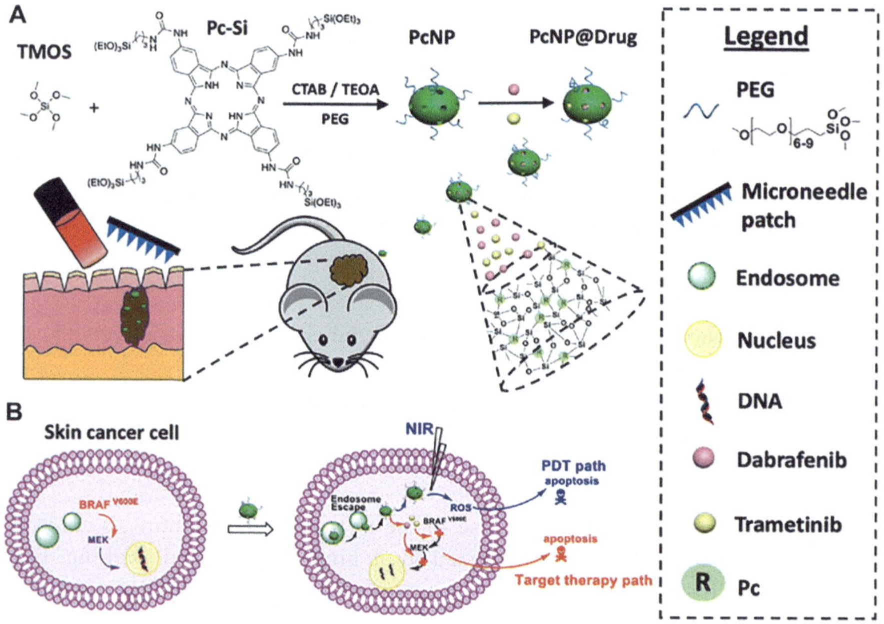

Fig. 7.4 **a** Schematic illustration of co-loaded photosensitizers and small-molecule inhibitors into microneedles for synergistic therapy of deep-seated melanoma. **b** Antitumor effects for these transdermally delivered drugs. The figure was reproduced with permission [41]

7.5 Advantages of PLA Microneedles

Polylactic acid (PLA) microneedles offer several significant advantages, including minimal invasiveness, enhanced patient compliance, controlled release of therapeutic agents, and the ability to be customized for specific dermatological needs. Their biodegradability ensures that there is no need for removal after application, which reduces the risk of infection and scarring. Moreover, the precise fabrication of these microneedles allows for the production of devices tailored to various clinical applications, making them highly versatile in the field of dermatology [42].

7.5.1 Minimal Invasiveness

One of the primary advantages of PLA microneedles is their minimal invasiveness. When applied to the skin, they create microchannels that penetrate the stratum corneum without causing significant pain or damage to deeper tissues. This characteristic makes them an ideal choice for patients who are needle-phobic or have a low tolerance for pain [43]. The small size of the microneedles and the precision of their application result in a procedure that is generally well-tolerated, even by those with sensitive skin.

7.5.2 Patient Compliance

The user-friendly nature and minimal discomfort associated with PLA microneedles significantly enhance patient compliance with treatment regimens. Patients are more likely to adhere to treatments that are painless and straightforward to administer. This increased compliance can lead to better therapeutic outcomes, as consistent application of treatment is crucial for the effectiveness of many dermatological therapies [44]. The ease of use also allows for self-administration, which can be particularly beneficial in chronic conditions requiring regular treatments.

7.5.3 Controlled Release

PLA microneedles can be engineered to provide controlled release of therapeutic agents over time, which is a substantial benefit in medical treatments. This controlled release can be achieved through the gradual degradation of the PLA matrix or by incorporating drug reservoirs within the microneedles [2]. Such mechanisms ensure that the therapeutic agents are released in a sustained manner, reducing the need for frequent dosing and maintaining consistent drug levels in the target tissue. This feature is particularly advantageous for treatments that require prolonged exposure to the drug, enhancing efficacy and reducing the risk of side effects associated with peak doses [45].

7.5.4 Customization

The fabrication techniques used to create PLA microneedles allow for extensive customization to meet specific dermatological needs. Microneedles can be designed with varying lengths, shapes, and drug loading capacities to target different skin conditions and treatment areas effectively [46]. This flexibility is critical in dermatology, where the type of skin condition and its location can significantly influence the choice of treatment. For instance, shorter microneedles may be used for more superficial treatments, while longer microneedles may be necessary for conditions that require deeper penetration. Additionally, the ability to load different drugs or multiple drugs into a single microneedle array can enhance the treatment's effectiveness [47].

7.6 Limitations and Challenges

Despite their considerable potential, PLA microneedles face several challenges that need to be addressed to fully realize their clinical application. Key issues include mechanical strength, manufacturing complexity, and drug stability. These challenges can impact the effectiveness, scalability, and practical implementation of PLA microneedles in dermatological treatments [48].

7.6.1 Mechanical Strength

The mechanical strength of PLA microneedles is a crucial factor in their effectiveness. If the microneedles lack sufficient strength, they may break during skin penetration, compromising the efficiency of drug delivery and potentially causing patient discomfort or injury. The mechanical properties of PLA can be a limiting factor, especially when deeper penetration is required for effective treatment. Ongoing research is focused on enhancing the mechanical strength of PLA microneedles by blending PLA with other polymers or incorporating reinforcement techniques. These approaches aim to create a more robust microneedle that can withstand the forces encountered during skin penetration without breaking [49, 50].

7.6.2 Manufacturing Complexity

The manufacturing process for PLA microneedles can be complex and costly, posing a significant barrier to large-scale production. Techniques such as 3D printing and micro-molding, while highly precise, require specialized equipment and expertise. These methods can be time-consuming and expensive, limiting the widespread adoption of PLA microneedles in clinical settings. Efforts are underway to develop more cost-effective and scalable fabrication methods that can produce high-quality microneedles at a lower cost. Innovations in automated manufacturing processes and the development of simpler fabrication techniques are essential to overcoming this challenge [26, 27].

7.6.3 Drug Stability

The stability of drugs incorporated into PLA microneedles is another critical challenge. Some therapeutic agents may degrade during the fabrication process or during storage, which can significantly reduce their efficacy. This degradation can result from exposure to heat, moisture, or other environmental factors that affect the stability of the drug. To address this issue, researchers are exploring various strategies, such as encapsulation, lyophilization, and the use of stabilizing agents. Encapsulation involves incorporating the drug into a protective matrix within the microneedle, which can shield it from degrading conditions. Lyophilization, or freeze-drying, is another technique that can improve the stability of sensitive drugs by removing moisture. Additionally, stabilizing agents can be added to the formulation to protect the drug from degradation during storage and use [24, 51].

7.7 Future Prospects

The future of polylactic acid (PLA) microneedles in dermatology looks promising, with ongoing research efforts aimed at enhancing their mechanical properties, drug loading capacities, and biodegradation rates. Advances in nanotechnology and materials science are expected to play a crucial role in developing next-generation PLA microneedles with improved performance and broader applications [1].

7.7.1 Nanotechnology Integration

Integrating nanotechnology with PLA microneedles could lead to significant advancements in their performance. By incorporating nanoparticles into the microneedles, researchers can enhance drug delivery efficiency, improve mechanical strength, and introduce additional functionalities such as imaging and diagnostics [52]. For example, nanoparticles can be designed to carry therapeutic agents and release them in response to specific stimuli, providing targeted and controlled drug delivery. This integration not only improves the therapeutic outcomes but also enables the microneedles to be used in diagnostic applications, such as detecting biomarkers or monitoring treatment progress. Research in this area is rapidly expanding, showing promising results that could revolutionize the capabilities of microneedles [25].

7.7.2 Advanced Materials

The development of new materials and composites for PLA microneedles is another critical area of focus. By blending PLA with other biodegradable polymers or incorporating reinforcing agents, researchers aim to enhance the mechanical properties and degradation rates of microneedles. These advancements are essential for creating microneedles that are both robust and effective in delivering drugs over extended periods [2]. For instance, incorporating materials like polycaprolactone (PCL) or silk fibroin can improve the mechanical strength and tailor the biodegradation profile of the microneedles, making them suitable for a wider range of dermatological applications. The use of advanced materials could also allow for the incorporation of multiple drugs within a single microneedle, enabling combination therapies that are more effective against complex skin conditions [13].

7.7.3 Clinical Trials

Translating PLA microneedles from the laboratory to clinical practice necessitates extensive clinical trials to establish their safety and efficacy. These trials are crucial for gathering data on the performance of PLA microneedles in various dermatological treatments, such as their effectiveness in drug delivery, patient comfort, and long-term outcomes [24, 53]. Ongoing and future clinical studies will provide the necessary evidence to support regulatory approval and ensure that the microneedles meet the required standards for medical use. These trials will also help identify any potential side effects or limitations, guiding further refinements and improvements. The successful completion of these trials will be a significant step towards making PLA microneedles a standard tool in dermatological practice, offering patients a safer, more effective, and less invasive treatment option.

7.8 Conclusion

PLA microneedles represent a significant advancement in dermatology, offering a versatile and minimally invasive tool for a wide range of applications, including drug delivery, cosmetic treatments, and therapeutic interventions. These microneedles have the potential to revolutionize dermatological practices by providing more effective and patient-friendly treatment options. Despite some challenges, such as enhancing mechanical strength, streamlining manufacturing processes, and ensuring drug stability, continued research and innovation are expected to address these issues. As advancements in nanotechnology and materials science progress, PLA microneedles are likely to see expanded clinical applications. This ongoing development promises to improve patient outcomes significantly, making PLA microneedles a crucial component in future dermatological treatments.

References

1. Donnelly, R.F., Singh, T.R.R., (eds.).: Microneedles for Drug and Vaccine Delivery and Patient Monitoring. Wiley (2015)
2. Prausnitz, M.R., Langer, R.: Microneedles: a new frontier in drug delivery. Nat. Biotechnol. **26**(11), 1261–1268 (2008)
3. Ita, K.: Transdermal delivery of drugs with microneedles—Potential and challenges. Pharmaceutics **7**(3), 90 (2017)
4. Quinn, H.L., Donnelly, R.F., McCrudden, M.T.C.: Microneedles: from vision to commercial products. Ther. Deliv. **5**(12), 1373–1380 (2014)
5. Kim, Y.C., Park, J.H., Prausnitz, M.R.: Microneedles for drug and vaccine delivery. Adv. Drug Deliv. Rev. **64**(14), 1547–1568 (2012)
6. Lee, J.W., Park, J.H., Prausnitz, M.R.: Dissolving microneedles for transdermal drug delivery. Biomaterials **29**(13), 2113–2124 (2008)
7. Bariya, S.H., Gohel, M.C., Mehta, T.A.: Microneedles: an emerging transdermal drug delivery system. J. Pharm. Pharmacol. **64**(1), 11–29 (2012)
8. Larrañeta, E., McCrudden, M.T.C., Courtenay, A.J.: Microneedles: a new frontier in nanomedicine delivery. Pharm. Res. **33**(5), 1055–1073 (2016)
9. Henry, S., McAllister, D.V., Allen, M.G.: Microfabricated microneedles: a novel approach to transdermal drug delivery. J. Pharm. Sci. **87**(8), 922–925 (1998)
10. Sullivan, S.P., Murthy, N., Prausnitz, M.R.: Minimally invasive protein delivery with rapidly dissolving polymer microneedles. Adv. Mater. **20**(5), 933–938 (2008)
11. Yan, G., Warner, K.S., Zhang, J., Sharma, G., Gale, B.K.: Evaluation of basic technologies for microneedle coating. J. Lab. Autom. **15**(1), 27–32 (2010)
12. Lee, K., Jung, H.: Drawing lithography for microneedles: a review of fundamentals and biomedical applications. Biomater. Res. **16**(4), 163–172 (2012)
13. Gill, H.S., Prausnitz, M.R.: Coating formulations for microneedles. Pharm. Res. **24**(7), 1369–1380 (2007)
14. Park, J.H., Allen, M.G., Prausnitz, M.R.: Polymer microneedles for controlled-release drug delivery. Pharm. Res. **22**(1), 14–19 (2005)
15. McAllister, D.V., Wang, P.M., Davis, S.P.: Microfabricated needles for transdermal delivery of macromolecules and nanoparticles: fabrication methods and transport studies. Proc. Natl. Acad. Sci. **100**(24), 13755–13760 (2003)

16. Ameri, M., Fan, S.C., Maa, Y.F.: Parenteral delivery of plasmid DNA by dissolvable microneedle arrays. Drug Deliv. Transl. Res. **4**(6), 525–534 (2014)
17. Kim, M., Jung, B., Park, J.H.: Hydrogel swelling as a trigger to release biodegradable polymer microneedles in skin. Biomaterials **33**(2), 668–678 (2012)
18. Miyano, T., Tobinaga, Y., Kanno, T.: Sugar micro needles as transdermic drug delivery system. Biomed. Microdev. **7**(3), 185–188 (2005)
19. Wang, C., Ye, Y., Hochu, G.M.: Enhanced cancer immunotherapy by microneedle patch-assisted delivery of anti-PD1 antibody. Nano Lett. **16**(4), 2334–2340 (2016)
20. Donnelly, R.F., Singh, T.R.R., Garland, M.J.: Hydrogel-forming microneedle arrays for enhanced transdermal drug delivery. Adv. Func. Mater. **22**(23), 4879–4890 (2012)
21. Kwon, S.Y., Kim, Y.C., Prausnitz, M.R.: Assessment of onset time of local anesthesia with lidocaine patch using a microneedle array: a comparative study. J. Control. Release **185**, 1–8 (2014)
22. Lee, B.M., Lee, C., Lahiji, S.F., Jung, U.W., Chung, G., Jung, H.: Dissolving microneedles for rapid and painless local anesthesia. Pharmaceutics **12**(4), 366 (2020)
23. Liu, S., Jin, M.N., Quan, Y.S.: Formulation development of a novel DNA vaccine delivery system using dissolving microneedles. J. Control. Release **161**(3), 610–618 (2012)
24. Kim, Y.C., Prausnitz, M.R.: Reducing the pain of vaccination: increasing the painless penetration depth of microneedles in the skin. Adv. Drug Deliv. Rev. **63**(4–5), 293–299 (2011)
25. Ma, G., Wu, C.: Microneedle, bio-microneedle and bioinspired microneedle: a review. J. Control. Release **251**, 11–23 (2017)
26. Wang, P., Zhang, J.: Microneedle-assisted transdermal delivery of hyaluronic acid: a mini-review. Mater. Today **4**(1), 23–28 (2017)
27. Jin, C.J., Han, J.W.: Microneedle-based transdermal drug delivery: a review. J. Control. Release **202**, 118–138 (2015)
28. van der Maaden, K., Jiskoot, W., Bouwstra, J.: Microneedle technologies for (trans)dermal drug and vaccine delivery. J. Control. Release **161**(2), 645–655 (2012)
29. Liu, F., Luo, Y., Chen, H., Xu, S., Zhang, D., Sang, H., Xu, C., Zhang, M.: Comparison of the efficacy of seven types of microneedles for treating a rabbit hypertrophic scar model. Nanoscale Adv. **5**(3), 927–933 (2022)
30. Du, H., Liu, P., Zhu, J.: Transforming growth factor beta (TGF-β) microneedles for scar reduction and skin remodeling. J. Control. Release **168**(2), 172–179 (2013)
31. Haq, M.I., Smith, E., John, D.N.: Clinical administration of microneedles: skin puncture, pain and sensation. Biomed. Microdevice **11**(1), 35–47 (2009)
32. Daugimont, L., Baron, N., Vandermeulen, G.: Hollow microneedle arrays for intradermal drug delivery and DNA vaccination. J. Control. Release **149**(3), 211–217 (2010)
33. Chen, H., Wu, Z.: Microneedle arrays for intradermal delivery of photosensitizers in photodynamic therapy. Adv. Drug Deliv. Rev. **65**(11–12), 1563–1570 (2015)
34. Li, G., Badkar, A.: Microneedle-mediated transdermal delivery of anti-acne agents. J. Control. Release **168**(2), 172–179 (2013)
35. Sinha, V.R., Kaur, M.P.: Microneedle-mediated transdermal delivery of acne treatment agents: a comparative study. J. Pharm. Sci. **89**(11), 1272–1278 (2000)
36. Wu, X., Chen, Y.: Microneedle-assisted transdermal delivery of skin-lightening agents. J. Cosmet. Dermatol. **15**(4), 445–454 (2016)
37. Yang, M., Li, J.: Microneedle arrays for transdermal delivery of hydroquinone: a review. J. Dermatol. Sci. **66**(1), 77–81 (2012)
38. Xu, J., Zeng, Y.: Microneedles for drug and vaccine delivery: a review. J. Control. Release **171**(2), 118–133 (2016)
39. Peng, S., Cheng, L., Wu, Q., Li, Y., Ran, L., Wang, W., Huang, K., Zhu, R., Xue, S., Zhou, C., Zhu, W., Cheng, B., Fu, X., Wang, R.: A modified hyaluronic acid-based dissolving microneedle loaded with Daphnetin improved the treatment of Psoriasis. Front. Bioeng. Biotechnol. **10**, 900274 (2022)
40. Lau, S., Fei, J.: Transdermal delivery of anti-inflammatory agents via microneedles for psoriasis treatment. J. Control. Release **272**, 199–207 (2018)

41. Chen, J., Ren, H., Zhou, P., Zheng, S., Du, B., Liu, X., Xiao, F.: Microneedle-mediated drug delivery for cutaneous diseases. Front. Bioeng. Biotechnol. **10**, 1032041 (2022)
42. Chen, H., Wu, Z.: Microneedle-mediated transdermal drug delivery: current status and future perspectives. J. Control. Release **165**(2), 120–131 (2014)
43. Kumar, S., Sahni, J.K.: Microneedles: a paradigm shift in transdermal drug delivery system. Biomed. Microdev. **13**(2), 129–140 (2011)
44. Levin, Y., Bachar, G.: Microneedles for drug and vaccine delivery: a review. J. Control. Release **153**(2), 71–76 (2011)
45. Ameri, M., Maa, Y.F.: Microneedles for drug and vaccine delivery. J. Control. Release **185**, 1–8 (2014)
46. Yan, G., Sharma, G.: Evaluation of basic technologies for microneedle coating. J. Lab. Autom. **15**(1), 27–32 (2010)
47. Wang, C., Ye, Y.: Enhanced cancer immunotherapy by microneedle patch-assisted delivery of anti-PD1 antibody. Nano Lett. **16**(4), 2334–2340 (2016)
48. Yang, M., Zhang, T.: Microneedle arrays integrated with pancreatic beta-cells for diabetes therapy. Biomed. Microdevice **12**(6), 1031–1038 (2010)
49. Kim, M., Jung, B.: Hydrogel swelling as a trigger to release biodegradable polymer microneedles in skin. Biomaterials **33**(2), 668–678 (2012)
50. Ameri, M., Fan, S.C.: Parenteral delivery of plasmid DNA by dissolvable microneedle arrays. Drug Deliv. Transl. Res. **4**(6), 525–534 (2014)
51. van der Maaden, K., Jiskoot, W.: Microneedle technologies for (trans)dermal drug and vaccine delivery. J. Control. Release **161**(2), 645–655 (2012)
52. Liu, S., Jin, M.N.: Formulation development of a novel DNA vaccine delivery system using dissolving microneedles. J. Control. Release **161**(3), 610–618 (2012)
53. McAllister, D.V., Wang, P.M.: Microfabricated needles for transdermal delivery of macromolecules and nanoparticles: fabrication methods and transport studies. Proc. Natl. Acad. Sci. **100**(24), 13755–13760 (2003)

Chapter 8
Diagnostic and Therapeutic Effects of Microneedles for Breast Cancer

Vaishali Y. Londhe and Gandhali Ghaisas

8.1 Introduction

Breast cancer is the most common cancer of all cancers in women. Despite a great deal of research on the detection and management of breast cancer, patient morbidity and death rates are still very high [1]. Consequently, the survival of patients will be greatly impacted by early detection of breast cancer as well as disease detection at all levels of care. Patients would have a 30% better chance of surviving if their cancer was discovered early on. As an example, a phase III ALLTO trial's central review found that there were up to 21.4% false negative instances for estrogen receptor (ER) and up to 14.5% false positive cases for a human epidermal growth factor receptor 2 (HER2) [2]. However, the difficulty in identifying breast cancer early on and the scarcity of diagnostic tools mean that treating breast cancer remains a problem. Furthermore, in spite of the wide range of methods available for detecting breast cancer, including histology, clinical examinations, magnetic resonance imaging (MRI), and mammography, the primary obstacle to diagnosing breast cancer is the absence of an affordable, non-invasive platform for early cancer detection. Chemotherapy is the most widely used treatment for advanced breast cancer; it can be administered intravenously or orally and is often coupled with other therapeutic modalities such as radiotherapy, surgery, and heat [3]. Obstacles include the gastrointestinal tract's enzymatic drug breakdown and the intestinal wall's inadequate absorption, which causes quick medication clearance, have led doctors to favour the cutaneous injection approach. However, patients' concerns about injection discomfort, needle pollution in the environment, and the requirement for recurrent, periodic injections pose challenges to subcutaneous or intravenous injections. However, patients' mental and physical well-being is negatively impacted by intricate and intrusive sample

V. Y. Londhe (✉) · G. Ghaisas
Shobhaben Pratapbhai Patel School of Pharmacy and Technology Management, SVKM's NMIMS, Mumbai, Maharashtra, India
e-mail: vaishali.londhe@nmims.edu; vaishali_londhe2000@yahoo.com

N. Talreja et al. (eds.), *Microneedles (MNs)-Based Technology*, Series in BioEngineering, https://doi.org/10.1007/978-981-96-3916-8_8

procedures used after the initial diagnosis and by recurring sampling to monitor the breast cancer's response to treatment [4]. Nevertheless, there are some non-invasive breast cancer treatments, such as radiofrequency and high-intensity focus ultrasound. Radiofrequency (RF) ablation is an additional technology that has been shown to be effective in treating incurable liver tumours. As a result, research on early breast cancer was conducted, and the results indicated a possible non-invasive treatment. Frictional heat produces radio frequency (RF) [5]. Electrode tips inserted into the lesion generate a high-frequency current that permeates the surrounding tissue and starts an ionic agitation process that results in heat generation and eventual cell death [5]. Usually, magnetic resonance imaging (MRI) is used to track the tumour ablation while the RF probe is positioned under ultrasound guidance [6]. High-intensity focused ultrasound (HIFU) is a significant emerging method that is totally non-invasive. In order to achieve high focal power levels during HIFU, the ultrasonic beam is concentrated into a tiny target volume. This causes temperature increases that cause cell death in the targeted tissue while sparing the surrounding structures. MRI (MR-HIFU) or traditional diagnostic ultrasonography (US-HIFU) can be used to guide HIFU. US-HIFU has treated hundreds of patients worldwide for bone tumours, kidney cancer, breast cancer, pancreatic cancer, liver cancer, and uterine fibroids [5].

8.2 Breast Cancer Pathophysiology

8.2.1 Risk Factors

As age advances, there is a greater likelihood of breast cancer. Based on statistics from the Surveillance, Epidemiology, and End Results (SEER) database, a woman in the United States has an 8% lifetime risk of having breast cancer; this risk increases to 1 in 202 from birth to age 39, 1 in 26 from age 40 to age 59, and 1 in 28 from age 60 to age 69 [7]. Breast cancer risk factors for premenopausal and postmenopausal women include early menarche age. Menarches that are delayed by two years are linked to a 10% decrease in risk. Women who menarched before the age of 13 had almost twice the incidence of hormone receptor-positive tumours in the European Prospective Investigation into Cancer and Nutrition cohort. An elevated risk of breast cancer has also been linked to a later onset of menopause [8, 9]. A 3% increase in risk arises for each year that the menopause is delayed, and a 17% increase in risk arises for every five years that the menopause is delayed. Worldwide, 21% of breast cancer fatalities are attributable to modifiable risk factors such as weight gain, lack of exercise, and excessive alcohol consumption [8–10]. Compared to parous women, nulliparous women have a higher chance of developing breast cancer. While comparatively older age at first birth confers a higher relative risk of breast cancer than that of a nulliparous woman, young age at first birth has an overall protective impact [11]. The cumulative incidence of breast cancer in women giving birth for the first time at ages 20, 25, and 35 was 20% lower, 10% lower, and 5% higher, respectively, than in nulliparous women [11]. Drinking alcohol has been linked to a statistically significant increase in the incidence of breast cancer even at levels as low as 5.0–9.9 g per day, or 3–6 drinks per week (RR = 1.15; 95%CI: 1.06–1.24; 333 cases/100000 person-years). After adjusting for cumulative alcohol intake, binge drinking was linked to an increased risk of breast cancer, but not frequency of drinking. Drinking alcohol at any stage of adulthood was linked to risk on its own [12]. Regular exercise has been demonstrated to lower the risk of breast cancer in a dose-dependent way, with moderate exercise resulting in a 2% risk reduction and severe exercise inducing a 5% risk reduction. It has also been demonstrated that obesity increases a woman's risk of breast cancer, particularly in postmenopausal women. The EPIC multicentre prospective cohort study found that postmenopausal women without hormone replacement therapy (HRT) had a higher risk of breast cancer in proportion to their weight, body mass index (BMI), and hip circumference [13]. When comparing women in the normal weight range to overweight (BMI 25.0–29.9) and obese (BMI > 30.0) women in this cohort, the multivariate relative risk was 1.28. Paradoxically, lean women on HRT had a higher risk of breast cancer (RR = 2.04) than their peers who were overweight (1.93) and obese (1.39). Studies have looked at insulin resistance and hyperinsulinemia as risk factors for diabetes and cardiovascular disease, two comorbidities linked to obesity [13]. Insulin receptor upregulation has been shown in human cancer cells, and insulin exerts anabolic effects on cellular metabolism. In postmenopausal women without diabetes, hyperinsulinemia has been demonstrated to be an independent risk factor for breast cancer, which may help to explain the link between obesity and breast cancer. A woman who has a family history of breast cancer is more likely to develop the disease herself [14, 15]. In the follow-up of the Nurses' Health Study, women without a family history of breast cancer had an adjusted relative risk of 1.37 and 1.69, respectively, depending on whether their mother was diagnosed before or after the age of 50. In addition, compared to patients without a family history, those who had a sister diagnosed with breast cancer had a higher relative risk of 1.66 if the diagnosis was made before the age of 50 and a higher related risk of 1.52 if it was made after that age. A rising number of first-degree relatives diagnosed with breast cancer at an early age (under 50) is linked to the highest risk. Women with one, two, or three affected first-degree relatives exhibited risk ratios of 1.80, 2.93, and 3.90, respectively, in comparison to women without any affected relatives [16]. For both premenopausal and postmenopausal women, elevated levels of endogenous sex hormones are associated with an increased risk of breast cancer. In postmenopausal women, high levels of circulating testosterone have been associated with an increased risk of breast cancer [17].

8.2.2 Breast Cancer Categories

Cancer arises from the mutation of normal genes. Mutated genes that cause cancer are called oncogenes [18]. Every cell, tissue, and organ has the potential to undergo carcinogenesis, which can result in pathological changes that give rise to a large range of malignancies. Evading apoptosis, having an infinite proliferative potential, increased angiogenesis, resistance to anti-growth signals, and metastization are the main factors that allow it to continue. Right now, breast cancer is the most common malignancy among women to receive a diagnosis [19]. Cooper's ligaments provide the framework that holds the glandular and fatty tissue that makes up the breast in place. Alveoli and ducts are parts of the glandular tissue. The alveoli, which are microscopic sacs bordered by mammary secretory epithelial cells, form the clusters that make up the lobes. The lobule is drained by the minuscule ducts that emerge from the alveoli and combine to generate more large ducts [20]. There are two types of breast cancer, invasive (infiltrating) carcinoma and in situ carcinoma. Breast cancer in situ can be further classified as ductal or lobular; the distinction between the two forms is based on cytological characteristics and growth patterns. There are further classifications for ductal carcinoma in situ, comedo, cribiform, micropapillary, papillary, and solid. The histological subtypes of invasive carcinomas, which include infiltrating ductal, invasive lobular, ductal lobular, mucinous (colloid), tubular, medullary, and papillary carcinomas, are a heterogeneous collection of tumours. At 80% of all cases of breast cancer, invasive ductal carcinoma, commonly referred to as infiltrating ductal carcinoma or IDC, is the most prevalent kind of disease [21]. Histopathological classification of breast cancer subtypes is shown in Fig. 8.1.

Breast cancer occurs in a total of five stages, from 0 to IV. At stage 0, the illness is limited to the breast's ducts or lobules. It's not infected the tissue around it. Another

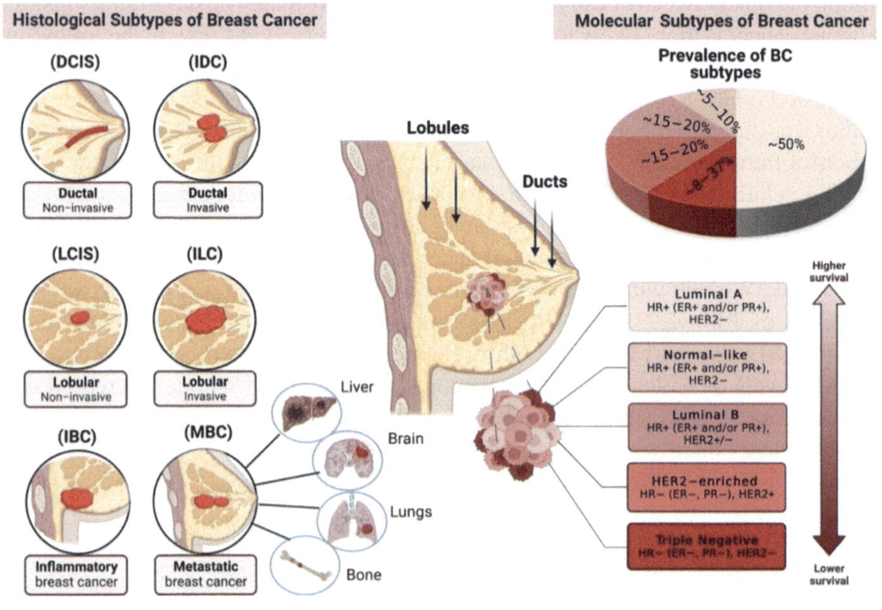

Fig. 8.1 Histological classification of breast cancer subtypes. The image was reproduced with permission under a common creative license [22]

name for it is non-invasive cancer. Stage I is subdivided into two further stages. Stage IA, there is not much tumour. The lymph nodes (T1, N0, and M0) have not been affected by it.

In stage IB, the tumour may be in the breast tissue in addition to the lymph nodes. Its size is less than 2 cm. Stage II describes invasive breast cancer. Two categories exist; stage IIA, the cancer cells have migrated to at least one to three lymph nodes. However, the tumour may not be present in the breast, and it may be 2 cm or smaller. Sometimes, a 2–5 cm tumour in the breast that hasn't migrated to the axillary lymph nodes may be seen in stage IIA.

Stage IIB, 1–3 axillary lymph nodes have been affected by the illness, and the tumour measures 2–5 cm. Alternatively, the tumour is more than 5 cm in size but hasn't reached the lymph nodes in the axis. Stage III also describes invasive breast cancer. Three categories exist; stage III is any size tumour that has metastasized to four to nine lymph nodes. Alternatively, the tumour is more than 5 cm in size and has only reached 1–3 lymph nodes. At stage IIIB, the cancer has progressed to the chest wall, and the tumour could be any size. It may affect up to nine lymph nodes and result in breast enlargement. Breast cancer that is inflammatory is classified as Stage IIIB. Stage IIIC is any size tumour that has progressed to ten or more lymph nodes. Stage IV is a metastatic stage. Any size tumour is possible, and the illness has progressed to other organs and tissues such as the brain, liver, bones, lungs, distant lymph nodes, and chest wall [23]. There are two major recognized molecular targets in the aetiology of breast cancer. One is the estrogenic receptor alpha (ERα), expressed in around 70% of aggressive breast tumours. A transcription factor and steroid hormone receptor, ERα, triggers oncogenic growth pathways in breast cancer cells when it is triggered by oestrogen. Another indicator of ERα signalling is the expression of the closely related steroid hormone progesterone receptor (PR) [24]. HR+ tumours are defined as those that express either the PR or the oestrogen receptor (ER) in at least 1% of their cells. The main systemic treatment for breast tumours that are ER-positive or PR-positive is the use of endocrine drugs to downregulate ER signalling. Epidermal growth factor 2 (ERBB2, formerly HER2 or HER2/neu) is a transmembrane receptor tyrosine kinase belonging to the epidermal growth factor receptor family [25]. It is overexpressed or amplified in about 20% of breast cancers and is linked to a poor prognosis if systemic therapy is not administered [26]. This makes it the second main molecular target. ERBB2+ tumours are those that have an overexpression or amplification of the ERBB2 gene [27]. ERBB2-targeted therapy, such as anti-ERBB2 antibodies (like trastuzumab and pertuzumab) and small-molecule tyrosine kinase inhibitors (like lapatinib and neratinib), is beneficial for patients with ERBB2-amplified or overexpressing breast cancer [28]. Three molecular targets ER, PR, or ERBB2 are not expressed in triple-negative breast cancer, which accounts for 15% of all breast cancers [29]. In the first three to five years after diagnosis, there is a significant chance of a distant relapse for triple-negative tumours. There is still much to learn about the precise molecular pathophysiology of triple-negative breast cancer. The three subtypes of breast cancer HR+, ERBB2+, or triple-negative, have different prevalences, prognoses, and possibilities for systemic therapy. While HR+ breast cancers are more common in older women, triple-negative breast tumours are more common in younger, black, or Hispanic women. Stages I–IV are used to categorize breast cancer, with IV meaning distant metastases. The 5-year breast cancer-specific survival rates for stage I breast cancers are at least 99%, 94%, and 85% for HR+, ERBB2+, and triple-negative subtypes, respectively. Anatomically, a stage I breast tumour is defined as one that is smaller than 2 cm and does not involve lymph nodes. For HR+ or ERBB2+ subtypes, the median overall survival of stage IV breast cancers is roughly 5 years; for triple-negative subtypes, it is 1 year [24].

8.2.2.1 Cancer Stem Cells

A tiny subset of tumour cells known as cancer stem cells (CSCs) are responsible for a variety of tumour behaviours, such as invasion, metastasis, recurrence, and resistance [30]. Malignant transformation of a self-renewing stem or precursor cell can cause cancer [31]. Breast cancer stem cells (BCSCs) have stem cell characteristics and are identified by the expression of surface markers such as CD44, EpCAM, CD133, CD49f, CD61 and the absence of surface markers such as CD24, CD2, CD3, CD10, CD16, CD18, CD31 and CD64 [32]. Different BCSCs are capable of producing diverse BCs with different behaviours. Several signalling pathways are involved in the maintenance and activity of CSCs that subsequently play a substantial role in tumour behaviour, such as growth, invasion, metastasis, and resistance to treatment [33]. Any type of cancer can be treated with surgery, chemotherapy, radiation therapy, targeted endocrine or molecular therapy, among many other options, depending on a number of different criteria [33]. Recent studies, however, have demonstrated that a single treatment is insufficient to fully remove tumours, stop cancer cells from returning, extend survival, and alleviate symptoms. Because of this, a multimodality approach combining many treatments is necessary. This technique includes systemic therapy, which may be administered preoperatively (neoadjuvant), postoperatively (adjuvant), or both. There are still many restrictions on one's ability to be physically, emotionally, socially, and psychologically well despite the advances in cancer therapies today [31].

8.3 Microneedles

Microneedles (MNs), or small needle-like structures, are used to administer drugs through the skin's layers. They are non-invasive and cause very little or no discomfort when they are applied to the skin. MNs are great at getting both big and small compounds to the patients who need them. There are a number of methods for delivering drugs via MNs, and each method has advantages and disadvantages [34]. Poke and patch, coat and poke, poke and release, and ultimately poke and flow are some of the most used tactics [35]. To construct microchannels, the MNs are punctured through the stratum corneum in poke and patch. After that, the MNs are removed, and a drug dosage form, such as a cream, gel, or solution, is applied. This serves as a drug reservoir (Fig. 8.2a) [36]. This technique has a drawback in that the microchannels close after a predetermined amount of time. Hence, the drug distribution is stopped, and it is a two-step process [36, 37]. The medication material is coated over the MNs in the second method, known as coat and poke. The procedure entails piercing the stratum corneum with drug-coated MNs, which allows the coated medication to diffuse into the skin's layers from the coating layers (Fig. 8.2b) [38]. This method is only applicable to medications with very low dosages. Moreover, medication coating dulls the MNs' sharpness. The poke and release method, in which the MNs usually disintegrate after insertion (Fig. 8.2c), skipping the removal step as a result, as demonstrated by the poke and patch technique. Additionally, the rapidly separating MNs are suggested as a cross between the dissolving MNs and the coat and poke technique. With this approach, the solid is removed but the drug-loaded soft matrix stays under the skin. Additionally, hydrogel-forming MNs work with interstitial fluid to create microchannels that can be used to deliver less potent medications [39–41]. Finally, the poke-and-flow approach causes much less pain, if any, at the administration site than the traditional hypodermic injection technique. Here, the drug solutions are placed into the MNs, which release the medication when inserted into the skin [42] (Fig. 8.2d).

8.4 Materials Used for Microneedle Fabrication

8.4.1 Silicon

Silicon gives MNs significant mechanical strength that enables them to puncture flesh effectively. Solid, hollow, and coated MNs can all be made with silicon. It can be applied to the fabrication of MNs with different heights and forms. The primary drawbacks of silicon are its expensive cost, complicated production specifications, protracted fabrication periods, and multistep, complex processing. One risk associated with silicon MNs is the potential for skin fractures that could compromise safety. On the other hand, silicon has been extensively employed to improve and ease transdermal medication administration [43, 44].

8.4.2 Metal

The most often utilized metals for making MNs include alloys, gold, palladium, titanium, nickel, platinum, and stainless steel. Metals are used to make coated MNs that are solid and hollow and serve as their foundation. Using metals has the benefit of their favourable mechanical characteristics and strong tensile strength, which allow for simple stratum corneum penetration. A lot of stainless steel is utilized in the creation of MNs. On the other hand, its rates of corrosion are higher than those of titanium alloys. Stainless steel is not as mechanically robust as titanium alloys. Less

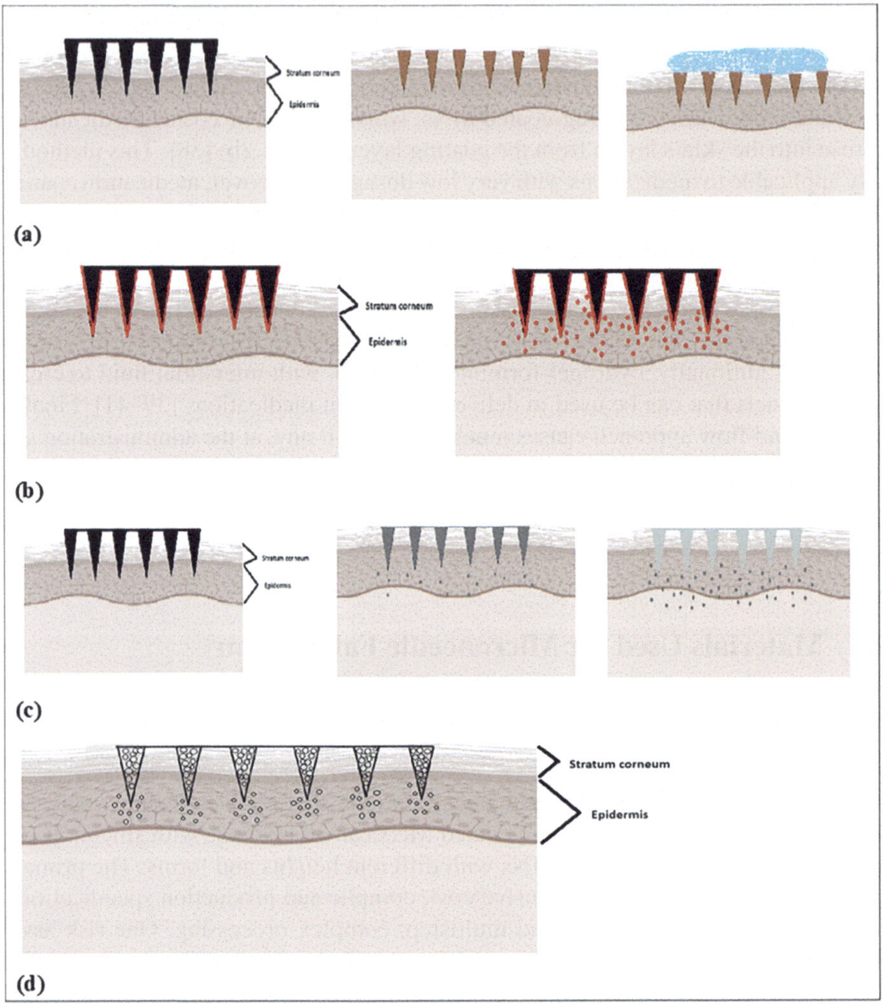

Fig. 8.2 **a** Coat and patch, **b** coat and poke, **c** poke and release, **d** poke and flow

research has been done on the use of palladium and platinum in the synthesis of MNs. Because of its problems with biocompatibility, nickel usage needs to be done carefully.

8.4.3 Glass

Glass MNs with a variety of shapes have been produced and have been shown to effectively puncture SC. The hollow variety is frequently made. The primary drawbacks of silica glass are its brittleness and the potential for needle tips to break through the skin, leading to granulomas and irritation. Nevertheless, borosilicate glass has demonstrated strong biocompatibility [45].

8.4.4 Polymers

Using polymers, MNs that dissolve, form hydrogels, or are biodegradable have been created.

Commonly used polysaccharides include hydroxypropyl methylcellulose, hyaluronic acid, carboxymethycellulose (CMC), alginates, and synthetic polymers like poly (methylvinylether/maleic anhydride) (e.g., GantrezVR), polystyrene, polyvinyl alcohol, polyvinylpyrrolidone (PVP), polylactic acid, polyglycolic acid, and their co-polymers (poly (lactic-co-glycolic acid) [PLGA]). The main benefits of polymers are their biodegradability and biocompatibility [46].

8.4.5 Sugars

MNs that can pierce SC are produced by sugars like maltose, trehalose, raffinose, mannitol, xylitol, and galactose. Nevertheless, they have additional disadvantages, such as instability, high processing temperature requirements, and quick pore resealing [47].

8.5 Fabrication Techniques

8.5.1 Laser Cutting

Metallic MNs are produced using this process. Stainless steel is the material that is used most commonly. Using an infrared laser, metallic sheets are cut into the shape of tiny needles in this process. The arrays' orientation and shape are then drawn using a software program, such as AutoCAD. The laser beam forms the needles in the plane of the sheet, ablates the metal sheet, and traces the required needle shape. The sheets are then cleaned, and the MNs are typically bent out of the sheet at a 90° angle. By sharpening the points, electropolishing establishes the final structure [48, 49].

8.5.2 Vapor Deposition

For the production process to be completed, vapour deposition alone is insufficient; additional processing steps, such as photolithography and etching, are needed. Its sole purpose is to cover the silicon substrate with a metal coating. Subsequently, these films are patterned by appropriate etching and photolithographic processes. When using procedures based on physical vapour deposition, heated raw materials emit vapours that are then deposited on the substrate surface from the source (the material to be coated). Thin films are created in chemical vapour deposition when a hot substrate and inert carrier gases in the chamber react chemically [50].

8.5.3 Photolithography

MNs are made from ceramics and polymers, such as poly (ethylene glycol) diacrylate, using photolithography. This method can create systems that are hollow or solid. This approach uses the spin coating technique to coat the substrate with a photosensitive polymer. The polymer, called photoresist, is thermally hardened, and the substrate is a silicon wafer. After that, the photoresist is exposed to UV light via a mask. UV radiation modifies the polymer's "crosslink" bonds, either by promoting or preventing crosslinking. As a result, the polymer region exposed to UV light has a significantly different solubility than the portion in the mask shade. By immersing the treated substrate in the developing solution, the exposed photoresist is removed. The last phase, commonly referred to as the etching step, involves using wet or dry chemical attack to replicate the photoresist drawing on the substrate [51].

8.5.4 Pulling Pipettes

The process of pulling pipettes is unique to the creation of hollow glass MNs. Using a micropipette puller, fire-polished type I borosilicate glass pipettes are pulled to create arrays. Reproducible and consistent MN manufacture is made possible by programmable puller parameters. After that, the extracted needles are bevelled at a 30° angle with a hollow MN beveller that has an oval-shaped aperture [52].

8.5.5 Droplet Born Air Blowing

This is a new method of creating polymeric MNs without the need for silicon moulds that have already been prepared. A polymer solution is dispensed at a controlled pace using a solution dispenser to manufacture the base array in this method. Drug-containing polymer solution is equally placed on the upper and lower plates after drying. The viscous solution elongates as contact is made between the solutions of the top and lower plates. After that, it is dried under carefully regulated airflow conditions. Finally, an isolation phase is carried out to isolate two MN arrays from each surface [53].

8.5.6 *Two-Photon Polymerization (2PP)*

This process, which works with ceramics and polymers, is a piece of very advanced equipment. Additionally, it can be utilized to create hollow MNs. The idea of photo-excitation of initiator molecules are used in two-photon polymerization. Multiphoton absorption starts with the polymerization of a resin. A femtosecond or picosecond laser, such as a titanium sapphire oscillator, is part of the 2PP configuration. It is concentrated and adapted so that it can move over the polymer resin in three dimensions (3D). Computer-aided design is used to connect the 3D design with spatial coordinates. Absorption of two photons: When a laser is concentrated on a small portion of the polymer resin, two-photon absorption causes nonlinear energy absorption that breaks the chemical bonds of photoinitiator molecules. In order to make the laser trace scan through the resin, the substrate is also polymerized. Following the completion of the MN's manufacture, the nonirradiated resin is removed using the proper solvent and dried with UV light [54].

8.5.7 *Dipping*

Using this technique, the intended medication formulation can be applied on premade solid MNs. The location of the coating solution reservoir eliminates base contamination by limiting the coating liquid's access to the MN shaft alone. Using micro positioners to help with 3D alignment and sinking MN rows or arrays into the dip holes, the reservoir is positioned and fixed. The MN-array holder and the solution reservoir are placed across from one another so that the coating-solution reservoir's cover plate has dip holes that match the MNs in the row or array. The MN-array holder is slid down to dip the array's MNs into the dip-holes in order to coat the MNs [49].

8.5.8 *Spraying*

MNs are coated using the traditional film coating technique. An atomized spray is created using a nozzle that is attached to a source of compressed air and coating solution. Every crucial parameter, including pressure and flow rate, is optimized. During the spray coating procedure, double-sided tape is used to secure the MNs to an adjustable stage. Coatings are applied and then let to dry for 12 h at room temperature before being analyzed [55].

8.6 Diagnostic Microneedle in Breast Cancer

The gold standard for detecting breast cancer is still mammography. Women who have a palpable lump or other breast disease symptoms get diagnostic mammography [56]. Another is breast MRI, which is becoming a crucial component of the detection and treatment of breast cancer. When a patient is diagnosed with breast cancer for the first time, a breast MRI can be used to assess them and determine the degree of disease at that point. When evaluating a patient's response to neoadjuvant chemotherapy, breast MRI imaging is helpful prior to, during, and following treatment [57, 58]. Ultrasound technology is an additional one. Suspected abnormalities on mammography or MRI, as well as issues with breast implants, are indications for this treatment. However, because dense tissues can obscure abnormalities, mammograms become more challenging to interpret when breast density is high. Mammograms may, therefore, be unable to correctly detect tumours in women whose breast tissue is thick. However, for the same reason, ultrasonography might overlook a solid tumour or little bump [58].

These days, a range of MNs, including point-of-care devices, have been developed for diagnostic procedures that can identify illnesses and malignancies [59]. Fixable on the skin, MNs are capable of collecting and evaluating interstitial fluid from the skin and blood vessels [60]. The presence of various biomarkers over a wider sampling area can be detected using modified MNs, increasing the accuracy of disease diagnosis by enabling the simultaneous identification of many analyte types [1]. Finding blood biomarkers for breast cancer is still a significant obstacle to patient prognosis, diagnosis, and screening [61]. The process of taking blood or tissue samples in order to identify breast cancer is frequently difficult, error-prone, and uncomfortable. Diagnosis of cancer by MNs has received a lot of attention after it was proved that interstitial fluid, mainly dermal and subcutaneous tissues, can detect biomarkers [62]. Nevertheless, several biomarkers from the early phases of carcinogenesis can be discovered in the interstitial fluid around the growing breast cancer. For breast cancer examination, tissue fluid around the breasts can reflect the real-time status of the cancer cells, which is timelier than serum. The changes in the level of tumour markers accompany the progression of breast cancer. In the process of carcinogenesis, at first, markers produced by tumour cells will accumulate in the tissue fluid around the tumours. These markers enter the circulation through exchange with tissue fluids. Consequently, the interstitial fluid likely has a higher concentration of breast cancer markers than blood, which could result in a notable increase in detection accuracy [1]. Consequently, using MNs to sample the interstitial fluid inside the breast may enable a more accurate detection of breast cancer. By removing carcinoembryonic antigen from skin tissue fluid, a unique technique for in situ breast cancer detection using MNs and nano-Ag/MBL film is proposed [63]. They discovered that the microneedle immunoassay biosensor not only permitted breast cancer detection in an animal model quicker than with traditional blood collection but it also enhanced the cutoff point for early breast cancer detection [63]. To extract enough skin interstitial fluid, a swellable microneedle patch composed of methacrylate hyaluronic acid and further crosslinked by UV light was produced. This MNs patch may extract enough interstitial fluid, an increasing source of biomarkers for disease diagnosis and prognosis, because of its exceptional water affinity [64]. Another example of MNs with good water affinity is made up of gelatine methacrylate. In addition, this microneedle was used for screening breast cancer in the genetically modified and 4T1 orthotopic mouse models. Using gelatine methacrylate microneedle in conjunction with other conventional screening techniques such as blood testing, ultrasound, or micro-CT, the cancer occurrence could be detected earlier by assessing the tumour markers carcinoembryonic antigen (CEA) and glycoantigen CA15-3 [65]. In another microneedle-based sampling of breast tissue, a two-port microneedle is used to extract lipids from tissue samples in situ. Technologies for silicon micromachining are used to create the microneedle. The two-port microneedle is integrated into a polymer microfluidic chip made of polymethyl methacrylate (PMMA), with each of the needle ports linked to a separate fluidic channel within the polymer chip. The polymer channel increases the overall fluidic volume that the needle can handle to more than 10 μL, even though the inner liquid capacity of the microneedle is just 5 μL. Hot embossing has been used to structure the polymer chip. At 400 μm in width and 100 μm in depth, the channels. The extracted sample is transferred from the needle to the mass spectrometer for the lipid analysis after the extraction procedure [66].

8.7 Therapeutic MNs in Breast Cancer

8.7.1 Dip-Coated MNs

An MNs is dipped into the coating liquid, and as soon as the microneedle leaves the coating liquid, a liquid film forms on its surface. Bhatnagar and colleagues reported MNs of zein (ZMN) for delivery of tamoxifen and gemcitabine, which are used in the treatment of cancer. Gemcitabine or tamoxifen was dip-coated on these blank ZMN. ZMN were carefully dipped into the medication solution by fastening the array to a texture analyzer setup's arm. Drug penetration through and into the skin, as well as ZMN implantation, were studied using excised pig ear skin as a model [67, 68].

8.7.2 Dissolving MN

The "poke and release" concept underpins the release of drugs from dissolving MNs (DMN). This indicates that the medication is released into the skin's pores by DMNs after they have been created. Drug molecules are delivered into the skin by DMs, which are made of water-soluble substances such as maltose, polyvinylpyrrolidone, chondroitin sulphate, dextran, hyaluronic acid, and albumin. DMs are applied to the skin with the tip of the finger. After use, they dissolve entirely in the skin and don't leave any biohazardous sharps tips behind because they are composed of biocompatible and water-soluble ingredients like sugars and cellulose derivatives. In another study, the preparation of dissolving MNs using polyvinyl pyrrolidone and polyvinyl alcohol for co-delivery of doxorubicin and docetaxel for the treatment of breast cancer was reported. A mixture of 10% w/v PVA and 11% w/v PVPK360 was used to prepare the base plate of the DMN array, while 21% w/v PVPK360 was used to prepare the needles (in a 50:50 ethanol: water mixture). The PVPK360 solution was supplemented with doxorubicin and docetaxel at the same concentrations as the individual preparations. For research on skin penetration, already shaved and excised mouse skin was employed. Immunodeficient mice with tumours were used to test the effectiveness of medication-loaded MNs [69].

MNs made up of maltose were prepared to deliver the anticancer compound 'Honokiol', obtained from the plant parts of *Magnolia glandiflora.* Gao and colleagues investigated the delivery of this natural anticancer agent employing skin penetration enhancers in addition to the MN-based administration of honokiol. Chemical enhancers such as isopropyl myristate, oleic acid, and propylene glycol were used in the study. In comparison to other compounds, oleic acid was proven to be an outstanding penetration booster. It was discovered that both the delivery methods (MNs and penetration enhancers) were progressive and promising in getting honokiol through the skin's layers. Furthermore, the anti-tumour action of honokiol was demonstrated by the inhibition of the Ki-67 protein and the reduced production of the cytokine interleukin-6. The methyl thiazolyl tetrazolium (MTT) assay was used to assess the cytotoxicity or irritation of the administered dose. The next step was to do the MTT assay to measure the cell viability, which would indicate whether or not the formulation was irritating. Three-dimensional psoriatic tissues that had been treated with and without honokiol underwent immunohistochemical examination. When compared to the negative control, the stained tissues revealed a morphological decrease in the amount of Ki-67 in the skin treated with honokiol [70]. Bevacizumab is a recombinant humanized monoclonal antibody used clinically to treat advanced-stage or metastatic breast cancers. To overcome the delivery challenges, such as structural changes resulting from variations in pH, alterations in ionic and osmotic pressure, and shear stresses exerted on formulation, delivery of biotherapeutics like bevacizumab using microneedles was studied [71]. Dissolving MNs loaded with cetuximab-anchored hyaluronic acid-oleyl amine and chitosan-oleic acid hybrid nanoparticles (HA-OA/CS-OA NPT) for the targeted delivery of cabazitaxel (CBT) in localized breast cancer tumours were reported [72]. Asha Matta and team studied the 5-fluorouracil loaded, matrix-forming microneedle-based medication delivery method that may be able to restrict systemic effects and localize drug doses to the application site. A composite dissolvable microneedle patch made of sodium carboxymethyl cellulose, sodium alginate, and PEG 400 was formulated for buccal mucosa insertion. Using a micromolding technique, the patch with a 20×20 array format and pyramid-shaped MNs was created [73].

8.7.3 Hollow MN

Glass, silicon, and metal combine, as was already indicated, an empty each needle has a bore on the tip and a cavity inside. As a result, skin injections of tiny amounts of medication solutions are possible. Higher medication delivery is one of the primary benefits of this type of MN. Capacity is in contrast to MN arrays that are solid, coated, or dissolved, where chemicals, proteins, and vaccination oligonucleotides can also be injected into the skin. Metallic MNs are also used to deliver breast cancer vaccines. In this work, using a mouse model, we investigate the immunological response elicited by a particulate breast cancer vaccine delivered via MNs. The particle vaccine was administered using the AdminPatch® 1200, a commercially available microneedle patch. Others have successfully assessed AdminPatch® microneedle patches for use in skin-based medication delivery applications. The 4T07 mouse breast cancer cell line was selected as a model because, in contrast to 67NR and 4T1 murine cell lines, it had intermediate metastatic capabilities [74].

8.7.4 Solid MN

The goal of this kind of microneedle shape is to promote drug administration to the dermis, hence enhancing bioavailability and kinetic transport across the skin by penetrating the stratum corneum. Mojeiko and the team reported that MNs can be used to augment the penetration of celecoxib microemulsion in the treatment of breast cancer. Using Franz diffusion cells and pig ear skin as model tissue, skin penetration and percutaneous administration of Celecoxib were evaluated. The impact of encapsulation on Celecoxib's cytotoxicity against cancer cells (MCF-7) was evaluated using cytotoxicity assays. Additionally, MNs could be considered for drug delivery via the transpapillary route (breast) for the treatment of breast cancer [75]. An effort was made to look into the possible therapeutic usage of MNs in conjunction with co-loaded liposomes containing doxorubicin HCl (DOX) and Celecoxib (CEL) for the treatment of melanoma. Co-loaded liposomes/gels containing DOX and CEL were made and described. The outcomes demonstrated that, in comparison to passive delivery, microneedle pretreatment with liposome gel enhanced DOX penetration into the skin by around a factor of two [76]. Some of these examples of drug-loaded MNs, along with the method of fabrication, are enlisted in Table 8.1.

Table 8.1 MNs for breast cancer

S. No	Loaded material	MNs type	Method of fabrication	Use	Preclinical/cell line studies	References
1	Tamoxifen and gemcitabine	Coated MNs	Moulding	Breast cancer treatment	Preclinical	[68]
2	Doxorubicin and Celecoxib co-loaded liposomes	Solid MNs	Reactive ion etching	Anticancer treatment	Preclinical	[76]
3	Hyaluronic acid-Oleyl amine and chitosan-oleic acid conjugate-based hybrid nanoparticles	Dissolving MNs	Moulding	Localized breast cancer	Preclinical	[72]
4	Doxorubicin and docetaxel	Dissolving MNs	Moulding	Treatment of breast cancer	Cell line	[69]
5	Honokiol	Solid MNs	Moulding	Anticancer activity	MTT assay	[70]
6	Breast cancer vaccine	Metallic MNs	Photolithography	Induce immunological response	Murine cell line	[74]
7	Celecoxib microemulsion	Solid MNs	Laser cutting	Breast cancer	Cell line	[75]

8.8 Clinical Trials in MNs

Currently, recruitment is underway for clinical trial research using MNs to test the Modi-1 vaccine for breast cancer. Assessing the safety, tolerability, immunological activity, and preliminary efficacy of the Modi-1 vaccination in patients with advanced triple-negative breast cancer is the primary goal of this study. Expectations for the development of MNs appear to have been enhanced, particularly in the field of cancer, by published and unreported clinical trials in the drug and vaccine delivery domain. Nevertheless, there are still few clinical trials investigating the effectiveness of MNs in the detection and management of cancer, particularly breast cancer, and the results of these trials have not yet been made public. Oleg E. Akilov created doxorubicin-containing MNs to treat cutaneous T-cell lymphoma. However, the report that can be accessed has not yet been released. Similarly, SkinJect, Inc. is conducting clinical trials to find the least harmful dose or non-toxic concentration of doxorubicin-loaded on dissolvable MNs at 25–200 μg for treating noduled or superficial basal cell carcinoma; however, no report has been released as of yet. The majority of microneedle clinical trials have been conducted on vaccinations, cancer medicine delivery, diabetes treatment, and microorganism identification [1].

8.9 Regulations of MNs Patches

There is an exponential increase in the number of medical products based on MN for therapeutic uses. However, because submissions must take the form of combination goods utilizing MNs, the FDA clearance process is not simple. Such submissions must include sufficient details regarding the product analysis, testing, and validations, including risk analysis, content uniformity, stability testing (mechanical properties, API migration, formulation), sterility validation, and manufacturing. According to the FDA, "the safety and effectiveness questions associated with each constituent and the product as a whole must be taken into account" while regulating combination products [78, 77]. The regulatory bodies' present licencing strategy for microneedle products involves product-specific approval rather than specific MN systems, which causes significant delays in approval and limits the commercialization of MNs. In order to facilitate the commercialization of MN goods, licencing laws pertaining to shape, formulation, sterilization, and packaging should be clearly established, and cGMP and quality control should be combined. However, the development of microneedle devices for clinical use might proceed independently from the formulation of drugs or vaccines. This may make the regulatory procedures much simpler and enable the technology to be included in the supply chains of specific medications more quickly. Even if only tiny amounts of a chemical may reach their desired delivery locations, the ease of use of this strategy may mitigate the regulatory issues that other complex formulation techniques confront. Strong guidelines are also needed in order to fully classify MN-based products; however, it has been suggested that in the case of monitoring diagnostic applications, this would probably fall under the medical device category, and in the case of drug or vaccine delivery, it would probably fall under the category of "combination product" (drug and device) or "drug product". After making this distinction, it might be able to modify current quality control protocols for MNs. The inherent variations between transdermal patches and hypodermic needles may mean that the present conventional quality control methodologies are not entirely appropriate for use with MN products. The objective of introducing MN-based products to the transdermal market could soon come to pass if all outstanding issues can be appropriately resolved to satisfy the requirements of both regulators and patients [79].

8.10 Challenges and Limitations

In the foreseeable future, approving the transfer of MNs from research labs to pertinent businesses is a thrilling but challenging process. Some important concerns and obstacles should be taken into consideration in order to transition this revolutionary technology from the lab bench to viable products in the relevant markets. One necessity for MN patches is that they must be able to pierce the skin enough. The properties of the skin, which might differ from person to person and throughout the body, should also be considered while handling this issue. The way MNs insert and penetrate the skin to overcome its elasticity depends on a number of factors, including length, interspace, base and tip diameters, and shape. When constructing MNs for clinical purposes, one parameter that needs to be considered early on is their geometry. According to a recent study, the geometric structure of microneedle arrays influences the mechanical strength and penetrating capabilities of MNs. The maximum amount of medication that a coated microneedle device can bolus dose is around 1 mg. While hollow MNs enable "as-needed/on-demand" dosing or continuous infusion, squeezed skin tissue following microneedle insertion may block central exits.

MNs can penetrate the skin's barrier qualities, but their effectiveness is mostly reliant on the biological formulation's passive diffusion into the skin. Large dosages may be challenging to deliver as a result, and a significant portion of the dose may be lost on the skin's surface. Because of this, there has been hesitation to employ this technology for some clinical applications due to the timing of application and the difficulty to monitor dose administration. Protein medications, such as insulin, erythropoietin, glucagon, growth hormones, and parathyroid hormones, are difficult to store and administer because bio-macromolecules quickly degrade and become inactive. Therefore, it is difficult to use MNs for the delivery of biomedicine. The skin is an extremely receptive organ to the MN administration of any medicinal substance due to its immunogenic nature. As a side effect, mild and transient erythema may appear based on the medication's size, composition, and kind. During clinical trials, skin irritation, sensitization, and immunological response need to be assessed as part of MN product safety evaluations. For microchip-based MNs to be fully used in therapeutic applications, large-scale production of MNs will require improvements to current manufacturing procedures. Although comprehensive economic evaluations of the technology have not yet been completed, it is easy to assume that, similar to any new technology, the clinical use of MNs may be relatively costly because of the intricate fabrication and storage processes as well as the drawn-out approval process. Another issue that needs to be considered early on when MN-based goods are intended for commercial use is MN patch sterilization. The most popular techniques, such as moist heat, gamma or microwave radiation, and ethylene oxide, may negatively impact any cargoes containing sensitive ingredients, such as biomolecules, vaccines, peptides, and the MNs themselves. This makes method selection crucial if sterilization is required [80]. Drug pharmacokinetics and side effects are among the many drawbacks of conventional therapy options for breast cancer, which include injectable or oral chemotherapeutic medicines. According to the findings that have been published in the literature thus far, using MNs as part of a chemotherapy regimen for breast cancer can be effective and have very few negative effects. According to reports compiled in this chapter, MNs may be used extensively in the future to treat breast cancer. There is proof that MNs can be used for medicinal, immunization, and diagnostic purposes. MNs have the potential to reduce medication metabolism by the liver, cause less pain and invasiveness, and enable continuous and regulated drug delivery into breast tissue. Depending on the patient's characteristics for personalized medicine, a variety of different drug devices will probably increase the choice of delivery mechanism due to the variety of materials available to be utilized in the manufacture of MNs. Drugs could be released into the skin one after the other in a sequential manner using a layer-by-layer manufacturing approach. In the future, simultaneous drug release might be offered to provide a more comprehensive therapy platform for breast cancer. Nevertheless, there are still not many thorough studies on the use of MNs in the treatment of breast cancer, and this topic needs more research.

References

1. Khan, S., Hasan, A., Attar, F., Babadaei, M.M., Zeinabad, H.A., Salehi, M., Alizadeh, M., Hassan, M., Derakhshankhah, H., Hamblin, M.R., Bai, Q.: Diagnostic and drug release systems based on microneedle arrays in breast cancer therapy. J. Control. Release **10**(338), 341–357 (2021)
2. McCullough, A.E., Dell'Orto, P., Reinholz, M.M., Gelber, R.D., Dueck, A.C., Russo, L., Jenkins, R.B., Andrighetto, S., Chen, B., Jackisch, C., Untch, M.: Central pathology laboratory review of HER2 and ER in early breast cancer: an ALTTO trial [BIG 2–06/NCCTG N063D (Alliance)] ring study. Breast Cancer Res. Treat. **143**, 485–492 (2014)
3. ElOuassif, B., Idri, A., Hosni, M., Abran, A.: Classification techniques in breast cancer diagnosis: a systematic literature review. Comput. Methods Biomech. Biomed. Eng.: Imaging Vis. **9**(1), 50–77 (2021)
4. Lin, C., Clark, R., Tu, P., Bosworth, H.B., Zullig, L.L.: Breast cancer oral anticancer medication adherence: a systematic review of psychosocial motivators and barriers. Breast Cancer Res. Treat. **165**, 247–260 (2017)
5. Hamza, A., Elrefaey, S.: Non-surgical treatment of early breast cancer: techniques on the way. Gland Surg. **3**(3), 149 (2014)
6. Hynynen, K.: MRI-guided focused ultrasound treatments. Ultrasonics **50**(2), 221–229 (2010)

7. Siegel, R., Naishadham, D., Jemal, A.: Cancer statistics, 2013. CA: Cancer J. Clin. **63**(1) (2013)
8. Hsieh, C.C., Trichopoulos, D., Katsouyanni, K., Yuasa, S.: Age at menarche, age at menopause, height and obesity as risk factors for breast cancer: associations and interactions in an international case-control study. Int. J. Cancer **46**(5), 796–800 (1990)
9. Ritte, R., Lukanova, A., Berrino, F., Dossus, L., Tjønneland, A., Olsen, A., Overvad, T.F., Overvad, K., Clavel-Chapelon, F., Fournier, A., Fagherazzi, G.: Adiposity, hormone replacement therapy use and breast cancer risk by age and hormone receptor status: a large prospective cohort study. Breast Cancer Res. **14**, 1–4 (2012)
10. Kelsey, J.L., Gammon, M.D., John, E.M.: Reproductive factors and breast cancer. Epidemiol. Rev. **15**(1), 36–47 (1993)
11. Rosner, B., Colditz, G.A., Willett, W.C.: Reproductive risk factors in a prospective study of breast cancer: the Nurses' Health Study. Am. J. Epidemiol. **139**(8), 819–835 (1994)
12. Chen, W.Y., Rosner, B., Hankinson, S.E., Colditz, G.A., Willett, W.C.: Moderate alcohol consumption during adult life, drinking patterns, and breast cancer risk. JAMA **306**(17), 1884–1890 (2011)
13. Lahmann, P.H., Hoffmann, K., Allen, N., Van Gils, C.H., Khaw, K.T., Tehard, B., Berrino, F., Tjønneland, A., Bigaard, J., Olsen, A., Overvad, K.: Body size and breast cancer risk: findings from the European prospective investigation into cancer and nutrition (EPIC). Int. J. Cancer **111**(5), 762–771 (2004)
14. Milazzo, G., Giorgino, F., Damante, G., Sung, C., Stampfer, M.R., Vigneri, R., Goldfine, I.D., Belfiore, A.: Insulin receptor expression and function in human breast cancer cell lines. Can. Res. **52**(14), 3924–3930 (1992)
15. Gunter, M.J., Hoover, D.R., Yu, H., Wassertheil-Smoller, S., Rohan, T.E., Manson, J.E., Li, J., Ho, G.Y., Xue, X., Anderson, G.L., Kaplan, R.C.: Insulin, insulin-like growth factor-I, and risk of breast cancer in postmenopausal women. J. Natl Cancer Inst. **101**(1), 48–60 (2009)
16. Colditz, G.A., Kaphingst, K.A., Hankinson, S.E., Rosner, B.: Family history and risk of breast cancer: nurses' health study. Breast Cancer Res. Treat. **133**, 1097–1104 (2012)
17. Sieri, S., Krogh, V., Bolelli, G., Abagnato, C.A., Grioni, S., Pala, V., Evangelista, A., Allemani, C., Micheli, A., Tagliabue, G., Schunemann, H.J.: Sex hormone levels, breast cancer risk, and cancer receptor status in postmenopausal women: the ORDET cohort. Cancer Epidemiol. Biomark. Prev. **18**(1), 169–176 (2009)
18. Kontomanolis, E.N., Koutras, A., Syllaios, A., Schizas, D., Mastoraki, A., Garmpis, N., Diakosavvas, M., Angelou, K., Tsatsaris, G., Pagkalos, A., Ntounis, T.: Role of oncogenes and tumor-suppressor genes in carcinogenesis: a review. Anticancer Res. **40**(11), 6009–6015 (2020)
19. Makhoul, I., Atiq, M., Alwbari, A., Kieber-Emmons, T.: Breast cancer immunotherapy: an update. Breast Cancer: Basic Clin. Res. **30**(12), 1178223418774802 (2018)
20. Shah, R., Sabir, S., Alhawaj, A.F.: Physiology, breast milk
21. Malhotra, G.K., Zhao, X., Band, H., Band, V.: Histological, molecular and functional subtypes of breast cancers. Cancer Biol. Ther. **10**(10), 955–960 (2010)
22. Sarvari, P., Sarvari, P., Ramírez-Díaz, I., Mahjoubi, F., Rubio, K.: Advances of epigenetic biomarkers and epigenome editing for early diagnosis in breast cancer. Int. J. Mol. Sci. **23**(17), 9521 (2022)
23. Koh, J., Kim, M.J.: Introduction of a new staging system of breast cancer for radiologists: an emphasis on the prognostic stage. Korean J. Radiol. **20**(1), 69–82 (2019)
24. Waks, A.G., Winer, E.P.: Breast cancer treatment: a review. Jama **321**(3), 288–300 (2019)
25. Hammond, M.E., Hayes, D.F., Dowsett, M., Allred, D.C., Hagerty, K.L., Badve, S., Fitzgibbons, P.L., Francis, G., Goldstein, N.S., Hayes, M., Hicks, D.G.: American Society of Clinical Oncology/College of American Pathologists guideline recommendations for immunohistochemical testing of estrogen and progesterone receptors in breast cancer. J. Clin. Oncol. **28**(16), 2784–2795 (2010)
26. Piccart-Gebhart, M.J., Procter, M., Leyland-Jones, B., Goldhirsch, A., Untch, M., Smith, I., Gianni, L., Baselga, J., Bell, R., Jackisch, C., Cameron, D.: Trastuzumab after adjuvant chemotherapy in HER2-positive breast cancer. N. Engl. J. Med. **353**(16), 1659–1672 (2005)

27. Wolff, A.C., Hammond, M.E., Hicks, D.G., Dowsett, M., McShane, L.M., Allison, K.H., Allred, D.C., Bartlett, J.M., Bilous, M., Fitzgibbons, P., Hanna, W.: Recommendations for human epidermal growth factor receptor 2 testing in breast cancer: American Society of Clinical Oncology/College of American Pathologists clinical practice guideline update. J. Clin. Oncol. **31**(31), 3997–4013 (2013)
28. Foulkes, W.D., Smith, I.E., Reis-Filho, J.S.: Triple-negative breast cancer. N. Engl. J. Med. **363**(20), 1938–1948 (2010)
29. Denkert, C., Liedtke, C., Tutt, A., von Minckwitz, G.: Molecular alterations in triple-negative breast cancer—The road to new treatment strategies. Lancet **389**(10087), 2430–2442 (2017)
30. Atashzar, M.R., Baharlou, R., Karami, J., Abdollahi, H., Rezaei, R., Pourramezan, F., Zoljalali Moghaddam, S.H.: Cancer stem cells: a review from origin to therapeutic implications. J. Cell. Physiol. **235**(2), 790–803 (2020)
31. Barzaman, K., Karami, J., Zarei, Z., Hosseinzadeh, A., Kazemi, M.H., Moradi-Kalbolandi, S., Safari, E., Farahmand, L.: Breast cancer: biology, biomarkers, and treatments. Int. Immunopharmacol. **1**(84), 106535 (2020)
32. Al-Hajj, M., Wicha, M.S., Benito-Hernandez, A., Morrison, S.J., Clarke, M.F.: Prospective identification of tumorigenic breast cancer cells. Proc. Natl. Acad. Sci. **100**(7), 3983–3988 (2003)
33. Chekhun, S.V., Zadvorny, T.V., Tymovska, Y.O., Anikusko, M.F., Novak, O.E., Polishchuk, L.Z.: CD44+/CD24? markers of cancer stem cells in patients with breast cancer of different molecular subtypes. Exp. Oncol. **37**(1), 58–63 (2015)
34. Ganeson, K., Alias, A.H., Murugaiyah, V., Amirul, A.A., Ramakrishna, S., Vigneswari, S.: Microneedles for efficient and precise drug delivery in cancer therapy. Pharmaceutics **15**(3), 744 (2023)
35. Prausnitz, M.R.: Engineering microneedle patches for vaccination and drug delivery to skin. Annu. Rev. Chem. Biomol. Eng. **8**(1), 177–200 (2017)
36. Guillot, A.J., Cordeiro, A.S., Donnelly, R.F., Montesinos, M.C., Garrigues, T.M., Melero, A.: Microneedle-based delivery: an overview of current applications and trends. Pharmaceutics **12**(6), 569 (2020)
37. Larrañeta, E., Lutton, R.E., Woolfson, A.D., Donnelly, R.F.: Microneedle arrays as transdermal and intradermal drug delivery systems: materials science, manufacture and commercial development. Mater. Sci. Eng. R. Rep. **1**(104), 1–32 (2016)
38. Chen, Y., Chen, B.Z., Wang, Q.L., Jin, X., Guo, X.D.: Fabrication of coated polymer microneedles for transdermal drug delivery. J. Control. Release **10**(265), 14–21 (2017)
39. Rodgers, A.M., McCrudden, M.T., Vincente-Perez, E.M., Dubois, A.V., Ingram, R.J., Larrañeta, E., Kissenpfennig, A., Donnelly, R.F.: Design and characterization of a dissolving microneedle patch for intradermal vaccination with heat-inactivated bacteria: a proof of concept study. Int. J. Pharm. **549**(1–2), 87–95 (2018)
40. Lee, W.J., Han, M.R., Kim, J.S., Park, J.H.: A tearable dissolving microneedle system for shortening application time. Expert Opin. Drug Deliv. **16**(3), 199–206 (2019)
41. Ito, Y., Yoshimitsu, J.I., Shiroyama, K., Sugioka, N., Takada, K.: Self-dissolving microneedles for the percutaneous absorption of EPO in mice. J. Drug Target. **14**(5), 255–261 (2006)
42. Norman, J.J., Choi, S.O., Tong, N.T., Aiyar, A.R., Patel, S.R., Prausnitz, M.R., Allen, M.G.: Hollow microneedles for intradermal injection fabricated by sacrificial micromolding and selective electrodeposition. Biomed. Microdev. **15**, 203–210 (2013)
43. Wilke, N., Mulcahy, A., Ye, S.R., Morrissey, A.: Process optimization and characterization of silicon microneedles fabricated by wet etch technology. Microelectron. J. **36**(7), 650–656 (2005)
44. O'Mahony, C.: Structural characterization and in-vivo reliability evaluation of silicon microneedles. Biomed. Microdev. **16**, 333–343 (2014)
45. van der Maaden, K., Trietsch, S.J., Kraan, H., Varypataki, E.M., Romeijn, S., Zwier, R., van der Linden, H.J., Kersten, G., Hankemeier, T., Jiskoot, W., Bouwstra, J.: Novel hollow microneedle technology for depth-controlled microinjection-mediated dermal vaccination: a study with polio vaccine in rats. Pharm. Res. **31**, 1846–1854 (2014)

46. Luangveera, W., Jiruedee, S., Mama, W., Chiaranairungroj, M., Pimpin, A., Palaga, T., Srituravanich, W.: Fabrication and characterization of novel microneedles made of a polystyrene solution. J. Mech. Behav. Biomed. Mater. **1**(50), 77–81 (2015)
47. Xie, L., Zeng, H., Sun, J., Qian, W.: Engineering microneedles for therapy and diagnosis: a survey. Micromachines **11**(3), 271 (2020)
48. Martanto, W., Davis, S.P., Holiday, N.R., Wang, J., Gill, H.S., Prausnitz, M.R.: Transdermal delivery of insulin using microneedles in vivo. Pharm. Res. **21**, 947–952 (2004)
49. Gill, H.S., Prausnitz, M.R.: Coated microneedles for transdermal delivery. J. Control. Release **117**(2), 227–237 (2007)
50. Dharadhar, S., Majumdar, A., Dhoble, S., Patravale, V.: Microneedles for transdermal drug delivery: a systematic review. Drug Dev. Ind. Pharm. **45**(2), 188–201 (2019)
51. Dardano, P., Caliò, A., Di Palma, V., Bevilacqua, M.F., Di Matteo, A., De Stefano, L.: A photolithographic approach to polymeric microneedles array fabrication. Materials **8**(12), 8661–8673 (2015)
52. Gupta, J., Park, S.S., Bondy, B., Felner, E.I., Prausnitz, M.R.: Infusion pressure and pain during microneedle injection into skin of human subjects. Biomaterials **32**(28), 6823–6831 (2011)
53. Kim, J.D., Kim, M., Yang, H., Lee, K., Jung, H.: Droplet-born air blowing: Novel dissolving microneedle fabrication. J. Control. Release **170**(3), 430–436 (2013)
54. Gittard, S.D., Ovsianikov, A., Monteiro-Riviere, N.A., Lusk, J., Morel, P., Minghetti, P., Lenardi, C., Chichkov, B.N., Narayan, R.J.: Fabrication of polymer microneedles using a two-photon polymerization and micromolding process. J. Diabetes Sci. Technol. **3**(2), 304–311 (2009)
55. McGrath, M.G., Vrdoljak, A., O'Mahony, C., Oliveira, J.C., Moore, A.C., Crean, A.M.: Determination of parameters for successful spray coating of silicon microneedle arrays. Int. J. Pharm. **415**(1–2), 140–149 (2011)
56. Barba, D., León-Sosa, A., Lugo, P., Suquillo, D., Torres, F., Surre, F., Trojman, L., Caicedo, A.: Breast cancer, screening and diagnostic tools: all you need to know. Crit. Rev. Oncol. Hematol. **1**(157), 103174 (2021)
57. Morris, E.A.: Breast cancer imaging with MRI. Radiol. Clin. **40**(3), 443–466 (2002)
58. Mann, R.M., Balleyguier, C., Baltzer, P.A., Bick, U., Colin, C., Cornford, E., Evans, A., Fallenberg, E., Forrai, G., Fuchsjäger, M.H., Gilbert, F.J.: Breast MRI: EUSOBI recommendations for women's information. Eur. Radiol. **25**, 3669–3678 (2015)
59. Sharifi, M., Hasan, A., Attar, F., Taghizadeh, A., Falahati, M.: Development of point-of-care nanobiosensors for breast cancers diagnosis. Talanta **1**(217), 121091 (2020)
60. Miller, P.R., Narayan, R.J., Polsky, R.: Microneedle-based sensors for medical diagnosis. J. Mater. Chem. B **4**(8), 1379–1383 (2016)
61. Hamam, R., Hamam, D., Alsaleh, K.A., Kassem, M., Zaher, W., Alfayez, M., Aldahmash, A., Alajez, N.M.: Circulating microRNAs in breast cancer: novel diagnostic and prognostic biomarkers. Cell Death Dis. **8**(9), e3045 (2017)
62. Tang, S., Zhou, F., Sun, Y., Wei, L., Zhu, S., Yang, R., Huang, Y., Yang, J.: CEA in breast ductal secretions as a promising biomarker for the diagnosis of breast cancer: a systematic review and meta-analysis. Breast Cancer **23**, 813–819 (2016)
63. Chen, L., Zhang, C., Xiao, J., You, J., Zhang, W., Liu, Y., Xu, L., Liu, A., Xin, H., Wang, X.: Local extraction and detection of early stage breast cancers through a microneedle and nano-Ag/MBL film based painless and blood-free strategy. Mater. Sci. Eng. C **1**(109), 110402 (2020)
64. Yu, M., Lu, Z., Shi, Y., Du, Y., Chen, X., Kong, M.: Systematic comparisons of dissolving and swelling hyaluronic acid microneedles in transdermal drug delivery. Int. J. Biol. Macromol. **30**(191), 783–791 (2021)
65. Huang, H., Qu, M., Zhou, Y., Cao, W., Huang, X., Sun, J., Sun, W., Zhou, X., Xu, M., Jiang, X.: A microneedle patch for breast cancer screening via minimally invasive interstitial fluid sampling. Chem. Eng. J. **15**(472), 145036 (2023)

66. Hokkanen, A., Stuns, I., Schmid, P., Kokkonen, A., Steinecker, A., Budczies, J., Heimala, P., Hakalahti, L.: Microneedle based sampling for breast cancer tissues. In: 18th International Conference on Miniaturized Systems for Chemistry and Life Sciences, MicroTAS 2014, pp. 1154–1156 (2014)
67. Bhatnagar, S., Chawla, S.R., Kulkarni, O.P., Venuganti, V.V.K.: Zein microneedles for transcutaneous vaccine delivery: fabrication, characterization, and in vivo evaluation using ovalbumin as the model antigen. Acs Omega **2**(4), 1321–1332 (2017)
68. Bhatnagar, S., Kumari, P., Pattarabhiran, S.P., Venuganti, V.V.: Zein microneedles for localized delivery of chemotherapeutic agents to treat breast cancer: drug loading, release behavior, and skin permeation studies. AAPS PharmSciTech **19**, 1818–1826 (2018)
69. Bhatnagar, S., Bankar, N.G., Kulkarni, M.V., Venuganti, V.V.: Dissolvable microneedle patch containing doxorubicin and docetaxel is effective in 4T1 xenografted breast cancer mouse model. Int. J. Pharm. **10**(556), 263–275 (2019)
70. Gao, X., Patel, M.G., Bakshi, P., Sharma, D., Banga, A.K.: Enhancement in the transdermal and localized delivery of honokiol through breast tissue. AAPS PharmSciTech **19**, 3501–3511 (2018)
71. Courtenay, A.J., McCrudden, M.T., McAvoy, K.J., McCarthy, H.O., Donnelly, R.F.: Microneedle-mediated transdermal delivery of bevacizumab. Mol. Pharm. **15**(8), 3545–3556 (2018)
72. Jha, A., Kumar, M., Goswami, P., Manjit, M., Bharti, K., Koch, B., Mishra, B.: Hyaluronic acid-oleylamine and chitosan-oleic acid conjugate-based hybrid nanoparticle delivery via. dissolving microneedles for enhanced treatment efficacy in localized breast cancer. Biomater. Adv. **160**, 213865
73. Matta, A., Janardhanam, L.S., Venuganti, V.V.: Dissolvable layered microneedle patch containing 5-fluorouracil for localized treatment of oral carcinoma. J. Chem. Sci. **135**(2), 23 (2023)
74. Chablani, L., Tawde, S.A., Akalkotkar, A., D'Souza, M.J.: Evaluation of a particulate breast cancer vaccine delivered via skin. AAPS J. **21**, 1–1 (2019)
75. Mojeiko, G., de Brito, M., Salata, G.C., Lopes, L.B.: Combination of microneedles and microemulsions to increase celecoxib topical delivery for potential application in chemoprevention of breast cancer. Int. J. Pharm. **5**(560), 365–376 (2019)
76. Ahmed, K.S., Shan, X., Mao, J., Qiu, L., Chen, J.: Derma roller® microneedles-mediated transdermal delivery of doxorubicin and celecoxib co-loaded liposomes for enhancing the anticancer effect. Mater. Sci. Eng. C **1**(99), 1448–1458 (2019)
77. US Department of Health and Human Services. Use of International Standard ISO 10993-1.: Biological Evaluation of Medical Devices–Part 1: evaluation and Testing Within a Risk Management Process. Silver Spring, MD: US Department of Health and Human Services. Food and Drug Administration, Center for Devices and Radiological Health (2016)
78. Donnelly, R.F., Mooney, K., Caffarel-Salvador, E., Torrisi, B.M., Eltayib, E., McElnay, J.C.: Microneedle-mediated minimally invasive patient monitoring. Ther. Drug Monit. **36**(1), 10–17 (2014)
79. Lutton, R.E., Moore, J., Larrañeta, E., Ligett, S., Woolfson, A.D., Donnelly, R.F.: Microneedle characterization: the need for universal acceptance criteria and GMP specifications when moving towards commercialization. Drug Deliv. Transl. Res. **5**, 313–331 (2015)
80. Avcil, M., Çelik, A.: Microneedles in drug delivery: progress and challenges. Micromachines **12**(11), 1321 (2021)

Chapter 9
Microneedles as Carriers: A Futuristics Delivery System for Medicines and Cosmetics

Fuad Saleh, Zakarya A. M. Hazaea, and Waleed Alahmad

9.1 Introduction

Drugs have been administered in several methods to enhance health and prolong life. Since eating medicinal leaves, drug delivery methods have drastically improved to include capsules, tablets, injectables, and implantable devices [1]. Transdermal drug delivery (TDD) is one of the drug delivery routes that focuses on administrating the drug through the layers of the skin. This technique attempts to regulate the skin's diffusion to transfer the medication molecules into the bloodstream [2]. Hypodermic needles and topical creams are the most common methods for delivering drugs through the skin. However, needles are often avoided by patients due to associated pain, while topical creams typically show limited bioavailability. This is largely due to the skin's natural structure, which presents a significant barrier to drug absorption. The skin comprises three main layers: the outermost stratum corneum, the middle epidermis, and the dermis, the thickest layer. The stratum corneum serves as the primary barrier, restricting passage to certain molecules, particularly those that are lipophilic and of low molecular weight. Its limited permeability poses a challenge for developing effective topical formulations, often requiring innovative approaches to enhance drug penetration [3, 4]. Microneedle (MNs) offers an advanced transdermal drug delivery system designed to replace conventional syringe injections. Developed to address limitations associated with hypodermic needles and transdermal patches, MNs serve as a hybrid solution that combines the benefits of both methods. By gently penetrating the stratum corneum, the microneedle array facilitates effective drug delivery with minimal invasiveness, offering a patient-friendly

F. Saleh · W. Alahmad (✉)
Department of Chemistry, Faculty of Science, Chulalongkorn University, Bangkok, Thailand
e-mail: waleed.alahmad.cu@gmail.com; waleed.al@chula.ac.th

Z. A. M. Hazaea
Department of Chemical Technology, Faculty of Science, Chulalongkorn University, Bangkok, Thailand

N. Talreja et al. (eds.), *Microneedles (MNs)-Based Technology*, Series in BioEngineering, https://doi.org/10.1007/978-981-96-3916-8_9

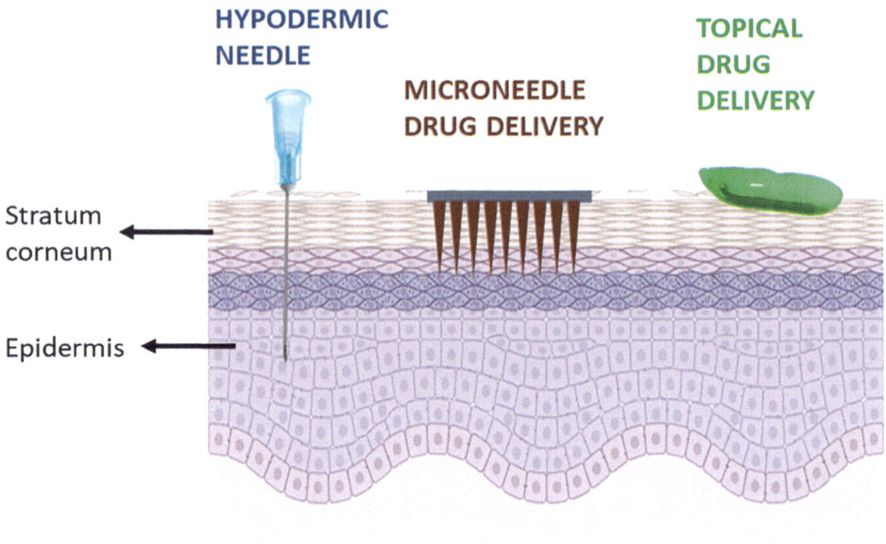

Fig. 9.1 Different types of transdermal drug delivery systems; this figure illustrates three major transdermal drug delivery systems: hypodermic needles, microneedles, and topical applications. The hypodermic needle penetrates deep into the dermal layers for direct injection, the microneedle system allows for minimally invasive delivery just beneath the skin surface, and topical drug delivery is applied on the outermost layer, the epidermis, providing non-invasive administration of medications

alternative for efficient medication administration (Fig. 9.1) [5]. MNs is regarded as one of the most efficient methods for transdermal drug delivery, as medications administered through this technique bypass essential organs like the liver [6].

Furthermore, it reduces the pain caused by hypodermic injection needles by giving a painless experience. As a consequence, it serves as the greatest option for persons with needle phobia (trypanophobia). The application of microneedle transdermal drug delivery eliminates the need for trained professionals, making it easy to use [7–9]. MNs arrays consist of tiny needles, generally measuring between 25 and 2000 μm in height, and have been utilized in various fields such as drug and vaccine delivery, cosmetics, and disease diagnostics. These MNs exhibit a variety of structural designs, shapes, forms, and materials, and are manufactured using different techniques. As highlighted by Donnelly et al., 30% of the latest scientific research on "transdermal delivery technology" centers on studies [7, 10]. Various types of MNs have been designed and studied for drug delivery applications, including solid, coated, dissolving, hollow, and hydrogel-based MNs, each with distinct properties (Fig. 9.2). These MNs employ different mechanisms to deliver drugs into the epidermis: some create microchannels in the stratum corneum, others are coated with drug solutions on their surface, dissolve upon insertion, or are pre-filled with the drug solution. Each type offers unique advantages tailored to specific drug delivery needs, enhancing therapeutic effectiveness and patient comfort [11–15].

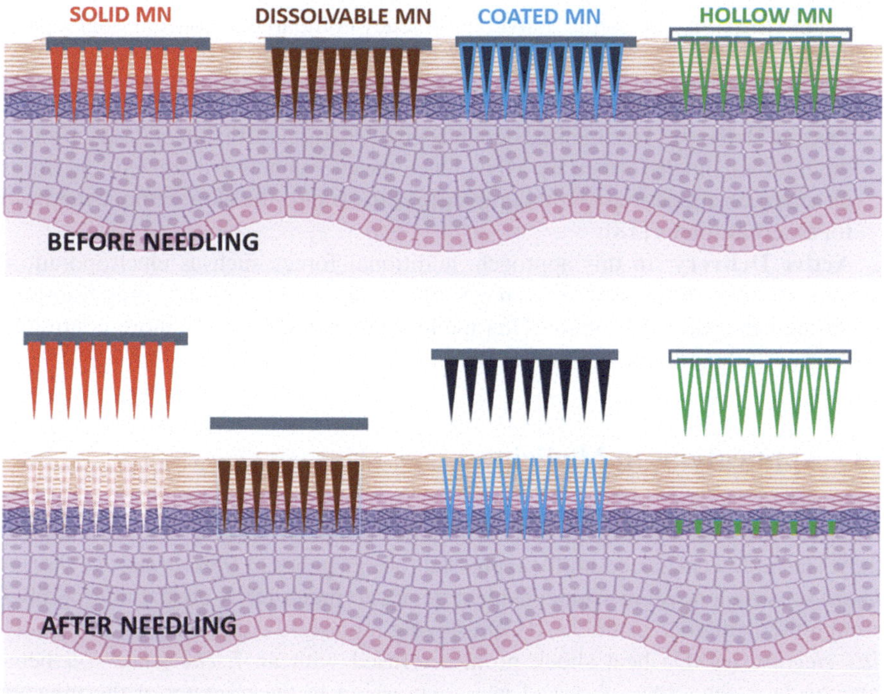

Fig. 9.2 Different types of microneedles; This figure illustrates the mechanisms of various microneedle types before and after skin insertion. **Solid microneedles (red)**: Use the "poke with patch" method to pre-treat the skin, creating microchannels for subsequent drug application. **Dissolving microneedles (brown)**: Made of biodegradable polymers that dissolve upon skin penetration, releasing the drug directly into the skin. **Coated microneedles (blue)**: Employ the "coat and poke" approach, where a drug solution is coated on the needle surface and released into the skin after insertion. **Hollow microneedles (green)**: Filled with drug solutions deposited into the dermis through the hollow channels after insertion

9.2 Mechanisms of Delivery

Topical drug delivery generally relies on diffusion, but microneedle technology temporarily disrupts the skin barrier to enhance drug absorption. A microneedle device typically consists of hundreds of MNs arranged in arrays on a compact patch, designed to deliver an effective drug dose for the desired therapeutic outcome. This approach enables efficient and controlled drug administration through the skin, offering a promising alternative to traditional methods [16, 17]. MNs create tiny, painless microchannels in the skin's outermost layer, the stratum corneum. Their length is precisely engineered to bypass this barrier without reaching deeper nerve endings, ensuring a minimally invasive and comfortable experience for the patient. This design allows for effective drug delivery while minimizing discomfort and potential side effects. These micro-channels facilitate the direct absorption of drugs or cosmetic agents into the skin, enhancing delivery efficiency compared to conventional topical methods.

Two mechanisms can be employed in microneedle technology:

1. **Passive Delivery**: This method depends on the natural diffusion of drugs through the microchannels formed by MNs. Once these channels are created, the drugs or active cosmetic ingredients gradually diffuse into the skin without the need for external forces [18].
2. **Active Delivery**: In this approach, additional forces such as electroporation, iontophoresis, or pressure-driven systems are applied to enhance drug transport through the micro-channels. This method enables a faster or more controlled release of the drug or substance [19].

9.3 Applications in Medicine

Historically, MNs have been utilized as effective drug carriers (Fig. 9.3). In a notable study, Hasmi S. et al. successfully expressed foreign genes in *Heterorhabditis bacteriophora* using a novel micromechanical device to inject DNA-coated microprobes through the nematode's cuticle. By employing a plasmid containing a *Caenorhabditis elegans* 16-kDa heat shock promoter fused with an *Escherichia coli* beta-galactosidase gene, they observed gene expression in the progeny of the injected nematodes. This approach highlights the versatility of MNs for genetic material delivery across biological systems. This approach highlights the versatility of MNs for genetic material delivery across biological systems. The microprobe tip penetrated the cuticle without injuring the nematode, with approximately 8% of the progeny exhibiting the foreign gene. This study establishes a new, efficient, and user-friendly genetic transformation method for nematodes [20].

Influenza vaccines are commonly delivered intramuscularly, often causing pain and requiring trained healthcare professionals to administer and safely dispose of biohazardous sharps. Microstructure patch technologies have recently been developed as a pain-free alternative for influenza vaccine delivery. This innovative method simplifies vaccine administration and leverages the skin's active immune cells to elicit a strong immune response [6, 21]. Amanda et al. evaluated the MicroCor transdermal system for delivering a cell culture-derived trivalent influenza vaccine. Mechanical strength tests conducted on excised pig skin demonstrated successful skin penetration and tip dissolution within just 5 min of insertion. Guinea pigs vaccinated with the influenza-loaded microstructures showed hemagglutinin inhibition (HI) and IgG titers comparable to those achieved with intramuscular injections. Following two rounds of immunization, serum HI titers in all vaccinated groups exceeded 40, marking a more than four-fold increase over the untreated group. These findings highlight the potential of skin-delivery technologies for effective administration of cell culture-derived influenza vaccines [22]. Anto Vardoljak et al. developed an innovative microfluidic, drop-dispensing method for producing dissolvable MNs, addressing the limitations associated with traditional dissolvable microneedle systems. Similarly,

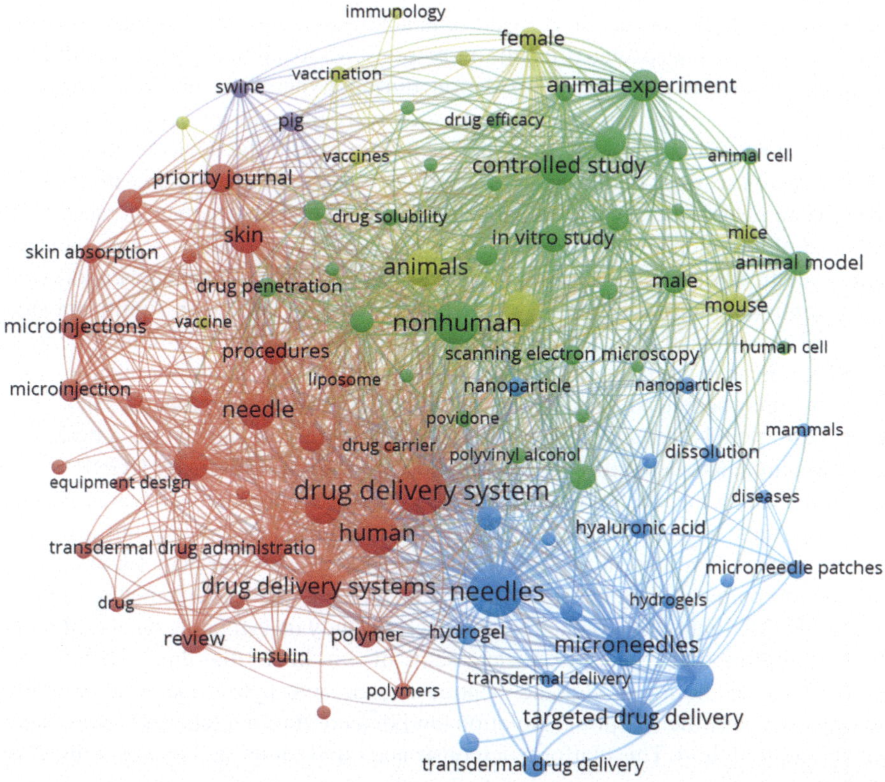

Fig. 9.3 Keywords visualization; Co-occurrence network visualization of author keywords from microneedles drug delivery research papers. This figure illustrates the interconnectedness of various keywords, highlighting the most prevalent themes and topics in the literature. The size of each keyword node reflects its frequency of occurrence, while the links between nodes indicate the strength of association between terms, providing insights into current research trends and focus areas within microneedle technology

Mei-Chin et al. addressed the challenge of complete microneedle insertion by developing a fully insertable MNs system composed of poly-γ-glutamic acid (γ-PGA) MNs supported by polyvinyl alcohol (PVA) and polyvinyl pyrrolidone (PVP) structures. These supporting materials were engineered to enhance insertion depth and mechanical strength, ensuring full skin penetration. Upon insertion, both the MNs and their support structures dissolved within 4 min, enabling rapid drug release. This system was tested for diabetes treatment by delivering insulin-loaded MNs to diabetic rats daily for two days. Results showed that the hypoglycemic effect in rats treated with insulin-loaded MNs was comparable to that achieved with subcutaneous insulin injections, underscoring the method's feasibility and effectiveness [23].

Jicheng Yu et al. developed an innovative glucose-responsive microneedle-array patch for rapid, biocompatible, and painless insulin delivery. This "smart insulin patch" utilizes hypoxia-sensitive, hyaluronic acid-based vesicles that rapidly disassociate to release encapsulated insulin in response to localized hypoxic conditions triggered by enzymatic glucose oxidation during hyperglycemia. This enzyme-triggered, glucose-responsive mechanism effectively regulates blood glucose levels in type 1 diabetic mice, swiftly restoring them to normal. The device demonstrates a faster response than traditional pH-sensitive formulations, reducing the risk of hypoglycemia and offering a promising advancement in diabetes management [24]. Xi Xie et al. broadened the application of microneedle technology to pain management, specifically targeting neuropathic pain, which is often inadequately addressed by conventional pharmacological treatments and is associated with significant side effects. They utilized dissolvable MNs to deliver calcitonin gene-related peptide (CGRP), a neuropeptide implicated in the pathophysiology of chronic neuropathic pain. The microneedle patches demonstrated effective analgesic properties in neuropathic pain models, achieving pain relief without disrupting normal nociception or motor function, a benefit attributed to the peptide's high specificity for CGRP receptors. Importantly, the patches did not elicit skin irritation or systemic adverse effects. These findings suggest that dissolvable microneedle patches delivering a CGRP antagonist peptide represent a safe, efficient, and user-friendly strategy for managing neuropathic pain, offering substantial advantages over existing therapeutic options [25]. D. Al Sulaiman et al. conducted a pioneering study investigating the use of MNs for the sampling and sensing of nucleic acids in skin interstitial fluid. They developed MNs coated with an alginate–peptide nucleic acid hybrid material, enabling the sequence-specific sampling, isolation, and detection of nucleic acid biomarkers from interstitial fluid. This platform demonstrated rapid sampling kinetics, achieving a volume capacity of approximately 6.5 μL within just 2 min. The technology allows for the detection of specific nucleic acid biomarkers either directly on the patch or in solution after light-triggered release from the hydrogel. Given the clinical significance of cell-free nucleic acids as biomarkers in bodily fluids, this minimally invasive and automated platform holds substantial promise for personalized diagnostics and the ongoing monitoring of disease progression tailored to individual patients [26]. SN Economidou et al. developed 3D-printed microneedle arrays using a biocompatible resin through stereolithography (SLA) for transdermal insulin delivery. The MNs were created by polymerizing successive layers of photopolymer resin, and thin layers of insulin were deposited onto the needle surfaces using inkjet printing, along with sugar alcohol or disaccharide carriers. This optimized printing process significantly enhanced the skin penetration capabilities of the 3D-printed MNs compared to traditional metal arrays. In vivo animal trials demonstrated rapid insulin action, effectively controlling hypoglycemia and reducing glucose levels within 60 min. Moreover, steady-state plasma glucose levels were maintained for up to 4 h, showcasing superior performance compared to subcutaneous injections. These findings highlight the potential of 3D-printed microneedle technology for improving insulin delivery methods [27–29]. Methotrexate (MTX) is a first-line treatment for moderate to severe psoriasis; however, its oral and injectable forms are often hindered by side effects. Transdermal drug delivery offers a promising alternative. H. Du et al. developed a dissolving microneedle (MN) patch using hyaluronic acid (HA) due to its excellent water solubility, biocompatibility, biodegradability, and mechanical strength. The encapsulation of MTX within the MNs was precisely controlled during fabrication to ensure accurate dosing. Importantly, the MTX-loaded MNs effectively penetrated the imiquimod (IMQ)-induced thickened epidermis in mice, allowing for intralesional drug delivery. The rapid dissolution of HA made the MNs user-friendly for patients. In vitro studies demonstrated that the MTX-loaded MNs had a strong inhibitory effect and effectively alleviated psoriasis-like skin inflammation in mice, proving significantly more effective than oral MTX at equivalent doses. In contrast, oral MTX required higher doses to achieve similar therapeutic outcomes, resulting in increased systemic toxicity. These findings underscore the potential of dissolving microneedle technology for enhancing the efficacy and safety of MTX in the treatment of psoriasis. Overall, the MTX-loaded dissolving MN patch offers a novel, efficient, and safer strategy for psoriasis treatment [30–33]. Existing therapies for bacterial biofilms frequently face challenges due to inadequate antibiotic penetration through biofilm barriers, unintended toxicity, and the emergence of antibiotic resistance [34, 35]. To address these challenges, J. Xu et al. created a microneedle patch-mediated treatment aimed at enhancing the effectiveness of biofilm elimination. This innovative system utilizes self-dissolving MNs, whose tips are infused with gelatin nanoparticles sensitive to gelatinase and loaded with chloramphenicol (CAM) (CAM@GNPs). This microneedle patch can penetrate biofilms and deliver antibiotics directly to regions of active bacterial growth. Notably, the microneedle-mediated treatment demonstrated greater effectiveness in treating *Vibrio vulnificus* biofilms compared to free drug solutions. This innovative approach could significantly enhance the targeted delivery of various antimicrobial agents to biofilm-infected areas, improving treatment outcomes [36]. A novel microneedle has been developed utilizing bioinspired backward-facing curved barbs to enhance tissue adhesion, employing digital light processing (DLP) 3D printing techniques. These backward-facing barbs are formed through a dissolving-induced deformation process that utilizes a cross-linking density gradient within a photocurable polymer. By adjusting printing parameters and material composition, researchers can precisely control the thickness and bending curvature of the barbs, optimizing their performance for improved tissue integration. This innovative design holds significant promise for advancing the efficacy of microneedle-based delivery systems in various biomedical applications. This bioinspired design significantly enhances tissue adhesion, with the barbed MN demonstrating 18 times stronger adhesion compared to barbless MNs. Furthermore, barbed MNs facilitate sustained drug release within tissues. The enhanced tissue adhesion establishes a more stable and reliable platform for drug delivery, biofluid collection, and biosensing applications. This improved stability not only facilitates effective therapeutic interventions but also ensures consistent performance in diagnostic and monitoring tasks, ultimately advancing the utility of microneedle technology across various biomedical fields [37]. The integration of sampling and immediate metabolite readout can significantly enhance patient compliance. An innovative all-in-one transdermal colorimetric microneedle patch was developed for real-time glucose monitoring in mice, eliminating the need for complex laboratory equipment. This patch detects hyperglycemia through a visible color change triggered by the enzymatic reactions of glucose oxidase (GOx) and horseradish peroxidase (HRP) in response to elevated glucose levels. To improve sensitivity and reduce nonspecific reactions, HRP in the upper layer of the patch is biomineralized with a calcium phosphate (CaP) shell, which also imparts pH responsiveness. This design not only simplifies glucose monitoring but also enhances the accuracy and reliability of metabolic assessments in real-time. The patch minimally invasively extracts interstitial fluid and immediately translates glucose concentrations into a color change. The color intensity can be quantitatively analyzed by scanning the patch and interpreting the red, green, and blue (RGB) values. This low-cost, portable sensor offers a non-invasive alternative for daily glucose monitoring, eliminating the need for traditional blood tests [38, 39]. MNs have revolutionized painless localized drug delivery through the skin, but their effectiveness is often hindered by the slow diffusion of molecules, frequently requiring external stimuli. To address this challenge, an innovative autonomous and degradable active microneedle delivery system has been developed. This platform integrates magnesium microparticles within the microneedle patch, which function as an internal propulsion mechanism, enabling deeper and faster intradermal drug delivery. This advancement enhances the efficiency of drug administration while ensuring that the system is both effective and environmentally friendly [40]. The magnesium particles react with interstitial fluid, rapidly generating hydrogen (H_2) bubbles that produce sufficient force to penetrate dermal barriers, thereby significantly enhancing drug penetration. The release kinetics of this active microneedle system were evaluated in vitro by assessing the passage of IgG antibodies (used as a model drug) through phantom tissue and pigskin barriers. In vivo studies conducted on a B16F10 mouse melanoma model demonstrated that the active delivery of anti-CTLA-4, a checkpoint inhibitor drug, elicited a markedly stronger immune response and significantly improved survival rates. These results highlight the potential of this innovative microneedle system for enhancing therapeutic efficacy in cancer treatment [40]. Additionally, the platform features spatially organized regions of active and passive MNs, facilitating a combination of rapid burst delivery and slow, sustained release. This versatile, autonomous microneedle delivery technology holds significant promise for various therapeutic applications, providing enhanced treatment outcomes, improved patient convenience, and cost-effectiveness. By integrating both delivery mechanisms, this innovative system can optimize therapeutic efficacy while minimizing the frequency of administration, ultimately leading to better patient adherence and overall treatment success [41, 42]. Microneedle arrays provide a promising minimally invasive solution for bypassing the skin barrier and achieving effective transdermal drug delivery. This study investigates the use of solid polymer-based ion-conductive porous MNs (PMN) with interconnected micropores to enhance iontophoresis, a technique that facilitates transdermal molecular transport via direct current. The PMN, modified with a charged hydrogel, offers several advantages for iontophoresis, including reduced transdermal resistance due to minimally invasive punctures in the resistant stratum corneum, improved transport of larger molecules through the interconnected micropores, and the generation of electroosmotic flow (EOF). The EOF produced by the PMN significantly enhances the transdermal penetration and extraction of molecules, effectively mimicking pressure-induced flow.

This increased efficiency has been demonstrated through EOF-assisted delivery of a model drug, dextran, and glucose extraction from pig skin samples. These findings highlight the potential of PMN technology to improve drug delivery and biomarker extraction in transdermal applications. Additionally, the PMN-based EOF system can be powered by a built-in enzymatic biobattery, such as a fructose/O_2 battery, representing a viable option for a fully organic iontophoresis patch suitable for transdermal applications [43–45]. Cell therapies targeting skin disorders can greatly benefit from innovative technologies designed for the transdermal delivery of therapeutic cells [46]. A novel design and proof-of-concept application for cryogenic microneedle patches have been developed to facilitate the transdermal delivery of living cells. This innovative approach leverages cryogenic technology to preserve cell viability during the delivery process, ensuring that the cells remain functional upon administration. By utilizing MNs for minimally invasive penetration of the skin, this method enhances the efficiency and effectiveness of cell delivery while minimizing discomfort for patients. This groundbreaking technique holds significant promise for advancing regenerative medicine and targeted therapies that rely on the precise delivery of live cells [47].

These MNs are fabricated using a stepwise cryogenic micro-molding technique, which incorporates pre-suspended cells within a cryogenic medium. They can be easily inserted into porcine skin and dissolve after delivering the cells. Studies conducted in mice demonstrated that the cells delivered via cryogenic MNs retained their viability and proliferative capacity. In a subcutaneous melanoma model, the administration of ovalbumin-pulsed dendritic cells through these MNs resulted in enhanced antigen-specific immune responses and reduced tumor growth compared to traditional intravenous and subcutaneous injection methods. These findings suggest that cryogenic microneedle technology could represent a significant advancement in the transdermal delivery of living cells for cancer immunotherapy and other therapeutic applications. This biocompatible cryogenic microneedle technology shows considerable potential for enabling minimally invasive cell delivery across various therapeutic applications [48]. The inherent properties of therapeutic proteins, including their large molecular size and susceptibility to degradation, pose significant challenges for transdermal delivery. MNs offer a solution by painlessly penetrating the stratum corneum to facilitate the direct translocation of protein drugs into the systemic circulation. Engineered MNs can incorporate suitable structural materials and therapeutics with tailored physicochemical properties, enabling local and systemic drug delivery across various applications, including vaccination, diabetes management, and cancer therapy. This chapter evaluates the design and application of polymeric MNs for transdermal protein delivery while discussing their clinical potential and future translation [49]. Eye diseases and injuries pose significant clinical challenges due to the complexities associated with effective ocular drug delivery, primarily hindered by various ocular barriers. To address this issue, an innovative eye patch equipped with detachable MNs has been developed. This patch is designed to penetrate the ocular surface tissue and function as micro-reservoirs for controlled drug delivery. By facilitating direct access to the targeted tissues, this advanced delivery system aims to enhance therapeutic efficacy and improve treatment outcomes for patients suffering from ocular conditions. The double-layered design allows for biphasic drug release, enhancing therapeutic efficacy. In a corneal neovascularization model, the eye patch demonstrated a substantial reduction in neovascular area through the delivery of an anti-angiogenic agent. Additionally, rapid release of an anti-inflammatory compound, followed by sustained release of the anti-angiogenic agent, produced a synergistic therapeutic effect. This minimally invasive eye patch represents a promising strategy for home-based treatment of various eye conditions, ensuring improved patient compliance [50].

9.4 Applications in Cosmetics

Healthy skin plays a crucial role in psychological well-being and social interactions. Research indicates a direct link between skin conditions and mental health, revealing that individuals with skin issues are often more vulnerable to depression, social isolation, loneliness, and a diminished quality of life. A range of strategies has been employed to improve skin conditions, including topical treatments, chemical peels, and both ablative and non-ablative laser photo-rejuvenation. These approaches are commonly utilized to address various skin concerns, such as wrinkles, signs of aging, acne scars, and hyperpigmentation. By enhancing the appearance and health of the skin, these treatments can contribute to improved mental health outcomes and overall quality of life [51–53]. Recently, the microneedling technique, commonly referred to as percutaneous collagen induction (PCI), has gained substantial popularity within the cosmetics industry (Fig. 9.4). This technique utilizes various microneedling devices, including the Dermaroller and Dermapen, as well as several types of fabricated MNs patches. These patches can be categorized into solid, hollow, dissolving, coated, and hydrogel MNs, each offering unique advantages for skin rejuvenation and treatment. By promoting collagen production and enhancing product absorption, microneedling has emerged as a versatile and effective method for addressing a range of skin concerns, further solidifying its position in aesthetic dermatology [54].

MNs technology has been effectively utilized to improve cutaneous scarring by stimulating the production of collagen and elastin within the dermis, thereby facilitating collagen remodeling and enhancing the thickness of both the epidermis and dermis. This technique is particularly advantageous for addressing acne scars, which often result in significant emotional and psychosocial distress for affected individuals. By promoting skin regeneration and improving overall texture, microneedling not only aids in the physical appearance of scars but also contributes to enhanced emotional well-being for those seeking treatment [55]. In a study conducted by J. Sharad et al., the efficacy of combining microneedling with glycolic acid (GA) for acne scar treatment was examined. Results showed a significant improvement in the healing of superficial and moderately deep scars, with enhanced skin texture and reduced acne pigmentation [56]. Saadawi et al. conducted a comparative study to evaluate the effectiveness of combining microneedling with glycolic acid (GA) peels versus administering each treatment individually for the management of acne

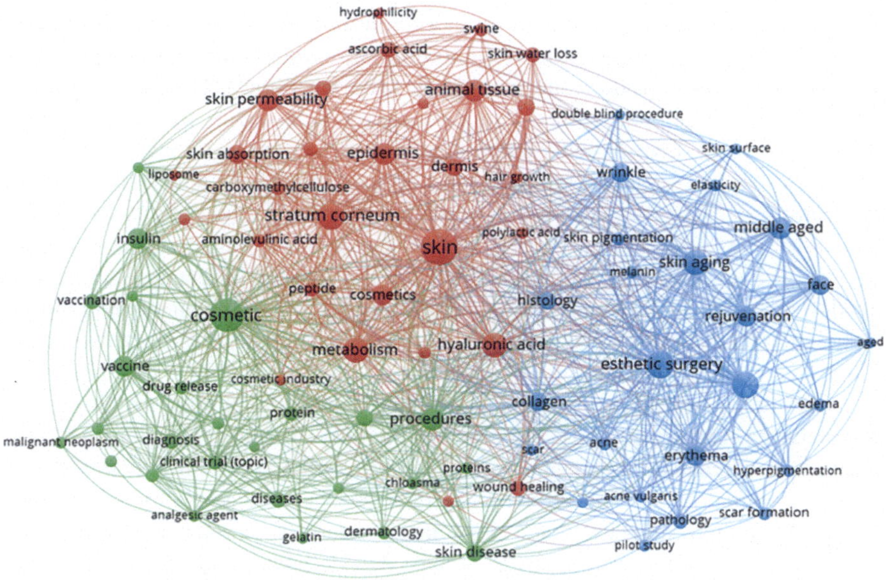

Fig. 9.4 Keywords visualization; Co-occurrence network visualization of author keywords from research papers on cosmetics microneedles. This figure depicts the relationships between various keywords, illustrating the most commonly discussed topics and themes in the field. The size of each keyword node corresponds to its frequency of occurrence, while the connections between nodes indicate the degree of association among the terms, offering insights into emerging trends and focal points in cosmetics applications of microneedle technology

scars. Thirty participants were randomly assigned to one of three groups, with each group undergoing six treatment sessions at two-week intervals. The study concluded that the combination therapy provided superior results in treating acne scars compared to using microneedling or GA peel alone [57]. Another study demonstrated that combining microneedling with the topical application of platelet-rich plasma (PRP) significantly accelerated wound healing in atrophic acne scars by enhancing nucleogenesis and promoting scar remodeling. Additionally, the authors observed that pairing microneedling with chemical peeling using 15% trichloroacetic acid (TCA) resulted in marked improvements in the treatment of severe atrophic acne scars. This synergistic approach not only enhances the efficacy of scar treatment but also offers patients a more comprehensive solution for addressing their skin concerns [58]. Moreover, studies have indicated that microneedling is effective in enhancing the appearance of varicella scars in teenagers with darker skin tones [59]. B. Ali et al. conducted a comparative study to evaluate the efficacy and safety of microneedling using a Dermapen versus superficial chemical peeling with Jessner's solution—a formulation containing salicylic acid, resorcinol, and lactic acid in 95% ethanol—for the treatment of atrophic acne scars. The results demonstrated that the combined treatment of Dermapen and Jessner's solution yielded the most significant clinical improvement, necessitating fewer treatment sessions compared to either microneedling or Jessner's solution used individually. This study underscores the potential benefits of combining these modalities to optimize treatment outcomes for atrophic acne scars [60]. Microneedling was performed using a Dermastamp® equipped with 2.1 mm stainless steel needles to address hypertrophic burn scars. Five patients underwent 8–12 treatment sessions at two-week intervals, with each session comprising 3–4 applications of the device, resulting in the creation of 200–300 micro-holes in the scar tissue. Evaluations based on the Vancouver Scar Scale and measurements of scar depth revealed substantial clinical improvement following the treatment. The enhancement in scar appearance was linked to microneedling, which promoted collagen fiber rearrangement and reduced scar height [61]. Scars are a common and challenging outcome of surgery. MNs have been utilized in the treatment of postsurgical scars, demonstrating collagen production in treated patients. Notably, the data revealed that initiating treatment during the early maturation phase, around 6–7 weeks postoperatively, enhanced collagen formation at a time when natural collagen synthesis was declining [62]. Initially introduced for skin rejuvenation, microneedling has broadened its applications to include acne scars, post-traumatic and burn scars, alopecia, hyperhidrosis, stretch marks, and drug delivery. Significant innovations over the past decade have improved the effectiveness of microneedling instruments. This technique can also be combined with other surgical methods to further improve treatment outcomes. Notably, microneedling is considered particularly safe for individuals with darker skin types, where the risk of post-inflammatory pigmentation is increased with other procedures. Numerous studies have shown the effectiveness of microneedling in addressing various types of scars, including surgical scars, highlighting its versatility and suitability for diverse patient populations [63–66].

Melasma is an acquired hyperpigmentation disorder that predominantly affects sun-exposed areas of the face, significantly diminishing the quality of life for those affected. Microneedling serves as an effective adjunctive therapy for melasma by optimizing the transdermal delivery of topical agents, thereby facilitating their penetration into the epidermis and dermis. This technique enhances therapeutic efficacy and promotes targeted treatment of hyperpigmented lesions [67]. EA Lima et al. assessed the effectiveness of microneedling with a Dermaroller in combination with depigmentation therapy and sunscreen in a study involving 22 patients with recalcitrant melasma that had proven resistant to conventional topical lightening treatments. The results suggested that skin lightening could be achieved through the alterations induced by the moderate injury inflicted by the Dermaroller. However, the exact mechanisms driving the depigmentation effects associated with microneedling have yet to be fully clarified [68]. S. Farshi et al. conducted a pilot study to compare the effectiveness of microneedling combined with a meso-depigmentation solution (mesoneedling) against standard microneedling over a four-month treatment period. In this investigation, 20 patients received microneedling on one side of their face and mesoneedling on the other side. The findings revealed that 2–4 sessions of mesoneedling were significantly more effective in treating melasma compared to standard microneedling alone [69]. S. Shamsi Meymandi et al. investigated the effectiveness of microneedling combined with 4% tranexamic acid in comparison to 4%

hydroquinone for treating melasma. In this study, 60 patients diagnosed with melasma were randomly assigned to two groups: Group A received monthly microneedling sessions paired with topical 4% tranexamic acid, while Group B was treated with nightly applications of 4% hydroquinone. The microneedling device utilized featured six fine needles, each measuring 1500 μm in length and 0.25 mm in width, allowing for skin penetration ranging from 0.1 to 1.3 mm. The results indicated that the efficacy of microneedling with 4% tranexamic acid was comparable to that of 4% hydroquinone in effectively managing melasma [70]. Microneedling demonstrates potential in addressing both melasma (hyperpigmentation) and vitiligo (hypopigmentation), employing distinct mechanisms for each condition. In the case of melasma, microneedling enhances the absorption of topical treatments and promotes skin remodeling, thereby reducing pigmentation. Conversely, for vitiligo, microneedling may stimulate melanocyte activity and improve skin regeneration, contributing to repigmentation. These varying approaches highlight microneedling's versatility as a therapeutic option for diverse pigmentation disorders [68, 71].

MNs has shown efficacy in treating both androgenetic alopecia and alopecia areata. Research by Faghihi et al. demonstrated that combining microneedling, performed with an adjustable-depth electrical pen device, with a 5% minoxidil lotion yielded superior therapeutic outcomes in terms of hair count, thickness, and growth compared to minoxidil used alone. This enhanced effect is attributed to microneedling's ability to release platelet-derived growth factors through platelet activation and to stimulate the overexpression of genes associated with hair growth. Furthermore, microneedling is believed to promote collagen deposition, enhance the release of growth factors, and improve the absorption of treatments like minoxidil and topical steroids [72–74]. Recent preclinical studies have highlighted the advantages of integrating MNs with therapeutic exosomes for promoting hair regeneration. Yang et al. developed detachable MNs utilizing a water-soluble hyaluronic acid base infused with therapeutic exosomes and the small-molecule drug UK5099. The MN matrix was constructed from natural hair protein keratin, which encapsulated hair follicle stem cells within the exosomes while incorporating UK5099 into poly lactic-co-glycolic acid (PLGA) nanoparticles. Both delivery systems were integrated into keratin-hydrogel-based MNs. The results demonstrated that this innovative approach significantly enhanced pigmentation and stimulated hair regrowth in treated mice within six days by facilitating the transition from the telogen to the anagen phase and acting as a depot for the sustained release of therapeutics. Furthermore, the MN-exosomal system exhibited superior clinical efficacy compared to the subcutaneous injection of exosomes and the topical application of UK5099. Additionally, the combination of therapeutic exosomes with MNs effectively activated dermal papilla cells, which are crucial for hair regeneration [75, 76]. MNs were engineered using a combination of hyaluronic acid (HA) and swellable polyvinyl alcohol (PVA) to create needles loaded with chitosan lactate and exosome-encapsulated dermal papilla cells (DPCs). The chitosan lactate component released L-lactate, which enhanced cell growth by activating lactate dehydrogenase. Concurrently, the exosomes enabled a sustained release of DPCs, stimulating cell proliferation through the activation of the Wnt/β-catenin signaling pathway, a vital mechanism in hair follicle regeneration.

The results indicated that the MNs effectively penetrated the cuticle, facilitating the delivery of human adipose mesenchymal stem cell (hAMSC) exosomes into the dermis, thereby stimulating hair follicle stem cells. Additionally, yellow-light irradiation was employed to reduce inflammation in the hair follicles, further promoting hair regeneration. This innovative approach resulted in significant hair growth within seven days while minimizing inflammation, highlighting its substantial potential for clinical applications [77, 78].

9.5 Clinical Efficacy and Safety

MNs have been validated for drug delivery in humans; however, additional clinical and safety evaluations are necessary for wider approval. A frequently employed metric for assessing pain intensity in clinical trials is the Visual Analog Scale (VAS). Research indicates that the application of 150-μm-long solid MNs yields significantly lower VAS scores compared to traditional 26-gauge hypodermic needles, measuring 1.7 mm versus 59.8 mm, respectively. This suggests a markedly reduced perception of pain associated with the use of MNs [79]. In terms of skin reactions, increasing the application velocity of solid MNs from 4.3 to 8.5 m/s resulted in the occurrence of small bleeding spots, while slower application speeds did not produce significant skin redness. Additionally, a greater tip radius (increasing from 10 to 20 μm) and an extended needle length (from 100 to 300 μm) were linked to heightened skin redness. These findings highlight the critical influence of microneedle design and application speed on the skin's physiological response [80]. Studies on pretreatment with solid MNs have primarily focused on their effectiveness when combined with topical pharmaceutical formulations, such as in anesthetic therapy. Pretreatment with 70 μm solid MNs has been shown to reduce the time required for topical anesthetics to take effect by threefold, with a significant increase in pain reduction (39% in the intradermal group compared to 19% without pretreatment). Additionally, in another study involving lidocaine application, MN pretreatment reduced the required incubation time from 60 to 30 min, highlighting the efficacy of MNs in accelerating anesthetic onset [81].

Participants in various studies involving dissolving MNs have consistently reported minimal to no pain during their application. For instance, in one study utilizing maltose-based dissolving MNs measuring 500 μm in length, participants experienced no discomfort during needle insertion. Another investigation that assessed dissolving MNs of varying lengths (300, 500, and 800 μm) found that although Visual Analog Scale (VAS) scores increased with needle length, they remained low, ranging from 10 mm to under 20 mm. Additionally, self-administered 600 μm MN patches were reported as pain-free, while participants using a 650 μm MN patch on the forearm similarly noted no pain. Additionally, in a separate study using 800 μm MN patches for influenza treatment, 96% of participants reported no pain. These findings highlight the comfort and low pain sensation associated with dissolving MNs [82, 83]. Erythema was observed immediately after the removal of an 800 μm long dissolving microneedle (MN) patch, MicroHyala®, which was loaded with all-trans retinoic acid for the treatment of seborrheic keratosis. The mild redness reaction disappeared within 7 days, and participants' skin returned to normal within 30 days post-application, demonstrating a temporary and mild skin response to the treatment [84]. Safety evaluations of dissolving MNs have shown favorable outcomes, with no significant local or systemic adverse events observed [81]. Dissolving MNs have gained attention as a promising method for drug delivery, particularly in vaccination. The MicroHyala® system, which uses 800 μm dissolving MNs, successfully delivered a 15 μg TIV influenza vaccine with equivalent immunogenicity to subcutaneous (SC) administration. Similar immune responses were observed between dissolving MN patches and intramuscular (IM) injections. Notably, no significant differences were found between self-administration and professional application in delivering identical vaccine doses. Research also indicated that larger MN patches (10 cm × 12 cm) were equally effective as smaller patches, supporting the potential for scaling up MN patch size for transdermal drug delivery [83, 85]. A phase II clinical trial evaluated a novel transdermal microneedle patch system (ZP-PTH) for administering parathyroid hormone (PTH) in the treatment of advanced osteoporosis in men and postmenopausal women. The 6-month outpatient study involved the daily application of 190-μm coated MNs, and no instances of skin infection or sensitization at the patch site were reported, demonstrating the safety and efficacy of this delivery method [86]. Another notable example of a coated microneedle system is the Nanopatch™, which features a high-density array of short silicon MNs, each with a height of 230 μm. This innovative system is specifically engineered for the transdermal delivery of vaccines, including those targeting influenza and human papillomavirus-like particles. Furthermore, the Nanopatch™ is currently undergoing clinical trials for its application in polio vaccination [87, 88]. The average Visual Analog Scale (VAS) score for hydrogel-forming MNs measuring 600 μm in length, with a density of 361 needles per 0.5 cm^2, was significantly higher compared to an array with only 121 needles per 0.5 cm^2. Additionally, VAS scores for both microneedle designs showed a gradual increase throughout the 24 h treatment period. In a separate study using the same microneedle system, all volunteers exhibited minor erythema, which occurred regardless of the duration of microneedle insertion, suggesting that this erythema may have been attributable to the initial insertion process. Notably, there were no reports of local skin reactions or infections [89, 90]. The ability of hydrogel-forming MNs to swell while maintaining their structural integrity has garnered significant interest for applications in fluid extraction from the skin, especially in diagnostic settings. A recent study demonstrated that fluid uptake through hydrogel-forming MN arrays could be accurately quantified, resulting in the extraction of approximately 0.9–2.7 μL of fluid after one hour. Additionally, the mass of the MNs increased by about 30% after six hours of insertion into the skin, highlighting their effectiveness in fluid absorption [90].

9.6 Challenges and Future Directions

MNs have generated considerable interest in various transdermal biomedical applications, including biosensing and drug delivery, owing to their painless, semi-invasive, and sustainable nature. However, challenges remain in optimizing the materials and fabrication methods to achieve specific shapes, configurations, and functionalities that align with targeted biomedical applications [91]. Additionally, there is a limited selection of suitable biomaterials, along with challenges such as inadequate mechanical strength, poor control over drug delivery, and restrictions on drug loading capacity [92]. Potent drugs requiring low doses and vaccines are among the most suitable candidates for delivery in therapeutically effective concentrations via MNs. However, the delivery of macromolecules, particularly biotechnological products, poses significant challenges due to their high molecular weight and hydrophilicity, complicating their transdermal absorption. Additionally, certain MNs made from silicon or various polymers may lack sufficient mechanical strength to effectively pierce the skin. The ideal MN design would combine low insertion force with high fracture force, streamlining the application process to avoid the cumbersome two-step method of skin portion followed by patch application. In this context, hydrogel-forming MNs present a compelling option. Furthermore, it is crucial to balance penetration enhancement with the need for a painless application.

9.7 Conclusion

In conclusion, MNs offer a highly promising platform for enhancing the efficacy and safety of drug and cosmetic delivery systems. Their unique ability to penetrate the skin with minimal pain or discomfort makes them an appealing alternative to conventional delivery methods. The diverse applications discussed highlight the potential of MNs to address significant needs across medical and cosmetic fields, from improving vaccine delivery to supporting skin health through nutrient administration. Despite these advantages, challenges remain—such as technical limitations and regulatory hurdles-that must be addressed to unlock their full potential. Continued research and innovation will be critical to overcoming these barriers and driving broader market acceptance of microneedle technologies. Future advancements are expected to refine their capabilities further, solidifying MNs as a cornerstone in the evolution of transdermal delivery systems.

Acknowledgements Waleed Alahmad would like to thank Department of Chemistry, Faculty of Science, Chulalongkorn University and the Second Century Fund (C2F), Chulalongkorn University, Thailand.

Declaration of Competing Interest The authors declare that they have no known competing financial interests or personal relationships that could have appeared to influence the work reported in this paper.

References

1. Ranade, V.V., Hollinger, M.A., Cannon, J.B.: Drug Delivery Systems. CRC Press (2003)
2. Han, T., Das, D.B.: Potential of combined ultrasound and microneedles for enhanced transdermal drug permeation: a review. Eur. J. Pharm. Biopharm. **89**, 312–328 (2015)
3. Bora, P., Kumar, L., Bansal, A.K.: Microneedle technology for advanced drug delivery: evolving vistas. Curr. Res. Inf. Pharm. Sci. **9**(1), 7–10 (2008)
4. Ita, K.: Transdermal delivery of drugs with microneedles—Potential and challenges. Pharmaceutics **7**(3), 90–105 (2015)
5. Donnelly, R.F., et al.: Microneedle-Mediated Transdermal and Intradermal Drug Delivery (2012)
6. Kim, Y.-C., Park, J.-H., Prausnitz, M.R.: Microneedles for drug and vaccine delivery. Adv. Drug Deliv. Rev. **64**(14), 1547–1568 (2012)
7. Museau, M., Butdee, S., Vignat, F.: Design and manufacturing of microneedles toward sustainable products. Appl. Sci. Eng. Progr. **4**(4), 41–51 (2011)
8. Giudice, E.L., Campbell, J.D.: Needle-free vaccine delivery. Adv. Drug Deliv. Rev. **58**(1), 68–89 (2006)
9. Amsden, B., Goosen, M.: Transdermal delivery of peptide and protein drugs: an overview. AIChE J. **41**(8), 1972–1997 (1995)
10. Donnelly, R.F., et al.: Microneedles for Drug and Vaccine Delivery and Patient Monitoring. Wiley
11. Cheung, K., Han, T., Das, D.B.: Effect of force of microneedle insertion on the permeability of insulin in skin. J. Diabetes Sci. Technol. **8**(3), 444–452 (2014)
12. Suzuki, M., Takahashi, T., Aoyagi, S.: 3D laser lithographic fabrication of hollow microneedle mimicking mosquitos and its characterisation. Int. J. Nanotechnol. **15**(1–3), 157–173 (2018)
13. Prausnitz, M.R.: Engineering microneedle patches for vaccination and drug delivery to skin. Annu. Rev. Chem. Biomol. Eng. **8**(1), 177–200 (2017)
14. Li, J., et al.: Microneedle patches as drug and vaccine delivery platform. Curr. Med. Chem. **24**(22), 2413–2422 (2017)
15. Waghule, T., et al.: Microneedles: a smart approach and increasing potential for transdermal drug delivery system. Biomed. Pharmacother. **109**, 1249–1258 (2019)
16. Akhtar, N.: Microneedles: an innovative approach to transdermal delivery-a review. Int. J. Pharm. Pharm. Sci. **6**(4), 18–25 (2014)
17. Sharma, D.: Microneedles: an approach in transdermal drug delivery: a Review. PharmaTutor **6**(1), 7–15 (2018)
18. Avcil, M., Çelik, A.: Microneedles in drug delivery: progress and challenges. Micromachines (Basel) **12**(11) (2021)
19. Charoo, N.A., et al.: Electroporation: an avenue for transdermal drug delivery. Curr. Drug Deliv. **7**(2), 125–136 (2010)
20. Hashmi, S., et al.: Genetic transformation of nematodes using arrays of micromechanical piercing structures. Biotechniques **19**(5), 766–770 (1995)
21. Prausnitz, M.R.: Microneedles for transdermal drug delivery. Adv. Drug Deliv. Rev. **56**(5), 581–587 (2004)
22. Bonificio, A., et al.: Fabrication of cell culture-derived influenza vaccine dissolvable microstructures and evaluation of immunogenicity in guinea pigs. Vaccine **33**(25), 2930–2938 (2015)
23. Chen, M.-C., Ling, M.-H., Kusuma, S.J.: Poly-γ-glutamic acid microneedles with a supporting structure design as a potential tool for transdermal delivery of insulin. Acta Biomater. **24**, 106–116 (2015)
24. Yu, J., et al.: Microneedle-array patches loaded with hypoxia-sensitive vesicles provide fast glucose-responsive insulin delivery. Proc. Natl. Acad. Sci. **112**(27), 8260–8265 (2015)
25. Xie, X., et al.: Analgesic microneedle patch for neuropathic pain therapy. ACS Nano **11**(1), 395–406 (2017)

26. Al Sulaiman, D., et al.: Hydrogel-coated microneedle arrays for minimally invasive sampling and sensing of specific circulating nucleic acids from skin interstitial fluid. ACS Nano **13**(8), 9620–9628 (2019)
27. Economidou, S.N., et al.: 3D printed microneedle patches using stereolithography (SLA) for intradermal insulin delivery. Mater. Sci. Eng. C **102**, 743–755 (2019)
28. Lee, I.-C., et al.: Fabrication of two-layer dissolving polyvinylpyrrolidone microneedles with different molecular weights for in vivo insulin transdermal delivery. RSC Adv. **7**(9), 5067–5075 (2017)
29. Yu, W., et al.: Polymer microneedles fabricated from alginate and hyaluronate for transdermal delivery of insulin. Mater. Sci. Eng. C **80**, 187–196 (2017)
30. Du, H., et al.: Hyaluronic acid-based dissolving microneedle patch loaded with methotrexate for improved treatment of psoriasis. ACS Appl. Mater. Interfaces **11**(46), 43588–43598 (2019)
31. Nguyen, H.X., Banga, A.K.: Delivery of methotrexate and characterization of skin treated by fabricated PLGA microneedles and fractional ablative laser. Pharm. Res. **35**(3), 68 (2018)
32. Abla, M.J., et al.: Transdermal delivery of methotrexate for pediatrics using silicon microneedles. Ther. Deliv. **4**(5), 543–551 (2013)
33. Vemulapalli, V., et al.: Synergistic effect of iontophoresis and soluble microneedles for transdermal delivery of methotrexate. J. Pharm. Pharmacol. **60**(1), 27–33 (2010)
34. Koo, H., et al.: Apigenin and tt-Farnesol with fluoride effects on S. mutans biofilms and dental caries. J. Dental Res. **84**(11), 1016–1020 (2005)
35. Koo, H., et al.: Inhibition of Streptococcus mutans biofilm accumulation and polysaccharide production by apigenin and tt-farnesol. J. Antimicrob. Chemother. **52**(5), 782–789 (2003)
36. Xu, J., et al.: Microneedle patch-mediated treatment of bacterial biofilms. ACS Appl. Mater. Interfaces **11**(16), 14640–14646 (2019)
37. Han, D., et al.: 4D printing of a bioinspired microneedle array with backward-facing barbs for enhanced tissue adhesion. Adv. Func. Mater. **30**(11), 1909197 (2020)
38. Wang, Z., et al.: Transdermal colorimetric patch for hyperglycemia sensing in diabetic mice. Biomaterials **237**, 119782 (2020)
39. Chua, B., et al.: Sensing contact between microneedle array and epidermis using frequency response measurement. IEEE Sens. J. **14**(2), 333–340 (2013)
40. Pilon-Thomas, S., et al.: Neutralization of tumor acidity improves antitumor responses to immunotherapy. Can. Res. **76**(6), 1381–1390 (2016)
41. Lopez-Ramirez, M.A., et al.: Built-in active microneedle patch with enhanced autonomous drug delivery. Adv. Mater. **32**(1), 1905740 (2020)
42. Li, J., et al.: Micromotors spontaneously neutralize gastric acid for pH-responsive payload release. Angew. Chem. Int. Ed. **56**(8), 2156–2161 (2017)
43. Kusama, S., et al.: Transdermal electroosmotic flow generated by a porous microneedle array patch. Nat. Commun. **12**(1), 658 (2021)
44. Li, J., Li, D.: Electroosmotic flow velocity in DNA modified nanochannels. J. Colloid Interface Sci. **553**, 31–39 (2019)
45. Samant, P.P., Prausnitz, M.R.: Mechanisms of sampling interstitial fluid from skin using a microneedle patch. Proc. Natl. Acad. Sci. **115**(18), 4583–4588 (2018)
46. MacNeil, S.: Progress and opportunities for tissue-engineered skin. Nature **445**(7130), 874–880 (2007)
47. Bhatnagar, S., Dave, K., Venuganti, V.V.K.: Microneedles in the clinic. J. Control. Release **260**, 164–182 (2017)
48. Chang, H., et al.: Cryomicroneedles for transdermal cell delivery. Nat. Biomed. Eng. **5**(9), 1008–1018 (2021)
49. Ye, Y., et al.: Polymeric microneedles for transdermal protein delivery. Adv. Drug Deliv. Rev. **127**, 106–118 (2018)
50. Than, A., et al.: Self-implantable double-layered micro-drug-reservoirs for efficient and controlled ocular drug delivery. Nat. Commun. **9**(1), 4433 (2018)
51. Huang, Y., et al.: Research progress on cosmetic microneedle systems: preparation, property and application. Eur. Polymer J. **163**, 110942 (2022)

52. Krasuska, M., et al.: The role of adult attachment orientation and coping in psychological adjustment to living with skin conditions. Br. J. Dermatol. **178**(6), 1396–1403 (2018)
53. Ganceviciene, R., et al.: Skin anti-aging strategies. Dermatoendocrinol **4**(3), 308–319 (2012)
54. McCrudden, M.T., et al.: Microneedle applications in improving skin appearance. Exp. Dermatol. **24**(8), 561–566 (2015)
55. Harris, A.G., Naidoo, C., Murrell, D.F.: Skin needling as a treatment for acne scarring: an up-to-date review of the literature. Int. J. Womens Dermatol. **1**(2), 77–81 (2015)
56. Sharad, J.: Combination of microneedling and glycolic acid peels for the treatment of acne scars in dark skin. J. Cosmet. Dermatol. **10**(4), 317–323 (2011)
57. Saadawi, A.N., et al.: Microneedling by dermapen and glycolic acid peel for the treatment of acne scars: comparative study. J. Cosmet. Dermatol. **18**(1), 107–114 (2019)
58. Garg, S., Baveja, S.: Combination therapy in the management of atrophic acne scars. J. Cutan. Aesthet. Surg. **7**(1), 18–23 (2014)
59. Costa, I.M., Costa, M.C.: Microneedling for varicella scars in a dark-skinned teenager. Dermatol. Surg. **40**(3), 333–334 (2014)
60. Ali, B., ElMahdy, N., Elfar, N.N.: Microneedling (Dermapen) and Jessner's solution peeling in treatment of atrophic acne scars: a comparative randomized clinical study. J. Cosmet. Laser Ther. **21**(6), 357–363 (2019)
61. Kim, S., et al.: Management of hypertrophic scar after burn wound using microneedling procedure (dermastamp). Burns **35**, S37 (2009)
62. Claytor, R.B., Sheck, C.G., Chopra, V.: Microneedling outcomes in early postsurgical scars. Plast. Reconstr. Surg. **150**(3), 557e–561e (2022)
63. Singh, A., Yadav, S.: Microneedling: Advances and widening horizons. Indian Dermatol. Online J. **7**(4), 244–254 (2016)
64. Juhasz, M.L., Cohen, J.L.: Microneedling for the treatment of scars: an update for clinicians. Clin. Cosmetic Investigat. Dermatol. 997–1003 (2020)
65. Ramaut, L., et al.: Microneedling: where do we stand now? A systematic review of the literature. J. Plast. Reconstr. Aesthet. Surg. **71**(1), 1–14 (2018)
66. Lima, E., et al.: Correcting post-surgical scar using PCI. In: Percutaneous Collagen Induction With Microneedling: a Step-by-Step Clinical Guide, pp. 133–139 (2021)
67. Wu, S.Z., Muddasani, S., Alam, M.: A systematic review of the efficacy and safety of microneedling in the treatment of Melasma. Dermatol. Surg. **46**(12), 1636–1641 (2020)
68. Lima, E.D.A.: Microneedling in facial recalcitrant melasma: report of a series of 22 cases. Anais brasileiros de dermatologia **90**, 919–921 (2015)
69. Farshi, S., Mansouri, P.: Study of efficacy of microneedling and mesoneedling in the treatment of epidermal melasma: a pilot trial. J. Cosmet. Dermatol. **19**(5), 1093–1098 (2020)
70. Shamsi Meymandi, S., et al.: Efficacy of microneedling plus topical 4% tranexamic acid solution vs 4% hydroquinone in the treatment of melasma: a single-blind randomized clinical trial. J. Cosmet. Dermatol. **19**(11), 2906–2911 (2020)
71. Speeckaert, R., et al.: The impact of antioxidants on vitiligo and melasma: a scoping review and meta-analysis. Antioxidants **12**(12), 2082 (2023)
72. Martinez-Lopez, A., et al.: Advanced medical therapies in the management of non-scarring alopecia: areata and androgenic alopecia. Int. J. Mol. Sci. **21**(21), 8390 (2020)
73. Faghihi, G., et al.: Microneedling in androgenetic alopecia; comparing two different depths of microneedles. J. Cosmet. Dermatol. **20**(4), 1241–1247 (2021)
74. Fertig, R., et al.: Microneedling for the treatment of hair loss? J. Eur. Acad. Dermatol. Venereol. **32**(4), 564–569 (2018)
75. Yang, G., et al.: A therapeutic microneedle patch made from hair-derived keratin for promoting hair regrowth. ACS Nano **13**(4), 4354–4360 (2019)
76. Shi, Y., et al.: A drug-free, hair follicle cycling regulatable, separable, antibacterial microneedle patch for hair regeneration therapy. Adv. Healthc. Mater. **11**(19), 2200908 (2022)
77. Hong, C., et al.: Hair grows hair: Dual-effective hair regrowth through a hair enhanced dissolvable microneedle patch cooperated with the pure yellow light irradiation. Appl. Mater. Today **25**, 101188 (2021)

78. Choi, B.Y.: Targeting Wnt/β-catenin pathway for developing therapies for hair loss. Int. J. Mol. Sci. **21**(14), 4915 (2020)
79. Kaushik, S., et al.: Lack of pain associated with microfabricated microneedles. Anesth. Analg. **92**(2), 502–504 (2001)
80. Watanabe, T., Hagino, K., Sato, T.: Evaluation of the effect of polymeric microneedle arrays of varying geometries in combination with a high-velocity applicator on skin permeability and irritation. Biomed. Microdev. **16**(4), 591–597 (2014)
81. Nguyen, T.T., Park, J.H.: Human studies with microneedles for evaluation of their efficacy and safety. Expert Opin. Drug Deliv. **15**(3), 235–245 (2018)
82. Hirobe, S., et al.: Development and clinical study of a self-dissolving microneedle patch for transcutaneous immunization device. Pharm. Res. **30**, 2664–2674 (2013)
83. Rouphael, N.G., et al.: The safety, immunogenicity, and acceptability of inactivated influenza vaccine delivered by microneedle patch (TIV-MNP 2015): a randomised, partly blinded, placebo-controlled, phase 1 trial. Lancet **390**(10095), 649–658 (2017)
84. Hiraishi, Y., et al.: Development of a novel therapeutic approach using a retinoic acid-loaded microneedle patch for seborrheic keratosis treatment and safety study in humans. J. Control. Release **171**(2), 93–103 (2013)
85. Ripolin, A., et al.: Successful application of large microneedle patches by human volunteers. Int. J. Pharm. **521**(1–2), 92–101 (2017)
86. Daddona, P.E., et al.: Parathyroid Hormone (1–34)-coated microneedle patch system: clinical pharmacokinetics and pharmacodynamics for treatment of osteoporosis. Pharm. Res. **28**(1), 159–165 (2011)
87. Fernando, G.J., et al.: Nanopatch targeted delivery of both antigen and adjuvant to skin synergistically drives enhanced antibody responses. J. Control. Release **159**(2), 215–221 (2012)
88. Corbett, H.J., et al.: Skin vaccination against cervical cancer associated human papillomavirus with a novel micro-projection array in a mouse model. Plos One **5**(10), e13460 (2010)
89. Donnelly, R.F., et al.: Hydrogel-forming microneedle arrays for enhanced transdermal drug delivery. Adv. Func. Mater. **22**(23), 4879–4890 (2012)
90. Donnelly, R.F., et al.: Hydrogel-forming microneedles increase in volume during swelling in skin, but skin barrier function recovery is unaffected. J. Pharm. Sci. **103**(5), 1478–1486 (2014)
91. Luo, X., Yang, L., Cui, Y.: Microneedles: materials, fabrication, and biomedical applications. Biomed. Microdev. **25**(3), 20 (2023)
92. Lee, J., et al.: Rapid and repeatable fabrication of high A/R silk fibroin microneedles using thermally-drawn micromolds. Eur. J. Pharm. Biopharm. **94**, 11–19 (2015)

Chapter 10
Dissolvable Microneedle-Based Patch: A Next-Generation Solution for Tumor Detection and Treatment

Bo Zhi Chen and Xin Dong Guo

10.1 Introduction

Tumors are neoplasms formed by abnormal proliferation of local tissue cells in the body. According to their biological behavior and their impact on the body, tumors can be classified into benign tumors and malignant tumors. Benign tumors lack invasive and metastatic capabilities, posing minimal harm to the body. Malignant tumors, commonly referred to as cancer, typically exhibit invasive and metastatic properties, posing significant challenges in treatment and severely threatening human life and health. According to the World Health Organization (WHO), nearly 10 million people died from cancer globally in 2020, accounting for about one-sixth of the global death toll [1].

Traditional tumor treatment methods mainly include surgical treatment, chemotherapy, and radiotherapy. Surgical treatment involves the direct removal of localized tumors, achieving curative effects for some benign tumors. However, it is ineffective for tumor cells that have spread throughout the body, and surgery itself may increase the risk of tumor cell dissemination to other sites. Chemotherapy or radiotherapy is usually required for malignant tumor patients. However, while chemotherapy drugs inhibit tumor cell growth, they also severely affect the normal growth of healthy cells, causing significant damage to the patient's body. Radiotherapy is a local treatment method that utilizes radiation to treat tumors, but it may damage surrounding normal tissues, leading to complications such as radiation necrosis [2]. Therefore, the development of treatment methods with high efficacy,

B. Z. Chen · X. D. Guo (✉)
State Key Laboratory of Organic-Inorganic Composites, Beijing University of Chemical Technology, Beijing, China
e-mail: xdguo@buct.edu.cn

B. Z. Chen
e-mail: chenbz@buct.edu.cn

N. Talreja et al. (eds.), *Microneedles (MNs)-Based Technology*, Series in BioEngineering, https://doi.org/10.1007/978-981-96-3916-8_10

minimal side effects, low cost, and the ability to effectively suppress tumor growth and recurrence is crucial.

In recent years, new methods such as photothermal therapy, photodynamic therapy, gene therapy, and immunotherapy have been developed for tumor treatment. These methods aim to safely kill tumor cells without affecting the growth of healthy cells and have shown excellent potential for tumor suppression [3]. However, the complex administration routes of these therapies may cause serious side effects. Hence, effective delivery of therapeutic agents to tumor tissues is of paramount importance.

Transdermal drug delivery systems not only offer advantages such as painlessness and convenience of administration but also avoid first-pass metabolism by the liver, making them one of the most acceptable administration routes for patients. Therefore, they are widely used in tumor treatment. However, due to the barrier function of the skin's stratum corneum, drugs find it challenging to penetrate into tumor tissues completely. The introduction of microneedle (MN) technology effectively overcomes the barrier function of the stratum corneum, enabling direct delivery of therapeutic agents to the subcutaneous tissue [4]. The combination of microneedle-assisted delivery of tumor therapeutic drugs is beneficial not only for increasing the amount of anti-tumor agents reaching deep tumor tissues but also for reducing the extent of drug diffusion to adjacent tissues, thereby reducing drug side effects. Moreover, microneedles can also load different therapeutic drugs and achieve controlled release, providing a new approach for selecting the optimal treatment combination during the drug development stage. Additionally, microneedle technology plays a significant auxiliary role in tumor detection and diagnosis [5].

10.2 Biomedical Microneedles for Tumor Detection

Early detection and prevention of some cancers can significantly improve the life expectancy of cancer patients. Therefore, screening and diagnosis of cancer are important means of preventing cancer, including in vivo imaging and biochemical testing, and so forth [6]. Microneedles can play a unique role in cancer detection and diagnosis with their unique characteristics [7]. On one hand, microneedles can help medical personnel complete tissue fluid extraction, avoiding the inconvenience of conventional blood tests. On the other hand, microneedles can assist in the efficient delivery of colorants and other substances, thereby helping doctors accurately identify tumor sites through medical image detection.

To date, blood is the most commonly tested body fluid. However, the interstitial fluid in the skin also contains a wide range of molecular biomarkers from cells and surrounding capillaries, which are often overlooked due to the lack of efficient means to extract tissue fluid [8]. Swelling hydrogel microneedles are an effective means of tissue fluid extraction. As inserted into the skin, hydrogel microneedles may form a gel network that serves as a "storage point", where the surrounding tissue can permeate. After the microneedles are removed, the stored tissue fluid inside the needles can be used for subsequent component analysis [9]. Figure 10.1 shows a coating microneedle with (alginate-peptide nucleic acid) hybrid material. Polylactic acid is used as the matrix material of the microneedle to ensure that the microneedle has sufficient mechanical strength. The outer layer of alginate gel enriched with peptide nucleic acids (PAN), which is applied to specifically screened RNA substances corresponding to the sequence from tissue fluid for diagnosis. This technology features fast sampling, large capacity (approximately 6.5 μL of tissue interstitial fluid can be extracted within 2 min), strong specificity, and can detect different types of mRNA by customizing PNA sequences. By combining fluorescence detection with a light sensor platform, this technology can more accurately assist doctors in diagnosis [10]. Similarly, Yang et al. [11] constructed a hydrogel microneedle system, in which the functionalized hydrogel contains DNA triple helix nanostructures and methylene blue ligands, enabling targeted recognition of the target DNA. Sampling on the skin followed by detection using a wearable analysis pool can be used for monitoring of nasopharyngeal carcinoma.

Compared to biochemical component detection, colorimetric aided imaging diagnosis analysis, as a rapid and intuitive, is another important way of cancer diagnostic. Users can benefit from microneedle delivery of color rendering regents due to the ability of microneedles to overcome the first-pass effect and the barrier of the skin's stratum corneum. For example, Kim [12] and his coworkers used dissolvable microneedles to deliver functional fluorescent materials to the skin. These fluorescent substances are sensitive to reactive oxygen species (ROS) in vivo, allowing real-time monitoring and quantitative analysis of physiological and pathological parameters through colorimetry. Researchers have demonstrated that this system is highly sensitive to ROS, and the microneedles have high biocompatibility and stable response signals (Fig. 10.2).

10.3 Design Principles and Strategies of Dissolvable Microneedles for Tumor Treatment

10.3.1 Dissolvable Microneedles for Tumor Chemotherapy

Chemotherapy is one of the main weapons against cancer. However, in recent years, people have gradually realized that when anti-cancer chemotherapeutic drugs circulate in the body along with the blood, they can cause many problems. For example, the poor tumor targeting ability of drugs can lead to high off-target toxicity, making the drugs unable to cope with cancer metastasis and severe side effects. The emerging microneedle technology has provided a new possibility for the application of such anti-cancer drugs.

Specifically, microneedle local drug delivery can greatly reduce the chances of drugs coming into contact with healthy tissues, fundamentally reducing the side effects of the drugs. In addition, compared to conventional direct drug delivery

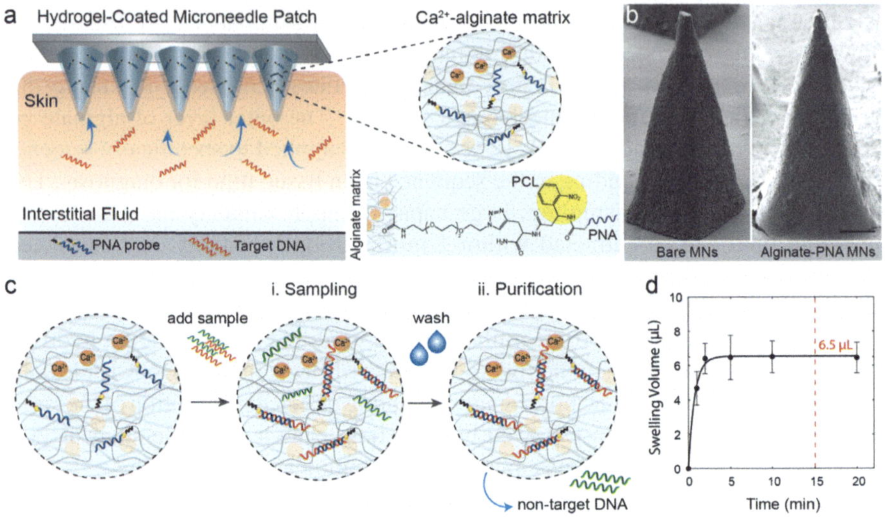

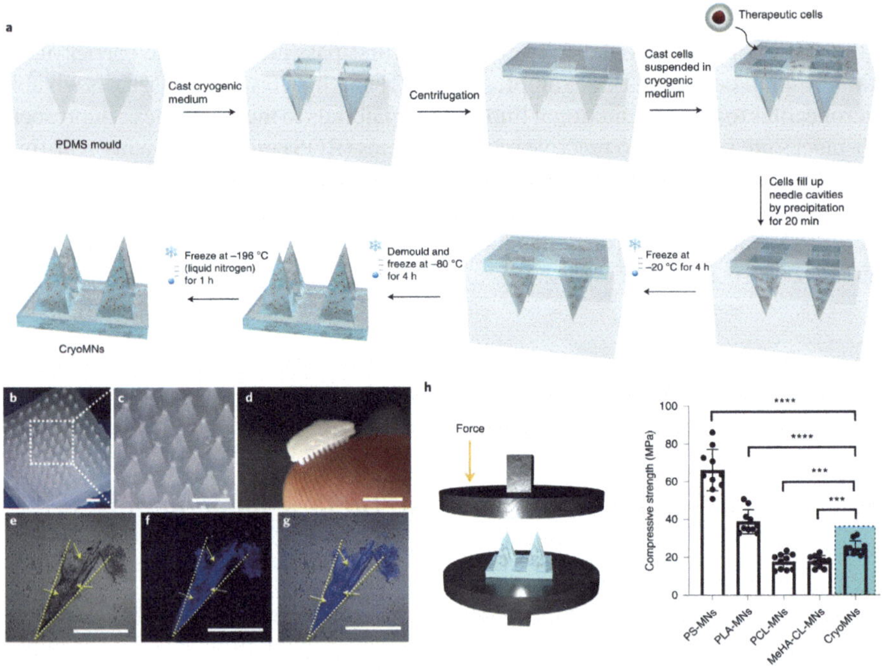

Fig. 10.1 Microneedle technique for organizing liquid extraction. **a** Microneedle arrays are functionalized with bespoke peptide nucleic acid (PNA) probes (blue) which are covalently bound to an alginate hydrogel matrix via a photocleavable linker (PCL, yellow). **b** Scanning electron micrograph of the bare MNs and alginate–PNA hydrogel coated MNs (scale bar = 100 μm). **c** Schematic illustration of the generic protocol for MN sampling of target biomarker (red) and purification to remove nontarget sequences (green). **d** Swelling kinetics of the hydrogel MNs fitted by the Spring and Dashpot Voight-based model (black solid line), showing an equilibrium swelling capacity of 6.5 ± 0.2 μL and a sampling rate constant of 0.74. The figure was reproduced with permission [10]

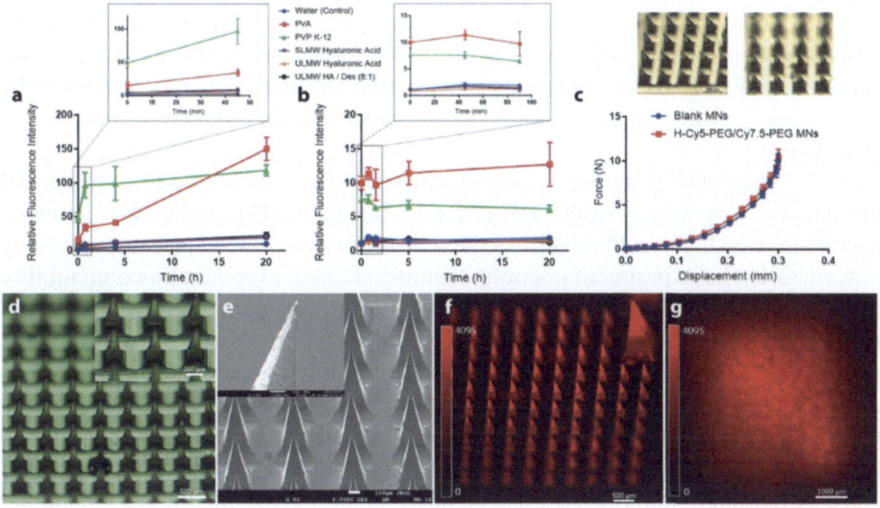

Fig. 10.2 *Soluble microneedles used for colorimetric detection.* **a** *Stability of free H-Cy5 in* common MN polymers. **b** Stability of H-Cy5- poly (ethylene glycol) (PEG) in common MN polymers. **c** Force–displacement curve of failure force of MNs with and without the presence of H-Cy5-PEG/Cy7.5-PEG. Inset shows MNs before and after deflection. Scale bar = 500 μm. **d** Brightfield image of MNs made from ULMW HA/dextran (8:1). **e** SEM of ULMW HA/Dextran MNs. Scale bar = 100 μm, or 10 μm for inset. **f** Fluorescent image of Cy5-PEG contained within MN tips. **g** Microneedle tattoo of Cy5-PEG delivered in vivo to rat skin using MNs. The figure was reproduced with permission [12]

into tumors, some special types of microneedles can achieve stronger sustained and controlled release of tumor drugs, as well as limiting the excessive penetration of drugs into adjacent healthy tissues.

10.3.1.1 Microneedle Delivery of Chemotherapy Drugs

Microneedle technology plays a crucial role in chemotherapy drug delivery strategies. Firstly, by utilizing solid microneedles to pretreat the skin, it effectively opens up microchannels, thereby promoting drug penetration and absorption. Secondly, drugs can be directly formulated into microneedle form, allowing them to penetrate the epidermis directly, achieving transdermal delivery and enhancing localized therapeutic effects. These two methods demonstrate the potential of microneedle technology in improving drug delivery efficiency and efficacy.

Non-drug-loaded solid microneedles are commonly used as tools for skin pretreatment. Compared to traditional surgical dressings, applying drugs to skin that has been pretreated with microneedles can greatly improve drug delivery efficiency. This treatment method is also used in cancer therapy. Sabri et al. [13] explored the mechanism of enhancing the delivery of hydrophobic molecules using solid microneedles combined with Aldara cream and a vibrating Dermapen, which improved the local intradermal delivery of imiquimod and enhanced the treatment of basal cell carcinoma.

On the other hand, drug-loaded soluble microneedles can be used for transdermal delivery of chemotherapy drugs. For example, Lee et al. [14] used double-stranded salmon DNA (sDNA) as the microneedle matrix material to load and transdermally deliver DOX. The experiment not only demonstrated the excellent biocompatibility and material stability of sDNA, but also found that strong interactions between DNA molecules and DOX allowed the drug to be preserved for a longer period of time in the microneedles. The enhancement of drug stability by microneedle matrix materials is not an isolated case, as similar findings exist between other drugs and microneedle materials. All-trans retinoic acid (ATRA) is a drug used to treat seborrheic keratosis (a benign skin tumor) with broad anti-proliferative and differentiation-promoting activities, able to inhibit malignant growth of various tumor cells, making it a commonly used differentiating agent. Yasuhiro et al. [15] utilized hyaluronic acid microneedles for transdermal delivery of ATRA, with experimental results showing significant therapeutic effects of the drug. Moreover, even after 24 weeks, 85% of ATRA remained active in the hyaluronic acid microneedles, demonstrating good stability of ATRA in the microneedles. These results all highlight the importance of selecting appropriate materials when developing microneedles for anticancer drug delivery to achieve better microneedle performance.

Microneedles can not only effectively deliver a single drug percutaneously, but can also realize the simultaneous delivery of multiple drugs to enhance the efficacy of anticancer therapy. Bhatnagar et al. [16] designed a polyvinylpyrrolidone/polyvinyl alcohol microneedle patch for the combined transdermal delivery of DOX and docetaxel to treat breast cancer tumors. DOX and docetaxel are two commonly used chemotherapy drugs for such cancers. DOX is a non-selective type I anthracycline drug that induces cell death by intercalating into DNA and inhibiting macromolecular biosynthesis. Docetaxel interferes with the normal function of microtubule growth, affecting cell cytoskeleton flexibility and cell mitosis. After the two drugs were directly dissolved in a polyvinylpyrrolidone solution, they were deposited in the conical needles using a micro-template method, followed by coating the surface of the microneedles with a polyvinylpyrrolidone/polyvinyl alcohol mixture to further enhance the mechanical strength of the patch. Subsequent experimental results showed that the microneedles had a compressive strength of over 5 N, and the drugs could be detected at a depth of approximately 140 μm under the skin of mice with the aid of microneedle delivery. Additionally, the microneedle patch completely dissolved within 33 min, releasing 90% of docetaxel within the first 15 min. Subsequent in vivo anticancer experiments on mice carrying breast tumors showed that compared to conventional drug delivery methods, the microneedle-treated group had significantly improved survival rates and long-term maintenance of body weight, indicating that chemotherapy delivery mediated by microneedles can significantly reduce the toxic side effects of drugs on the body.

Additionally, some new forms of microneedles have been attempted for drug delivery. Jonas et al. [17] developed a micro-needle containing multiple drug reservoirs, each reservoir containing a unique single drug or drug combination that can be implanted inside tumors. This cylindrical device, with a diameter of 820 μm and a length of 4 mm, has up to 16 drug reservoirs on its surface, which can be implanted into the tumor using a biopsy needle, allowing the loaded drugs to diffuse into the tumor area within a range of 200–300 μm. By adjusting the size of the reservoirs, the ratio of polymer matrix, and hydrogel, drug diffusion (slowing or accelerating) can be controlled. This device can not only independently control the release of each drug reservoir but also load different drugs in the same reservoir for combination therapy. This excellent drug release capability allows various drugs to reach tumor-matching drug concentrations for cancer treatment and is expected to be rapidly translated into clinical applications. Additionally, this device can be used to study the therapeutic effects of different drugs on the same tumor. Results show that paclitaxel has the highest apoptosis index for triple-negative breast cancer tumors (54%), followed by doxorubicin (36%), cisplatin (25%), gemcitabine (12%), and lapatinib (4%). Furthermore, using these microneedles can determine the most effective combination for a specific tumor, for example, in BT474 human breast cancer tumor cells, the combination therapy of doxorubicin and lapatinib increased the apoptosis index from approximately 3% to about 5.5%, and when combined with docetaxel, the index further increased to 13.45%.

Research has also been conducted using insoluble photocrosslinkable dextran methacrylate hydrogels (DexMa hydrogels) to prepare microneedles loaded with both DOX and trametinib, as shown in Fig. 10.3 [18]. DOX can directly act on DNA, inhibiting DNA polymerase activity to suppress the synthesis of DNA and RNA. Trametinib can inhibit the growth of BRAFV600 mutant positive melanoma cells and can also antagonize the efflux of DOX by the P-glycoprotein transporter (DOX is a substrate of the P-glycoprotein transporter), thereby increasing the intracellular concentration of DOX and enhancing its cytotoxic effects on cancer cells. DOX and trametinib can synergistically treat melanoma, with the best inhibitory effects on tumors.

In conclusion, microneedles serve as a tool for direct delivery of chemotherapy drugs, providing a safer and more reliable new approach to cancer treatment in clinical settings. Currently, many researchers are focusing on microneedle development, using innovative biomedical materials as the substrate material for microneedles to improve their applicability and the storage time of chemotherapy drugs. Research results indicate that microneedles are not only suitable for carrying different types of molecules (regardless of their molecular weight, charge, and hydrophobicity) but can also enhance delivery efficiency and promote uniform distribution of drugs in tissues. Furthermore, microneedles can not only release single anticancer drugs but can also achieve synergistic antitumor effects by releasing multiple anticancer drugs simultaneously.

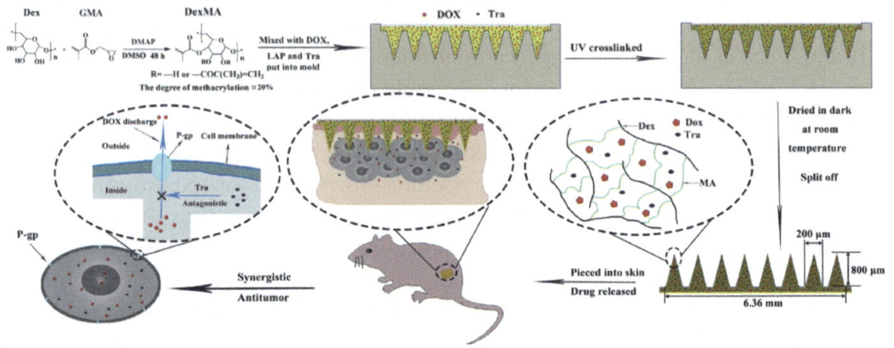

Fig. 10.3 Microneedles used for delivering chemotherapy drugs to combat tumors. The figure was reproduced with permission [18]

10.3.1.2 Microneedle Delivery of Chemotherapy Drug Carriers

Nanoparticle-based drug carriers and microneedle-based drug delivery technology are pivotal components in the transdermal delivery system. Nanoparticle technology, being one of the most commonly employed passive strategies for enhancing penetration, has found extensive application in delivering various small molecule drugs, proteins, vaccines, and nucleotides. Diverse forms of nanoparticles such as solid lipid nanoparticles (SLN), polymer nanoparticles, liposomes, nano-lipid carriers (NLC), and nanocrystals have been developed. In the field of tumor chemical drug delivery, well-designed nanoparticle carriers can effectively enhance release rate, drug targeting precision, in vivo distribution uniformity and bioavailability of chemotherapeutic drugs while safeguarding their active ingredients and improving solubility for hydrophobic drugs. The integration of nanoparticle carriers with microneedles, a powerful physically enhanced penetration technology, significantly enhances the efficiency of transdermal drug delivery. In the current report, the combination of the two technologies for the delivery of chemotherapy drugs has achieved promising therapeutic effects.

Gill et al. [14] utilized a coated microneedle for the delivery of DOX, which was initially encapsulated in polylactic acid-glycolic acid copolymer (PLGA) nanoparticles and then coated onto solid microneedles. In subsequent diffusion experiments using a porcine skin model, it was observed that the drug exhibited a uniform distribution within the porcine skin. Furthermore, the microneedles demonstrated significant reduction in DOX diffusion into non-diseased tissues compared to subcutaneous injection. Shivaprasad et al. [19] encapsulated resveratrol (RVT) with NLC and combined it with a macromolecular microneedle for the treatment of breast cancer, as shown in Fig. 10.4. Resveratrol, derived from plants, exhibits anti-cancer properties by regulating gene expression and tumor-related signaling pathways to inhibit cancer cell proliferation and reduce drug resistance in patients. However, its limited bioavailability poses challenges for effective utilization. NLCs are formulated based on solid lipid nanoparticles but incorporating liquid lipids enhances drug encapsulation efficiency while minimizing leakage. NLC is developed on the basis of solid lipid nanoparticles. Adding liquid lipids to solid lipids can greatly improve the drug encapsulation efficiency and reduce the leakage of drugs. In the experiment, NLC was prepared using the melt emulsification method, and the appropriate size was achieved through phacoemulsification (the lipid particle size optimized by independent variables was 104.47 ± 4.35 nm). The optimized NLCs formulation was incorporated into the carboxyvinyl polymer as a thickener, along with the addition of benzoic acid (0.1%, w/v) for preservation and ascorbic acid (0.3%) for antioxidation. Under ambient conditions of 5 ± 3 °C and 60% ± 5% humidity, the drug in microneedles can be stored for a period of 4 months. By comparing the cytotoxicity and cell migration assays of RVT and RVT-NLCs, the researchers observed that the sustained release of RVT from NLCs effectively inhibited cancer cell metastasis, leading to a superior anticancer effect compared to the lipid-free carrier. In vivo antitumor experiments in rats with breast tumors further demonstrated that microneedle delivery of RVT-NLCS, as compared to oral pure RVT and subcutaneous injection of RVT-NLCs, circumvents systemic drug release-associated side effects, thereby effectively enhancing drug concentration in the blood and breast tissue.

In recent years, the incidence of skin cancer has been on the rise due to the rapid aging of the population. Hao Ying et al. [20] designed a near-infrared light-responsive 5-fluorouracil (5-Fu) and indocyanine green (ICG) loaded monomethoxy-poly (ethylene glycol)-polycaprolactone (MPEG-PCL) nanoparticle (5-Fu-ICG-MPEG-PCL), and then 5-Fu-ICG-MPEG-PCL was integrated with a hyaluronic acid dissolvable microneedle system (HA MN) to get 5-Fu-ICG-MPEG-PCL loaded HA MN for treating skin cancers. In this system, HA is a microneedle carrier with good skin penetration ability. 5-Fu is used as a chemotherapy drug in the treatment of skin cancer;

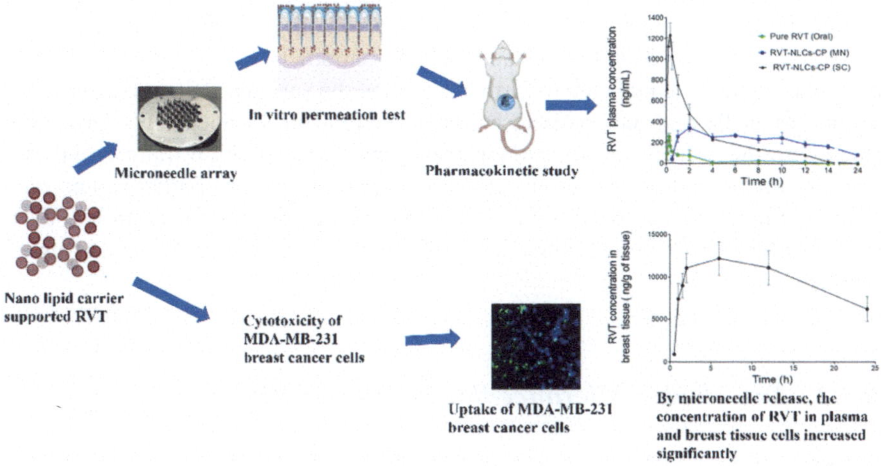

Fig. 10.4 Transcutaneous delivery of resveratrol by microneedles for breast cancer therapy. The figure was reproduced with permission [19]

ICG is a photothermal agent. The microneedle exhibits excellent photothermal efficacy, achieving a rapid temperature increase to 60 °C within 5 min under an 808 nm NIR laser. When the pork tissue is treated with microneedles, it can rise to 65 °C and reach a depth of 1.5 cm in 5 min, indicating that heat could be effectively transferred to the tumor center. The experimental results demonstrated that the combination of chemotherapy and photothermal therapy, along with laser treatment using 5-Fu-ICG-MPEG-PCL@HA MN, exerted a remarkable inhibitory effect on tumor growth, leading to complete remission in all mice without any signs of recurrence. Furthermore, minimal changes in body weight during treatment validate its safety profile. Collectively, these findings highlight the immense potential of this microneedle system for skin cancer management.

Roller microneedles, as a commonly used commodity, have found wide applications in various fields such as cosmetics and scientific research. Ahmed et al. [21] utilized the thin-film hydration method to prepare DOX/celecoxib (CEL) co-loaded liposomes for the treatment of cutaneous melanoma. These drug-loaded liposomes were subsequently applied topically on roller microneedle-prepared skin surfaces. The results demonstrated a twofold increase in the delivery efficiency of DOX in skin-liposome pretreated with roller microneedles compared to the untreated experimental group. Furthermore, transdermal administration of DOX/CEL co-loaded liposomes effectively suppressed subcutaneous melanoma in female mice, with the efficacy of DOX/CEL co-loaded liposomes surpassing that of single loaded liposomes and exhibiting a significantly enhanced anti-tumor effect.

However, the concurrent utilization of nanoparticle carriers and microneedle technology presents certain limitations. For instance, the nanoparticle carrier within the soluble MNs may be influenced by the MNs matrix material, leading to potential damage to the carrier structure or entrapment of the drug by the MNs material, thereby hindering normal release. Additionally, there is a possibility that the combined effect of nanoparticle carrier and MNs could exacerbate skin irritation. However, these issues do not represent inherent contradictions, and numerous subsequent studies can be conducted to further optimize the various challenges associated with microneedle and nanoparticle systems. Significant advancements have been achieved in the delivery of anti-tumor drugs, including the high transdermal efficiency of microneedles, the strong targeting capability of nanoparticles, and their ability to protect drug activity. It has been demonstrated that the integration of these two technologies holds great potential for broader applications in the field of tumor treatment.

10.3.2 Dissolvable Microneedles for Tumor Biologic Therapy

10.3.2.1 Immunotherapy

As early as the Ming Dynasty in China, there is solid evidence of the Chinese adopting the "Nose Vaccine Method" to prevent smallpox. This represents the earliest large-scale implementation of human pox inoculation, marking an empirical stage in the emergence of immunology. With the rapid development of cell biology, molecular biology, and other fields, immunology has become a cutting-edge field of life sciences. In 2021, the first CAR-T cell product was launched in China, and its efficacy caught the public's attention once again to oncology immunotherapy.

Immunology can abstractly be understood as the organism's function to distinguish between "self" and "non-self" recognizing and eliminating "non-self" to protect "self". The study of tumor immunology aims to understand the organism's immune response to tumors and their effector mechanisms to achieve tumor immunodiagnosis and prevention. In cases where traditional methods such as surgery, radiotherapy, and chemotherapy have shown poor efficacy in treating tumors, tumor immunotherapy has become a major research hotspot, leading to the development of many types of immunotherapy modalities. Table 10.1 summarizes several major types of tumor immunotherapy to date, including tumor vaccines, cytokine therapy, adoptive immunotherapy, and antibody-directed therapy. Immunotherapies for tumors can generally be divided into active and passive immunity, or specific and nonspecific immunity. As the third generation of percutaneous immunotechnology, microneedle technology has greatly improved the efficiency of percutaneous drug delivery and can even challenge traditional methods such as subcutaneous injection. It also offers advantages such as low pain, ease of operation, and low cost. Therefore, microneedles also play an important role in tumor immunotherapy. The following section will introduce the important applications of microneedles in tumor vaccine therapy, cytokine therapy, immune checkpoint inhibitors, etc.

Vaccines are immune formulations made from antigens, cells, or related molecules, providing patients with immunity against specific diseases including

Table 10.1 Main types of tumour immunotherapy

Common therapies	Coverage aspects
Tumour vaccine (active immunity)	(1) Tumour cell vaccine (2) Tumour peptide vaccine (3) Tumour nucleic acid vaccine (4) Others……
Cytokine therapy (passive immunity)	(1) Interleukin IL-2, IL-4, IL-12…… (2) Interferon IFN-α, IFN-β, IFN-γ…… (3) Tumour necrosis factor TNF……
Relay immunotherapy (passive immunity)	(1) Lymphokine activated killer cell, **LAK** (2) Tumor activated killer cell, **TAK** (3) Tumor-infiltration lymphocytes, **TIL** (4) Cytokine-introduced killer cell, **CIK** (5) Chimeric antigen receptor T-cell, **CAR-T**
Antibody-directed therapy (passive immunity)	(1) Monoclonal antibodies, polyclonal antibodies, bispecific antibodies (2) Antibody coupling (3) Immune checkpoint inhibitors PD-1, CTLA-4

tumor. These vaccine categories include many new vaccines based on recombinant technology, such as recombinant bacterial vaccines and recombinant antigen vaccines. Nucleic acid vaccines encompass DNA vaccines and RNA vaccines.

The application of microneedles for vaccination didn't originate with oncology vaccines. In 2009, Prausnitz et al. [22] first attempted influenza vaccine delivery using coated microneedles, yielding promising immunological results. Nowadays, the fusion of oncology vaccines with microneedles is gaining momentum with increasing studies. In 2017, Ali et al. [23] synthesized RALA nanoparticle complexes carrying E6 and E7 molecules and delivered them using PVP (polyvinylpyrrolidone) microneedles for cervical cancer treatment. Cervical cancer ranks as the second most common cancer among females, with over 90% of cases attributed to HPV subtypes 16 and 18. E6 and E7 are HPV viral genes capable of disrupting the tumor suppressor functions of p53 and retinoblastoma protein (pRb), respectively. RALA, a synthetic amino acid sequence, serves as a vaccine carrier, facilitating DNA passage through the plasma membrane and shielding it from intracellular degradation. In vivo experiments demonstrated that mice in the microneedle group had higher levels of E6 and E7 specific IgGs (immunoglobulins) than the injection group, exhibiting stronger T-cell mediated TC-1 cytotoxicity. Additionally, the microneedle group showed delayed TC-1 tumor onset, effectively retarding tumor growth. In 2018, McCarthy et al. [24] enhanced PVP microneedles' drug loading capacity to 50 μg by pre-lyophilizing the aforementioned nanoparticles. In 2022, Joo et al. [25] explored the use of dissolvable self-locking microneedle patches integrated with immunomodulators for cancer immunotherapy. They designed microneedles using a digital light processing (DLP) chip-based projection micro-stereolithography 3D printer, which allowed for the precise insertion and locking of microneedles within skin tissue. This innovative design improves drug delivery accuracy compared to conventional microneedles, which often fail to fully penetrate the skin.

The study specifically investigated the efficacy of combining SD-208, a transforming growth factor-β (TGF-β) inhibitor, with anti-PD-L1 (aPD-L1 Ab), an immune checkpoint inhibitor, loaded into these microneedles. This combination was tested on a B16F10 melanoma-bearing mouse model. The results showed that the self-locking microneedles significantly enhanced the delivery and effectiveness of the immunotherapy agents compared to traditional intratumoral injections, leading to improved anti-tumor responses and therapeutic outcomes [26]. As depicted in Fig. 10.5, PVP microneedle delivery of DNA vaccine encapsulated in RALA particles proved most immunologically effective, leading to delayed tumor generation time, minimal tumor volume growth, and the highest survival rate among all subgroups. These consecutive microneedle vaccine research efforts underscore the efficacy of antigen release facilitated by microneedles, effectively enhancing the antibody response and resulting in a robust immune effect on the organism.

Cytokine therapy is another common approach in tumor immunotherapy. Cytokines are low molecular weight glycoproteins secreted by various cells in the body, typically ranging in molecular weight from 10 to 25 kDa, and they have direct effector functions. Cytokines regulate gene expression in cells by binding to corresponding receptors on the cell membrane and initiating signaling pathways. For

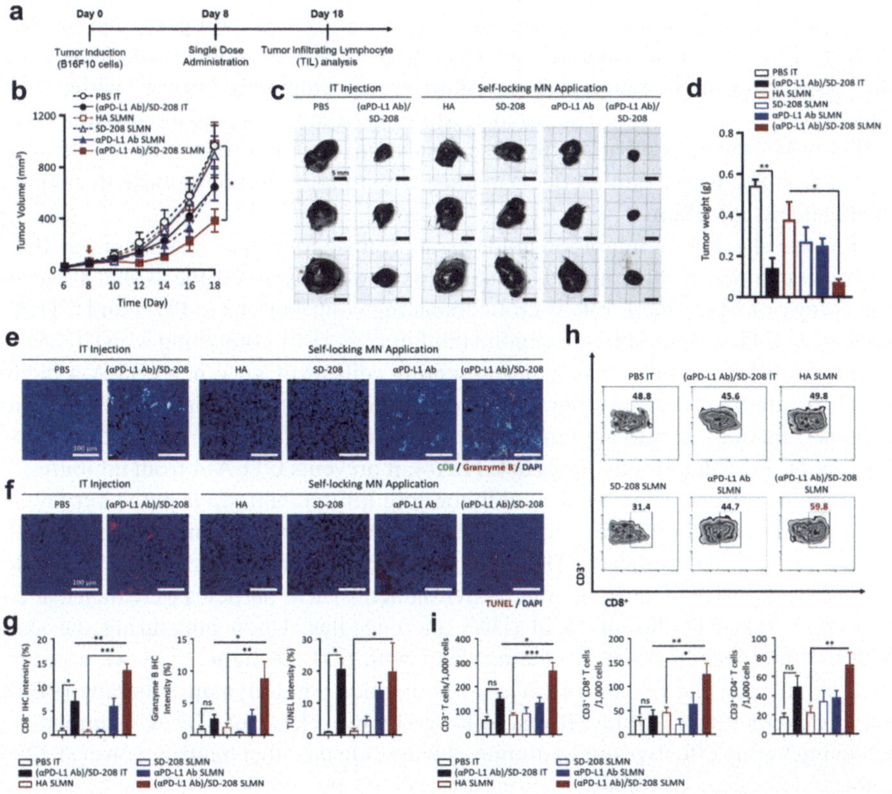

Fig. 10.5 Anti-tumoral and immunomodulation efficacies of (αPD-L1 Ab)/SD-208 SLMNs in orthotopic melanoma mouse model. **a** Timeline overview of treatment and evaluation of C57BL/6 melanoma mouse model inoculated with B16F10 cells. **b** The tumor volume measurement for 18 days post-treatment. **c** Representative images of tumors harvested on the 18th day and **d** tumor weight measurements. **e** Immunofluorescent staining in tumor sections. **f** TUNEL assay in tumor sections. **g** The ratio of positive fluorescence intensity in immunofluorescent staining and TUNEL assay. **h** Representative plots of CD3+ and CD8+ T cells in tumor tissues analyzed using flow cytometry. **i** Total tumor tissue was harvested and stained for anti-CD3, anti-CD4, and anti-CD8 antibodies. The ratio of CD3+ T cells, CD3+ CD8+ T cells, and CD3+ CD4+ T cells in 1 × 104 cells was measured using flow cytometric analysis. The figure was reproduced with permission [26]

instance, tumor necrosis factor (TNF) can stimulate cells to release other cytokines, enhance the expression of MHC-like molecules, and improve the body's immune response.

Korkmaz et al. [25] at the University of Pittsburgh utilized microneedles to deliver antibodies specific to the cytokine TNF-α for treating diseases associated with skin inflammation. Interferon (IFN) is another common cytokine, with IFN-α being a well-known drug for treating chronic hepatitis C and various human malignancies. However, interferon, being a proteinaceous drug, exhibits poor permeability and stability in the bloodstream or gastrointestinal tract. In 2016, Kusamori et al. [27]

developed a novel interferon-coated microneedle array based on polyvinyl alcohol (IFN-α-MNs) for transdermal delivery of interferon. Microneedle treatment induced slight erythema and irritation in mouse skin, which completely resolved within 24 h after removal of the IFN-α-MNs patches. Moreover, the pharmacokinetic parameters of IFN-α-MNs were comparable to those of subcutaneous administration and demonstrated significant anti-tumor effects in mice with tumors similar to those treated via subcutaneous injection.

Immune checkpoint inhibition holds great promise for cancer treatment. In 2018, Professors James P. Allison and Tasuku Honjo were awarded the Nobel Prize in Physiology or Medicine for their groundbreaking contributions to PD-1 and CTLA-4 research. CTLA-4 immune checkpoint inhibitors work by competing with CD28 for binding to B7 molecules on antigen-presenting cells (APCs). When CTLA-4 binds to B7, it inhibits T cell proliferation and secretion, thereby suppressing systemic immune function, akin to applying "brakes" on the immune system. When human-delivered CTLA-4 antibody binds to CTLA-4, it prevents CTLA-4 from inhibiting T cell immune effects. The CTLA-4 inhibitor drug Ipilimumab (trade name Yervoy) is currently FDA (US Food and Drug Administration)-approved for marketing. In 2020, Chen et al. [28] encapsulated CTLA-4 antibodies and photosensitizers in self-soluble particles responsive to the tumor microenvironment. These particles were then loaded into cross-linked hyaluronic acid (HA) microneedles. Upon puncturing the skin with the microneedles and irradiating them with 660 nm light, ROS were generated, stimulating the release of CTLA-4 antibodies. Photodynamic therapy (PDT) initially exerts tumor-killing effects, followed by the release of CTLA-4 antibodies, achieving highly effective tumor immunotherapy. On the other hand, Kwon et al. [29] employed an array of hollow microneedles with a tip length of 350 μm combined with a microfluidic structure to percutaneously inject CTLA-4 antibodies. Compared to the intraperitoneal injection group, the microneedle transdermal injection group exhibited a 40% reduction in cancer cell metastasis to the lungs, lymph nodes, or bone marrow, along with significant reduction in tumor volume in mice. Immune cell infiltration into the tumor was significantly higher in the microneedle injection group, indicating more effective immune activation and stronger tumor-killing effects.

PD-1 serves as an "immune checkpoint" similar to CTLA-4, acting as another regulatory "brake" for T cells. In response to the cytotoxic effects of T cells, tumor cells express PD-L1 on their cellular surfaces, which binds to PD-1, thereby inhibiting the normal function of T cell immunity. Blocking this interaction by delivering PD-1 antibody or PD-L1 antibody can effectively restore the normal initiation of T cell immune responses. In 2016, Gu et al. [30] developed crosslinked hyaluronic acid microneedles delivering PD-1 antibody for the treatment of cutaneous melanoma, as depicted in Fig. 10.6. The PD-1 antibody was not merely loaded into the microneedle; instead, it was encapsulated within dextran nanoparticles (NPs) together with glucose oxidase (Gox) enzymes. These nanoparticles exhibit pH sensitivity, allowing for drug degradation and release in acidic environments. In the presence of oxygen, GOx converts glucose to gluconic acid, creating an acidic milieu conducive to the release of PD-1 antibodies from the nanoparticles. Upon treatment of melanoma in mice with microneedle-delivered nanoparticles carrying GOx-modified PD-1 antibody,

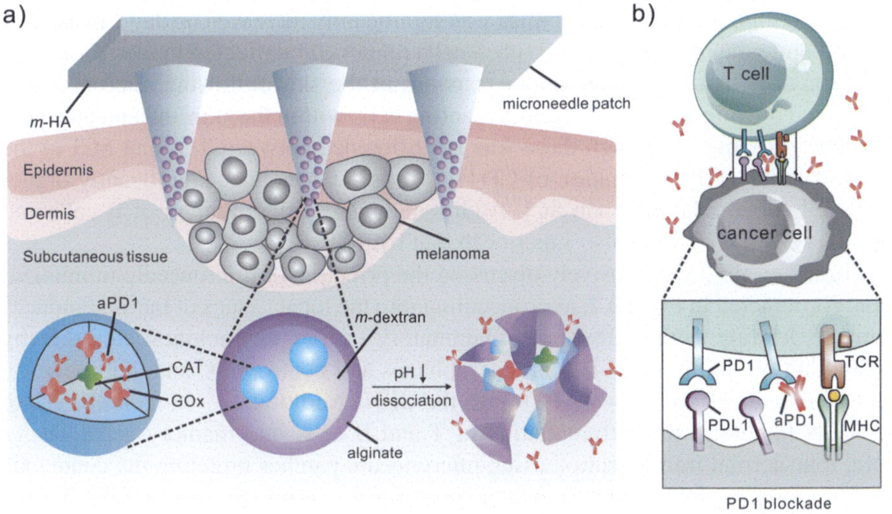

Fig. 10.6 Schematic of microneedle-assisted delivery of PD-1 antibody for skin cancer treatment. **a** Schematic of MN patch loaded with physiologically self-dissociating NPs for delivery of PD-1 antibody; **b** PD-1 antibody blocks PD-1, activates the immune system, and destroys skin tumour cells. The figure was reproduced with permission [30]

significant reduction in tumor cell signals was observed after 10 days, indicating effective tumor cell elimination. Moreover, the survival rate of this mouse group exceeded 40% after 40 days, the highest among all experimental groups. Furthermore, the T-cell infiltration ratio in this group was more than five times that of the control group, surpassing even the effects of intratumoral injection at the same dosage. Utilizing microneedles to deliver immune checkpoint inhibitors for superficial tumors such as melanoma allows for local drug aggregation within the tumor and rapid binding with immune cells.

The immune-enhancing effects of transdermal techniques have been under investigation for quite some time. In 2015, Hirobe et al. [31] demonstrated that transdermal immunization elicited stronger immune responses and exhibited dose-saving properties compared to subcutaneous immunization in both animal and human experiments. Earlier, in 2010, Henri et al. [32] elucidated that this unique immune function of the skin stems from the abundance of lymphocytes, such as Langerhans cells, within the dermis of the epidermis, from a fundamental standpoint. In 2021, Hirobe et al. [33] conducted a more comprehensive study on transcutaneous immunization (TCI) combined with microneedling using ovalbumin (OVA) as a model vaccine. The experimental data reaffirmed the utility of microneedle technology for TCI in several aspects. The advantages of microneedle technology for transdermal immunization were reaffirmed in various ways: OVA-specific antibodies reached stable antibody titers after five subcutaneous immunization (SCI) administrations of the vaccine, whereas the same stable levels were achieved in only three transdermal immunizations of the same dose via microneedles. The proportion of total CD11c+ cells in the draining lymph nodes (dLNs) of mice was significantly increased on day 1 in the TCI group, while the proportion of CD11c+ cells remained unaffected in the SCI group. This suggests that the inflammatory response of the skin following microneedling facilitated the migration of antigen-presenting cells within the skin and enhanced the immune response. Although there was no difference between TCI and SCI in the level of induced differentiation of CD4+ T cells, TCI induced significantly higher Tem cell subpopulations than SCI. Additionally, the differentiation of B cells and plasma cells by TCI was also superior to that by SCI.

Hirobe et al. [33] extensively discussed the principle of microneedle immunization. As depicted in Fig. 10.7, microneedling into the local tissues of the skin induces damage, leading to the release of inflammatory factors, chemokines, etc., thereby promoting the infiltration of local macrophages and neutrophils, and the migration and activation of cutaneous APCs into the dLN. Additionally, antigen entry into the dLN facilitates the differentiation of T and B cells into memory cells. Therefore, transdermal immunization using microneedle patches targeting the cutaneous immune system yields a more potent antigen-specific immune response effect.

Yang et al. [34] also confirmed that microneedles can enhance the immune response through a combination of computer simulation and experimentation. Experiments utilizing gel microneedles to deliver OVA revealed that the intramuscular injection group only induced the production of IgG1, whereas the crosslinked gel microneedle group induced the production of both IgG1 and IgG2, with a wider array of antibody classes and an IgG1 titer more than six times that of the injection group.

CAR-T cell therapy, short for chimeric antigen receptor-modified T-cell therapy, is a type of adoptive cellular immunotherapy listed in Table 10.1. As illustrated in Fig. 10.8, cellular immunotherapy typically involves isolating lymphocytes from human peripheral blood, processing them in vitro, and reintroducing them into the patient. This process expands T lymphocytes in vitro and enhances their tumor-killing capabilities. CAR-T cells acquire specific anti-tumor capabilities by genetically engineering a receptor segment that recognizes a specific protein on the tumor surface onto the surface of T cells during in vitro processing.

Currently, CAR-T cell therapy primarily targets lymphohematopoietic malignancies, with investigations ongoing for solid tumors. A 2020 article in The Lancet Haematology analyzed 35 clinical studies applying CAR-T cells for B-lymphoblastic leukemia treatment, revealing an overall complete remission rate of 80%, reaching 83% with autologous T cells. Despite impressive outcomes in tumor therapy, CAR-T cell therapy may entail side effects such as cytokine release syndrome (CRS) and neurotoxicity.

In 2021, Gu et al. [36] reported using PLGA matrix microneedles preloaded with CaCO3 particles and acid-washed to create pores (5–20 μm) for CAR-T cell loading. These microneedles achieved a 20% loading rate at a concentration of 10^7 cells/ml, with each needle holding about 22,000 cells. This method does not affect CAR-T cell proliferation or killing ability. The porous structure allows CAR-T cells to better

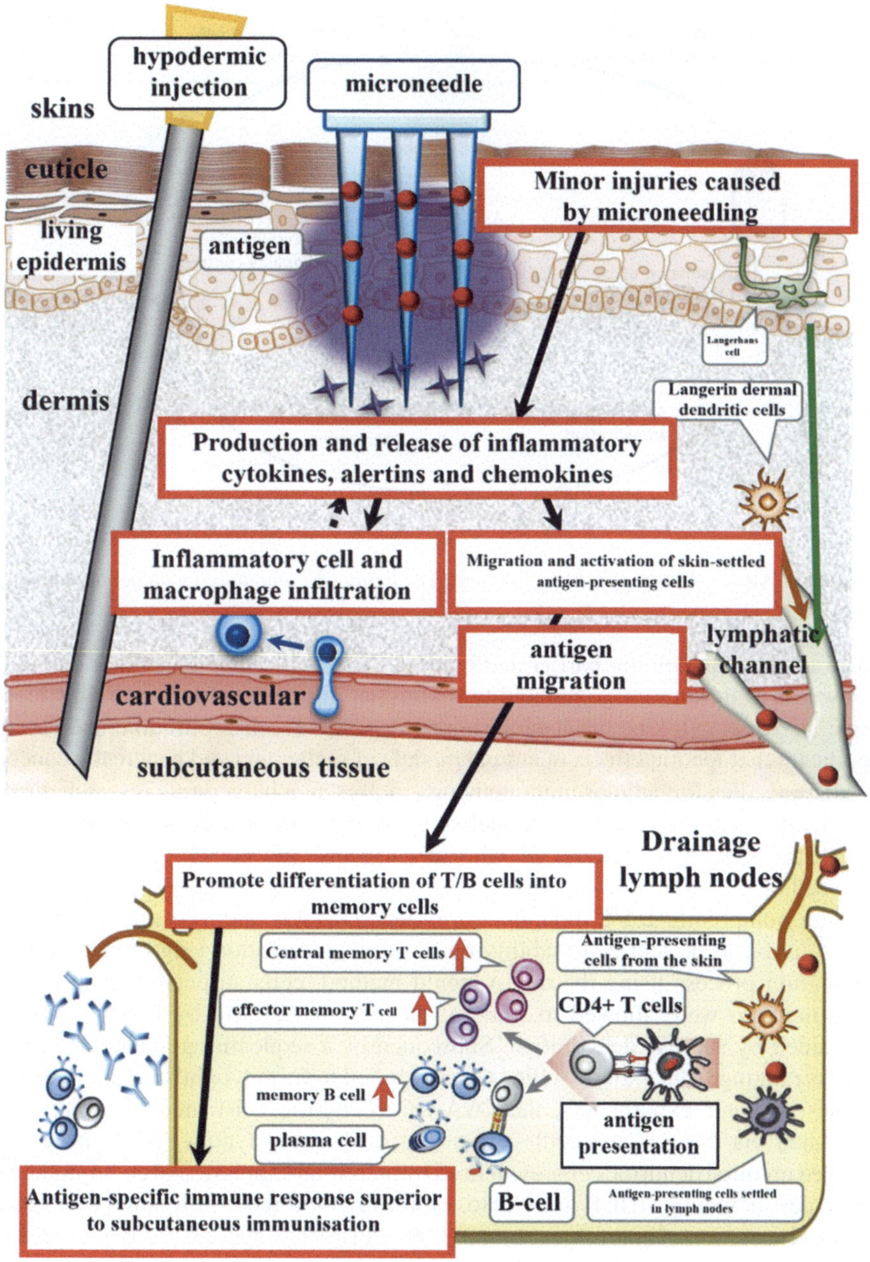

Fig. 10.7 Mechanisms of microneedle transdermal induction of immunity. The figure was reproduced with permission [33]

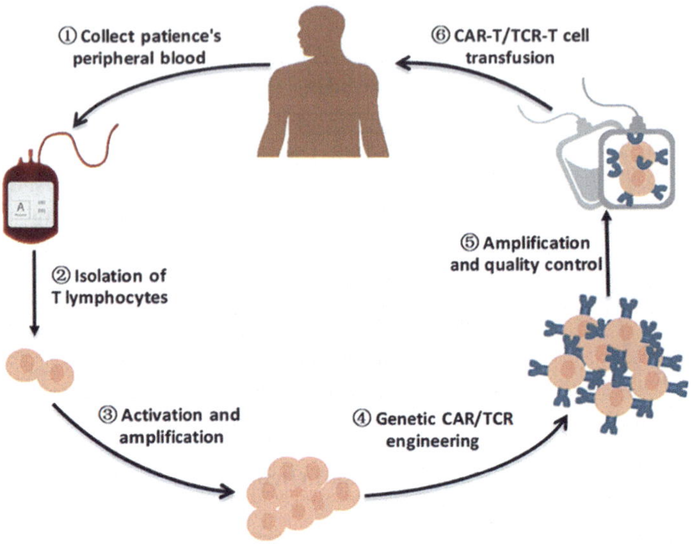

Fig. 10.8 CAR-T cell therapy simple flowchart. The figure was reproduced with permission [35]

penetrate the extracellular barrier and bind to tumor cells, enhancing their distribution within solid tumors. Compared to traditional intratumoral injection, this technique increases cytokine production, improves CAR-T cell infiltration, and shows significant therapeutic effects in animal models of melanoma and pancreatic cancer.

Microneedles for cellular immunotherapy represent a burgeoning research direction, facilitating the percutaneous delivery of living cells. Recent advancements include the delivery of various cell types such as pancreatic cells [37], mesenchymal stem cells [38], and cardiomyocytes [39]. Chang et al. [40] introduced a freezing medium comprising 2.5% dimethyl sulfoxide (DMSO) and 100 mM sucrose in PBS solution to protect cell viability during cryopreservation (Fig. 10.9). In their preparation process using the micro-mould method, cells suspended in the cryogenic medium were allowed to settle at the microneedle tip by gravity over 20 min, aided by sucrose dehydration. Subsequently, a sequential gradient cryogenic process ranging from −20 to −196 °C was applied to form lyophilized microneedle patches. In later experiments, the OVA-DC (dendritic cell) model vaccine was inoculated via these microneedles. Removal of draining lymph nodes on day 28 revealed matured dendritic cells, evidenced by increased expression of co-stimulatory molecules including MHCII and CD86. Moreover, elevated proliferation of splenocytes and secretion of interferon-gamma (IFN-γ) in mice suggested that twice-weekly vaccination with OVA-DC-cryoMNs effectively induced an immune response.

Preliminarily, this freeze-dried microneedle technology enables multiple rounds of conventional cellular immunotherapy in a single session by harvesting blood and loading cells into multiple batches of microneedles, thereby significantly reducing patient suffering and economic costs.

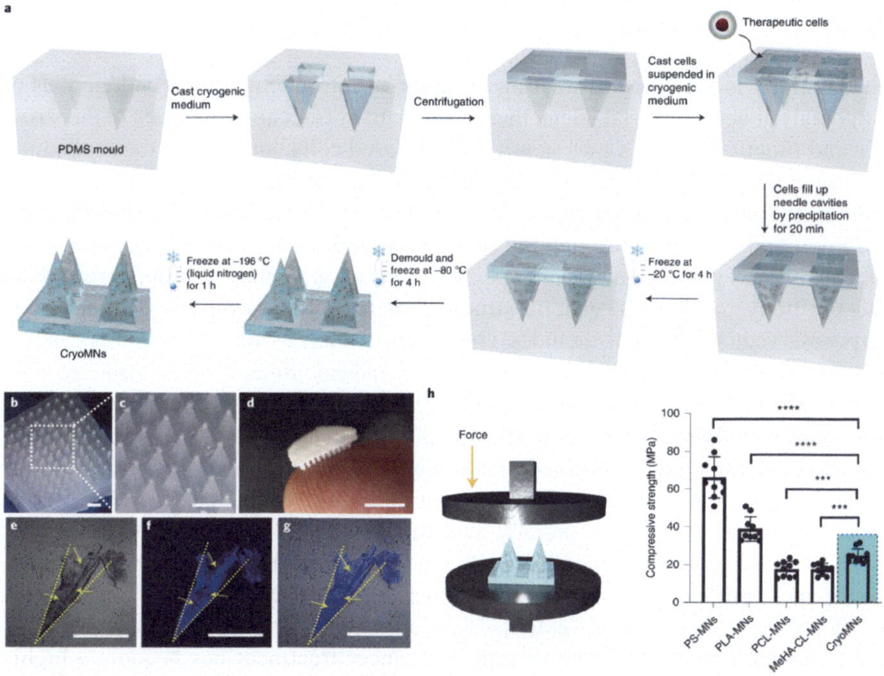

Fig. 10.9 Schematic diagram of the preparation process and characterization of cryo-microneedling (cryoMN), **a** schematic of cryoMN fabrication. **b** Digital microscopy image of the cryoMNs. Scale bar, 1,000 μm. **c** Magnified view of (**b**). Scale bar, 1,000 μm. **d** Photograph of the cryoMNs patch. Scale bar, 5 mm. **e–g** Images of a single MN loaded with cells stained with NucBlue Live ReadyProbes reagent. Yellow arrows indicate individual cells. Image **g** is an overlay of the bright-field image in **e** and a fluorescent image (**f**). Scale bars, 500 μm. **h** Compressive strength of cryoMNs (right) tested by an Instron tensile meter (left). The figure was reproduced with permission [40]

The utilization of microneedles in tumor immunotherapy remains in a phase of rapid development, with tumor immunotherapy technology itself evolving and undergoing optimization. Current research findings demonstrate the distinctive advantages of microneedles in tumor immunotherapy, including the capacity to elicit stronger immune responses and notably reduce patient discomfort and economic burdens. Consequently, it is clinically imperative to persist in exploring the integration of microneedles with tumor immunotherapy.

10.3.2.2 Gene Therapy

Gene therapy is a method of treating diseases by introducing either endogenous or exogenous genetic material. Gene therapy aims to correct abnormalities in the structure and function of one's own genes, kill diseased cells, and inhibit the replication of exogenous pathogenic genetic material, as shown in Fig. 10.10.

Gene therapy approaches include ex vivo and in vivo methods. Microneedle gene therapy mainly operates through the in vivo method. In vivo gene therapy involves introducing exogenous genes into specific expression vectors and delivering them into the body to alter the genetic structure of cells for preventive or therapeutic purposes. Vector selection includes viral vectors such as retroviral vectors, adenoviral vectors, herpes simplex virus vectors, and adeno-associated virus vectors, as well as non-viral vectors such as naked DNA/plasmid DNA, liposome vectors, and cationic polymer vectors, as shown in Fig. 10.11. Viral vector gene therapy systems have high gene transduction efficiency and long expression times, but they may induce immunogenicity, mutations, and limitations in the length of the inserted gene. Non-viral vector gene therapy systems are safer during treatment and have no size restrictions for the inserted gene, but they suffer from low transduction efficiency and short expression times. In studies of microneedle-assisted gene therapy for tumors, mainly non-viral vectors are used for delivering therapeutic genes.

Microneedle-mediated gene therapy for cancer treatment has become a highly popular research direction in recent years. In the field of oncology, microneedle-mediated gene therapy is primarily used for treating or preventing cancer. Microneedles can directly deliver DNA vectors for treating cancer or DNA vaccines for preventing tumor occurrence. Microneedles can also be used in combination with drugs or photothermal therapy for synergistic tumor treatment based on DNA delivery. Wu et al. [41] used dissolvable microneedles to deliver siRNA for treating melanoma. They chose polyethyleneimine (PEI) cationic polymers as carriers to enhance cell uptake of siRNA. PEI, rich in amino groups and with abundant positive charges, can bind to DNA fragments through electrostatic interactions and compress DNA molecules into stable spherical structures of 50–200 nm, forming PEI/DNA

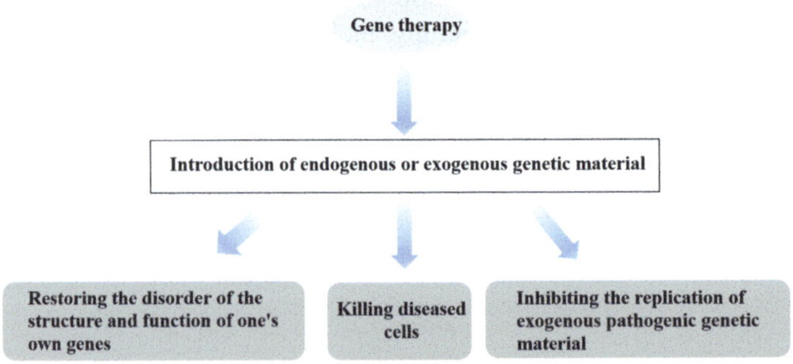

Fig. 10.10 Functions of gene therapy

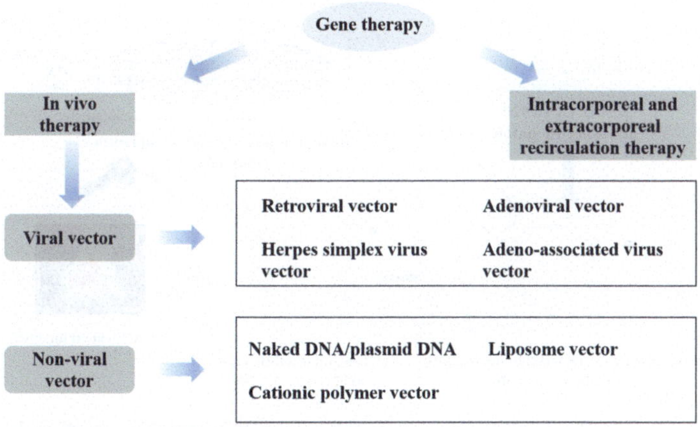

Fig. 10.11 Therapeutic approaches of gene therapy

complex particles. These particles can enter cells through endocytosis. Because every third atom in the PEI monomer (–CH_2–CH_2–NH–) contains a protonatable nitrogen atom, the pKa of the primary, secondary, and tertiary amino groups formed by nitrogen atoms differs, allowing PEI to absorb protons under any pH condition, known as the "proton sponge effect". This characteristic helps to avoid DNA degradation by lysosomes and enables the release of DNA from lysosomes into the cytoplasm and nucleus, facilitating plasmid DNA expression. Research results show that microneedle-mediated siRNA can effectively transfect cells, enhancing gene silencing efficiency and effectively inhibiting melanoma growth.

Wang et al. [42] devised soluble microneedles carrying both pDNA and the photothermal agent IR820 for combined therapy of subcutaneous tumors, as depicted in Fig. 10.12. HA was selected as the material for the needle body, being one of the predominant constituents in skin tissue and an FDA-approved soft tissue filler with outstanding biocompatibility and in vivo degradability. Additionally, HA can serve as a carrier for targeted tumor drugs, primarily mediated through its interaction with receptors to enhance drug delivery specificity. Notably, HA binds specifically to receptors such as CD44, IVd4, RHMAA, and LEC, which are frequently overexpressed on the surface of tumor cells. Upon entering the body, HA selectively binds to these receptors, facilitating the targeted delivery of drugs. For instance, CD44, a receptor significantly overexpressed in pancreatic cancer cells compared to normal breast cells, serves as a compelling example. When HA is employed as a carrier for delivering therapeutic drugs to pancreatic cancer cells, it binds specifically to the CD44 receptor, thereby concentrating the therapeutic drugs within the pancreatic cancer cell area, achieving targeted therapy. Studies have demonstrated that HA microneedles rapidly release pDNA and IR820 within 5 min of skin penetration. While pDNA effectively transfects target cells, IR820 exhibits high photothermal conversion efficiency under near-infrared irradiation, rapidly elevating the temperature at the tumor site to eradicate cancer cells. Consequently, HA microneedles

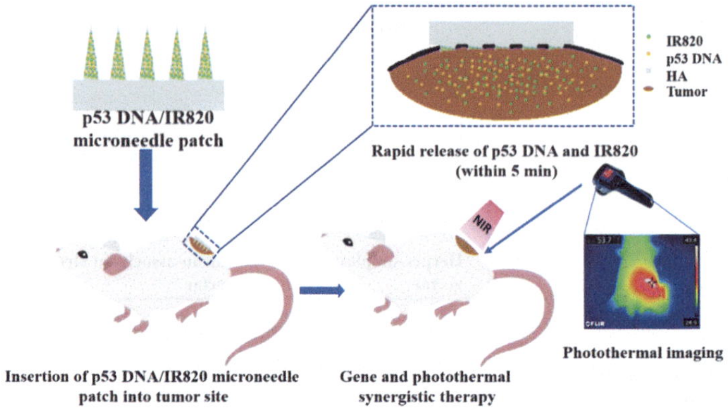

Fig. 10.12 Treatment of subcutaneous tumors with soluble microneedles loaded with pDNA and photothermal agent IR820. The figure was reproduced with permission [42]

for combined pDNA and photothermal therapy represent a promising approach for effectively inhibiting tumor growth.

In addition, recent studies have revealed the potential of DNA vaccines in both tumor prevention and effective treatment. Ali et al. [23] employed a strategy where DNA targeting HPV16 E6 or E7 was complexed with cationic peptide RALA nanoparticles and delivered via soluble microneedles. E6 and E7 viral genes, being the target antigens for therapeutic vaccination, play crucial roles in tumor malignancy transformation and growth. The research findings indicate that mice vaccinated with DNA vaccines showed higher levels of E6/E7-specific IgG in their serum, leading to a more efficient suppression of tumors.

10.3.3 Dissolvable Microneedles for Tumor Phototherapy

Tumor phototherapy refers to the use of near-infrared light of different wavelengths to achieve a special response of materials, and then cause the death of tumor cells. Compared with surgery, chemotherapy and radiotherapy, phototherapy is more controllable, less invasive and easy to operate. The photo-responsive materials used have good biocompatibility in the unexcited state, and use targeted recognition technology to accumulate near the tumor tissue, thereby reducing some side effects and systemic toxicity during tumor treatment. Based on the above advantages, phototherapy has been more and more widely used in tumor treatment, and some have also been applied to clinical treatment. However, for local treatment of some superficial tumors, due to the barrier effect of the skin, the penetration depth of drugs is limited, and it is difficult to achieve high concentration enrichment of drugs by injection, resulting in insufficient therapeutic effect. As a novel transdermal drug delivery method, microneedles can efficiently deliver drugs to the dermis to achieve local drug enrichment. In previous studies, microneedle has been successfully applied to the efficient delivery and local treatment of insulin, vaccines and chemotherapy drugs, which provides a reliable basis for the combination of phototherapy and microneedle technology.

10.3.3.1 Photothermal Therapy

Photothermal therapy usually uses some photothermal conversion materials (such as precious metal nanoparticles, carbon materials, metal and non-metallic compounds, organic dye substances, etc.) to achieve efficient thermal conversion under the excitation of near-infrared light, and kill cells by the heat generated. Hood et al. [43] first combined photothermal therapy with microneedle technology in 2013, by using hollow silicon microneedles to deliver single-walled carbon nanoangle to porcine bladder tissue to achieve efficient photothermal conversion in the deep tissue. Although silicon as the matrix material of microneedles has sufficient mechanical properties to penetrate the skin, it is brittle and has the risk of subcutaneous fracture, and it is difficult to process and shape. Carcamo et al. [44] fixed gold nanorods by using a light-curing resin coating on the surface of microneedles. After the microneedle was inserted into the tumor, the gold nanorods in the surface coating of the microneedle produced a thermal effect under the excitation of near-infrared light, which increased the tissue temperature. After removal of the MNs array, no polymer or gold nanorods were left behind, confirming the high safety of this method and its potential for future biomedical minimally invasive applications of deep skin hyperthermia. As shown in Fig. 10.13, Lin et al. [45] selected PVP as the matrix material to deliver Nb_2C nanosheets. After penetration into the skin, PVP can be rapidly dissolved. The tumor-bearing mice were selected as the model. After the micro-needle was inserted into the tumor, the temperature of the tumor site was quickly increased to 70 °C under the 1064 nm near-infrared (1 W/cm^2) irradiation, thereby killing the tumor cells. After two weeks of treatment, the tumor volume decreased significantly. Wei et al. [46] prepared NIR950 encapsulated methoxy polyethylene glycol-poly (β-amino ester) (mPEG-PAE) micelle by nanodeposition method to improve the solubility of photothermal therapy drugs and make them uniformly distributed in solution, which is conducive to the preparation of microneedles. When coumarin 6 was used as the mimic drug and co-cultured with cells, it was easier to observe the mimetic drug in the cells under acidic environment, proving that the charge conversion characteristics of mPEG-PAE in acidic tumor microenvironment could improve the uptake and retention time of NIR950@PMs in tumor tissues. In the animal experiment, the drug was concentrated in the tumor by in vivo imaging, and the local temperature of the tumor rose to 55 °C after 4 min of near-infrared irradiation. Compared with other drug delivery methods, the microneedle-treated group was given only once within two weeks, and the tumor was almost completely removed without recurrence.

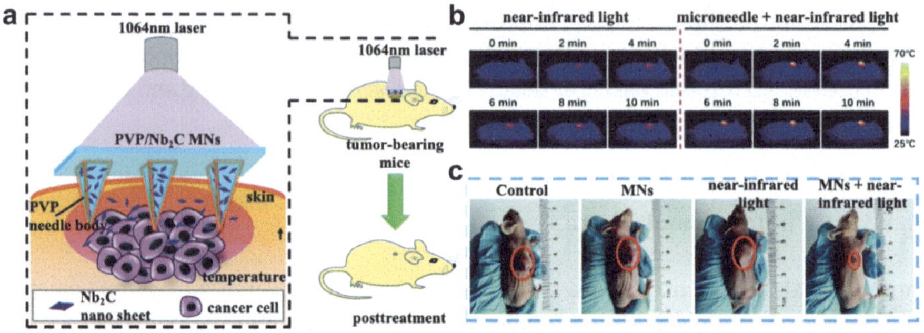

Fig. 10.13 Microneedle-patch (PVP/Nb_2C MN) for delivery of drugs to treat superficial tumors. **a** Schematic representation of the microneedle system. **b** Thermal images of tumor sites in NIR group and MNs + NIR group irradiated with 1064 nm NIR (1 W/cm^2) for 0, 2, 4, 6, 8, and 10 min. **c** images of tumor sites of tumor mice after treatment. The figure was reproduced with permission [45]

10.3.3.2 Photodynamic Therapy

Photodynamic therapy usually uses some porphyrin molecular photosensitizers, which can achieve energy conversion under irradiation of light of specific wavelengths to produce highly cytotoxic reactive oxygen species that kill tumor cells. Jain et al. [47] used microneedle technology in combination with photodynamic therapy for the treatment of superficial cutaneous tumors in 2016, where stainless steel microneedle arrays were prepared by wet etching, and then soaked in 5-aminolevulinate drug solution to prepare drug coating on the surface of microneedles. After 11 days of treatment of skin tumors, the volume of skin tumors in the coated microneedle group (1.75 mg 5-aminolevulinic acid) was reduced to 43%, whereas the topical ointment group (5 mg 5-aminolevulinic acid) did not inhibit the tumors but rather was similar to the blank group, with the tumors increasing in volume to 5–6 times their initial volume. Hamdan et al. [48] compared the diffusion in vivo of microneedles and intravenous photosensitizers, and found that the photosensitizers injected by tail vein rapidly spread to the whole body and maintained a high intensity within 7 days. However, after the use of microneedles, the photosensitive drugs concentrated on the site of action, and no large area of diffusion occurred within 7 days, which proves that microneedles can avoid the adverse effects caused by the systemic diffusion of photosensitizer and make the treatment more accurate and efficient. Zhao et al. [49] delivered 5-aminolevulinic acid by preparing dissolving microneedles of hyaluronic acid. As a water-soluble polymer, hyaluronic acid is a high-level polysaccharide composed of D-glucuronic acid and n-acetylglucosamine, and has a large number of hydroxyl groups on its side bonds, so it has good water solubility and can be processed and modified. Based on the use of sugars as matrix materials, they have been widely used in the medical cosmetic industry due to their high biocompatibility. Because of its high biocompatibility, it has been widely used in medical cosmetology industry. By comparing microneedle administration with tail vein injection administration, it was found that for the subcutaneous tumor model in BALB/c nude mice, the photodynamic therapeutic effect of HA microneedle-loaded 5-aminolevulinic acid was significantly better than that of the 5-aminolevulinic acid injection group, even though the HA microneedle-loaded 5-aminolevulinic acid dosage was comparatively low, and the results further verified the experimental conclusions of Hamdan et al. In addition, the stability of the photosensitizer is also an important factor affecting the therapeutic effect, which is prone to hydrolysis, photodegradation and thermal degradation during storage and application. To solve this problem, Zhu et al. [50] found that the microneedle matrix material has a protective effect on photosensitizers, and the hyaluronic acid microneedle provides an acidic, oxygen-free environment for 5-aminolevulinic acid, which reduces the dimerization of the 5-aminolevulinic acid molecule to form an inactive pyrazine derivative activity through Schiff base bonding, thus maintaining its chemical structure and biological activity. The characteristic peaks of 5-aminolevulinic acid, as determined by ultraviolet–visible spectrophotometry, remained essentially unchanged after 9 months of storage in the microneedle and still showed significant efficacy in in vitro cellular assays and in vivo tumor therapy.

In addition, the strong hydrophobicity and poor tumor selectivity of photosensitizers, as well as the oxygen consumption characteristics of PDT, lead to unsatisfactory therapeutic effect. As shown in the figure, Huang et al. [51] has developed a tumor acidic microenvironment activatable dissolving microneedle (DHA@HPFe-MN) to realize controlled drug release and excellent chemo-photodynamic therapy of melanoma via oxidative stress amplification (Fig. 10.14). The acidic pH triggered the degradation of microneedles, enabling Fe-activated and DHA-mediated oxidation treatment, as evidenced by abundant ROS generation. Moreover, under light irradiation, a combined chemo-photodynamic therapeutic effect was achieved with amplified ROS generation. Importantly, the Fecatalyzed ROS production of DHA was oxygen-independent, which work in synergy with the oxygen-dependent PDT to effectively destroy tumor cells.

10.3.4 Dissolvable Microneedles for Tumor Combination Therapy

Due to the heterogeneity and complexity of the tumor, it is difficult to achieve significant therapeutic effects with a single therapy, so in recent years there has been a gradual shift from traditional monotherapy to a combination of multiple therapies. As a newly developed transdermal drug delivery technology, microneedles can not only penetrate the stratum corneum for efficient drug delivery, but also can avoid contact with the pain nerve tissue below the dermis due to its unique micron-scale needle structure, showing painless penetration [3]. Different types of microneedles have been designed and prepared for tumor therapy. Microneedle synergistic therapy, as a strategy to enhance tumor therapy, has attracted more and more

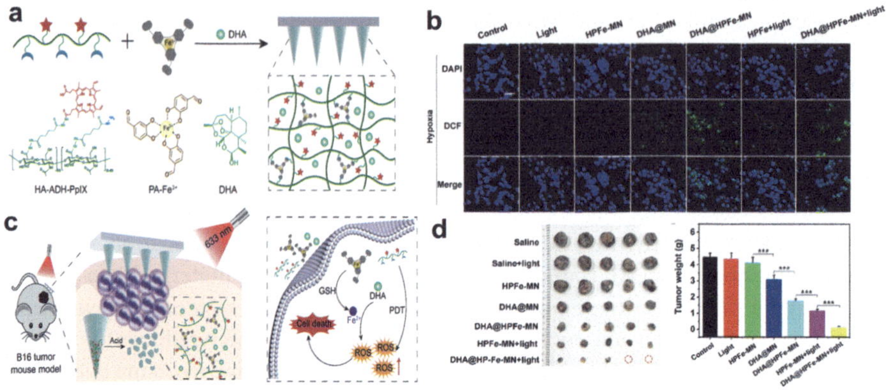

Fig. 10.14 **a** Schematic illustration of the fabrication of hydrogel-based DHA@HPFe-MN microneedles. **b** Schematic illustration of the in vivo antitumor mechanism of DHA@HPFe-MN via oxidative stress amplification. **c** The ROS generation capacity of different microneedles formulations under hypoxic conditions. Scale bar = 20 μm. **d** Photograph of the excised tumors and average tumor weights after different treatments. The figure was reproduced with permission [51]

attention, mainly including chemotherapy and photothermal therapy (PTT) synergistic therapy, chemotherapy and PDT synergistic therapy, light therapy and gene synergistic therapy, light therapy and immunosynergistic therapy.

10.3.4.1 Microneedles-Based Synergistic Treatment of Chemotherapy and Photothermal Therapy

Microneedle-assisted chemotherapy and photothermal synergistic therapy mainly involve loading photosensitizers and chemotherapeutic drugs into microneedles, microneedles are administered at the tumor site, photosensitizers and chemotherapeutic drugs are released locally into the tumor, and controlling the time and intensity of the photothermal therapy can lead to thermal ablation of tumor cells. At the same time, the blood vessels in the tumor area are destroyed, and the chemotherapeutic drugs can play a better role in inhibiting the tumor. In addition, the thermal effect produced by photothermal therapy can enhance the absorption of chemotherapeutic agents by tumor cells, accelerate the release of chemotherapeutic agents loaded into nanocarriers, and increase the concentration of drugs in tumor cells, thus achieving better therapeutic effects [52].

Dissolving microneedle-assisted therapy is most common in chemotherapy and photothermal synergistic therapy, and the materials used are generally polyvinyl alcohol (PVA), PVP, and HA. All three materials have been approved by the FDA and have excellent biocompatibility. The prepared microneedles have sufficient mechanical properties to pierce the skin and have strong water solubility, which can dissolve quickly in the skin and deliver the carried drugs without any sharp substances harmful to organisms remaining. PVP is a dissolvable material used as a binder in many pharmaceutical tablets, and it has a long history of clinical use as a plasma expander. HA is a major component of the extracellular matrix, with anti-aging and anti-inflammatory properties, skin repair, tissue regeneration, and wound recovery properties [53]. HA has good biocompatibility, high moisturizing properties, and adjustable viscoelasticity. It is a naturally unbranched-chain polymer consisting of repeating disaccharide units of D-glucuronic acid and N-acetylglucosamine linked by glucuronic acid β (1–3) bonds. The molecular weight of HA directly affects its properties. For example, low molecular weight HA (LMW-HA) is associated with the promotion of angiogenesis, whereas high molecular weight HA (HMW-HA) inhibits angiogenesis. The proper balance between HA synthesis and degradation is essential for the regulation of a variety of biological functions including cell proliferation, migration, differentiation, angiogenesis, and the modulation of cellular adhesion and motility, properties that are also dictated by its molecular weight [54].

Dong et al. [55] developed gold nanocages (AuNCs) and chemotherapeutic drug adriamycin (DOX)-loaded HA microneedles, which were uniformly administered without DOX leakage, and the body weights of the mice in the DOX/AuNC microneedle + NIR group were stable, suggesting that this delivery method can improve the therapeutic effect of DOX on tumors while ensuring biosafety. Chen et al. [56] incorporated graphene oxide (GO) and HA15 into microneedles prepared with PVP for the treatment of melanoma. The presence of GO increased the mechanical strength of the microneedles, improved the moisture resistance, bacteriostatic and anti-inflammatory properties, prevented microbial contamination during storage and transportation, and reduced the risk of infection in clinical applications.

Researchers have also developed multifunctional nano-carrier systems to enhance the bioavailability and solubility of chemotherapy drugs due to their lack of targeting and low bioavailability. These carrier systems include inorganic, organic, polymer, and metal nanoparticles. Pei et al. [57] synthesized mesoporous silica nanoparticles (MSN) using a sol-gel method, then covalently bound indocyanine green ICG to MSN to form MSN@ICG, which was used to improve the stability of ICG and avoid the reduction of fluorescence quantum efficiency. They prepared microneedles (PVP@DOX/MSN@ICG MN) by centrifuging a mixture of MSN@ICG and PVP solution. In vitro experiments using MG-63 osteosarcoma cells showed that PVP@DOX/MSN@ICG microneedle patches combined with chemotherapy and PTT maximally induced cell death. After the drug-loaded microneedles were inserted into the tumor site of mice, local tumor temperatures reached 48°C within 2 min of 808 nm light irradiation, demonstrating optimal therapeutic tumor effects. Qin et al. [58] encapsulated paclitaxel (PTX) and IR-780 SLNs (PTX/IR-780 SLN), achieving pulse drug release profiles through "ON" and "OFF" heating cycles at intervals of 10 min, 1, and 2 h, with each PTX release ranging from 2 to 4%. They centrifuged the PTX/IR-780 SLN suspension three times to position it at the tip of the needle, used 400 mg/mL HA solution as the needle material, and finally centrifuged PVP ethanol solution to form microneedle baseplates (PTX/IR-780 SLNs @DMNs). Compared to intravenous and intratumoral injections, PTX/IR-780 microneedles showed minimal distribution to other organs, improving the bioavailability at the tumor site with good safety. In vivo experimental results showed that PTX/IR-780 SLNs microneedles had significant anti-tumor effects, eliminating primary tumors within 30 days, with a 100% cure rate and the highest 100-day survival rate among all groups, reaching 66.67%. Moreira et al. [59] coated the surface of soluble microneedles with chitosan loaded with DOX and polyvinyl alcohol coated with gold nanocages (Dox@MicroN) for the delivery of DOX and AuNCS to tumor cells. The Dox@MicroN patch exhibited good photothermal capacity, with a temperature increase of 12 °C under near-infrared (808 nm, 1.7 W/cm^2, 5 min) irradiation. Combined with the pH sensitivity of chitosan, it could be used to control DOX release, mediating both chemotherapy and photothermal therapy, showing significant cytotoxicity to cervical cancer cells. Hao et al. [60] designed a near-infrared-responsive polyethylene glycol-modified gold nanorod (GNR-PEG) and embedded it in poly(D, L-lactic acid) microneedles (PDLLA MN) to form a near-infrared-responsive microneedle delivery system (GNR-PEG@MN) to enhance the antitumor efficacy of methoxy poly(ethylene glycol)-poly(D,L-lactic acid)-paclitaxel (MPEG-PDLLA-DTX) micelles in treating A431 human squamous cell carcinoma. PEG, as a hydrophilic polymer-modified nanocarrier, can achieve long circulations in vivo and spatially avoid the destruction of the reticular endothelial system by carriers. Under 2 W/cm^2 808 nm laser irradiation, the temperature of GNR-PEG@microneedles quickly reached 55 °C, and the temperature returned to room temperature after stopping irradiation. After the microneedle was inserted into the tumor site, it quickly heated up with laser irradiation, reaching 50 °C in 1 min, $\Delta T = 23.8$ °C, and the heating depth could reach 1.5 cm, causing irreversible damage to the tumor tissue. The study ultimately concluded that the combined therapy of PEG-PDLLA-DTX (5 mg/kg) + GNR-PEG@MNs with laser had a better therapeutic effect, rapidly reducing tumor volume, disappearing within 14 days. GNR-PEG@MN enhances the anti-tumor effect of MPEG-PDLLA-DTX micelles on superficial tumors, and is expected to achieve clinical translation in the treatment of squamous cell carcinoma in humans, as shown in Fig. 10.15.

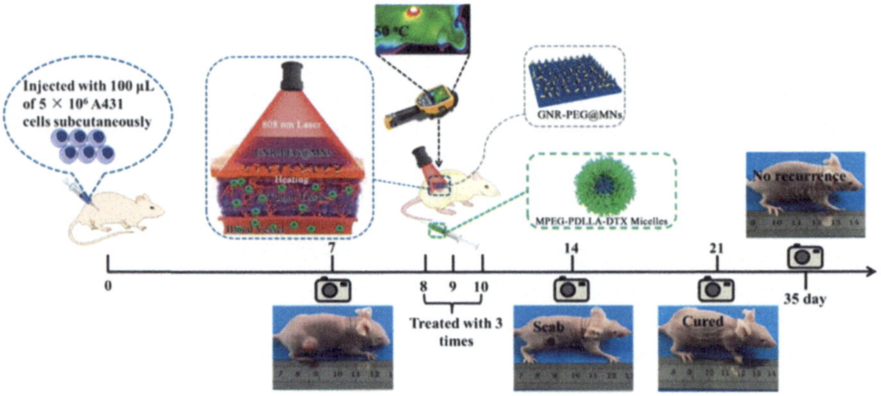

Fig. 10.15 MPEG-PDLLA-DTX and GNR-PEG@MNs synergistic treatment of tumor. The figure was reproduced with permission [60]

PCL is an FDA-approved biodegradable polyester material that is biocompatible and degradable, and also improves drug stability and is often used as a drug carrier. Its melting point is 55 ~ 60 °C, and controlled release of drugs can be achieved by controlling the temperature at which it is exposed. Hao et al. [20] designed a near-infrared light-responsive 5-fluorouracil (5-Fu) and ICG-loaded monomethoxy-poly (ethyleneglycol)-poly (caprolactone) (MPEG-PCL) nanoparticles (5-Fu-ICG-MPEG-PCL); and then the 5-Fu-ICG- MPEG-PCL was then integrated with a hyaluronic acid soluble microneedle system to obtain 5-Fu-ICG-MPEG-PCL-loaded hyaluronic acid microneedles (5-Fu-ICG-MPEG-PCL@HA MN) for the treatment of skin tumors including squamous cell carcinoma and melanoma, which exhibited good skin-piercing ability and heat transfer. At the same time, it also has a good photothermal effect, which can both thermally ablate tumors and control the release behavior of 5-Fu in the nanoparticles, realizing synergistic chemotherapy and photothermal therapy for skin tumors. PCL can not only be made into nanorods, but also be used as a microneedle needle body material.

Chen et al. [5, 61, 62] chose PVP and PVA as the matrix material, PCL as the material of tip site, chemotherapeutic drug DOX, and photosensitizer lanthanum hexaboride (LaB6), ICG, and Prussian blue nano-cubes (PB NCs) were used to prepare MN to achieve synergistic treatment of chemotherapy and photothermal therapy, which has a very good anti-tumor effect, as shown in Fig. 10.16. In addition, the details of the materials for microneedle preparation and the used chemotherapeutic drug and photosensitizer are listed in Table 10.2.

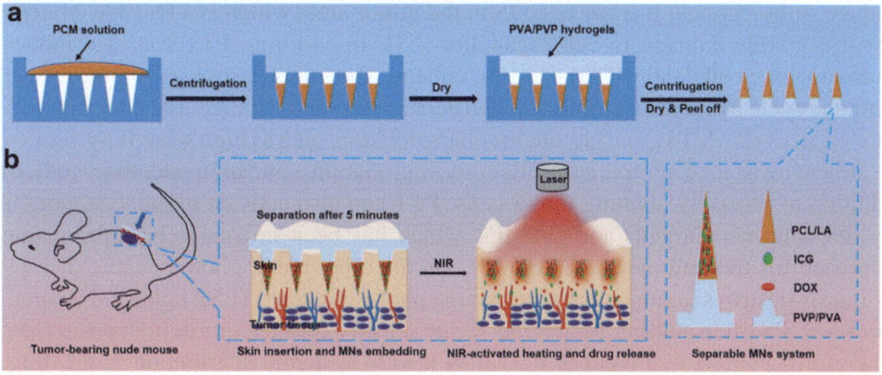

Fig. 10.16 The fabrication of the separable MN system (**a**), and ICG/DOX-loaded MN system for synergistic chemo-photothermal therapy against SSTs (**b**). The figure was reproduced with permission [61]

Table 10.2 Microneedle materials, chemotherapy drugs and photosensitizers used in the synergistic treatment of chemotherapy and photothermal therapy

Microneedle type	Microneedle material	Chemotherapeutic drug	Photothermic drug	References
Dissolving MN	HA	DOX	AuNCs	[55]
Dissolving MN	PVP	HA15	GO	[56]
Dissolving MN	PVP	MSN@ICG	ICG	[57]
Dissolving MN	HA + PVP	PTX	IR-780	[58]
Coated MN	PLLA	MPEG-PDLLA-DTX	GNR-PEG	[60]
Dissolving MN	PVP	DOX	AuMSS	[59]
Dissolving MN	HA	5-Fu	ICG	[20]
Dissolving MN	PCL + PVA/PVP	DOX	LaB6	[5]
Dissolving MN	PCL + PVA/PVP	DOX	ICG	[61]
Dissolving MN	PCL + PVP/PVA	DOX	PB NCs	[62]

10.3.4.2 Microneedles-Based Synergistic Treatment of Phototherapy and Immunotherapy

Photothermal therapy utilizes photosensitizers accumulated in tumor tissues to induce sufficient heat therapy or ROS in the tumor area, which is a feasible strategy for irreversible damage to cell organelles [63]. In addition, PTT can also induce immunogenic cell death (ICD) patterns, enhance the immune response of the cancer immune cycle by releasing tumor-related neoantigens and damage-associated molecular patterns (DAMPs), and has unique characteristics such as high selectivity, limited treatment resistance, remote controllability, high stability, and high selectivity [64]. In addition to precisely ablating tumor cells, PTT can also activate immune responses by releasing neoantigens and DAMPs, which has been proven to be a promising approach for treating metastatic tumors. On the other hand, photodynamic therapy primarily destroys tumor cells by combining photosensitizers (PS), light, and oxygen. However, PS often exhibits poor target tissue selectivity and requires high doses [65]. Therefore, a single treatment of PDT may lead to systemic distribution of the drug, resulting in severe long-term phototoxicity to the skin. Although researchers have made great efforts in creating novel photosensitizers and overcoming tumor hypoxia, the effectiveness of PDT in killing cells remains an important challenge.

Tumor immunotherapy is a more targeted approach to tumor cells by modulating the function of specific immune cells, with the ultimate goal of initiating or restoring the self-sustaining cycle of tumor immunity and amplifying and delivering it. However, due to the strong inhibitory nature of tumor immune regulation, the immune system often cannot induce a strong immune response to combat tumors. Some typical immunosuppressive molecules [66], such as programmed cell death receptor-1 (PD-1), PD-1 ligand (PD-L1), and cytotoxic T-lymphocyte antigen-4 (CTLA-4), involve receptor-ligand interactions that severely inhibit immune responses.

To further improve the effectiveness of treatment, the development of combination therapy has received increasing attention. So far, most combination therapies follow programmable steps, including: (I) inducing tumor cell necrosis by phototherapy/chemotherapy to enhance tumor immune-related responses; (II) systemic administration of immunomodulatory drugs to boost immune responses. Wu et al. [67] designed a core-shell structured microneedle array for localized co-delivery of the photosensitizer ICG and indoleamine 2, 3-dioxygenase (IDO) inhibitor to achieve synergistic immune responses. Indocyanine green was encapsulated into chitosan nanoparticles (ICG-NPs) concentrated at the tip of the microneedles. 1-methyltryptophan was loaded into crosslinked polyvinylpyrrolidone and polyvinyl alcohol hydrogels as the microneedle core layer. 1-methyltryptophan helps inhibit the action of IDO, reversing the immune suppressive effects caused by overexpression of IDO regionally, ensuring its effectiveness and durability. Ye et al. [68] developed a melanin-mediated tumor immunotherapy strategy. Melanin-containing B16F10 whole tumor lysate was loaded into polymer microneedles, with the lysate continuously released by the insertion of the microneedles into the skin. Combined with near-infrared light irradiation, melanin in the tumor lysate mediated heat generation, further promoting dendritic cell recognition and uptake of tumor antigens, enhancing the capability of anti-tumor vaccine to activate immune responses. Compared to solo microneedle inoculation, simultaneous inoculation increased BP by 2.5 times, and 4T1 model by 4 times. Additionally, combination therapy can induce the production of pro-inflammatory cytokines, leading to enhanced DC activation, as shown in Fig. 10.17. He et al. [69] prepared soluble microneedles of hyaluronidase-modified semiconducting polymer nanoparticles (SPNs) co-delivered in the microneedles with SPNs containing poly(cyclopentadithiophene-bithiophene) and the adjuvant immunologique acide polyinosinique-polycytidylique (PIC). Through hyaluronidase-mediated degradation of the extracellular matrix, SPNs and PIC can penetrate deep into the tumor, activating immune cells, enhancing T cell immune responses, and inhibiting the growth and metastasis of tumors in synergy with photothermal therapy.

10.3.4.3 Microneedles-Based Synergistic Treatment of Phototherapy, Photothermal Therapy, and Immunotherapy

The triple combination of chemotherapy, photothermal therapy, and photodynamic therapy can be achieved through a nanocarrier platform. Under the irradiation of an exogenous excitation light source, the heat generated by photothermal therapy can achieve thermal ablation, enhance cellular uptake of photo-sensitizers, increase local oxygen content, enhance cytotoxic effects on tumor cells, significantly increase drug/ROS-induced DNA damage. Therefore, photothermal/photodynamic/immunotherapy will produce a synergistic effect of triple-peaked therapy.

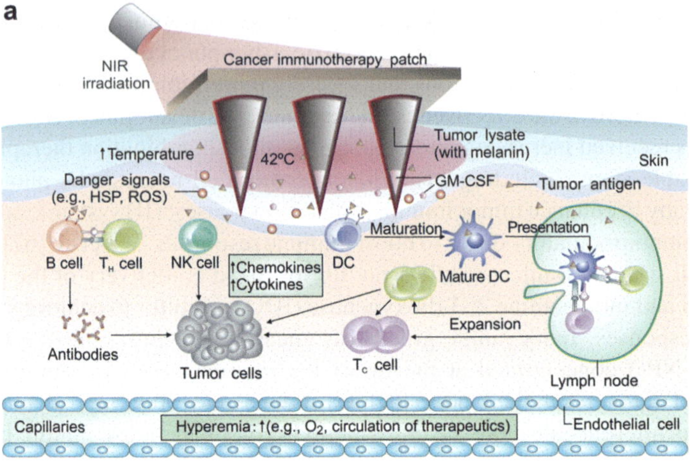

Fig. 10.17 Schematic of microneedle transdermal vaccination. The figure was reproduced with permission [69]

Chen et al. [70] prepared in-situ self-assembled nanomicelle dissolving microneedle patches for targeted delivery of immunogenic cell death inducer (IR780) and autophagy inhibitor (chloroquine, CQ) co-encapsulated micelles (C/IMil) in the lesion tissue, which can effectively conduct anti-tumor therapy. After self-assembly, C/I-Mil can electrostatically bind with the matrix material hyaluronic acid of the microneedle and be released along with the dissolution of the needle upon insertion into the skin. Subsequently, the size-controllable C/I-Mil mediated by photothermal can effectively penetrate deep into the tumor tissue, be internalized in large quantities through CD44 receptor-mediated endocytosis, precisely ablate the tumor with the help of autophagy inhibition, and promote the release of damage-associated molecular patterns. In addition, CQ can also act as an immunomodulator, reshaping tumor-associated macrophages towards the M1 phenotype through NF-κB activation. In vivo results demonstrate that localized photoimmunotherapy in synergy with autophagy inhibition can effectively eliminate primary and distant tumors, with a relapse-free survival period of over 40 days following reshaping.

10.4 Outlook and Challenges

Microneedles, as a novel drug delivery route, have significantly broadened the spectrum of drug administration in tumor diagnosis and treatment, enhancing delivery efficiency while also bolstering drug stability and reducing reliance on skilled operators. Nonetheless, several challenges persist in the clinical adoption and product realization of microneedle-based tumor therapy. In the realm of tumor diagnosis, microneedles encounter limitations in the volume of tissue fluid they can access or extract, directly impacting diagnostic accuracy. Similarly, in tumor treatment, the constrained drug-carrying capacity of microneedles can compromise therapeutic efficacy, while prolonged and frequent use may pose safety concerns. Furthermore, microneedle technology exhibits a preference for addressing superficial tumors, limiting its applicability to deep-seated tumors.

Presently, much of the research in microneedle-based tumor diagnosis and treatment centers on drug development, with microneedles serving merely as delivery tools. To advance the efficacy of tumor diagnosis and treatment, microneedle structures can be tailored, encompassing material selection and fabrication techniques to enable multistage drug release and intelligent detection and treatment strategies. Moreover, in tumor treatment, activating the subcutaneous immune system alongside superficial tumor cell eradication is paramount, given the pivotal role immune cells play in combating metastatic tumor cells.

Although microneedle-based tumor diagnosis and treatment face hurdles in clinical translation, their attributes of convenience, minimal invasiveness, and cost-effectiveness render them a promising therapeutic modality. As research progresses, microneedles hold the potential for widespread clinical application in tumor diagnosis and treatment, offering newfound convenience and optimism to this domain.

References

1. Ferlay, J., Colombet, M., Soerjomataram, I., Parkin, D.M., Piñeros, M., Znaor, A., Bray, F.: Cancer statistics for the year 2020: an overview. Int. J. Cancer **149**(4), 778–789 (2021)
2. Kanda, T., Wakabayashi, Y., Zeng, F., Ueno, Y., Sofue, K., Maeda, T., Nogami, M., Murakami, T.: Imaging findings in radiation therapy complications of the central nervous system. Jpn. J. Radiol. **36**(9), 519–527 (2018)
3. Lin, S., Cao, Y., Chen, J., Tian, Z., Zhu, Y.: Recent advances in microneedles for tumor therapy and diagnosis. Appl. Mater. Today **23**, 101036 (2021)
4. Alimardani, V., Abolmaali, S.S., Tamaddon, A.M., Ashfaq, M.: Recent advances on microneedle arrays-mediated technology in cancer diagnosis and therapy. Drug Deliv. Transl. Res. **11**(3), 788–816 (2021)
5. Chen, M.-C., Lin, Z.-W., Ling, M.-H.: Near-infrared light-activatable microneedle system for treating superficial tumors by combination of chemotherapy and photothermal therapy. ACS Nano **10**(1), 93–101 (2016)
6. Xu, C., Feng, Q., Yang, H., Wang, G., Huang, L., Bai, Q., Zhang, C., Wang, Y., Chen, Y., Cheng, Q., Chen, M., Han, Y., Yu, Z., Lesniak, M.S., Cheng, Y.: A light-triggered mesenchymal stem cell delivery system for photoacoustic imaging and chemo-photothermal therapy of triple negative breast cancer. Adv. Sci. (Weinh) **5**(10), 1800382 (2018)
7. Khan, S., Hasan, A., Attar, F., Babadaei, M.M.N., Zeinabad, H.A., Salehi, M., Alizadeh, M., Hassan, M., Derakhshankhah, H., Hamblin, M.R., Bai, Q., Sharifi, M., Falahati, M., ten Hagen, T.L.M.: Diagnostic and drug release systems based on microneedle arrays in breast cancer therapy. J. Control. Release **338**, 341–357 (2021)
8. Miller, P.R., Taylor, R.M., Tran, B.Q., Boyd, G., Glaros, T., Chavez, V.H., Krishnakumar, R., Sinha, A., Poorey, K., Williams, K.P., Branda, S.S., Baca, J.T., Polsky, R.: Extraction and biomolecular analysis of dermal interstitial fluid collected with hollow microneedles. Commun. Biol. **1**(1), 173 (2018)
9. Mandal, A., Boopathy, A.V., Lam, L.K.W., Moynihan, K.D., Welch, M.E., Bennett, N.R., Turvey, M.E., Thai, N., Van, J.H., Love, J.C., Hammond, P.T., Irvine, D.J.: Cell and fluid

sampling microneedle patches for monitoring skin-resident immunity. Sci. Transl. Med. **10**(467) (2018)
10. Al Sulaiman, D., Chang, J.Y.H., Bennett, N.R., Topouzi, H., Higgins, C.A., Irvine, D.J., Ladame, S.: Hydrogel-coated microneedle arrays for minimally invasive sampling and sensing of specific circulating nucleic acids from skin interstitial fluid. ACS Nano **13**(8), 9620–9628 (2019)
11. Yang, B., Fang, X., Kong, J.: In situ sampling and monitoring cell-free DNA of the Epstein-Barr virus from dermal interstitial fluid using wearable microneedle patches. ACS Appl. Mater. Interfaces **11**(42), 38448–38458 (2019)
12. Babity, S., Couture, F., Campos, E.V.R., Hedtrich, S., Hagen, R., Fehr, D., Bonmarin, M., Brambilla, D.: A naked eye-invisible ratiometric fluorescent microneedle tattoo for real-time monitoring of inflammatory skin conditions. Adv. Healthc. Mater. **11**(6), 2102070 (2022)
13. Sabri, A., Ogilvie, J., McKenna, J., Segal, J., Scurr, D., Marlow, M.: Intradermal Delivery of an immunomodulator for basal cell carcinoma; expanding the mechanistic insight into solid microneedle-enhanced delivery of hydrophobic molecules. Mol. Pharm. **17**(8), 2925–2937 (2020)
14. Ma, Y., Boese, S.E., Luo, Z., Nitin, N., Gill, H.S.: Drug coated microneedles for minimally-invasive treatment of oral carcinomas: development and in vitro evaluation. Biomed. Microdevice **17**(2), 44 (2015)
15. Hiraishi, Y., Nakagawa, T., Quan, Y.-S., Kamiyama, F., Hirobe, S., Okada, N., Nakagawa, S.: Performance and characteristics evaluation of a sodium hyaluronate-based microneedle patch for a transcutaneous drug delivery system. Int. J. Pharm. **441**(1), 570–579 (2013)
16. Bhatnagar, S., Bankar, N.G., Kulkarni, M.V., Venuganti, V.V.K.: Dissolvable microneedle patch containing doxorubicin and docetaxel is effective in 4T1 xenografted breast cancer mouse model. Int. J. Pharm. **556**, 263–275 (2019)
17. Jonas, O., Landry, H.M., Fuller, J.E., Santini, J.T., Baselga, J., Tepper, R.I., Cima, M.J., Langer, R.: An implantable microdevice to perform high-throughput in vivo drug sensitivity testing in tumors. Sci. Transl. Med. **7**(284), 284ra57–284ra57 (2015)
18. Huang, S., Liu, H., Huang, S., Fu, T., Xue, W., Guo, R.: Dextran methacrylate hydrogel microneedles loaded with doxorubicin and trametinib for continuous transdermal administration of melanoma. Carbohydr. Polym. **246**, 116650 (2020)
19. Gadag, S., Narayan, R., Nayak, A.S., Catalina Ardila, D., Sant, S., Nayak, Y., Garg, S., Nayak, U.Y.: Development and preclinical evaluation of microneedle-assisted resveratrol loaded nanostructured lipid carriers for localized delivery to breast cancer therapy. Int. J. Pharm. **606**, 120877 (2021)
20. Hao, Y., Chen, Y., He, X., Yang, F., Han, R., Yang, C., Li, W., Qian, Z.: Near-infrared responsive 5-fluorouracil and indocyanine green loaded MPEG-PCL nanoparticle integrated with dissolvable microneedle for skin cancer therapy. Bioact. Mater. **5**(3), 542–552 (2020)
21. Ahmed, K.S., Shan, X., Mao, J., Qiu, L., Chen, J.: Derma roller® microneedles-mediated transdermal delivery of doxorubicin and celecoxib co-loaded liposomes for enhancing the anticancer effect. Mater. Sci. Eng. C **99**, 1448–1458 (2019)
22. Kim, Y.-C., Quan, F.-S., Yoo, D.-G., Compans, R.W., Kang, S.-M., Prausnitz, M.R.: Improved influenza vaccination in the skin using vaccine coated microneedles. Vaccine **27**(49), 6932–6938 (2009)
23. Ali, A.A., McCrudden, C.M., McCaffrey, J., McBride, J.W., Cole, G., Dunne, N.J., Robson, T., Kissenpfennig, A., Donnelly, R.F., McCarthy, H.O.: DNA vaccination for cervical cancer; a novel technology platform of RALA mediated gene delivery via polymeric microneedles, nanomedicine: nanotechnology. Biol. Med. **13**(3), 921–932 (2017)
24. Cole, G., Ali, A.A., McCrudden, C.M., McBride, J.W., McCaffrey, J., Robson, T., Kett, V.L., Dunne, N.J., Donnelly, R.F., McCarthy, H.O.: DNA vaccination for cervical cancer: strategic optimisation of RALA mediated gene delivery from a biodegradable microneedle system. Eur. J. Pharm. Biopharm. **127**, 288–297 (2018)
25. Korkmaz, E., Friedrich, E.E., Ramadan, M.H., Erdos, G., Mathers, A.R., Burak Ozdoganlar, O., Washburn, N.R., Falo, L.D.: Therapeutic intradermal delivery of tumor necrosis factor-alpha antibodies using tip-loaded dissolvable microneedle arrays. Acta Biomater. **24**, 96–105 (2015)

26. Joo, S.-H., Kim, J., Hong, J., Fakhraei Lahiji, S., Kim, Y.-H.: Dissolvable self-locking microneedle patches integrated with immunomodulators for cancer immunotherapy. Adv. Mater. **35**(10), 2209966 (2023)
27. Kusamori, K., Katsumi, H., Sakai, R., Hayashi, R., Hirai, Y., Tanaka, Y., Hitomi, K., Quan, Y.S., Kamiyama, F., Yamada, K., Sumida, S., Kishi, K., Hashiba, K., Sakane, T., Yamamoto, A.: Development of a drug-coated microneedle array and its application for transdermal delivery of interferon alpha. Biofabrication **8**(1), 015006 (2016)
28. Chen, S.-X., Ma, M., Xue, F., Shen, S., Chen, Q., Kuang, Y., Liang, K., Wang, X., Chen, H.: Construction of microneedle-assisted co-delivery platform and its combining photodynamic/immunotherapy. J. Control. Release **324**, 218–227 (2020)
29. Kwon, S., Velasquez, F.C., Rasmussen, J.C., Greives, M.R., Turner, K.D., Morrow, J.R., Hwu, W.-J., Ross, R.F., Zhang, S., Sevick-Muraca, E.M.: Nanotopography-based lymphatic delivery for improved anti-tumor responses to checkpoint blockade immunotherapy. Theranostics **9**(26), 8332–8343 (2019)
30. Wang, C., Ye, Y., Hochu, G.M., Sadeghifar, H., Gu, Z.: Enhanced cancer immunotherapy by microneedle patch-assisted delivery of Anti-PD1 antibody. Nano Lett. **16**(4), 2334–2340 (2016)
31. Hirobe, S., Azukizawa, H., Hanafusa, T., Matsuo, K., Quan, Y.-S., Kamiyama, F., Katayama, I., Okada, N., Nakagawa, S.: Clinical study and stability assessment of a novel transcutaneous influenza vaccination using a dissolving microneedle patch. Biomaterials **57**, 50–58 (2015)
32. Henri, S., Guilliams, M., Poulin, L.F., Tamoutounour, S., Ardouin, L., Dalod, M., Malissen, B.: Disentangling the complexity of the skin dendritic cell network. Immunol. Cell Biol. **88**(4), 366–375 (2010)
33. Hirobe, S., Susai, R., Takeuchi, H., Eguchi, R., Ito, S., Quan, Y.-S., Kamiyama, F., Okada, N.: Characteristics of immune induction by transcutaneous vaccination using dissolving microneedle patches in mice. Int. J. Pharm. **601**, 120563 (2021)
34. Yang, H.-W., Ju, S.-P., Chen, H.-Y., Cheng, Y.-C., Hsu, W.-L.: Ovalbumin-loaded gelation microneedles made of predictive formulation by molecular dynamics simulation for enhancement of skin immunization. ACS Biomater. Sci. Eng. **5**(11), 6012–6021 (2019)
35. Zhao, L., Cao, Y.J.: Engineered T cell therapy for cancer in the clinic. Front. Immunol. **10**, 2250–2250 (2019)
36. Li, H., Wang, Z., Ogunnaike, E.A., Wu, Q., Chen, G., Hu, Q., Ci, T., Chen, Z., Wang, J., Wen, D., Du, H., Jiang, J., Sun, J., Zhang, X., Dotti, G., Gu, Z.: Scattered seeding of CAR T cells in solid tumors augments anticancer efficacy. Natl. Sci. Rev. **9**(3), nwab172 (2022)
37. Ye, Y., Yu, J., Wang, C., Nguyen, N.-Y., Walker, G.M., Buse, J.B., Gu, Z.: Microneedles integrated with pancreatic cells and synthetic glucose-signal amplifiers for smart insulin delivery. Adv. Mater. **28**(16), 3115–3121 (2016)
38. Lee, K., Xue, Y., Lee, J., Kim, H.-J., Liu, Y., Tebon, P., Sarikhani, E., Sun, W., Zhang, S., Haghniaz, R., Çelebi-Saltik, B., Zhou, X., Ostrovidov, S., Ahadian, S., Ashammakhi, N., Dokmeci, M.R., Khademhosseini, A.: A patch of detachable hybrid microneedle depot for localized delivery of mesenchymal stem cells in regeneration therapy. Adv. Func. Mater. **30**(23), 2000086 (2020)
39. Tang, J., Wang, J., Huang, K., Ye, Y., Su, T., Qiao, L., Hensley, M.T., Caranasos, T.G., Zhang, J., Gu, Z., Cheng, K.: Cardiac cell–integrated microneedle patch for treating myocardial infarction. Sci. Adv. **4**(11), eaat9365 (2018)
40. Chang, H., Chew, S.W.T., Zheng, M., Lio, D.C.S., Wiraja, C., Mei, Y., Ning, X., Cui, M., Than, A., Shi, P., Wang, D., Pu, K., Chen, P., Liu, H., Xu, C.: Cryomicroneedles for transdermal cell delivery. Nat. Biomed. Eng. **5**(9), 1008–1018 (2021)
41. Pan, J., Ruan, W., Qin, M., Long, Y., Wan, T., Yu, K., Zhai, Y., Wu, C., Xu, Y.: Intradermal delivery of STAT3 siRNA to treat melanoma via dissolving microneedles. Sci. Rep. **8**(1), 1117 (2018)
42. Xu, Q., Li, X., Zhang, P., Wang, Y.: Rapidly dissolving microneedle patch for synergistic gene and photothermal therapy of subcutaneous tumor. J. Mater. Chem. B **8**(19), 4331–4339 (2020)

43. Hood, R.L., Carswell, W.F., Rodgers, A., Kosoglu, M.A., Rylander, M.N., Grant, D., Robertson, J.L., Rylander, C.G.: Spatially controlled photothermal heating of bladder tissue through single-walled carbon nanohorns delivered with a fiberoptic microneedle device. Lasers Med. Sci. **28**(4), 1143–1150 (2013)
44. Cárcamo-Martínez, Á., Mallon, B., Domínguez-Robles, J., Cordeiro, A.S., Celentano, M., Larrañeta, E., Bell, S.E.J., Donnelly, R.F.: Plasmonic photothermal microneedle arrays and single needles for minimally-invasive deep in-skin hyperthermia. J. Mater. Chem. B **8**(25), 5425–5433 (2020)
45. Lin, S., Lin, H., Yang, M., Ge, M., Chen, Y., Zhu, Y.: A two-dimensional MXene potentiates a therapeutic microneedle patch for photonic implantable medicine in the second NIR biowindow. Nanoscale **12**(18), 10265–10276 (2020)
46. Wei, S., Quan, G., Lu, C., Pan, X., Wu, C.: Dissolving microneedles integrated with pH-responsive micelles containing AIEgen with ultra-photostability for enhancing melanoma photothermal therapy. Biomater. Sci. **8**(20), 5739–5750 (2020)
47. Jain, A.K., Lee, C.H., Gill, H.S.: 5-Aminolevulinic acid coated microneedles for photodynamic therapy of skin tumors. J. Control. Release **239**, 72–81 (2016)
48. Hamdan, I.M.N., Tekko, I.A., Matchett, K.B., Arnaut, L.G., Silva, C.S., McCarthy, H.O., Donnelly, R.F.: Intradermal delivery of a near-infrared photosensitizer using dissolving microneedle arrays. J. Pharm. Sci. **107**(9), 2439–2450 (2018)
49. Zhao, X., Li, X., Zhang, P., Du, J., Wang, Y.: Tip-loaded fast-dissolving microneedle patches for photodynamic therapy of subcutaneous tumor. J. Control. Release **286**, 201–209 (2018)
50. Zhu, J., Dong, L., Du, H., Mao, J., Xie, Y., Wang, H., Lan, J., Lou, Y., Fu, Y., Wen, J., Jiang, B., Li, Y., Zhu, J., Tao, J.: 5-Aminolevulinic acid-loaded hyaluronic acid dissolving microneedles for effective photodynamic therapy of superficial tumors with enhanced long-term stability. Adv. Healthc. Mater. **8**(22), 1900896 (2019)
51. Huang, Y., Lai, H., Jiang, J., Xu, X., Zeng, Z., Ren, L., Liu, Q., Chen, M., Zhang, T., Ding, X., Zhao, C., Cui, S.: PH-activatable oxidative stress amplifying dissolving microneedles for combined chemo-photodynamic therapy of melanoma. Asian J. Pharm. Sci. **17**(5), 679–696 (2022)
52. Li, Z., Hu, Y., Howard, K.A., Jiang, T., Fan, X., Miao, Z., Sun, Y., Besenbacher, F., Yu, M.: Multifunctional bismuth selenide nanocomposites for antitumor thermo-chemotherapy and imaging. ACS Nano **10**(1), 984–997 (2016)
53. Chen, Z.: What's new about the mechanism of methotrexate action in psoriasis? Br. J. Dermatol. **179**(4), 818–819 (2018)
54. How, K.N., Yap, W.H., Lim, C.L.H., Goh, B.H., Lai, Z.W.: Hyaluronic acid-mediated drug delivery system targeting for inflammatory skin diseases: a mini review. Front. Pharmacol. **11** (2020)
55. Dong, L., Li, Y., Li, Z., Xu, N., Liu, P., Du, H., Zhang, Y., Huang, Y., Zhu, J., Ren, G., Xie, J., Wang, K., Zhou, Y., Shen, C., Zhu, J., Tao, J.: Au nanocage-strengthened dissolving microneedles for chemo-photothermal combined therapy of superficial skin tumors. ACS Appl. Mater. Interfaces **10**(11), 9247–9256 (2018)
56. Chen, Y., Yang, Y., Xian, Y., Singh, P., Feng, J., Cui, S., Carrier, A., Oakes, K., Luan, T., Zhang, X.: Multifunctional graphene-oxide-reinforced dissolvable polymeric microneedles for transdermal drug delivery. ACS Appl. Mater. Interfaces **12**(1), 352–360 (2020)
57. Pei, P., Yang, F., Liu, J., Hu, H., Du, X., Hanagata, N., Zhao, S., Zhu, Y.: Composite-dissolving microneedle patches for chemotherapy and photothermal therapy in superficial tumor treatment. Biomater. Sci. **6**(6), 1414–1423 (2018)
58. Qin, W., Quan, G., Sun, Y., Chen, M., Yang, P., Feng, D., Wen, T., Hu, X., Pan, X., Wu, C.: Dissolving microneedles with spatiotemporally controlled pulsatile release nanosystem for synergistic chemo-photothermal therapy of melanoma. Theranostics **10**(18), 8179–8196 (2020)
59. Moreira, A.F., Rodrigues, C.F., Jacinto, T.A., Miguel, S.P., Costa, E.C., Correia, I.J.: Poly (vinyl alcohol)/chitosan layer-by-layer microneedles for cancer chemo-photothermal therapy. Int. J. Pharm. **576**, 118907 (2020)

60. Hao, Y., Dong, M., Zhang, T., Peng, J., Jia, Y., Cao, Y., Qian, Z.: Novel approach of using near-infrared responsive PEGylated Gold nanorod coated poly(l-lactide) microneedles to enhance the antitumor efficiency of docetaxel-loaded MPEG-PDLLA micelles for treating an A431 Tumor. ACS Appl. Mater. Interfaces **9**(18), 15317–15327 (2017)
61. Song, G., Jiang, G., Liu, T., Zhang, X., Zeng, Z., Wang, R., Li, P., Yang, Y.: Separable microneedles for synergistic chemo-photothermal therapy against superficial skin tumors. ACS Biomater. Sci. Eng. **6**(7), 4116–4125 (2020)
62. Yu, W., Li, X., Huang, Y., Chen, Y., Gao, Q., Wang, Y., Ji, J.: Build an implanted "arsenal": detachable microneedles for NIR-triggered cancer photothermo-chemotherapy. Biomater. Sci. **9**(13), 4737–4745 (2021)
63. Luo, L., Zhu, C., Yin, H., Jiang, M., Zhang, J., Qin, B., Luo, Z., Yuan, X., Yang, J., Li, W., Du, Y., You, J.: Laser immunotherapy in combination with perdurable PD-1 blocking for the treatment of metastatic tumors. ACS Nano **12**(8), 7647–7662 (2018)
64. Chen, Q., Hu, Q., Dukhovlinova, E., Chen, G., Ahn, S., Wang, C., Ogunnaike, E.A., Ligler, F.S., Dotti, G., Gu, Z.: Photothermal therapy promotes tumor infiltration and antitumor activity of CAR T cells. Adv. Mater. **31**(23), 1900192 (2019)
65. Calzavara-Pinton, P., Venturini, M., Sala, R.: Photodynamic therapy: update 2006 Part 1: photochemistry and photobiology. J. Eur. Acad. Dermatol. Venereol. **21**(3), 293–302 (2007)
66. Binnewies, M., Roberts, E.W., Kersten, K., Chan, V., Fearon, D.F., Merad, M., Coussens, L.M., Gabrilovich, D.I., Ostrand-Rosenberg, S., Hedrick, C.C., Vonderheide, R.H., Pittet, M.J., Jain, R.K., Zou, W., Howcroft, T.K., Woodhouse, E.C., Weinberg, R.A., Krummel, M.F.: Understanding the tumor immune microenvironment (TIME) for effective therapy. Nat. Med. **24**(5), 541–550 (2018)
67. Chen, M., Quan, G., Wen, T., Yang, P., Qin, W., Mai, H., Sun, Y., Lu, C., Pan, X., Wu, C.: Cold to hot: binary cooperative microneedle array-amplified photoimmunotherapy for eliciting antitumor immunity and the abscopal effect. ACS Appl. Mater. Interfaces **12**(29), 32259–32269 (2020)
68. Ye, Y., Wang, C., Zhang, X., Hu, Q., Zhang, Y., Liu, Q., Wen, D., Milligan, J., Bellotti, A., Huang, L., Dotti, G., Gu, Z.: A melanin-mediated cancer immunotherapy patch. Sci. Immunol. **2**(17), eaan5692 (2017)
69. He, T., Luo, Y., Zhang, Q., Men, Z., Su, T., Fan, L., Chen, H., Shen, T.: Hyalase-mediated cascade degradation of a matrix barrier and immune cell penetration by a photothermal microneedle for efficient anticancer therapy. ACS Appl. Mater. Interfaces **13**(23), 26790–26799 (2021)
70. Chen, M., Yang, D., Sun, Y., Liu, T., Wang, W., Fu, J., Wang, Q., Bai, X., Quan, G., Pan, X., Wu, C.: In situ self-assembly nanomicelle microneedles for enhanced photoimmunotherapy via autophagy regulation strategy. ACS Nano **15**(2), 3387–3401 (2021)

Chapter 11
Microneedles-Based Electrochemical Sensing of Emerging Biomarkers

Mani Arivazhagan, S. Lokeswara Reddy, and Raheel Akram

11.1 Introduction

Microneedles' promise for use in biosensing, medication delivery, and fluid extraction has piqued the interest of many in the biomedical industry in recent years [1]. The microneedle points can penetrate the epidermal layer of the skin and reach the uppermost stratum of the dermis, situated above the blood capillaries. The microneedles can create microchannels on the skin that allow molecules to flow through, or they can bring devices, such as electrodes, to the skin to obtain biological signals [2]. The microneedle method of drug delivery has the following benefits: it circumvents the impact of gastrointestinal digestion, speeds up drug absorption, extends the duration of action, and allows patients to self-administer medication without the assistance of medical professionals [3]. The microneedle method makes it feasible to continuously monitor the biomarkers in interstitial fluid for biosensing.

Over time, the notion of microneedles has progressed from the utilization of huge needles to the present-day advanced design of microneedles. In 1905, Dr. Ernst Kromayer, a German dermatologist, employed various sizes of motor-powered dental burs [4] to address scarring, hyperpigmentation, and other skin conditions. The earliest recorded instance of microneedle usage in literature dates to 1921, when Chambers inserted the needle into the nucleus of an egg [5].

In the 1960s, pharmacological injections into the stratum corneum gained attention [6]. Although the microneedle concept was suggested in the 1970s, it was not experimentally proved until the 1990s [7]. The first transdermal solution for treating motion sickness, a three-day patch delivering scopolamine, was approved in 1979 [3]. Orentreich performed a sub-cision operation in 1994, inserting a tri-beveled

M. Arivazhagan · S. L. Reddy · R. Akram (✉)
Research Laboratory for Analytical Instrument and Electrochemistry Innovation, Department of Chemistry, Faculty of Science, Chiang Mai University, Chiang Mai, Thailand
e-mail: raheel.akram19@yahoo.com

N. Talreja et al. (eds.), *Microneedles (MNs)-Based Technology*, Series in BioEngineering, https://doi.org/10.1007/978-981-96-3916-8_11

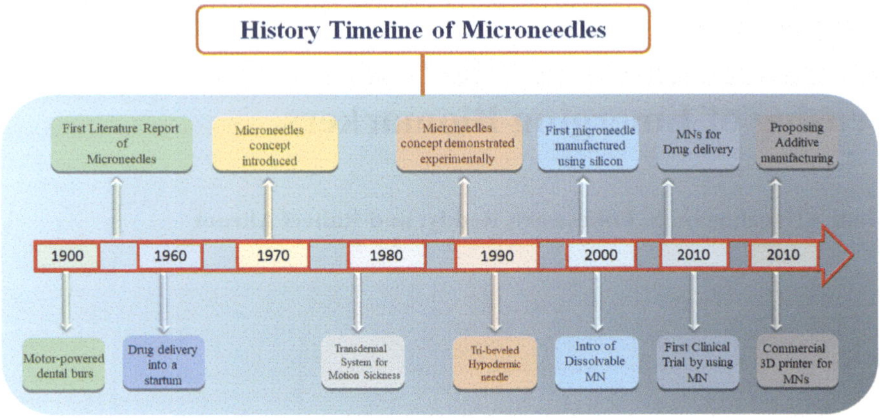

Fig. 11.1 An overview of the historical development of microneedles technologies

hypodermic needle into the epidermis to remove fibrous strands [8]. This procedure addressed cutaneous abnormalities under the skin, causing depressed scars and wrinkles. Transdermal microneedles were first proposed in 1998 and made from silicon wafers using ion etching and photolithography [9]. The study explored the use of microfabricated needles to improve drug administration across the skin. The microneedle domain was extensively researched after this paper. Different materials like glass, ceramic, metal, and polymers were used to create microneedles. In 2004, a microneedle array was employed to distribute transdermal drugs through skin piercings [10], leading to the exploration of various construction methods and materials for TDD. Different forms of MNs include solid, coated, hollow, dissolvable, and hydrogel-forming. Additionally, production processes like laser ablation, photolithography, and micro-injection molding are available. Recent additive manufacturing processes offer low-cost options for micro-mold fabrication. Several reports suggest that using commercially available 3D printers to create MN master molds offers new opportunities for device fabrication and large-scale manufacture [11, 12] (Fig. 11.1).

11.2 Materials for Microneedles

The main reason why MNs are made is that they can go through skin without breaking or bending. The MN manufacturing problem has been looked at from different angles, including the material, the way it is made, and the design. Different kinds of MNs have been made from a range of materials (Fig. 11.2). Silicon, metals, ceramics, polymers, and plastics are some examples of these types of materials. A mix of different types of materials have been used in biomedical uses like tissue engineering, biomedical implants, and drug delivery.

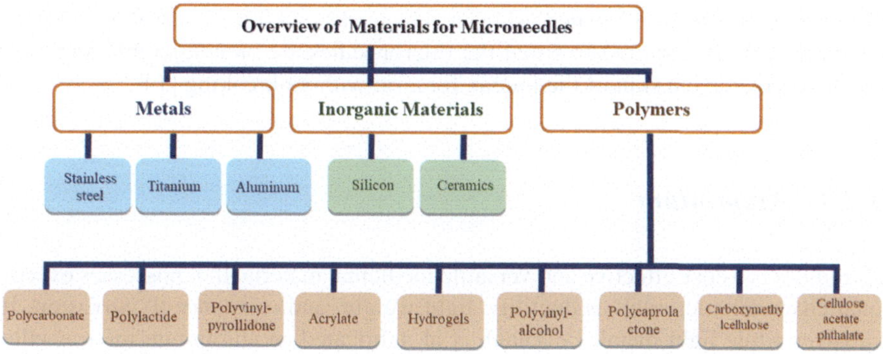

Fig. 11.2 An overview of the materials used for the microneedle's fabrications

11.2.1 Metals

Metal materials offer numerous advantages for use as structural materials in microneedles. These include exceptional mechanical and physical qualities, high durability, cost-effectiveness, and excellent biocompatibility. The metal microneedles have several applications, including hollow microneedles, coated microneedles, and solid-state microneedles. The metal microneedles consist of stainless steel, titanium, aluminum, and nickel. Nevertheless, the drawbacks of metal microneedles lie in their non-degradability and lack of flexibility.

11.2.2 Silicon

The initial MN, made from silicon material, was produced in the 1990s. Silicon has several advantages compared to other materials, such as its inherent flexibility, which enables convenient manufacturing of desired forms and sizes of micro/nanostructures. Silicon has been employed in the production of solid, hollow, and coated micro/nanoparticles. However, there are drawbacks when employing silicon, including the laborious process of manufacture, the high expense, and the potential for skin fractures [13].

11.2.3 Titanium

The silver-white transition metal titanium finds widespread use in the medical field. It possesses numerous exceptional characteristics, including being lightweight, having high strength, stable chemical properties, great temperature resistance, strong resistance to acid and alkaline substances, and low density. Li, JY presented a titanium porous microneedle array created using an enhanced metal injection molding (MIM) technique. This method ensures that the microneedles are biocompatible and may effortlessly pierce the skin of the human forearm without breaking [14].

11.2.4 Aluminium

Aluminium is a cost-effective and versatile medicinal material that possesses exceptional characteristics such as ductility, light weight, and resistance to corrosion. It is a silver-white light metal. Le Thanh et al. fabricated an aluminium microneedle sample using computer numerical control (CNC) processing. The procedure was distinguished by its cost-effectiveness and capacity to be scaled up for large-scale production [15].

11.2.5 Inorganic Materials

11.2.5.1 Silicon

The initial MN, made from silicon material, was produced in the 1990s. Silicon has several advantages compared to other materials, such as its inherent flexibility, which enables convenient manufacturing of desired forms and sizes of micro/nanostructures. Silicon has been employed in the production of solid, hollow, and coated micro/nanoparticles. However, there are drawbacks when employing silicon, including the laborious process of manufacture, the high expense, and the potential for skin fractures [13].

11.2.5.2 Ceramics

Ceramic materials, like as alumina, have been utilized to create an MN due to its exceptional chemical characteristics and ability to withstand compression. Nevertheless, alumina has a lower tensile strength in comparison to other materials. Calcium sulphate dihydrate and calcium phosphate dihydrate are other forms of ceramics that are used in the production of MNs [16]. A micro-molding process can be employed to manufacture a micro-needle utilizing ceramic material. This method provides the opportunity for large-scale production at a minimal expense.

11.2.5.3 Polymeric Materials

Polymers present a good material substitute for MN. Their biocompatibility is high, with minimal toxicity and cost. Nevertheless, they exhibit reduced strength in comparison to silicon and metals. Polymers are commonly used to make dissolvable and hydrogel-forming arrays of microneedles (MN), as well as solid, coated, and hollow MN arrays [17]. Biodegradable MNs have been used to provide several medications topically. The polymers utilized in the production of MNs encompass polylactic acid (PLA), poly (methyl methacrylate) (PMMA), polystyrene, poly (carbonate), and SU-8 photoresist [16].

11.3 Fabrication of Microneedles

Silicon, sugar, polymers, and stainless steel are among the materials that have been used to make microneedles, which can be solid, coated, hollow, or soluble. Figure 11.3 demonstrates the type of microneedles and the interaction of microneedles with the skin before and after needling. Furthermore, Fig. 11.4 illustrates the hierarchical division of the microneedle fabrication process.

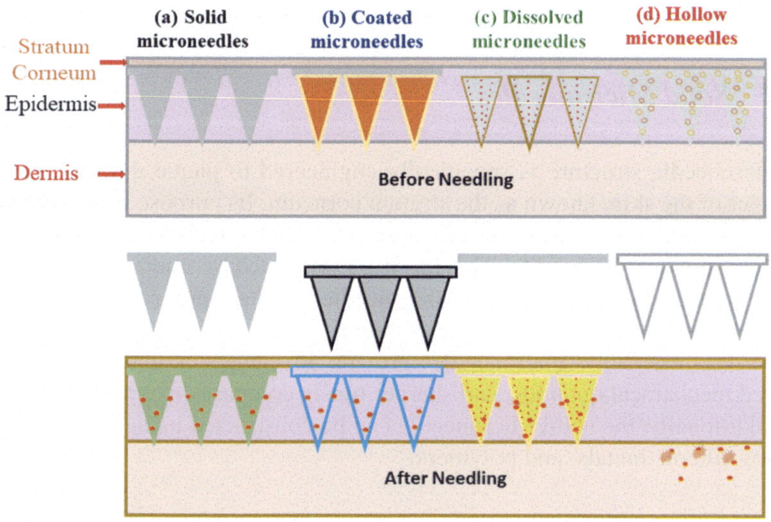

Fig. 11.3 An illustration of the contact and working principal between different types of microneedles and human skin. **a** Prior to skin treatment, solid microneedles are inserted using a poke-with-a-patch technique. **b** Coated microneedles are used for the coat and poke method, where a drug solution coating is put on the needle surface. **c** Dissolving microneedles are made of biodegradable polymers. **d** The medicine is transferred in the dermis using hollow microneedles that are filled with a medication solution [16]

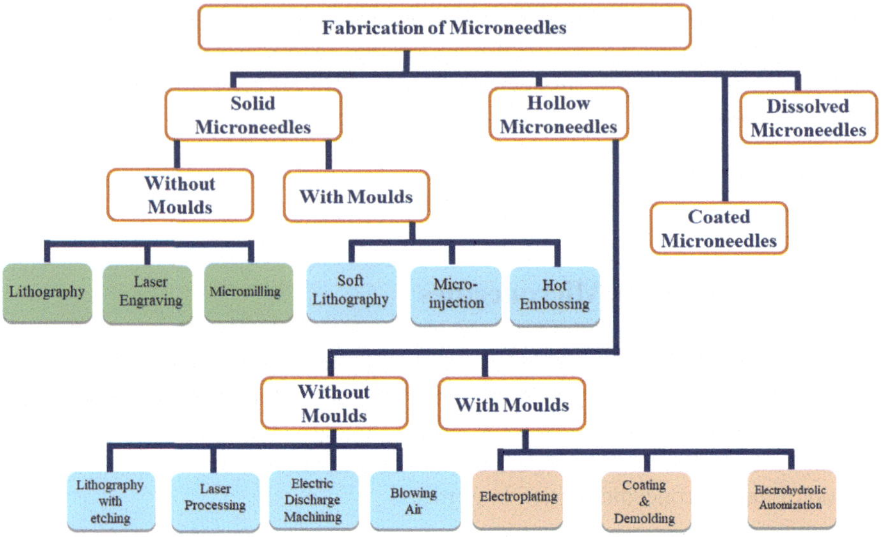

Fig. 11.4 An illustration of the types of fabrication technologies used for microneedles formation

11.3.1 Solid Microneedles

This microneedle structure is specifically engineered to pierce through the outermost layer of the skin, known as the stratum corneum. Its purpose is to optimize the delivery of drugs to the dermis, hence improving their bioavailability and facilitating their movement across the skin. This has been supported by studies [18]. Compared to intramuscular delivery, the solid Microneedle is better appropriate for delivering vaccinations because it has a longer duration of action and produces a stronger immune response. Solid microneedles provide advantageous manufacturing ease, enhanced mechanical qualities, and sharper tips in comparison to hollow microneedles. Additionally, the solid Microneedle can be constructed using many materials, including silicon, metals, and polymers.

11.3.2 Solid Microneedles Without Moulds

The most populous method for the solid microneedles without moulds is the lithography with selective etching. This classification is further subdivided based on the base of radiation source used to perform lithography. The employed light source may be either UV and/-or infrared light or soft x-ray source.

The production of the microneedles involved the use of photolithography, metal deposition, and selective etching techniques. The manufacturing process involves the use of photolithography to create a pattern at a microscale. This is followed by the application of metals, such as chromium, as a layer of material to serve as a mask [19]. Conversely, X-rays with short wavelengths have been effectively utilized and are necessary to penetrate deeper microneedle gully formations. The X-ray dosage energy is closely correlated with both the exposure depth of the Microneedle and the size transition.

11.3.3 Solid Microneedles with Moulds

Another method of solid Microneedle's fabrication is employing moulds. Its two important types are soft lithography and micro-injection moulding method. In soft lithography, the transfer process typically involves pouring polymeric ingredients into a mold. After the materials have hardened and been removed from the mold, the appropriate microneedles are obtained. Conventional machining, microfabrication, and 3D printing were used to construct the first designs of microneedles or cavities. Microneedles or their negative analogues, called cavities, can make up the first mould.

Pattern transfer and microneedle creation are achieved by the use of micro-injection moulding and heat embossing processes. Using a micro-injector and a compression technique, the solution was injected into the mold's cavities [20]. The secondary clamping pressure, injection speed, and melt temperature are critical parameters to consider. At every stage of injection, these variables significantly affect microneedle quality.

11.3.4 Hollow Microneedles

Hollow microneedles offer the advantage of continuous drug delivery, unlike solid microneedles. Nevertheless, the production procedure and expenses associated with hollow microneedles are frequently intricate and substantial. Various techniques have been examined for the production of hollow microneedles.

11.3.4.1 Hollow Microneedles Without Moulds

Likewise, for solid microneedles, the most widely acceptable method for the fabrication process of hollow microneedles is lithography, which is further subdivided into the following four types: lithography with etching, laser lithography, electrical discharge machining, and the air blowing method.

Utilizing excimer (UV) and infrared (IR) laser processing techniques, metal electrodeposition molds can be prepared. Additionally, solid needles made of titanium can be sliced straight through to create hollow microneedles. Reliable hollow three-dimensional microstructures, including microneedles for drug delivery, are manufactured with nanoscale resolution, and are not constrained by subtractive processing or etching techniques thanks to the programming of the laser focus to move in three-dimensional space.

Electric-discharge machining (EDM) is a method used to create intricate forms. The functional concept involves utilizing electrical discharges, commonly known as sparks, to eliminate materials and degrade hardened steel with small characteristics. Vinaya Kumar, KB, and other researchers developed a technique for manufacturing stainless steel microneedles using a mix of electrical discharge machining (EDM) and focused ion beam (FIB) technologies. Initially, a planar solid SS microneedle array was created using the EDM approach. Subsequently, a focused ion beam was employed to produce erosion-formed microstructures by making holes [21].

For air blowing, the underlying principle involves the formation of polymer droplets into microneedles through the application of air pressure, followed by solidification of the droplets to achieve the desired microneedle shape. This approach offers a combination of gentle temperature and rapid manufacturing conditions. The precise techniques involve: (1) distributing the polymer droplets onto a level surface; (2) contacting the dispersed droplets using the top plate; (3) regulating the length of the microneedles by vertically moving the higher plate. (4) Air purging Ensure that the droplets are slender and have a distinct shape; (5) displace the two plates to create two rows of microneedles on each plate.

11.3.4.2 Hollow Microneedles with Moulds

Hollow microneedles have been manufactured by a variety of pattern transfer techniques, such as electroplating with etching, coating, and demolding, and electro-hydraulic atomization. In order to create hollow metal microneedles, a process known as electroplating with etching combines two photon polymerizations (2PP) with laser direct writing (LDW). Solid microneedles with internal bosses were first prepared using LDW and 2 pp, and subsequently they were made using PDMS. Then, using pulse deposition, metal was electroplated on the PDMS to create metal hollow microneedles [22].

Aydin Sadeqi et al. employed coating and demolding method to prepare the 3D microneedles using a CO_2 laser by carving a cross-line pattern on an acrylic plate [23]. After the acrylic plate underwent salinization and plasma treatment, the PDMS was cast onto the acrylic mold to create a PDMS mold. Lastly, chitosan solid microneedles can be produced by pouring a chitosan solution onto the PDMS mold.

The process of creating smart microneedle coatings using the electro-hydraulic atomization (EHDA) principle entails connecting stainless steel microneedles (MNs) to ground electrodes. Then, by adjusting the flow rate and applying voltage, MNs are coated with particles and fibers that are between 100 and 3 μM and 400 and 1 μM in size, respectively.

11.3.5 Coated Microneedles

The coated Microneedle refers to a solid-type microneedle that has been covered with a medication solution, as seen in Fig. 11.8. Usually, the amount of the medicine it contains is determined by the thickness of the coating layer, which might vary. The efficacy of drug delivery utilizing a coated microneedle (MN) relies on the consistent application of a controlled drug layer onto the MNs. A coated magnetic nanoparticle possesses the capability to transport proteins and DNA in a manner that causes minimal damage or intrusion [24]. One benefit of using a coated MN is that it allows for fast medication administration to the skin. However, there is a risk that the remaining drug at the needle's tip could potentially infect other patients. The outcomes of the vaccine distribution utilizing coated MN were comparable to those achieved using intradermal and intramuscular methods.

11.3.6 Dissolvable Microneedles

The dissolvable MN, which was first introduced in 2005, is a technology that shows promise due to its distinctive features. These properties encompass the ability to quickly release large molecules, and the convenience of applying drugs in a single step, making drug administration easier. The "poke-and-release" method of applying dissolvable MNs has shown significant improvement, making it a superior strategy compared to other methods [25]. The dissolvable MN tip can be efficiently loaded using a two-step casting procedure. When the dissolvable MN is inserted into the skin, the medicine is released and spreads quickly by the dissolving of the needle tip. The dissolvable MN is best produced using water-soluble ingredients [26]. Similarly, the micro-mold fabrication approach is highly appropriate for producing the dissolvable MN. Technical competence is necessary for the design and manufacture of a dissolvable MN array. Nevertheless, this particular type of MN necessitates full insertion, a task that is frequently challenging to achieve, and also experiences a delay in dissolution [3].

11.4 Microneedles for the Sensing of Emerging Biomarkers

The primary consideration for designing a microneedle-based biosensor is the location of the analyte that needs to be reached and detected [27]. This involves understanding the depth and distribution of the target analyte within the skin or other tissues. For effective sensing, the microneedles must be designed to penetrate to the appropriate depth where the analyte is present in sufficient concentration. Additionally, factors such as the biochemical environment, potential interference from other substances, and the mechanical properties of the tissue must be considered to ensure accurate and reliable detection. A medical analytical device with a thin, sharp tip used to measure and report the concentration of an interesting analyte in biofluids beneath the skin's surface is called a "microneedle-based biosensor." The idea is to access the biofluid flowing beneath the skin by inserting the needle which incorporates a sensing function-through the dermis' outermost layer. Compared to conventional blood sample procedures, this method offers a convenient and possibly less painful option for the minimally invasive monitoring of multiple biomarkers [28].

The development of microneedle sensors has fundamentally transformed invasive blood-centered diagnostics by providing direct and continuous access to interstitial fluid (ISF). These sensors penetrate the skin minimally, reaching the ISF where they can measure various biomarkers in real time. This approach reduces the need for frequent blood draws, lowers the risk of infection, and increases patient comfort, making it a promising advancement in medical diagnostics and patient monitoring. A wide range of materials and fabrication methods have been established to construct microneedle sensors with diverse geometries, mechanical flexibility, and biocompatibility. These materials include metals, polymers, ceramics, and composites, each selected based on the specific application and desired properties of the microneedles. Fabrication techniques such as photolithography, micro moulding, 3D printing, and laser cutting allow for precise control over the size, shape, and structure of the microneedles. This versatility enables the development of customized sensors tailored to various biomedical applications, enhancing their performance and safety [29].

11.5 Microneedles for Glucose Detection

In order to physically damage the skin's outer layer and facilitate the delivery of medications or other therapeutic agents, microneedles are tiny devices. Pain and discomfort are reduced because they are tiny enough to pierce the stratum corneum, the skin's outermost layer, but not large enough to reach deeper tissues. Numerous uses of this technology have been investigated, including vaccinations, cosmetic procedures, and transdermal drug delivery. Most applications of microneedles have focused on minimally-invasive pain-free transdermal delivery of drugs. More recently, sensing applications of microneedles have been developed, focusing primarily on the minimally-invasive voltametric monitoring of emerging biomarkers like glucose, lactic acid, dopamine, cysteine etc.

For instance, Joseph Wang et al. [30] fabricated biofuel cell sensor makes use of the integration of modified carbon pastes into hollow microneedle devices. The system displays defined dependence of the power density vs glucose concentration in artificial interstitial fluid. An excellent selectivity against common electroactive interferences and long-term stability are obtained. The attractive performance of the device indicates considerable promise for subdermal glucose monitoring.

The current fabricated Microneedle self-powered biofuel cells (BFC) based sensor system consists of mating the bioanode and cathode into a hollow-microneedle device

(Fig. 11.5). Figure 11.5 illustrates the self-powered microneedle sensor, including an optical micrograph of nine-element hollow microneedle array (a) and two of these microneedle electrodes, serving as a bioanode and cathode, along with the reaction scheme involved in the self-powered sensor BFC operation (b) While the carbon-paste Microneedle BFC can harvest energy from artificial interstitial fluid under physiological conditions over the relevant 5–25 mM glucose range, further improvements in the power generation and extension of the upper glucose concentration range (relevant to hyperglycaemia) have been achieved using a metalized carbon-fiber BFC microneedle configuration (Fig. 11.5c). Figure 11.5d shows the dependence of the power density on the cell voltage for different glucose concentrations (5–25 mM), relevant to diabetes monitoring. The maximum power density displays a defined for glucose quantification by amperometry, and using phosphate buffer solutions.

In addition, Hiroyuki Kai et al. [31] Porous microneedle electrodes with pores of ~1 μm diameter were fabricated by electroless plating of nickel followed by gold on a polymer monolith of poly(glycidyl methacrylate). Electrochemical glucose sensors were then fabricated by immobilizing glucose oxidase on the gold-plated microneedle electrodes. The sensitivity of the porous microneedle glucose sensor was $\sim 22.99 \pm 0.72\ \mu A\ mM^{-1}$ and the linear ranges between 0 and 15 mM. Particularly, amperometric sensing of glucose with the applied potential of 0.3 V (vs. Ag/AgCl) by the porous microneedle electrode with an electron mediator dissolved in

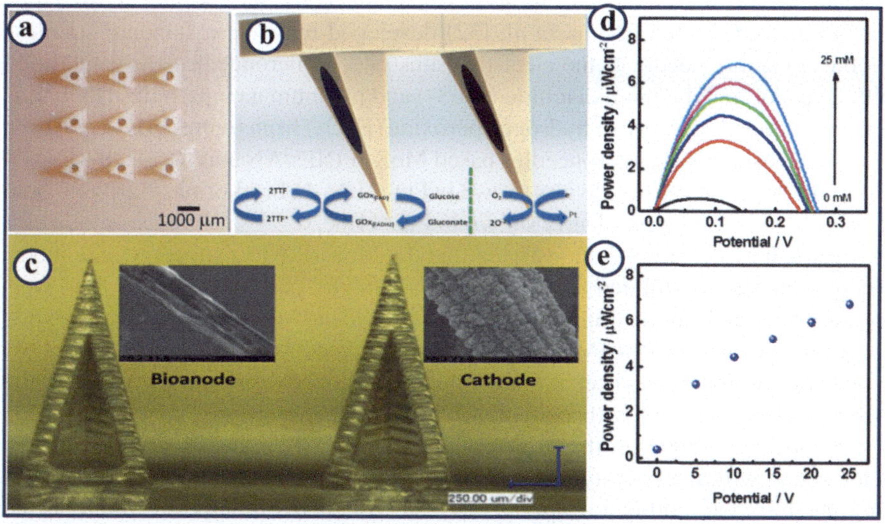

Fig. 11.5 **a** Optical micro graphical traits of the microneedle-based array. **b** Schematic representation of the carbon paste based bioanode and cathode electrodes mated on 2 of the hollow microneedles located within the array of electrode. Schematic reaction delineating the operational principle of the self-powered sensor device using a GO_x/TTF-based bioanode and Pt-based cathode, **c** Metalized-carbon-fiber BFC microneedle configuration, **d** power density obtained for varying glucose concentrations from 0 to 25 mM. **e** Power density vs glucose concentration illustrating the linear correlation between the two variables. The figure was reproduced with the permission [30]

solution. Conventional Ag/AgCl reference electrode and Pt counter electrode were used. The porous microneedle glucose electrode was achieved the successive addition of glucose from 0 to 20 mM (each step is 1.667 mM increase).

11.6 Microneedles for Hydrogen Peroxide Detection

As one of the relatively stable and well-known reactive oxygen species (ROS) in living organisms, hydrogen peroxide (H_2O_2) production and elimination are closely correlated with normal physiological processes. H_2O_2 plays a crucial role in various cellular functions, including signalling, defense against pathogens, and regulation of gene expression. However, imbalances in its production and elimination can lead to oxidative stress, contributing to the pathogenesis of numerous diseases, such as cancer, cardiovascular diseases, and neurodegenerative disorders. Monitoring H_2O_2 levels in biological systems is therefore essential for understanding and managing these conditions. Normal H_2O_2 levels are generally crucial for cellular signalling of growth factors and second messengers. According to the current study, high concentrations of H_2O_2 cause oxidative stress and damage in cells and tissues, which can lead to a variety of illnesses like Parkinson's disease (PD), Alzheimer's disease (AD), myocardial infarction, cancer, etc. In addition, H_2O_2 has been identified as a cancer biomarker, with a greater concentration of tumour cells than normal tissue cells. For instances, Guo-Jun Zhang. et al. [32] developed by using a standard stainless steel acupuncture needle as the electrode substrate, a microneedle sensor modified with layers of molybdenum disulfide (MoS_2) and platinum nanoparticles (PtNPs) is created to track the release of hydrogen peroxide (H_2O_2) from living cells in real time [32]. The fabrication of microneedles based MoS_2/PtNPs/AN and the electrodes can acts as the acupuncture needle was shown in Fig. 11.6a, b. In addition, morphological traits of MoS2/Pt NPs are clearly shown in Fig. 11.6c, d.

Electrochemical study demonstrated that the MoS_2/PtNPs nanocomposite modified microneedles exhibited excellent electrocatalytic performance and low overpotential toward the reduction of H_2O_2. Not only did the Microneedle achieve a wide linear range from 1 to 100 $\mu M\ L^{-1}$ with a limit of detection down to 0.686 $\mu M\ L^{-1}$.

In addition, James Davis et al., fabricated the conductive composite Microneedle (MN) patches modified with palladium clusters for the electrocatalytic detection of peroxide [32]. Through the use of micro molding techniques, carbon nanoparticles bound within a polystyrene matrix produce a 10×10 array of 700 micron-long needles. The ability to rapidly fabricate conductive composite microneedle patches by micro-molding in silicone templates has proven to be significantly advantageous over alternative micro-machining techniques. This method allows for the efficient and precise production of microneedles with desirable properties. Additionally, the ability to directly chelate palladium (Pd) to the anodized microneedle electrode surface has been demonstrated as a highly effective and sensitive approach for peroxide sensing. This technique enhances the detection capabilities of the

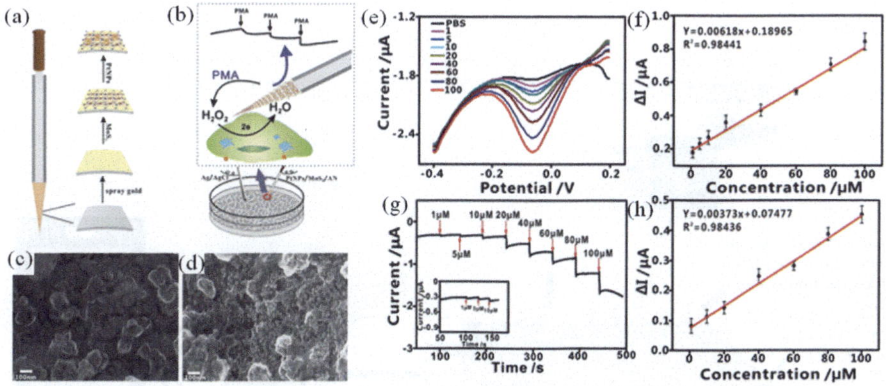

Fig. 11.6 **a** Fabrication process for the preparation of MoS_2/PtNPs/AN. **b** Schematic illustration of the acupuncture needle based microsensor for real-time monitoring of H_2O_2 release from living cells; SEM images of (C) MoS_2/AN and **d** MoS_2/PtNPs/AN; **e** The DPVs curves for the MoS_2/PtNPs/AN biosensor in 0.01 M PBS with increasing H_2O_2 concentrations of 0–100 μM L^{-1} (from top to bottom). **f** The corresponding calibration curves. **g** Amperometric response and **h** the corresponding calibration curves of the MoS_2/PtNPs/AN biosensor in N_2-saturated PBS (0.01 M, pH 7.4) at −0.15 V with successive addition of 1–100 μM L^{-1} H_2O_2. The figure was reproduced with the permission [31]

microneedle sensors, making them more reliable for monitoring hydrogen peroxide and potentially other analytes in biological systems.

11.7 Microneedles for Lactic Acid Detection

Because L-lactate concentrations are crucial for patient clinical diagnosis and early lactate concentration monitoring in intensive care units (ICUs), it is imperative to monitor serum lactate concentrations. An infection, ischemia, or traumatic shock can also result in elevated lactate levels in the blood or urine. Bacterial overproduction in the gut is the usual cause. Anaerobic metabolism can result from low oxygen delivery to tissues or stress on the body, which can be indicated by elevated lactate levels. The diagnosis and treatment of several illnesses, such as sepsis, severe infections, and respiratory or cardiovascular failure, can be greatly aided by lactate level monitoring. In this concern, Mugo et al. [33] fabricated multiplexed sensors were fabricated via layer-by-layer (LbL) assembly on a polydimethylsiloxane (PDMS) microneedle platform coated with a conductive PDMS/carbon nanotube (CNT)/cellulose nanocrystal (CNC) composite is denoted as (PDMS/CNT/CNC@PDMS) Fig. 11.7. The as prepared composite materials can be useful for the detection of lactic acid. In addition, the other biomolecules like epinephrine and dopamine also detected in the same electrode materials. The limits of detection (LOD) for the lactic acid is 0.07 ± 0.07 mM.

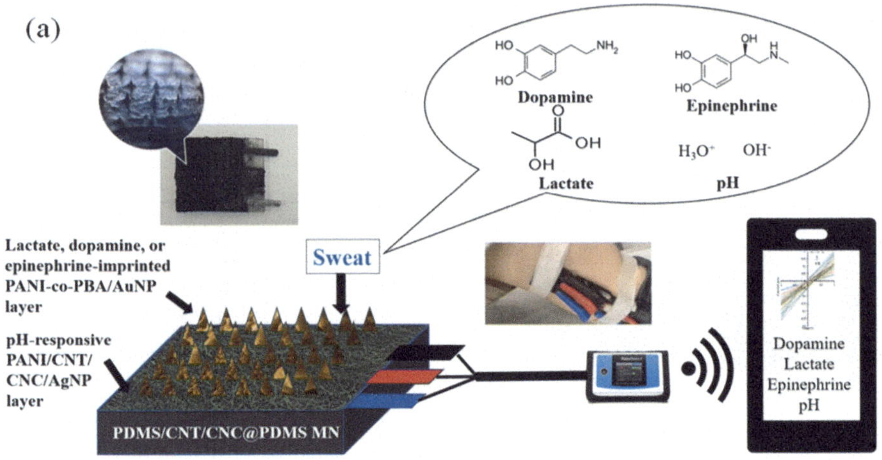

Fig. 11.7 **a** Pictorial representation of the layer-by-layer fabricated sensor patches for the detection of biomarkers. The figure was reproduced with the permission [33]

11.8 Microneedles for Dopamine Detection

A progressive neurodegenerative disease that affects millions of people worldwide is Parkinson's disease (PD). Degeneration of dopamine-producing neurons in the brain, especially in the substantia nigra region, is its defining feature. Tremors, stiffness, bradykinesia (slowness of movement), and postural instability are the symptoms that result from this. Non-motor symptoms like mood swings, autonomic dysfunction, and cognitive impairment may also appear as the disease worsens. Although there is not a known cure for Parkinson's disease, there are several treatments that can help control the symptoms, such as medication and surgery.

For instance, Joseph Wang et al., developed new multimodal microneedle sensing platform relies on parallel simultaneous independent enzymatic amperometric and nonenzymatic voltammetric detection of L-Dopa using different microneedles on the same sensor array patch [34]. With both enzymatic and nonenzymatic L-Dopa microneedle sensors providing distinct informational dimensions, high sensitivity (with a low detection limit), high selectivity in the face of possible interferences, and good stability in artificial interstitial fluid (ISF), the new wearable microneedle sensor device exhibits an attractive analytical performance shown in Fig. 11.8. The microneedle sensor array's appealing analytical performance and possible wearable uses have been shown in a phantom gel that mimics human skin and after penetrating mice's skin. The new orthogonal wearable microneedle sensor array's attractive analytical performance and robust design offer a lot of promise for continuous, dependable, minimally invasive L-Dopa monitoring in the ISF, which can help optimize drug dosage regimens and improve Parkinson disease management.

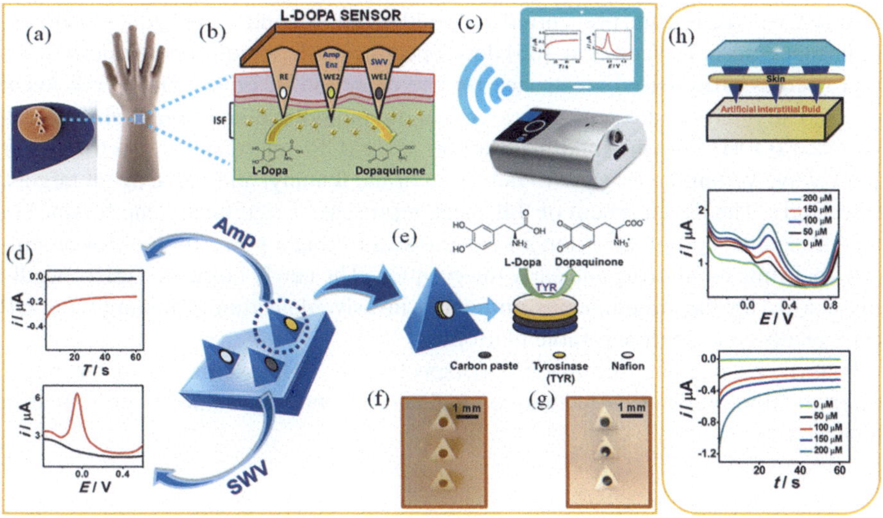

Fig. 11.8 Microneedle based electrochemical sensing platform for L-Dopa detection; schematic representation of **a** mannequin hand wearing the microneedle-based sensor and **b** the microneedle sensor for L-Dopa monitoring in interstitial fluid (ISF). **c** Portable wireless electrochemical workstation enabled with wireless data transmission to the smart device; **d** schematics of the microneedle sensor platform, illustrating three electrodes used for L-dopa sensing using SWV and amperometry; **e** microneedle sensor with the corresponding reagent layers, including the CP, tyrosinase, and nafion layer. **f**, **g** Actual optical images of microneedles before and after packing with CP (scale bar, 1 mm), **h** Real-time L-Dopa detection in artificial ISF through the mice skin-penetrated Microneedle. The figure was reproduced with the permission [34]

The new orthogonal microneedle sensing relies on the simultaneous L-Dopa sensing by SWV and chronoamperometry measurements at unmodified and enzyme-modified carbon paste electrodes, respectively. The resulting microneedle sensing platform displays an attractive analytical performance in skin-mimicking phantom gel as well as upon penetration through mice skin with high sensitivity, a wide linear dynamic range, and good stability along with high selectivity. The ability to combine several distinct independent processes onto a single wearable platform holds considerable potential for enhancing the power of a wide range of on-body chemical sensing devices in general and could be readily extended to the monitoring of other key biomarkers in different biofluids.

11.9 Microneedles for Opioid and Nerve Agents Detection

Sensing technology that can differentiate between episodes of opioid overdose and nerve agent poisoning is urgently needed. By addressing the challenges associated with dermal absorption and developing robust sensing technologies, it is possible to enhance the detection and prevention of opioid overdoses and nerve agent poisoning. For instance, Joseph Wang et al. [35] developed the modern multimodal micro structured needles array-based electrodes for the detection of opioid and nerve agent targets. The microneedle sensor array relies on unmodified and organophosphorus hydrolase (OPH) enzyme-modified carbon paste (CP) microneedle electrodes for square wave voltammetric (SWV) detection of the fentanyl and nerve agent targets, respectively. The development of this patch represents a significant leap forward in the fight against opioid addiction and overdose, offering a powerful tool for continuous, real-time monitoring and early intervention. The newly fabricated microneedle sensor array for the simultaneous minimally invasive detection of fentanyl and OP nerve agents on a single wearable platform.

To further enhancing the reader's perspective of MN, a summary of the major design of MN, preparation technique, applications and importance is presented in Table 11.1.

11.10 Future Outlooks

The microneedle derivatives can combine the outstanding qualities of the sensors in higher the sensitivity, lower the LOD, sharp edged microneedles can give more electrochemical active sites and it can increase the electrochemical active surface area (ECSA) of the microneedle-based electrode materials with the features of their precursors. Microneedles are minuscule instruments that pierce the stratum corneum, the outermost layer of skin, to directly administer medications or vaccines to the dermal layers. By avoiding the skin's natural barrier, this minimally invasive technique helps to improve the delivery and effectiveness of medicinal agents. Microneedles are available in solid, hollow, coated, and dissolvable varieties, and they can be fabricated from a wide range of materials, including silicon, metals, polymers, and ceramics. This technology has the potential to improve patient compliance, lower the risk of infection, and administer medication without pain. As a result, Microneedle based electrochemical sensing technology demonstrated the major benefits in overcoming the invasive methods in the point of care (POC) diagnostics for the biomedical application. Microneedles hold great promise for electrochemical applications; however, their practical implementation and commercialization require overcoming obstacles and exploring new opportunities. Investigating the clever creation, synthesis, characterization, and application of electrocatalysts derived from microneedles will yield novel concepts and increase the sustainability of electrochemical technologies. We firmly believe that the promising developments in this area, along with continued commitment and research, will open the door for electrodes derived from microneedles to be used in a variety of industrial applications. In this chapter we concluded the various methods of preparation of microneedles and their electrochemical sensing of different emerging biomarkers. Future efforts will thus involve a detailed animal study of the new microneedle patch and its integration with timely therapeutic action. This study will focus on evaluating the efficacy, safety,

Table 11.1 Summary of the major design of MN, preparation technique, applications and importance

Design/shape	Preparation method	Application	Importance	References
Porous	Template replication	To wound heal	Microneedle porosity allows for regulated drug release and high drug-carrying capacity	[36]
Mushroom	Micromolding	COVID 19 vaccine delivery	The structure can be detached from the underlying substrate, allowing for easy vaccination count recording without further storage and convenient readout	[37]
Bee sting	Micromolding	Drug delivery	Separate needle tips allow for drug administration to the gingival sulcus without disrupting oral function	[38]
Core-shell	Micromolding	Vaccination	Preprogrammed burst release of vaccination payloads for long-term distribution in one administration	[39]
Iced	Freeze-template molding	Drug delivery	Effective penetration with ability to administer diverse medication compositions	[40]
Faceted	3D printing	Vaccination	Coloading several vaccine components	[41]
Tuneable merged-tip	Photolithography	Drug delivery	Simple, adaptable fabrication. This customizable and modular platform may meet unique functional demands and adapt to various design requirements	[42]

and potential side effects of the microneedle patch in delivering therapeutic agents in a controlled and timely manner. By conducting these animal trials, researchers aim to gather critical data that will inform the subsequent steps in the development and potential clinical application of the microneedle patch.

After years of study and laboratory-based manufacture, microneedles are now ready for clinical applications with increased certainty. To achieve viable uses for microneedles, numerous crucial criteria must be addressed. Balancing microneedle preparation parameters, such as needle shape, mechanical insertion performance, and fabrication platform, is challenging. All microneedle platform parameters, from design to production, need robust justification. Integrating microneedle array devices with cameras or AI-assisted equipment for real-time monitoring of therapeutic or biosensing processes enhances efficiency [43–46]. As microneedles are small and used in the body, dependable medication delivery and sensing platforms are crucial.

11.11 Conclusion

This book chapter discusses the advancement of microneedles preparation methods and which can be applicable in electrochemical sensing platforms for the detection of various biomarkers. The microneedles and their nanocomposites materials to enrich the sensing property of biomarkers. In addition, this chapter discusses the conceptual background as well as evolutionary developments of electrochemical sensing in microneedles based nanostructured materials with high sensitivity and lower the detection limit of biomarkers. In the pursuit of more extensive industrial uses for these intriguing microneedles-based electrode materials, this constraint must be addressed. In order to facilitate the production of these materials with the necessary morphology and catalytic properties for large-scale industrial application, considerable effort must be put into establishing thorough guidelines and technical prerequisites that can improve stable and effective manufacturing processes. In this chapter highlights the recent developments and controlled preparation of microneedles of various techniques and their cutting-edge electrochemical sensing applications.

Authors' Contributions Mani Arivazhagan: Writing, review and editing.
S. Lokeswara Reddy: Writing, review and editing.
Raheel Akram: Methodology, investigation, data curation, writing original draft.

References

1. Chen, M.C., Chan, H.A., Ling, M.H., Su, L.C.: Implantable polymeric microneedles with phototriggerable properties as a patient-controlled transdermal analgesia system. J. Mater. Chem. B **5**, 496–503 (2017)
2. Kalluri, H., Banga, A.K.: Microneedles and transdermal drug delivery. J. Drug Deliv. Sci. Technol. **19**, 303–310 (2009)
3. Prausnitz, M.R., Langer, R.: Transdermal drug delivery. Nat. Biotechnol. **26**, 1261–1268 (2008)
4. Parry, A., Nursing, A., Sciences, H.: Surgical Instrument Specialists. J. Aesthet Nurs. **8**, 2019–2020 (2019)
5. Chambers, R.: Microdissection studies, Iii. Some problems in the maturation and fertilization of the echinoderm egg. Biol. Bull. **41**, 318–350 (1921)
6. Larrañeta, E., Lutton, R.E.M., Woolfson, A.D., Donnelly, R.F.: Microneedle arrays as transdermal and intradermal drug delivery systems: materials science, manufacture and commercial development. Mater. Sci. Eng. R Rep. **104**, 1–32 (2016)
7. Reed, M.L., Lye, W.K.: Microsystems for drug and gene delivery. Proc. IEEE **92**, 56–75 (2004)
8. Orentreich, D.S., Orentreich, N.: Subcutaneous Incisionless, vol. 21, pp. 543–49 (1995)
9. Henry, S., McAllister, D.V., Allen, M.G., Prausnitz, M.R.: Microfabricated microneedles: a novel approach to transdermal drug delivery. J. Pharm. Sci. **87**, 922–925 (1998)
10. Prausnitz, M.R.: Microneedles for transdermal drug delivery. Adv. Drug Deliv. Rev. **56**, 581–587 (2004)
11. Caudill, C.L., Perry, J.L., Tian, S., Luft, J.C., DeSimone, J.M.: Spatially controlled coating of continuous liquid Interface production microneedles for transdermal protein delivery. J. Control. Release **284**, 122–132 (2018)
12. Krieger, K.J., et al.: Simple and customizable method for fabrication of high-aspect ratio microneedle molds using low-cost 3D printing. Microsyst. Nanoeng. **5** (2019)

13. O'Mahony, C.: Structural characterization and in-vivo reliability evaluation of silicon microneedles. Biomed. Microdevices **16**, 333–343 (2014)
14. Li, J., et al.: Fabrication of a Ti porous microneedle array by metal injection molding for transdermal drug delivery. Plos One **12**, 1–15 (2017)
15. Thanh, H.L., et al.: Low-cost fabrication of hollow microneedle arrays using CNC machining and UV lithography. J. Microelectromech. Syst. **24**, 1583–1593 (2015)
16. Waghule, T., et al.: Microneedles: a smart approach and increasing potential for transdermal drug delivery system. Biomed. Pharmacother. **109**, 1249–1258 (2019)
17. Donnelly, R.F., et al.: Hydrogel-forming and dissolving microneedles for enhanced delivery of photosensitizers and precursors. Photochem. Photobiol. **90**, 641–647 (2014)
18. Kim, Y.C., Park, J.H., Prausnitz, M.R.: Microneedles for drug and vaccine delivery. Adv. Drug Deliv. Rev. **64**, 1547–1568 (2012)
19. Henry, S., McAllister, D.V., Allen, M.G., Prausnitz, M.R.: Micromachined needles for the transdermal delivery of drugs. In: Proceedings IEEE Micro Electro Mechanical Systems, pp. 494–498. https://doi.org/10.1109/memsys.1998.659807
20. Mirza, K.B., et al.: Injection moulded microneedle sensor for real-time wireless pH monitoring. In: Proceedings of the Annual International Conference of the IEEE Engineering in Medicine and Biology Society EMBS, pp. 189–192 (2017). https://doi.org/10.1109/EMBC.2017.8036794
21. Vinayakumar, K.B., et al.: A hollow stainless steel microneedle array to deliver insulin to a diabetic rat. J. Micromech. Microeng. **26**, 65013 (2016)
22. Simfukwe, J., Mapasha, R.E., Braun, A., Diale, M.: Biopatterning of keratinocytes in aqueous two-phase systems as a potential tool for skin tissue engineering. MRS Adv. **357**, 1–8 (2017)
23. Sadeqi, A., Nejad, H.R., Kiaee, G., Sonkusale, S.: Cost-effective fabrication of chitosan microneedles for transdermal drug delivery. In: Conference on Proceedings of the Annual International Conference of the IEEE Engineering in Medicine and Biology Society IEEE Engineering in Medicine and Biology Society Annual Conference, pp. 5737–5740 (2018)
24. Duong, H.T.T., et al.: Microneedle arrays coated with charge reversal pH-sensitive copolymers improve antigen presenting cells-homing DNA vaccine delivery and immune responses. J. Control. Release **269**, 225–234 (2018)
25. Guillot, A.J. et al.: Microneedle Based Delivery Review.Pdf
26. González-Vázquez, P., et al.: Transdermal delivery of gentamicin using dissolving microneedle arrays for potential treatment of neonatal sepsis. J. Control. Release **265**, 30–40 (2017)
27. Vora, L.K., et al.: Microneedle-based biosensing. Nat. Rev. Bioeng. **2**, 64–81 (2024)
28. Tehrani, F., et al.: An integrated wearable microneedle array for the continuous monitoring of multiple biomarkers in interstitial fluid. Nat. Biomed. Eng. **6**, 1214–1224 (2022)
29. Azizi Machekposhti, S., Khanna, S., Shukla, S., Narayan, R.: Microneedle fabrication methods and applications. MRS Commun. **13**, 212–224 (2023)
30. Valdés-Ramírez, G., et al.: Microneedle-based self-powered glucose sensor. Electrochem. Commun. **47**, 58–62 (2014)
31. Kai, H., Kumatani, A.: A porous microneedle electrochemical glucose sensor fabricated on a scaffold of a polymer monolith. J. Phys. Energy **3** (2021)
32. Zhou, J.X., et al.: MoS2/Pt nanocomposite-functionalized microneedle for real-time monitoring of hydrogen peroxide release from living cells. Analyst **142**, 4322–4329 (2017)
33. Mugo, S.M., Robertson, S.V., Lu, W.: A molecularly imprinted electrochemical microneedle sensor for multiplexed metabolites detection in human sweat. Talanta **259**, 124531 (2023)
34. Goud, K.Y., et al.: Wearable electrochemical microneedle sensor for continuous monitoring of levodopa: toward Parkinson management. ACS Sens. **4**, 2196–2204 (2019)
35. Mishra, R.K., et al.: Continuous opioid monitoring along with nerve agents on a wearable microneedle sensor array. J. Am. Chem. Soc. **142**, 5991–5995 (2020)
36. Yao, S., et al.: Porous MOF microneedle array patch with photothermal responsive nitric oxide delivery for wound healing. Adv. Sci. **9**, 1–11 (2022)
37. Li, Q., et al.: Smart mushroom-inspired imprintable and lightly detachable (MILD) microneedle patterns for effective COVID-19 vaccination and decentralized information storage. ACS Nano **16**, 7512–7524 (2022)

38. Song, C., Zhang, X., Lu, M., Zhao, Y.: Bee sting-inspired inflammation-responsive microneedles for periodontal disease treatment. Research **6**, 0119 (2023)
39. Tran, K.T.M., et al.: Transdermal microneedles for the programmable burst release of multiple vaccine payloads. Nat. Biomed. Eng. **5**, 998–1007 (2021)
40. Zhang, X., Fu, X., Chen, G., Wang, Y., Zhao, Y.: Versatile ice microneedles for transdermal delivery of diverse actives. Adv. Sci. **8**, 1–9 (2021)
41. Caudill, C., et al.: Transdermal vaccination via 3D-printed microneedles induces potent humoral and cellular immunity. Proc. Natl. Acad. Sci. U. S. A.**118** (2021)
42. Lim, J., Tahk, D., Yu, J., Min, D.-H., Jeon, N.L.: Design rules for a tunable merged-tip microneedle. Microsyst. Nanoeng. **4**, 29 (2018)
43. Tarar, C., Aydın, E., Yetisen, A.K., Tasoglu, S.: Machine learning-enabled optimization of interstitial fluid collection via a sweeping microneedle design. ACS Omega **8**, 20968–20978 (2023)
44. Tarar, C., Aydın, E., Yetisen, A.K., Tasoglu, S.: Bayesian machine learning optimization of microneedle design for biological fluid sampling. Sens. Diagn. **2**, 858–866 (2023)
45. Biswas, A.A., et al.: Advancements in microneedle fabrication techniques: artificial intelligence assisted 3D-printing technology. Drug Deliv. Transl. Res. **14**, 1458–1479 (2024)
46. Lori Zoudani, E., Nguyen, N.T., Kashaninejad, N.: Microneedle optimization: toward enhancing microneedle's functionality and breaking the traditions. Small Struct. **2400121** (2024)

Chapter 12
Microneedles-Based Interstitial Fluid Extraction for Systemic Lupus Erythematosus Monitoring

Nurul Asmak Md. Lazim, Syaidah Md. Saleh, and Ida Idayu Muhamad

12.1 Introduction

Drugs or therapeutically active ingredients are delivered mostly via the skin through transdermal drug delivery systems (TDDS). This is because the physical characteristics of medications, such as their polarity and molecular size, frequently limit their ability to deliver the specific drug transdermally. Therefore, only a few pharmaceuticals are administered transdermally. When applied to the skin, a transdermal patch releases a dose of medication gradually into the bloodstream. This patch is medicated and has sticky qualities. It can be utilized with patients who have trouble swallowing pills, capsules, or parenteral drug delivery. As a result, patients occasionally forget when to take their medications on time [1].

Currently, microneedles are commonly utilized for transdermal drug delivery systems and have been effectively used to extract interstitial fluid (ISF) by various methods to make further analysis easier. The combination of microneedles with sensors enables in situ ISF analysis and particular compound monitoring, while the integration of monitoring and delivery functions in wearable devices provides real-time dose modification [2]. Since ISF can also have biomarkers like those in blood, it can be extracted for drug or biomarker analysis to track the state of a disease and adjust the dose for individualized care [3]. However, trustworthy and user-friendly ISF extraction techniques must be developed because the current procedures are laborious, need specialized handling, and involve large equipment [4, 5].

N. A. Md. Lazim (✉) · I. I. Muhamad
Department of Bioprocess & Polymer Engineering, Faculty of Chemical and Energy Engineering, Skudai, Malaysia
e-mail: nurulasmak89@gmail.com

S. Md. Saleh · I. I. Muhamad
IJN-UTM Cardiovascular Engineering Centre, iHuman, Universiti Teknologi Malaysia, Johor Bahru, Johor, Malaysia

N. Talreja et al. (eds.), *Microneedles (MNs)-Based Technology*, Series in BioEngineering, https://doi.org/10.1007/978-981-96-3916-8_12

These days, transdermal drug delivery methods are applied in a number of therapeutic domains, including the treatment of heart disease, motion sickness, hormone replacement, pain management, and quitting smoking. Drug potency is increased, and relative adverse effects are reduced using TDDS [1]. However, the authors found that the use of microneedles for patients with systemic lupus erythematosus (SLE) is very limited. Almost fewer than 10 articles have been published on this issue.

An autoimmune condition that affects several systems is called systemic lupus erythematosus, or SLE. Ecology, environment, genetics, and epigenetics are all the reasons involved in this complex mechanism. First and foremost, it triggers the activation of both innate and adaptive immunity. This, in turn, triggers the activation of autoreactive B cells by T cells and the deposition of immune complexes in tissues, which sets off an autoimmune cascade that may involve only one organ or may affect the entire system. The current goals of SLE treatment are organ damage prevention, remission, and general quality of life enhancement, all of which call for a heavy dose of medication to maintain a healthy lifestyle [6, 7]. Combination medication therapy, which includes glucocorticoids, immunosuppressive agents, anti-malarial medications, and non-steroidal anti-inflammatory drugs, is currently a successful treatment for SLE [8–10]. However, while these medications can somewhat control some clinical symptoms, many patients suffer from serious side effects, including infections, which are often far more severe than the disease's symptoms [11, 12]. Therefore, in this chapter, the authors reviewed the possible ways to minimize the severe illness of SLE by monitoring using microneedle-based and current technology utilized for the possible treatment. It covers every major facet of the subject, such as material selection, design philosophy, operating mechanisms, and gadget utility. Regarding the transfer of these cutting-edge innovations from the bench to the bedside, the authors also discuss the past, the present, and the future.

12.2 Transdermal Drug Delivery Systems (TDDS) and Interstitial Fluid (ISF)

The development of novel drug delivery systems for pharmaceuticals that are already on the market has garnered increased attention in recent years. The development of a novel delivery method for well-known medications greatly improves patient adherence, overall therapeutic effect, efficacy, and safety. When patches are applied to intact skin, they are sometimes referred to as "patches" [13]. However, transdermal means through the skin or dermis. In contrast, transdermal drug delivery (TDD) refers to the application of a medication to the skin, typically in the form of an ointment or patch. However, using a transdermal drug delivery system (TDDS) allows the drug's active components to be administered through the skin's outer layer. The TDDS is necessary for controlled drug release in order to attain the drug's efficacy in a way that is safe, predefined, and sufficient for systematic circulation [14] (Fig. 12.1).

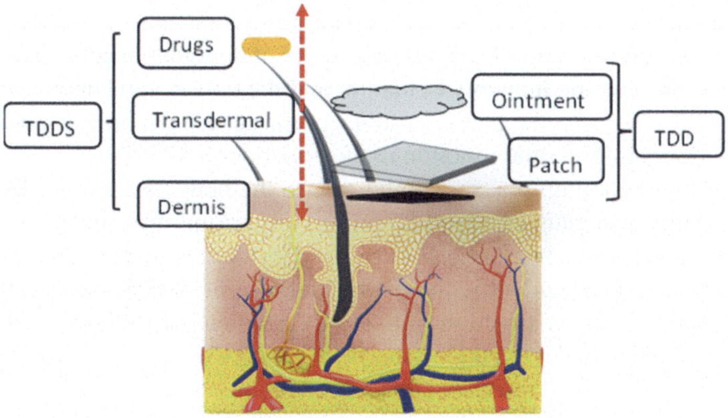

Fig. 12.1 Illustration of patches, TDD, and TDDS

Transdermal medication delivery is the introduction of an active ingredient for therapeutic use through the skin to produce a systemic impact. There are very few products that are used on the top since the physical characteristics of medications, such as their polarity and molecular size, frequently prevent them from being applied topically. Likewise, inadequate bioavailability and skin irritation have been issues [1]. Since TDDS does not require transit via the gastrointestinal tract, medications can be administered without interference from pH, enzymes, or intestinal microbes. Additionally, there is no loss resulting from first-pass metabolism. Furthermore, TDDS can be utilized to regulate medication release in accordance with consumption limitations, which adds to the method's high persistence. Most importantly, medication can be safely and conveniently given to youngsters or the elderly using TDDS, a non-invasive administration technique that causes the patient little discomfort or strain [15]. Because of the skin's natural barrier, TDDS still does not reach its full potential. The purpose of the skin, which is the outermost organ and has multiple layers, is to shield the body from harmful external elements, including heat, chemicals, and poisons. These layers of skin can be separated into the dermis, which creates skin cells and contains blood arteries, and the epidermis, which serves as protection. Each layer contains components that prevent transdermal distribution [14, 15]. The reason is that, firstly, the stratum corneum, the outermost layer, has the skin barrier effect of the epidermis, which is the ability to prevent substances from the outside. When it comes to the transportation of chemicals with a high molecular weight, the barrier effect is crucial. It is commonly acknowledged that the intracellular pathway is used in TDDS to deliver compounds with modest molecular weights. However, techniques and different mechanisms utilizing the intracellular system in addition to the intercellular pathway are introduced and used for drugs with a significant molecular weight [15–18]. This is caused by the way the skin is structured, as the lipid layer, which contains both cells and substances that are both hydrophilic and hydrophobic, does not have a totally regular position but rather exists regularly [19]. The physicochemical qualities that are tried to improve drug transport through the skin can explain these structural aspects. Next, transdermal delivery may be inhibited by the vascular system in the skin layer. The layer of endothelial cells that makes up the interface between the human vasculature and the tissues surrounding the skin is one cell thick and ends in the papillary loops of the superficial arteriovenous plexus close to the dermal-epidermal junction in the upper dermis [15].

The skin's endothelium plays a similar role to the body as a whole. By controlling permeability and causing vasodilation or constriction in response to pressure, shear, osmotic pressure, heat, chemokines, and cytokines, it actively responds to these stimuli [20]. Consequently, the main challenge of TDDS is to overcome the stratum corneum's barrier function, provide the medication to the skin tissue, and get past the cellular and vascular tissue to the target tissue. The medication can only be absorbed via the skin tissue in small amounts, which is difficult [21, 22]. Numerous cutting-edge TDDS solutions have been thoroughly designed as appealing administration methods to address this issue. Furthermore, these advancements may provide a competitive edge over alternative drug delivery techniques concerning the administered dosage, economic viability, and medicinal effectiveness [22–25]. When compared to topical drug treatment on the skin, external stimuli, including electrical, mechanical, or physical stimulation, are known to increase the permeability of medicines and biomolecules through the skin [26]. Active transdermal delivery, or TDDS enhanced by suitable equipment, is a fast and dependable method of delivering medications to the skin. Furthermore, this improved TDDS mode can hasten the therapeutic efficacy of medications that are administered. As for passive delivery, the drugs should have a low molecular weight (<1 kDa), an affinity for lipophilic and hydrophilic phases, a short half-life, and little skin irritation in order to maximize transdermal distribution and therapeutic efficacy [27]. Numerous factors, including species differences, skin age and site, skin temperature, skin condition, area of application, length of exposure, skin moisture content, pretreatment techniques, and physical properties of the penetrant, will influence how well a medicine penetrates the skin. There are many types of TDDSs [15]. However, in this chapter, the authors will only focus on microneedle-based delivery (Fig. 12.2).

12.2.1 Microneedle-Based Drug Delivery System

A needle is used in the microneedle drug delivery system, a revolutionary drug delivery method, to administer medication to the circulatory system [28]. This is one of the most widely used transdermal medication delivery techniques, and research on it is still ongoing. This involves a technique that diffuses medication across the epidermal layer by puncturing the skin's superficial layer with needles of a micron size. Normally, the length of microneedles ranges from 50 to 110 mm. These microneedles release the medicine contained in the reservoir upon penetrating the stratum corneum, the main transdermal delivery barrier [1]. The medication within the reservoir could be in the form of a gel, solution, or solid. These tiny, thin microneedles assist in preventing discomfort by delivering medication straight to the blood

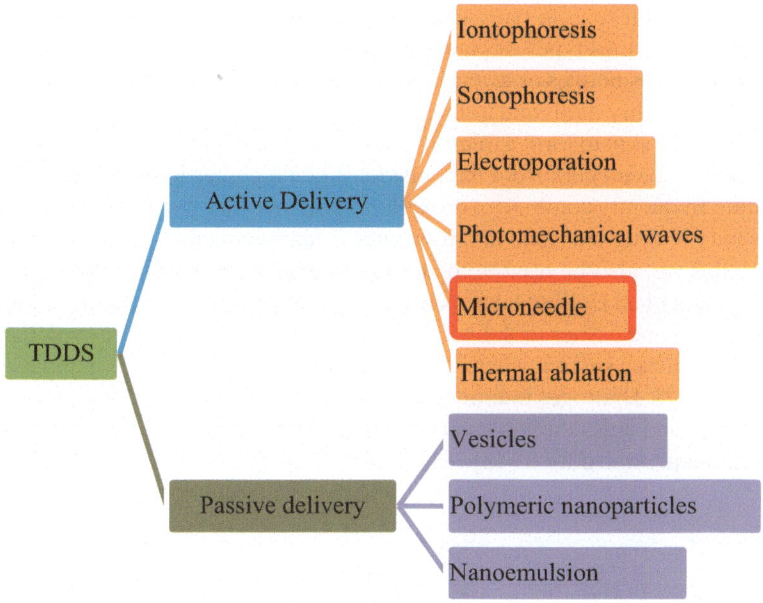

Fig. 12.2 The different types of TDDSs

capillary area for active absorption [29]. Researchers have endeavoured to employ several methodologies to achieve the suitable optimization and geometric measures necessary for the successful insertion of microneedles into human skin, which aligns with the overarching goal of microneedle research.

Numerous studies have been conducted on the creation of microneedle systems, taking into account the intended application, drug type and dosage, and goal [30]. Up until recently, photolithography and laser-mediated methods have been used to produce microneedles. Metal or polymer microneedles can be produced using laser-mediated fabrication processes. By utilizing a laser to cut or ablate a flat metal or polymer surface, a microneedle's 3D structure is created [31, 32]. The process of intricately creating microneedles is called photolithography, and it offers the benefit of allowing the production of needles in a variety of shapes and materials. This process is mostly used to create silicon or dissolving hydrogel microneedles by etching photoresist and creating an inverse mold based on the microneedle structure [33]. Two-photon polymerization [34], microstereolithography [35], and 3D printing [36] are all being researched to prepare different microneedle systems. A variety of types of prepared microneedles are available, including solid microneedles that act as a physical conduit for drug absorption, drug-coated microneedles that facilitate the delivery of drugs coated on the needles' surfaces as they enter the skin, dissolving microneedles composed of drug formulations that dissolve in the body, naturally delivered melting needles that involve the storage of drugs in hollow needles and then administration (e.g., a particular type of injection), and microneedle patches combined with various types of patches [37–41].

The operating principles of several kinds of microneedles are shown in Fig. 12.3. In order to gather bodily fluids and transport actives to the site of action, solid microneedles puncture the skin to form a microincision [42]. Meanwhile, coated microneedles have a surface-coated core from the outside. When it comes to medication delivery, this coating frequently combines a drug with a polymer. The coating is used to collect bodily fluids in drug analysis, and it mimics a semipermeable barrier [43]. Coated microneedles can also be used with sensors to gather and monitor medication and biomarker data [44]. Hollow microneedles' interior passageways are used to collect fluid from the same tissues or deliver drugs to them [45]. ISF exits through the pores created on the tissue surface when soluble microneedles are inserted into tissues and dissolve [46, 47]. Hydrogel microneedles' material has good swelling properties; it absorbs water and gels, but it does not dissolve, so it can be used for fluid aspiration or medication administration [48].

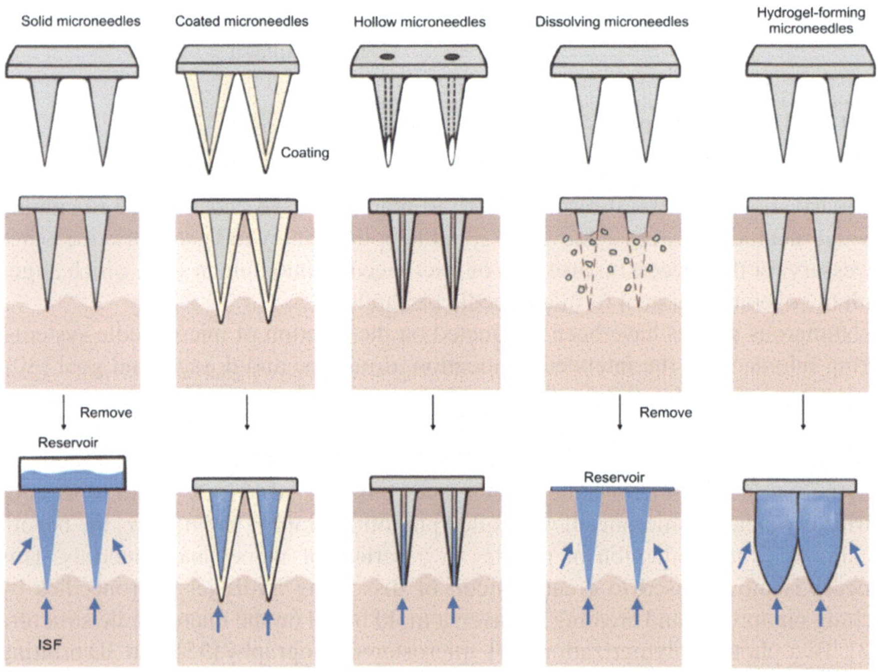

Fig. 12.3 Different types of microneedle design. Figure was reproduced with permission [2]

12.2.2 Interstitial Fluid (ISF) Profile

Pharmaceutical monitoring and quantification are highly practical undertakings, given the safety risks associated with incorrect drug dosage [49]. Blood samples are often used for drug testing, and because there is a relationship between blood concentration and treatment results, the blood concentration range was chosen to use as a reference [50, 51]. When it comes to minimizing the likelihood of improper dosage-induced side effects and enhancing therapeutic efficacy, this administration technique outperforms empirical dosing since it enables prompt dose and dosing interval adjustment. However, intrusive blood collection is fraught with risk and can be uncomfortable and difficult to perform in certain situations. Additionally, the procedures of collecting, transporting, and managing blood are costly and time-consuming. Due to these limitations, researchers have been looking for different body fluids to replace blood in drug tests, such as saliva, urine, and tear fluid. However, among these, interstitial fluid (ISF) is currently showing the most potential [52–55]. Since ISF has biomarkers that are similar to those in blood, it can be extracted for drug or biomarker analysis to track the state of the disease and adjust the dose for individualized care [3]. However, as the current methods are laborious, need specialized handling, and involve large equipment, ISF extraction strategies that are dependable and user-friendly must be developed [4, 5]. ISF sampling can be completed using acupuncture, a commonly used traditional Chinese medicine treatment. Filiform and plum blossom needles are only two of the various types of needles that can be used for this procedure [56, 57]. Furthermore, recent reports have shown that ISF extraction can be accomplished with microneedles [58]. In terms of real-time monitoring, a study that assessed the amount of glucose in microneedle-extracted insulin-stimulant fluid (ISF), for example, demonstrated that diabetic individuals could more readily monitor their blood glucose levels using this method. The rapid (within 3 min) detection of severe acute respiratory syndrome coronavirus 2 immunoglobulin G and M antibodies in ISF was made possible by a patch sensor that combined porous poly(lactic acid) microneedles with an immunochromatographic assay [59]. This operation also allowed for the quick diagnosis of Corona Virus Disease 2019 (COVID-19) previously [60]. Since ISF is not able to leave the body, it needs to be gathered via physiological barriers like the skin. In the past, ISF was commonly extracted and collected from the skin and tissues using suction blisters and microdialysis. These techniques have the added disadvantages of a small sample volume and a time-consuming nature. Also, the corresponding sampling process causes pain and/or discomfort and may result in scarring. These techniques also call for specific handling and instruments. The limitations mentioned earlier restrict the applicability of ISF in drug analysis and diagnosis. However, ISF is extracted quickly and easily with the use of microneedles, which are also used to gather ISF via the skin. In particular, nerve fibres and blood arteries found in the dermis are normally avoided by microneedles as they enter the stratum corneum of the epidermis. Microneedle-based ISF extraction is gaining a lot of attention in the drug detection sector since it is less invasive, safe, and painful than other approaches. It also has the added advantages of simple operation and easy-to-use equipment. Additionally, biological fluid collection and medication monitoring, including the collection of blood and lymphatic fluid, can be accomplished with microneedles [2].

ISF is absorbed differently by different microneedles. For example, the hydrogel microneedles' substance swells to permit ISF to permeate into it. Microneedle size and swelling capacity limit the amount of retrieved ISF at which point this diffusion slows down. By suctioning under capillary action, hollow microneedles aspirate ISF; nevertheless, when utilizing this technique, one should prevent congestion and be aware of the possibility of microneedle breakage [2]. In order to extract more ISF more quickly, researchers have made full use of the previously mentioned microneedle extraction techniques while purposefully utilizing some external forces (Fig. 12.4a). 3.82 μL of ISF from isolated mouse skin and 7.90 μL of ISF from isolated pig skin were extracted in just three minutes thanks to the accelerated extraction rate in an excellent work that involved creating a microneedle patch by combining an osmotic agent with a hydrogel to provide osmotic pressure during extraction [61]. Likewise, microneedles that create sorbitol hydrogel were engineered to expedite the extraction process [62]. ISF can be extracted through these micropores by using a vacuum pump as a convective driving force [63]. Microneedles that operate on the skin create micropores. However, this method is not well suited for widespread clinical application due to the requirement for specialized training, the use of specialized equipment, and the possibility of blood contamination of the tested ISF. The ISF extraction volume and rate are influenced by the microneedle material. It is also important to choose a material that fully takes into account the mechanical and swelling properties of hydrogel microneedles when using them for ISF extraction. A previous study found that in 10 min, a methacrylated hyaluronic acid microneedle patch (Fig. 12.4b) could extract approximately 2.3 μL of ISF [64]. Graphene oxide-GelMA microneedles, cross-linked GelMA microneedle patches, sponge-forming poly(vinyl acetal) microneedle patches, and gelatin methacrylate (GelMA) microneedle patches have all been created and utilized for ISF extraction in recent years [65–68]. The ISF extraction capabilities of microneedles made of different materials are contrasted in Table 12.1 [64–68].

The skin somewhat deforms as microneedles puncture it during ISF extraction, and variations in the force applied to the microneedle may have an impact on the extraction outcome. Precise punctures hold far greater significance. A bionic rotating microneedle inspired by a bee stinger allowed for a 45.7% reduction in the maximum penetration force and a 2.73-fold increase in the rate of force decay, as shown in Fig. 12.4c. This design made it possible to penetrate the skin with greater precision and reduced resistance [69]. The goal of several studies has been to improve ISF collection after microneedle penetration. For instance, after inserting the microneedle into the skin, two strips of filter paper were affixed to both sides of the microneedle base and used to absorb the effluent ISF (Fig. 12.4d) [70, 71], enabling the collection of approximately 4 μL of ISF in just one minute. Ultrafine nanoneedles were arranged on a microneedle array holder by Tran et al. [72], and each needle was connected to a 1–5 μL calibrated pipette capillary. ISF was collected into the capillary (Fig. 12.4e)

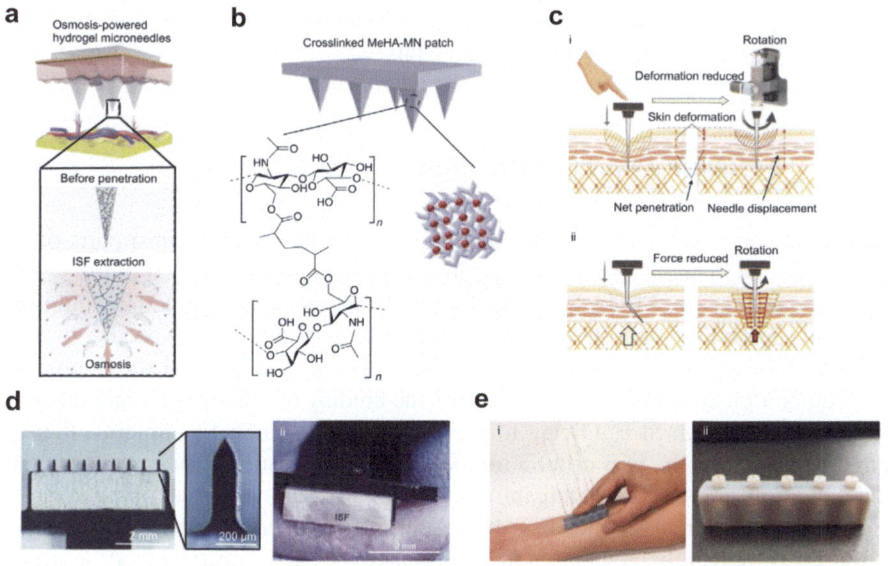

Fig. 12.4 Different types of microneedles with different mechanisms of ISF. Figure was reproduced with permission [2], **a** hydrogel microneedle with an osmotic agent, **b** methacrylated hyaluronic acid (MeHA)-microneedle (MN) patch, **c** rotating microneedle, **d** filter paper strips, **e** microneedle device with capillary

Table 12.1 Performance of interstitial fluid (ISF) extraction in microneedles made of various materials. The table was adapted with permission [2]

Microneedle material	Sampling volume (μL)	Time (min)	Sample tested
Methacrylate	2.3 ± 0.4	10	Mice
GelMA	2.5	10	Rats
poly(vinyl acetal)	1.6	1	Rats
c-GelMA	3.5 ± 0.1	30	Agarose gel system model
Graphene oxide-GelMA	21.34	30	Human skin

GelMA: gelatin methacrylate; c-GelMA: cross-linked GelMA

after the built microneedle array was inserted into the skin, and within 30 min, 10 μL of ISF was collected [72, 73].

The processes mentioned above necessitate the formation of micropores in the skin using microneedles to extract ISF for in vitro analysis. With the use of external forces, optimal material selection, and optimized collection schemes, significant progress has been made in improving the ISF extraction volume and minimizing the extraction time. However, it still takes a while to get the final analytical results because the extracted ISF needs to be transferred to another analytical instrument, and sample stability might not be sufficient. Furthermore, the likelihood of errors rises with the quantity of operational procedures.

12.2.3 Systemic Lupus Erythematosus (SLE) Profile

Despite having a variety of symptoms, humoral autoimmunity against parts of the nucleus defined patients with systemic lupus erythematosus (SLE), a multi-system autoimmune disease. Clinical observations and laboratory data, including antiphospholipid antibodies, complement protein levels, and antibodies specific to SLE (such as anti-dsDNA and anti-sm antibodies). Tissue damage and the deposition of immunological complexes are caused by the binding of homologous autoantigens by autoantibodies in SLE. Owing to the wide range of clinical symptoms, there is ongoing debate on the best criteria for classifying SLE patients, as well as whether the condition is really an amalgam of diseases with a shared pathophysiological route or a collection of distinct disorders. Additionally, overlap with other associated disorders is frequently included in the classification of SLE, since up to 40% of SLE patients also have antiphospholipid antibodies or overlap with other autoimmune disease syndromes, with secondary Sjogren's syndrome being the most common (up to 25%) [74–78].

Clinical symptoms of this illness range widely, from moderate cutaneous involvement to severe organ damage, including heart failure, pulmonary hypertension, and kidney failure. Based on clinical and test results, SLE is diagnosed. The most sophisticated and accurate categorization criteria to date are those employed by the American College of Rheumatology (ACR) [79] and the European League Against Rheumatism (EULAR). SLE management is difficult and calls for a multidisciplinary strategy. The organs affected and the severity of the disorders determine the therapy plan. Even though the illness affects several systems, each person's course will differ based on the severity, frequency of flare-ups, and length of remission. The degree of organ involvement, such as in the kidneys, lungs, and heart, may lower life expectancy. If not, 80–90% of SLE patients may have an average life expectancy with close monitoring [80]. Yohannan et al. [81] reported that an elective left-hip hemiarthroplasty was performed at an outside facility on a forty-year-old white woman who had systemic lupus erythematosus (SLE), positive lupus anticoagulant results, and left hip avascular necrosis. During the intraoperative phase, no problems were noted. A surgical bed hematoma measuring 6.9 by 4.3 cm on the left hip complicated the postoperative phase. There was also evidence of soft tissue hematoma formation across the pelvis. Peroneal nerve palsy resulted from the largest hematoma in the left gluteal area, hip, and thigh. The patient needed two units of fresh frozen plasma and five units of red blood cells due to a considerable decline in hemoglobin. She was admitted to the intensive care unit due to her activated partial thromboplastin time (aPTT) of 150 s. Upon physical assessment, the patient exhibited pallor and tachycardia. There was visible bleeding coming from the surgery site. She lacked feeling in her left foot when lightly touched, and her left foot dropped [81].

Geographically speaking, North America reports the highest incidence and prevalence of SLE worldwide, Africa has the lowest incidence, and Australia has the lowest prevalence. These differences are due to differences in reporting methods. Ethnicity, age, and gender all have a big impact on the clinical result and treatment of the illness. Although SLE is more common in women, men experience a more severe and rapid course of the disease, which leads to a poor prognosis. Genetic variations and environmental factors can be blamed for this discrepancy. The incidence rate for the Caucasian population is currently 6.73 cases per 100,000 people per year, while the African-American population has 31.4 cases per 100,000 people annually. The prevalence rate in the United States is 517 per 100,000 for the black population and 134 per 100,000 for the Caucasian and European populations [6, 74]. Epidemiological studies have clearly demonstrated the link between the occurrence of SLE and exposure to silica, cigarette smoke, oral contraceptives, ultraviolet B (UVB), specific medications, and Epstein-Barr virus (EBV) infection. These correlations may be caused by physiological factors such as elevated oxidative stress, systemic inflammation, elevation of inflammatory cytokines, and epigenetic changes. Here is a review of particular environmental exposure and its possible impact on the pathophysiology of SLE [6]. A broad range of clinical symptoms and an extensive autoantibody profile are seen in SLE patients. The variability in both clinical and serological features presents a significant obstacle to correct diagnosis. The simultaneous consideration of many clinical characteristics, serological results, imaging, and histology makes physician acumen crucial in the diagnosis of SLE. Here, the authors have covered the commonly used SLE categorization criteria as well as the biomarkers that are essential to the diagnostic process [6].

An interdisciplinary approach is necessary for the care of systemic lupus erythematosus since it is a disease with heterogenic manifestations that affect numerous organs. As a result, the disease's severity and organ involvement differ from patient to patient. The goal of treatment is to minimize drug side effects while preventing relapse, maintaining and promoting remission, and preventing disease flare-ups [82]. The disease's activity determines which medications are used to treat it. There have been fantastic improvements in the management of SLE since the 2008 publication of the EULAR guidelines for the condition. The Systemic Lupus Erythematosus Disease Activity Index-2000 (SLEDAI-2K), the Systemic Lupus Activity Questionnaire (SLAQ), Physician Global Assessment (PGA), the British Isles Lupus Assessment Group (BILAG 2004 index), and the Lupus Foundation of America Rapid Evaluation of Activity in Lupus (LFA-REAL) [83] are some of the widely accepted scoring systems used to assess the disease activity. Additionally, because the inflammation caused by SLE causes irreversible damage to many organs, a number of indices are employed to measure the damage, including the American College of Rheumatology Damage Index and the Brief Index of Lupus Damage [84].

The SLEDAI score of zero indicates complete remission or the absence of any active inflammation, the SLEDAI score of 1–5 indicates mild disease activity, the SLEDAI score of 6–10 indicates moderate disease activity, the increase of the SLEDAI score of 3 or more than 3 indicates a flare-up of the disease, and the decrease of the score of 3 or more indicates a response to the treatment and improvement in the disease activity. These scoring systems are important in helping to determine the choice of drugs and their effectiveness in managing the disease [84].

12.3 Transdermal Drug Delivery Systems (TDDS) and Interstitial Fluid (ISF) and Their Applications

The invention of transdermal drug delivery systems (TDDS) is a significant advancement in the field of controlled drug administration (Fig. 12.5). Transdermal dose forms are quickly becoming more and more popular as an alternative to conventional dosage forms due to their unique features. Such as a straightforward form of administration, zero-ordered regulated absorption, and the capability to terminate the activity in the event of an adverse effect are all excellent examples [13]. The TDDS increases their attractiveness for usage in the treatment of chronic diseases because those conditions necessitate long-term care. The Centers for Disease Control and Prevention (CDC) report that four out of ten Americans have two or more chronic diseases, and nearly six out of ten have at least one chronic disease [13].

Cardiovascular (CV) diseases such as atherosclerosis, angina pectoris, and acute myocardial infarction are the top causes of death globally as a result of our fast-paced, high-stress lifestyle. According to WHO data, CVDs are the world's largest

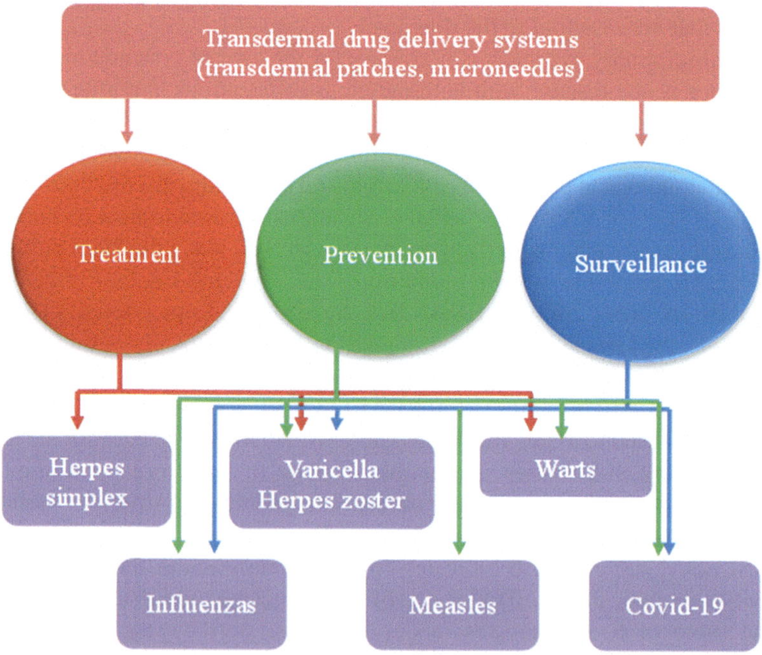

Fig. 12.5 Applications of TDDS

cause of death, accounting for an estimated 17.9 million deaths annually. Despite the fact that many commercially available medications are available for their treatment, the typical pills and capsules are insufficient to address these situations. Consequently, novel drug delivery strategies are being employed to deliver medications to certain regions of the cardiovascular system and hold them there for extended periods of time. The first step in creating innovative medication delivery techniques was applying transdermal patches to the skin that contained drug particles. They are believed to provide several benefits as compared to conventional oral medications. With transdermal therapy, a number of anti-hypertensive drugs bypass significant first-pass metabolism. Transdermal drug delivery devices are, therefore, anticipated to take off quickly, supporting the market's growth. Based on the kind, there are two types of transdermal medicine delivery systems: transdermal patches and transdermal semisolids. The transdermal patch market is divided into three sub-segments: drug-in-adhesive, reservoir membrane, and microneedle patches [13].

On the other hand, the application of microneedle-assisted drug monitoring in the COVID-19 epidemic is quite promising. For the COVID-19 immunization, microneedle technologies have been employed, and it has been demonstrated that storing vaccines in microneedles lowers transportation costs and lessens reliance on the cold chain. Additionally, fluorescence dyes are used to keep the microneedle array arrangement's pattern in the skin, allowing for the quick and easy recording of immunization data without the need for storage devices and the conservation of labour and materials. In addition, many people find it difficult to enter hospitals due to operational restrictions, including isolation. Since many patients with chronic illnesses, such as cancer, need to have their relevant biomarkers monitored over an extended period of time in order to optimize their therapy, research and development of microneedle-based home testing devices is essential [2]. Example of microneedle-ISF home-based applications is in Fig. 12.6. Figure 12.6a shows microneedle combined with a smartphone for glucose monitoring (smartphone-based glucose electrochemical detection platform (GEDP) (i), (ii) is referred to wireless chemical detector connected to a touch-type glucose sensor for electrochemical glucose detection and (iii) is showed schematic diagram of a smartphone-based GEDP consisting of a touch-sensitive glucose sensor, radio chemical detector, and smartphone application. Figure 12.6b shows a microneedle patch integrated with a lateral flow cassette for the home testing of chronic kidney disease, and lastly, Fig. 12.6c shows a diagram of the application process of a four-in-one test paper microneedle patch, where T1: test 1; T2: test 2; C: control; Cys C: cystatin C [2].

Figure 12.7 shows the current method of ISF sampling through microneedle. A small sample of ISF was extracted using a painless extraction microneedle, and the sample was then analyzed at home from a virtual reaction on the resultant sensor with a biomarker. From here, users can self-monitor their health status accordingly. The governing equations for the diverse kinds of drug and interstitial fluid movement in and out of the epidermal tissues, circulatory systems, and microneedles make up the modelling framework. The stratum corneum is a biphasic lipid-protein structure composed of layers of corneocytes shut off from the majority of things, including small molecule medicines, by densely packed lipids. Water molecules can move

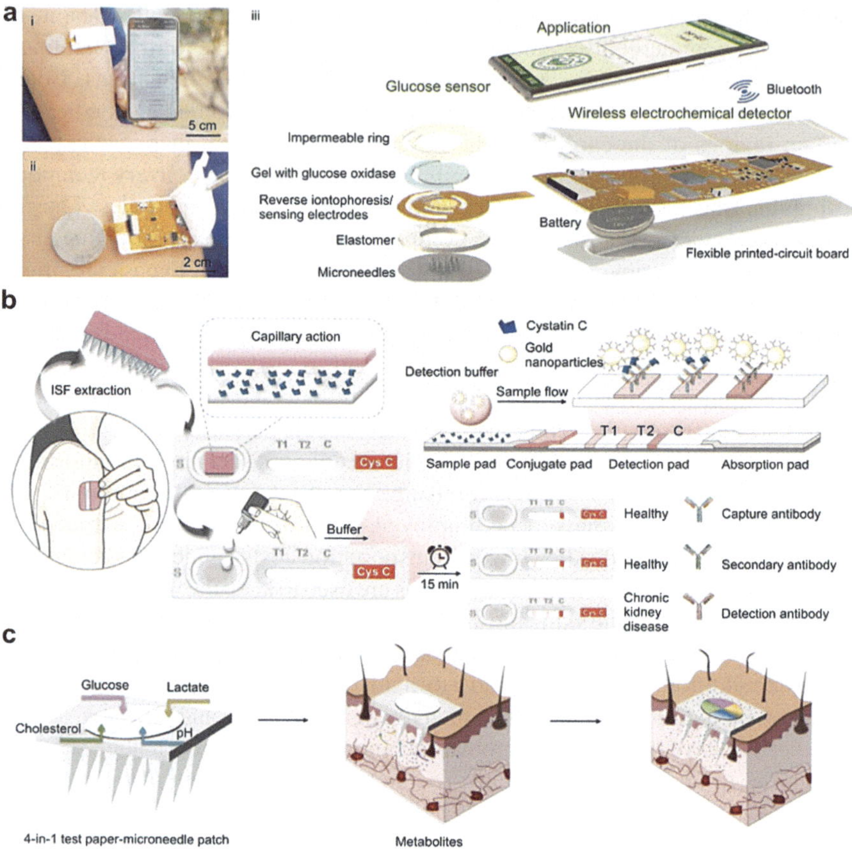

Fig. 12.6 Microneedle-ISF home-based applications. Figure was reproduced with permission [2]

across the stratum corneum, but they do so via different transport processes than those found in living skin tissue. As a result, the primary focus of this study is on live skin tissues, and trans-epidermal water loss is taken into account when analyzing the water flux across the stratum corneum. The details of the mathematical model are well explained in [86].

Figure 12.8 Shows the illustration for a microneedle that can fit with suggested mathematical models. Figure 12.8a refers to an example of a microneedle array on the supporting patch. Figure 12.8b shows the aggregation of representative elementary volume (REVs), and Fig. 12.8c is referred to the simulation for microneedles. The microneedle and skin tissue are marked in dark green and grey, respectively. Meanwhile, the values r and l are the radius and length of the microneedles. Given that microneedles are evenly distributed in the patch and have the same distance between each other, a REV can be applied to represent the delivery outcome. The

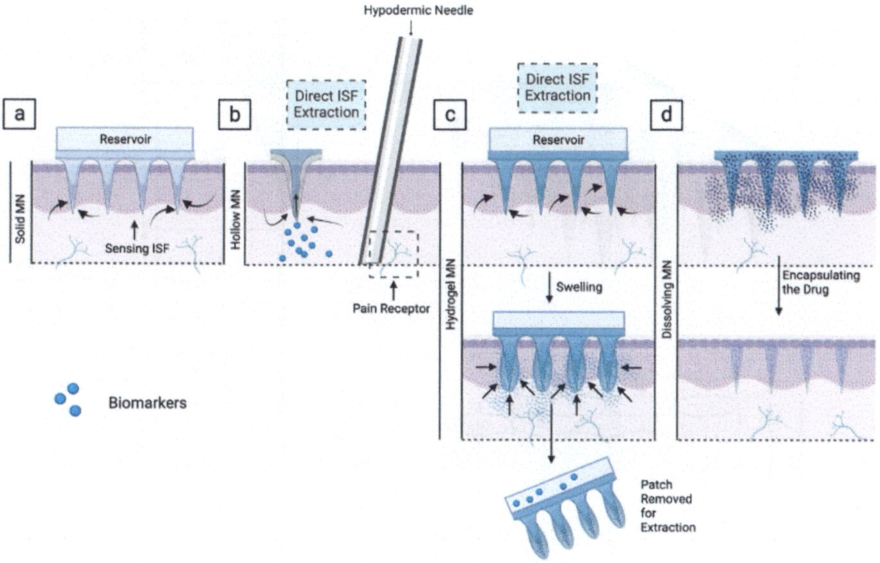

Fig. 12.7 Analysis flow of microneedle-ISF. Figure was reproduced with permission [85]

REV is axially symmetric with a radius, s equal to half the distance between two adjacent microneedles [86].

12.4 Microneedle-ISF Suggestion for SLE

The authors suggested the study by referring to [87] with little modification (Fig. 12.9). The drugs suggested are to prevent autoimmune systems that attack the main organs like the heart, kidneys, lungs, and so on, which produce immunity materials. The biomarker used can be detected by a small chemical sensor in microfluidic devices [88] and can be self-monitored. However, the detailed design is still an ongoing study.

12.5 Drug Monitoring

The use of non-invasive drug monitoring for recreational and medicinal substances, particularly addictive illicit drugs, has been studied in great detail. When it comes to prescription medications like levodopa, vancomycin, theophylline, and others, different people can require different amounts in order to get the best possible therapeutic results. In addition, there is a serious risk to social order and public health when recreational medications like morphine and opioids are abused [89]. Because

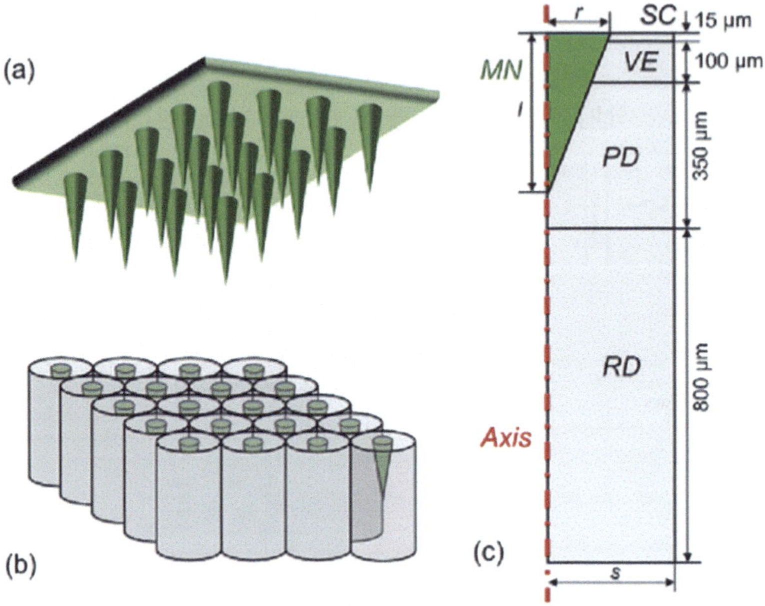

Fig. 12.8 Microneedle mechanism illustration for mathematical models' application. Figure was reproduced with permission [86]

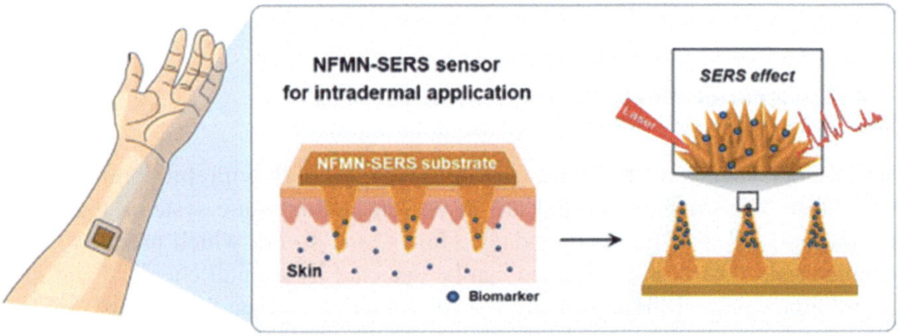

Fig. 12.9 Microneedle-ISF illustration for SLE patient. The figure was reproduced with permission [87]

of this, a great deal of work has gone into creating technology for real-time drug monitoring, which covers anything from adjusting drug dosages to meet specific needs to improving therapeutic efficacy and preventing drug misuse. In therapeutic drug monitoring, ISF-based sensors have the advantage of only detecting bioavailable drugs that are not bound to proteins, which gives a more accurate reflection of the active drug level. This is in contrast to blood-based methods, which typically detect the concentration of drugs that are bound to proteins [90].

12.6 Safety, Security, Data Privacy, Real-Time Monitoring

In the absence of a comprehensive federal law, states are racing to pass consumer data privacy rules that will give people greater control over how businesses collect and use their personal information. There are currently 12 states with complete data privacy laws in place: California, Virginia, Delaware, Connecticut, Colorado, Utah, Iowa, Indiana, Tennessee, Oregon, Montana, and Texas. In the 2022–2023 legislative year, privacy legislation was introduced in at least 16 states. These laws covered a wide range of topics, such as safeguarding health information and biometric identifiers. Companies with multi-state operations may face compliance and liability concerns as a result of this patchwork approach to privacy legislation [91, 92] (Table 12.2).

Cybersecurity for medical devices equals patient safety. Cybersecurity incidents have the potential to influence patients by affecting the functionality and availability of medical devices. To guarantee that everyone has access to safe and reliable medical devices, cybersecurity risks must be managed. Through its medical device cybersecurity team, the Center for Devices and Radiological Health (CDRH) of the U.S. Food and Drug Administration (FDA) supports this important focal area [91].

Table 12.2 Law status related to data privacy. The table was adapted from [90]

Law	Enacted date	Effective date
California Consumer Privacy Act (CCPA), as amended by the California Privacy Rights Act (CPRA)	CCPA: June 28, 2018 CPRA: November 3, 2020	CCPA: January 1, 2020 CPRA: January 1, 2023
Virginia Consumer Data Protection Act (VCDPA)	March 2, 2021	January 1, 2023
Colorado Privacy Act (CPA)	July 7, 2021	July 1, 2023
Connecticut Data Privacy Act (CTDPA)	May 10, 2022	July 1, 2023
Utah Consumer Privacy Act (UCPA)	March 24, 2022	December 31, 2023
Florida Digital Bill of Rights (FDBR)	June 6, 2023	July 1, 2024
Texas Data Privacy and Security Act (TDPSA)	June 18, 2023	July 1, 2024
Oregon Consumer Privacy Act (OCPA)	July 18, 2023	July 1, 2024
Montana Consumer Data Privacy Act (MTCDPA)	May 19, 2023	October 1, 2024
Iowa Consumer Data Protection Act (ICDPA)	March 29, 2023	January 1, 2025
Delaware Personal Data Privacy Act (DPDPA)	September 11, 2023	January 1, 2025
New Jersey Data Privacy Act (NJDPA)	January 16, 2024	January 15, 2025
Tennessee Information Protection Act (TIPA)	May 11, 2023	July 1, 2025
Indiana Consumer Data Protection Act (INCDPA)	May 1, 2023	January 1, 2026

Physiologic monitoring (RPM) refers to the monitoring of physiologic data such as weight, blood glucose, or blood pressure, while remote therapeutic monitoring (RTM) refers to the monitoring of patients' self-reported nonphysiologic data, such as pain levels or medication adherence. Currently, the Centers for Medicare & Medicaid Services (CMS) limits RTM reimbursement to cases involving the respiratory system, musculoskeletal system, and cognitive behavioural therapy. Medicare participants in 2021 accounted for 594 monthly claims, a very modest percentage of patients utilizing RPM; yet, between 2018 and 2021, RPM use grew among beneficiaries by more than six times. This rise can be partially attributed to the extended coverage rules implemented by CMS during the COVID-19 pandemic. As of March 2023, thirty-four state Medicaid plans covered RPM services, yet a number of Medicaid programs have limitations on the usage of RPMs. Since its inception in 2022, RTM uptake has also gradually increased; however, the requirements for billing and paperwork may prevent its general use [93].

12.7 Conclusion

The use of microneedles has gained a lot of interest and developed quickly since it was initially suggested, particularly as an alternate treatment for SLE. Microneedles are currently used in pharmaceutical analysis and monitoring in addition to cosmetic and delivery therapies, albeit considerable work needs to be done in the latter area. In certain studies, medications or biomarkers were extracted from ISF using microneedles, and advanced analysis was then carried out utilizing a variety of tools and techniques. Consequently, microneedle-sensor in situ extraction and monitoring systems have been developed. In light of the COVID-19 pandemic, for instance, the integration of sensors-equipped microneedles, smartphones, and lateral flow boxes facilitates self-testing by patients with chronic illnesses, optimizes therapy by monitoring results, and meets people's heightened expectations for improved health. However, for SLE in particular, over the past 20 years, there has been a noticeable improvement in the care of lupus and its long-term effects; nonetheless, many uncertainties remain about this illness. Targeted therapy development will require further randomized control trials in addition to the ongoing studies. Chronic illness management, particularly lupus management, calls for exceptional adherence and dedication. Poor results, relapses, and resistance instances are typically caused by noncompliance with medication regimens, neglect of patients, and failure to monitor and reevaluate disease activity. In order to control weight and lipid levels and prevent cardiovascular complications, patients should be counselled about the disease pathology, signs and symptoms, the need for regular monitoring, medication compliance, and preventive measures like dietary changes, regular exercise, and lifestyle modifications. Since it was first proposed in our study, studies on microneedle design and detail plan are still ongoing.

However, subsequent research endeavours may concentrate on enduring, dependable, and precise detection, particularly concerning macromolecules like proteins, cortisol, and cytokines. Personalized diagnosis and therapeutics could be revolutionized by advancements in ISF-based sensing, especially in immune system-related disorders. Furthermore, better physically and biochemically resilient sensing membranes that can resist both human activity and complicated biological settings are required. It is crucial to develop functional membranes that are strain-insensitive and anti-biofouling. Finally, to increase confidence in the use of ISF for biosensing applications, more in vivo research comparing sensor-detected ISF concentrations to blood levels is required.

References

1. Gulati, P., Pannu, S., Kumar, M., Bhatia, A., Mandal, U.K., Chopra, S.: Microneedles based drug delivery systems: an updated review. Int. J. Health Sci. **6**(7), 209–242 (2022). https://doi.org/10.53730/ijhs.v6n7.10813
2. Ma, S., Li, J., Pei, L., Feng, N., Zhang, Y.: Microneedle-based interstitial fluid extraction for drug analysis: advances, challenges, and prospects. J. Pharm. Anal. **13**, 111–126 (2023). https://doi.org/10.1016/j.jpha.2022.12.004
3. Rebrin, K., Steil, G.M.: Can interstitial glucose assessment replace blood glucose measurements? Diabetes Technol. Ther. **2**, 461–472 (2000)
4. Altendorfer-Kroath, T., Schimek, D., Eberl, A., et al.: Comparison of cerebral open flow microperfusion and microdialysis when sampling small lipophilic and small hydrophilic substances. J. Neurosci. Methods **311**, 394–401 (2019)
5. Ulrich, J.D., Burchett, J.M., Restivo, J.L., et al.: In vivo measurement of apolipoprotein E from the brain interstitial fluid using microdialysis. Mol. Neurodegener. **8**, 13 (2013)
6. Ameer, M., Chaudhry, H., Mushtaq, J., et al.: An overview of systemic lupus erythematosus (SLE) pathogenesis, classification, and management. Cureus **14**(10), e30330 (2022). https://doi.org/10.7759/cureus.30330
7. Nie, M., Chen, G., Zhao, C., Gan, J., Alip, M., Zhao, Y., Sun, L.: Bio-inspired adhesive porous particles with human MSCs encapsulation for systemic lupus erythematosus treatment. Bioact. Mater. **6**, 84–90 (2021). https://doi.org/10.1016/j.bioactmat.2020.07.018
8. Frangou, E., Vassilopoulos, D., Boletis, J., Boumpas, D.T.: An emerging role of neutrophils and NETosis in chronic inflammation and fibrosis in systemic lupus erythematosus (SLE) and ANCA-associated vasculitides (AAV): implications for the pathogenesis and treatment. Autoimmun. Rev. **18**(8), 751–760 (2019)
9. Kato, Y., Park, J., Takamatsu, H., Konaka, H., Aoki, W., Aburaya, S., et al.: Apoptosis-derived membrane vesicles drive the cGAS-STING pathway and enhance type I IFN production in systemic lupus erythematosus. Ann. Rheum. Dis. **77**(10), 1507–1515 (2018)
10. Flores-Mendoza, G., Sanson, S.P., Rodriguez-Castro, S., Crispin, J.C., Rosetti, F.: Mechanisms of tissue injury in lupus nephritis. Trends Mol. Med. **24**(4), 364–378 (2018)
11. Carter, E.E., Barr, S.G., Clarke, A.E.: The global burden of SLE: prevalence, health disparities and socioeconomic impact. Nat. Rev. Rheumatol. **12**(10), 605–620 (2016)
12. Anders, H.J., Saxena, R., Zhao, M.H., Parodis, I., Salmon, J.E., Mohan, C.: Lupus nephritis. Nat. Rev. Dis Primers. **6**(1), 7 (2020)
13. Otia, M.: A review of nanoformulations for transdermal delivery challenges and opportunities. J. Emerg. Technol. Innov. Res. **8**(11), A26-a43 (2021)
14. Anis, S.N.S., Muhamad, I.I., Marsin, A.M., Lazim, N.A.M., Kadiman, S., Khudzari, A.Z.M.: Transdermal Drug Delivery Systems: principles, Design, and Applications: Novel Formulations and Future Trends. Academic Press, United Kingdom & United States (2024)
15. Jeong, W.Y., Kwon, M., Choi, H.E., Kim, K.S.: Recent advances in transdermal drug delivery systems: a review. Biomater. Res. **25**, 24 (2021). https://doi.org/10.1186/s40824-021-00226-6

16. Hutton, A.R., McCrudden, M.T., Larrañeta, E., Donnelly, R.F.: Influence of molecular weight on transdermal delivery of model macromolecules using hydrogel-forming microneedles: potential to enhance the administration of novel low molecular weight biotherapeutics. J Mater Chem B. **8**(19), 4202–4209 (2020). https://doi.org/10.1039/D0TB00021C
17. Andrews, S.M., Jeong, E.H., Prausnitz, M.R.: Transdermal delivery of molecules is limited by full epidermis, not just stratum corneum. Pharm. Res. **30**(4), 1099–1109 (2013)
18. Chaulagain, B., Jain, A., Tiwari, A., Verma, A., Jain, S.K.: Passive delivery of protein drugs through transdermal route. Artif. Cells Nanomed. Biotechnol. **46**(1), 472–487 (2018). https://doi.org/10.1080/21691401.2018.1430695
19. Uchechi, O., Ogbonna, J., Attama, A.A.: Nanoparticles for dermal and transdermal drug delivery. In: Application of nanotechnology in drug delivery. Sezer AD: InTech C, pp. 193–235 (2014)
20. Zhou, X., Hao, Y., Yuan, L., Pradhan, S., Shrestha, K., Pradhan, O.: Nanoformulations for transdermal drug delivery: a review. Chin. Chem. Lett. **29**(12), 1713–1724 (2018). https://doi.org/10.1016/j.cclet.2018.10.037
21. Kováčik, A., Kopečná, M., Vávrová, K.: Permeation enhancers in transdermal drug delivery: benefits and limitations. Expert Opin. Drug Deliv. **17**(2), 145–155 (2020). https://doi.org/10.1080/17425247.2020.1713087
22. Pawar, P.M., Solanki, K.P., Mandali, V.A.: Recent advancements in transdermal drug delivery system. Int. J. Pharm. Clin. Res. **10**(3), 65–73 (2018)
23. Mujoriya, R., Dhamande, K.A.: Review on transdermal drug delivery system. Res. J. Sci. Tech. **3**(4), 227–231 (2011)
24. Patel, R., Patel, A., Prajapati, B., Shinde, G., Dharamsi, A.: Transdermal drug delivery systems: a mini review. Int. J. Adv. Res. **6**(5), 891–900 (2018). https://doi.org/10.21474/IJAR01/7109
25. Kakar, S., Singh, R., Rani, P.: A review on transdermal drug delivery. Innoriginal Int. J. Sci. **3**(4), 1–5 (2016)
26. Benson, H.A., Grice, J.E., Mohammed, Y., Namjoshi, S., Roberts, M.S.: Topical and transdermal drug delivery: from simple potions to smart technologies. Curr. Drug Deliv. **16**(5), 444–460 (2019). https://doi.org/10.2174/1567201816666190201143457
27. Stanekzai, A., Sudhakar, C.K., Zhakfar, A.M., Karan, V.S.: Recent approaches in transdermal drug delivery system. Res. J. Pharm. Technol. **12**(9), 4550–4558 (2019). https://doi.org/10.5958/0974-360X.2019.00783.2
28. Agrawal, S., Gandhi, S.N., Gurgar, P., Saraswathy, N.: Microneedles: an advancement to transdermal drug delivery system approach. J. Appl. Pharm. Sci. **10**(3), 149–159 (2020)
29. Zhao, Z., Chen, Y., Shi, Y.: Microneedles: a potential strategy in transdermal delivery and application in the management of psoriasis. RSC Adv. **10**(24), 14040–14049 (2020). https://doi.org/10.1039/D0RA00735H
30. Jung, J.H., Jin, S.G.: Microneedle for transdermal drug delivery: current trends and fabrication. J. Pharm. Investig. (2021). https://doi.org/10.1007/s40005-021-00512-4
31. Shakya, A.K., Ingrole, R.S.J., Joshi, G., Uddin, M.J., Anvari, S., Davis, C.M., et al.: Microneedles coated with peanut allergen enable desensitization of peanut sensitized mice. J. Control. Release **314**, 38–47 (2019). https://doi.org/10.1016/j.jconrel.2019.09.022
32. Lim, J., Tahk, D., Yu, J., Min, D.H., Jeon, N.L.: Design rules for a tunable merged-tip microneedle. Microsyst. Nanoeng. **4**(1), 1–10 (2018)
33. Dardano, P., Caliò, A., Palma, V.D., Bevilacqua, M.F., Matteo, A.D., Stefano, L.D.: A photolithographic approach to polymeric microneedles array fabrication. Materials **8**(12), 8661–8673 (2015). https://doi.org/10.3390/ma8125484
34. Han, D., Morde, R.S., Mariani, S., Mattina, A.A.L., Vignali, E., Yang, C., et al.: 4D printing of a bioinspired microneedle array with backward-facing barbs for enhanced tissue adhesion. Adv. Funct. Mater. **30**(11), 1909197 (2020). https://doi.org/10.1002/adfm.201909197
35. Xiao, Z., Li, X., Zhang, P., Wang, Y.: Research of polymeric microneedles for transdermal drug delivery. Prog. Chem. **29**(12), 1518 (2017)
36. Balmert, S.C., Carey, C.D., Falo, G.D., Sethi, S.K., Erdos, G., Korkmaz, E., et al.: Dissolving undercut microneedle arrays for multicomponent cutaneous vaccination. J. Control. Release **317**, 336–346 (2020). https://doi.org/10.1016/j.jconrel.2019.11.023

37. Kim, H.M., Lim, Y.Y., An, J.H., Kim, M.N., Kim, B.J.: Transdermal drug delivery using disk microneedle rollers in a hairless rat model. Int. J. Dermatol. **51**(7), 859–863 (2012). https://doi.org/10.1111/j.1365-4632.2011.05343.x
38. Li, K., Yoo, K.H., Byun, H.J., Lim, Y.Y., Kim, M.N., Hong, H.K., et al.: The microneedle roller is an effective device for enhancing transdermal drug delivery. Int. J. Dermatol. **51**(9), 1137–1139 (2012). https://doi.org/10.1111/j.1365-4632.2010.04703.x
39. Bariya, S.H., Gohel, M.C., Mehta, T.A., Sharma, O.P.: Microneedles: an emerging transdermal drug delivery system. J. Pharm. Pharmacol. **64**(1), 11–29 (2012)
40. Cheung, K., Das, D.B.: Microneedles for drug delivery: trends and progress. Drug Deliv. **23**, 2338–2354 (2016)
41. Lee, Y.S., Kang, T.G., Cho, H.R., Lee, G.J., Park, O.K., Kim, S.Y., et al.: Localized delivery of Theranostic nanoparticles and high-energy photons using microneedles on-bioelectronics. Adv. Mater. **33**(24), 2100425 (2021). https://doi.org/10.1002/adma.202100425
42. Du, H., Liu, P., Zhu, J., Li, Y., Zhang, L., Zzhu, J., et al.: Hyaluronic acid-based dissolving microneedle patch loaded with methotrexate for improved treatment of psoriasis. ACS Appl. Mater. Interfaces **11**(46), 43588–43598 (2019). https://doi.org/10.1021/acsami.9b15668
43. Sato, T., Okada, S., Hagino, K., et al.: Measurement of glucose area under the curve using minimally invasive interstitial fluid extraction technology: evaluation of glucose monitoring concepts without blood sampling. Diabetes Technol. Ther. **13**, 1194–1200 (2011)
44. Jakka, D., Matadh, A.V., Shivakumar, H.N., et al.: Polymer Coated Polymeric (PCP) microneedles for sampling of drugs and biomarkers from tissues. Eur. J. Pharmaceut. Sci. **175**, 106203 (2022)
45. Liu, F., Lin, Z., Jin, Q., et al.: Protection of nanostructures-integrated microneedle biosensor using dissolvable polymer coating. ACS Appl. Mater. Interfaces **11**, 4809–4819 (2019)
46. Miller, P.R., Taylor, R.M., Tran, B.Q., et al.: Extraction and biomolecular analysis of dermal interstitial fluid collected with hollow microneedles. Commun. Biol. **1**, 173 (2018)
47. Ito, Y., Taniguchi, M., Hayashi, A., et al.: Application of dissolving microneedles to glucose monitoring through dermal interstitial fluid. Biol. Pharm. Bull. **37**, 1776–1781 (2014)
48. Ito, Y., Inagaki, Y., Kobuchi, S., et al.: Therapeutic drug monitoring of vancomycin in dermal interstitial fluid using dissolving microneedles. Int. J. Med. Sci. **13**, 271–276 (2016)
49. He, R., Niu, Y., Li, Z., et al.: A hydrogel microneedle patch for point-of-care testing based on skin interstitial fluid. Adv. Healthc. Mater. **9**, e1901201 (2020)
50. Gershlick, A.H., Syndercombe Court, Y.D., Murday, A.J., et al.: Adverse effects of high dose aspirin on platelet adhesion to experimental autogenous vein grafts. Cardiovasc. Res. **19**, 770–776 (1985)
51. Carlsson, H., Hjorton, K., Abujrais, S., et al.: Measurement of hydroxychloroquine in blood from SLE patients using LC-HRMS-evaluation of whole blood, plasma, and serum as sample matrices. Arthritis Res. Ther. **22** 125 (2020)
52. Siddiqi, A., Khan, D.A., Khan, F.A., et al.: Therapeutic drug monitoring of amikacin in preterm and term infants. Singapore Med. J. **50**, 486–489 (2009)
53. Domes, C., Domes, R., Popp, J., et al.: Ultrasensitive detection of antiseptic antibiotics in aqueous media and human urine using deep UV resonance Raman spectroscopy. Anal. Chem. **89**, 9997–10003 (2017)
54. Franco, V., Gatti, G., Mazzucchelli, I., et al.: Relationship between saliva and plasma rufinamide concentrations in patients with epilepsy. Epilepsia **61**, e79–e84 (2020)
55. Nakajima, M., Sato, S., Yamato, S., et al.: Assessment of tear concentrations on therapeutic drug monitoring. III. Determination of theophylline in tears by gas chromatography/mass spectrometry with electron ionization mode. Drug Metabol. Pharmacokinet. **18**, 139–145 (2003)
56. Wen, J., Chen, X., Yang, Y., et al.: Acupuncture medical therapy and its underlying mechanisms: a systematic review. Am. J. Chin. Med. **49**, 1–23 (2021)
57. Wang, X., Han, Y., Jin, J., et al.: Plum-blossom needle assisted photodynamic therapy for the treatment of oral potentially malignant disorder in the elderly. Photodiagnosis Photodyn. Ther. **25**, 296–299 (2019)

58. Wang, Z., Luan, J., Seth, A., et al.: Microneedle patch for the ultrasensitive quantification of protein biomarkers in interstitial fluid. Nat. Biomed. Eng. **5**, 64–76 (2021)
59. Wang, P.M., Cornwell, M., Prausnitz, M.R.: Minimally invasive extraction of dermal interstitial fluid for glucose monitoring using microneedles. Diabetes Technol. Therapeut. **7**, 131–141 (2005)
60. Bao, L., Park, J., Qin, B., et al.: Anti-SARS-CoV-2 IgM/IgG antibodies detection using a patch sensor containing porous microneedles and a paper-based immunoassay. Sci. Rep. **12**, 10693 (2022)
61. Zheng, M., Wang, Z., Chang, H., et al.: Osmosis-powered hydrogel microneedles for microliters of skin interstitial fluid extraction within minutes. Adv. Healthc. Mater. **9**, e1901683 (2020)
62. Sabri, A.B.Q., Anjani, K., Donnelly, R.F.: Synthesis and characterization of sorbitol laced hydrogel-forming microneedles for therapeutic drug monitoring. Int. J. Pharm. **607**, 121049 (2021)
63. Samant, P.P., Niedzwiecki, M.M., Raviele, N., et al.: Sampling interstitial fluid from human skin using a microneedle patch. Sci. Transl. Med. **12**, eaaw0285 (2020)
64. Chang, H., Zheng, M., Yu, X., et al.: A swellable microneedle patch to rapidly extract skin interstitial fluid for timely metabolic analysis. Adv. Mater. **29**, 1702243 (2017)
65. Zhu, J., Zhou, X., Kim, H.J. et al., Gelatin methacryloyl microneedle patches for minimally invasive extraction of skin interstitial fluid. Small **16**, e1905910 (2020)
66. Chen, J., Wang, M., Ye, Y., et al.: Fabrication of sponge-forming microneedle patch for rapidly sampling interstitial fluid for analysis. Biomed. Microdevices **21**, 63 (2019)
67. Fonseca, D.F.S., Costa, P.C., Almeida, I.F., et al.: Swellable gelatin methacryloyl microneedles for extraction of interstitial skin fluid toward minimally invasive monitoring of urea. Macromol. Biosci. **20**, e2000195 (2020)
68. Qiao, Y., Du, J., Ge, R., et al.: A sample and detection microneedle patch for psoriasis MicroRNA biomarker analysis in interstitial fluid. Anal. Chem. **94**, 5538–5545 (2022)
69. Cai, Y., Huang, S., Zhang, Z., et al.: Bioinspired rotation microneedles for accurate transdermal positioning and ultraminimal-invasive biomarker detection with mechanical robustness. Research (Wash. D C) **2022**, 9869734 (2022)
70. Kolluru, C., Williams, M., Yeh, J.S., et al.: Monitoring drug pharmacokinetics and immunologic biomarkers in dermal interstitial fluid using a microneedle patch. Biomed. Microdevices **21**, 14 (2019)
71. Kolluru, C., Williams, M., Chae, J., et al.: Recruitment and collection of dermal interstitial fluid using a microneedle patch. Adv. Healthc. Mater. **8**, e1801262 (2019)
72. Tran, B.Q., Miller, P.R., Taylor, R.M., et al.: Proteomic characterization of dermal interstitial fluid extracted using a novel microneedle-assisted technique. J. Proteome Res. **17**, 479–485 (2018)
73. Iqbal, T., Elahi, A., Wijns, W., Shahzad, A. Cortisol detection methods for stress monitoring in connected health. Health Sci. Rev. **6**, 100079 (2023). https://doi.org/10.1016/j.hsr.2023.100079
74. Harley, I.T.W., Sawalha, A.H.: Systemic lupus erythematosus as a genetic disease. Clin. Immunol. **236**, 108953 (2022). https://doi.org/10.1016/j.clim.2022.108953
75. Echevarria, S., Johnson, E., Dhanraj, O., Provoast, K., Rizvi, S.H., Baksh, R.J., et al.: A case report on systemic lupus erythematosus-associated antiphospholipid syndrome and hyperhomocysteinemia. Int. J. Clin. Stud. Med. Case Rep. **40**(5) (2024). https://doi.org/10.46998/IJCMCR.2024.40.000997
76. Justiz-Vaillant, A.A., Gopaul, D., Soodeen, S., ArozarenaFundora, R., Barbosa, O.A., Unakal, C., et al.: Neuropsychiatric systemic lupus erythematosus: molecules involved in its imunopathogenesis. Clinical features, and treatment. Molecules **29**, 747 (2024). https://doi.org/10.3390/molecules29040747
77. Emerson, J.S., Gruenewald, S.M., Gomes, L., Lin, M.W., Swaminathan, S.: The conundrum of neuropsychiatric systemic lupus erythematosus: current and novel approaches to diagnosis. Front. Neurol. **14**, 1111769 (2023). https://doi.org/10.3389/fneur.2023.1111769
78. Askanase, A.D., Dall'Era, M., Almaani, S.: Insights into future management of lupus nephritis. Front. Lupus. **2**, 1334932 (2024). https://doi.org/10.3389/flupu.2024.1334932

79. Aringer, M., Costenbader, K., Daikh, D., et al.: European League Against Rheumatism/American College of Rheumatology classification criteria for systemic lupus erythematosus. Arthritis Rheumatol. **71**, 1400–1412 (2019). https://doi.org/10.1002/art.40930
80. Stratta, P., Mesiano, P., Campo, A., et al.: Life expectancy of women with lupus nephritis now approaches that of the general population. Int. J. Immunopathol. Pharmacol. **22**, 1135–1141 (2009). https://doi.org/10.1177/039463200902200432
81. Yohannan, B., Montanez, N.A., Escobar, M.A.: Simultaneous hemorrhage and venous thrombosis in a patient with systemic lupus erythematosus. Clevel. Clin. J. Med. **91**(5), 308 (2024). https://doi.org/10.3949/ccjm.91a.23059
82. Fanouriakis, A., Kostopoulou, M., Alunno, A., et al.: 2019 update of the EULAR recommendations for the management of systemic lupus erythematosus. Ann. Rheum. Dis. **78**, 736–745 (2019). https://doi.org/10.1136/annrheumdis2019-215089
83. Gladman, D.D., Ibañez, D., Urowitz, M.B.: Systemic lupus erythematosus disease activity index 2000. J. Rheumatol. **29**, 288–291 (2002)
84. Arora, S., Isenberg, D.A., Castrejon, I.: Measures of adult systemic lupus erythematosus: disease activity and damage. Arthritis Care Res. (Hoboken) **72**(Suppl 10), 27–46 (2020). https://doi.org/10.1002/acr.24221
85. Saifullah, K.M., Rad, Z.F.: Sampling dermal interstitial fluid using microneedles: a review of recent developments in sampling methods and microneedle-based biosensors. Adv. Mater. Interfaces **10**, 2201763 (2023)
86. Newell, B., Zhan, W.: Mathematical modelling of microneedle-mediated transdermal delivery of drug nanocarriers into skin tissue and circulatory system. J. Control. Release **360**, 447–467 (2023). https://doi.org/10.1016/j.jconrel.2023.07.011
87. Lu, H., Zada, S., Yang, L., Dong, H.: Microneedle-based device for biological analysis. Front. Bioeng. Biotechnol. **10**, 851134 (2022). https://doi.org/10.3389/fbioe.2022.851134
88. Saha, T., Mukherjee, S., Dickey, M.D., Velev, O.D.: Harvesting and manipulating sweat and interstitial fluid in microfluidic devices. Lab Chip **24**, 1244 (2024)
89. Minnesota Department of Employment & Economic Development (DEED) and Lathrop GPM. A Legal Guide to privacy and data security (2024)
90. Gibson Dunn.: U.S. Cybersecurity and Data Privacy Review and Outlook 2024. gibsondunn.com (2024)
91. FDA.: Medical Device Cybersecurity HELPING TO KEEP PATIENTS AND MEDICAL DEVICES SAFE (2024)
92. Harris, J., Curtis, M., Sandalow, M., Hoagland, G.W., Serafini, M.: The future of remote patient monitoring (2024)
93. Wu, Z., Qiao, Z., Chen, S., Fan, S., Liu, Y., Qi1, J., Lim, C.T.: Interstitial fluid-based wearable biosensors for minimally invasive healthcare and biomedical applications. Commun. Mater. **5**, 33 (2024)

Chapter 13
Microneedle-Based Delivery of Metal Oxide Nanoparticles

Jaison Jeevanandam, Tracy Ann Bruce-Tagoe, and Michael K. Danquah

13.1 Introduction

Microneedles have recently captivated researchers as a potential transdermal drug delivery system that utilizes micrometer-sized needles for the formation of microscopic pores in the skin [1]. These microneedles are minimally invasive, self-administrable and a painless approach for the delivery of drugs and other compounds via dermal administration [2]. They are synthesized through several methods and are classified based on distinct materials, such as sodium hyaluronate, polyvinyl pyrrolidone (PVP), polyvinyl pyrrolidone/vinyl acetate (PVPVAc), laser ablation, polydimethylsiloxane (PDMS) and micro-molding using ceramics. Sodium hyaluronate is typically dissolved in deionized water to prepare blank microneedles, after which the solution is injected into a mold, subjected to vacuum treatment to eliminate bubbles, and subsequently dried before being removed from the mold [3]. Microneedles were similarly fabricated using PVP, dissolved in a solution and processed through a two-step centrifugation and molding technique [4]. In another approach, laser ablation was employed for the creation of microneedle molds [5], whereas PDMS was used as a coating material, combined with an elastomer base and curing agent, to develop particle-attached microneedles [6]. Additionally, ceramic microneedles were produced using dual-replicated PDMS molds and ceramic micromolding technology. [7]. Microneedles offer several advantages, including reduced degradation in the gastrointestinal tract [8], avoidance of first-pass metabolism in the

J. Jeevanandam
Department of Experimental Neurobiology, Preclinical Research Program, National Institute of Mental Health, Topolová 748, 250 67 Klecany, Czechia

T. A. Bruce-Tagoe · M. K. Danquah (✉)
Department of Chemical and Biomolecular Engineering, University of Tennessee, Knoxville, TN, USA
e-mail: mdanquah@utk.edu

N. Talreja et al. (eds.), *Microneedles (MNs)-Based Technology*, Series in BioEngineering, https://doi.org/10.1007/978-981-96-3916-8_13

liver during oral delivery [9], painless intravenous injections [10], improved patient compliance [11], low microbial invasion risk [12] and reduced hazardous waste [13].

Metal oxide nanoparticles (NPs) are a class of emerging nanomaterials, which have unique physico-chemical characteristics and high stability, compared to metals and other NPs [14]. These NPs are synthesized via conventional physical [15], chemical [16] and biological approaches [17]. Physical and chemical methods can yield highly stable NPs [18], however they can lead to toxicity issues [19], whereas biological methods yield less/nontoxic NPs with poor stability [20]. These nanomaterials are highly beneficial as diagnostic and therapeutic agents for treating a range of diseases, such as cardiovascular ailments [21], diabetes [22] and cancer [23]. Although metal oxide NPs offer significant benefits in biomedical applications, proper formulation is essential to address their limitations [24]. Various carriers, including dendrimers [25], micelles [26], liposomes [27], and bio-polymers are widely utilized for the formulation of metal oxide NPs [28]. These formulations, while advantageous, are predominantly utilized for nasal, intravenous, and oral administration of metal oxide NPs in medical applications [29]. Recently, microneedles have been introduced as an innovative approach for formulating metal oxide NPs for dermal administration [30]. This chapter presents a comprehensive summary of various standalone microneedles and those formulated with metal oxide NPs for drug delivery applications. Additionally, it explores the antimicrobial, anticancer, and targeted bioimaging applications of microneedle-formulated metal oxide NPs.

13.2 Microneedles in Drug Delivery Applications

In recent times, microneedles are widely used in various drug delivery applications. Teo et al. [31] assessed the effectiveness of microneedles in transdermal drug delivery to maintain prolonged drug release and attain optimal blood concentrations. Microneedles are smaller in size which can eventually help in penetrating the protective, 10–20 μm thick stratum corneum barrier in the transdermal region. Silicon, metal, silicon dioxide, glass, polymer, and titanium-based microneedles were identified in this study as potential agents for transdermal drug delivery [31]. Martin et al. [32] fabricated biodegradable sugar glass microneedles under low temperature conditions for specific transdermal drug delivery applications. The study focused on the preparation of solid amorphous sugar glasses with low water residual quantities via trehalose dehydration and combinatorial solution of sucrose sugar. Later, biodegradable sugar glass microneedles were optimized to operate at low temperature via vacuum deposition micro-moulding methodology. The results showed that the microneedles possess absolute morphological fidelity of structures towards silicon master and displayed adequate rigidity to successfully penetrate human breast skin [32]. Bodhale et al. [33] developed an out-of-plane hollow polymeric microneedle featuring a cylindrical side-open hollow design for transdermal drug delivery. The study demonstrated that these microneedles, under specific boundary conditions, exhibited enhanced strength and flow rates at applied inlet pressures, making them advantageous for micro-electromechanical system (MEMS) applications [33]. Furthermore, Tsioris et al. [34] fabricated a microneedle system using silk fibroin, a biomaterial known for its enhanced biodegradability, excellent mechanical properties, biocompatibility, ease of processing, and ability to maintain and preserve biological activity within its material matrix. This study demonstrated the fabrication of high-aspect-ratio silk microneedles with enhanced mechanical properties at room temperature, utilizing micromoulding in an aqueous medium for the bulk loading of sensitive drugs. The rate of drug release was reduced by approximately 5.6-fold through adjustments in the microneedles' post-processing conditions, particularly through modulation of the secondary structure of the silk protein [34].

Recently, Aldawood et al. [35] developed six microneedle designs via stereolithography-mediated three-dimensional (3D) printing approach with three aspect ratios along with an array of conical-shaped 100 needles (10 × 10 needle/array). The study emphasized that microneedles possess higher aspect ratio with deformation properties, that are suitable for deeper penetration beyond stratum corneum in the transdermal region. The results revealed that the 0.3 mm and 0.4 mm diameters of microneedles can exhibit consistent behavior of force-displacement during skin-equivalent penetration analysis [35]. Nguyen et al. [36] prepared a innovative polyvinyl alcohol (PVA) microneedles using a micromoulding approach to encapsulate the anticancer agent doxorubicin for transdermal delivery. The efficacy of these microneedles in creating micropores in dermatomed human cadaver skin was assessed through histology, dye binding, confocal laser microscopy, pore uniformity, and skin integrity studies. The results indicated that the microneedles were strong, uniform, and sharp, effectively treating skin at a rate of 4351.55 ± 560.87 ng/cm^2, with enhanced drug permeability [36]. Zoudani and Soltani [37] recently introduced an innovative computational modeling approach to study the dissolution dynamics of microneedles in drug delivery applications. The model demonstrated that an initial increase in the drug load within the microneedle accelerates dissolution, while variations in pitch size affect drug concentration without altering the dissolution rate. According to the computational predictions, a microneedle with an array-in-array arrangement can deliver a drug concentration of 4.85×10^{-3} kg/m^3 to the tissue, achieving complete dissolution within 8 s [37]. Thakur Singh et al. [38] summarized various minimally invasive microneedles for the delivery of ocular drugs. These microneedles were determined to be highly advantageous for ocular drug delivery, offering reduced trauma to ocular tissues, lower drug dosages, and precise localization of medication. Additionally, they facilitate targeted injections within thin ocular tissues, such as the cornea and sclera, enabling effective localized drug delivery applications [38]. Similarly, Bodhale et al. [39] demonstrated the fabrication of hollow cylindrical silicon microneedle array with tapered tip for easy insertion into the skin. Initially, silicon wafers were synthesized with deep reactive ion etching and mask layout design, which was later combined with various anisotropic and isotropic etching processes via inductively coupled plasma etching technology. These microneedles were highly beneficial to fabricate drug delivery devices for targeted transdermal delivery applications [39]. Likewise, Sawon and Samad [40] designed and optimized microneedles with skin insertion benefits for transdermal drug delivery applications. In this study, single, out-of-plane silicon carbide microneedle was fabricated with six lumens for delivering drugs. The microneedle was identified to possess a total length of 350 μm with 52 μm of outer diameter and 30 μm of inner diameter. The resultant microneedle tip was identified to reduce pain during insertion process of ibuprofen drug with low velocity distribution, high dynamic viscosity, 0.01 and 0.9 μL/min of maximum flow rate as well as 10–100 kPa of inlet pressure [40]. Additionally, Larraneta et al. [41] designed an innovative hydrogel-based microneedle array by a microwave-assisted approach for effective transdermal drug delivery applications as shown in Fig. 13.1. In this study, polymeric microneedle arrays were fabricated from poly(methyl vinyl ether-alt-maleic acid) crosslinked with poly(ethylene glycol) at 80 °C for 24 h. The results revealed that the microwave-assisted preparation method produced microneedle arrays with properties comparable to those of conventional microneedles, while being 30 times faster than the traditional synthesis approach. These microneedles were shown to effectively facilitate in vitro caffeine diffusion through excised porcine skin, delivering 3000–3500 μg of caffeine within 24 h [41]. Novel maltose sugar microneedles were also fabricated by Nguyen and Banga [42], for the transdermal delivery of doxorubicin anticancer drug. These microneedles are synthesized via micromoulding technique to achieve effective microporation of human cadaver skin, preserved histological structure, high dye binding, maintained skin integrity and, pore uniformity. The resultant microneedles possess enhanced ability to dissolve rapidly with high sharpness and mechanical uniformity for effective permeation across human skin [42]. Furthermore, Sabri et al. [43] summarized various novel 'poke and patch' solid microneedles, that are prepared using stainless steel and silicon for lymphatic delivery. The author emphasized that the materials, design and geometry of the microneedles critically affect the performance of the microneedles [43].

13.3 Synthesis of Metal Oxide NPs with Microneedles

Recently, metal oxide NPs are formulated with microneedles to enhance their stability and enable targeted, controlled delivery Seeni et al. [44] developed a novel iron oxide nanoparticle-powered microscale optical coherence tomography system for in situ imaging of the swelling and penetration behavior of polymeric microneedles in the skin. This study introduced a real-time, noninvasive imaging approach by incorporating iron oxide NPs into polystyrene-block-poly(acrylic acid)-based microneedles to enhance the image contrast of coherence tomography. The results revealed that microneedles containing ~4–5 wt.% iron oxide NPs improved the contrast-to-noise ratio during the bioimaging process, with effective skin tissue swelling observed within ~30–35 s [44]. Cui et al. [45] designed a novel ferroferric oxide-infused, near-infrared-activated photothermal microneedle patch for regulated drug delivery. The fabrication process involved two steps: back layer preparation and needle formation. Iron oxide NPs were initially mixed with the microneedle material, dissolved in 2 g of deionized water, and stirred for 30 min at 60 °C, followed by an additional

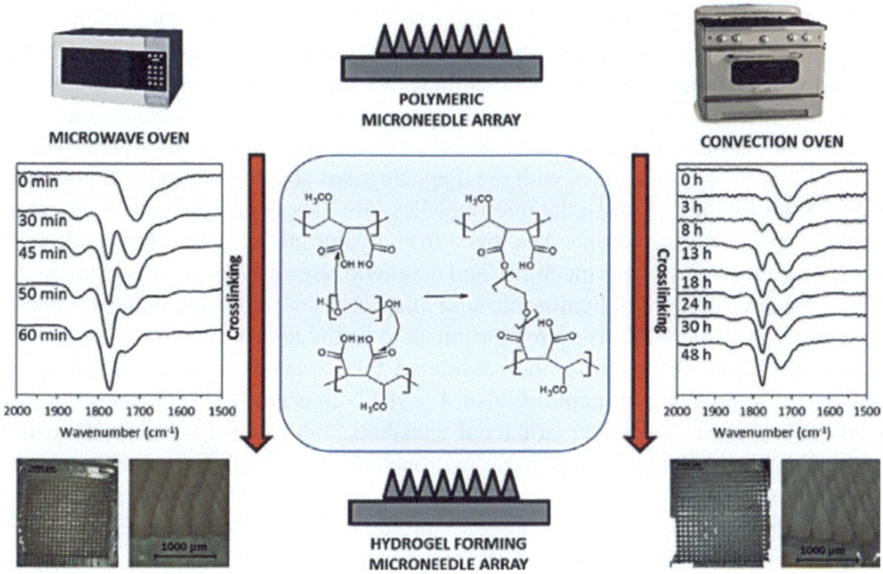

Fig. 13.1 Schematic showing results of microwave-assisted synthesis of polymer microneedle array prepared via crosslinking approach, compared to convection oven-mediated synthesis. Reproduced with permission from Larraneta et al. [41], ©Wiley, 2015

5 min of ultrasonic stirring. Subsequently, 0.5 g of poly(ethylene glycol) diacrylate (PEGDA) monomer was incorporated into the nanoparticle-microneedle mixture and stirred for 10 min at 60 °C. The back layer solution was prepared using Irgacure 2959, with polydimethylsiloxane (PDMS) molds utilized for microneedle formation. The PDMS molds were placed at 20 °C for 2 h, after which the nanoparticle-microneedle patches were peeled off. The study revealed that these nanoparticle-loaded microneedles possessed adequate structural integrity to pierce the skin and achieved a rapid temperature increase of 40 °C within 1 min under near-infrared irradiation. Additionally, the microneedles facilitated the release of approximately 80% of the drug within 20 min, including a ~70% release rate of doxycycline hydrochloride during the same irradiation period [45]. Tort et al. [30] developed novel microneedles coated with micro- or NPs using an electrospray technique for medical skin treatments. In this study, polyethylene oxide (PEO) of high and low molecular weights were dissolved in a water-ethanol mixture to facilitate the electrospraying process. Different dermarollers were employed to create microneedles with heights of either 0.25 mm or 2 mm, depending on the applied voltage during fabrication. The results showed that the microneedles with NPs, that are coated with drug or insulin possess a particle size of 515 ± 286 and 522 ± 261 nm, respectively. These microneedles were identified to possess the ability for transferring >70% of the loaded drug into porcine skins [30]. Haghighi and Haghighatzadeh [46] developed unique actinomorphic zinc oxide microneedles decorated with tin dioxide nanospheres. The fabrication process

involved a hydrothermal method for synthesizing needle-like zinc oxide microstructures, utilizing zinc nitrate tetrahydrate dissolved in hexamethylenetetramine (HMT). Subsequently, a co-precipitation deposition technique was employed to decorate the zinc oxide microneedles with tin dioxide nanospheres, using tin chloride dihydrate to form zinc oxide-tin dioxide heterojunctions, as shown in Fig. 13.2. The results revealed that the lattice fringes with the d-spacing was about 0.17 nm for tin dioxide and 0.28 nm for zinc oxide in the microneedles [46]. Choi et al. [47] also fabricated exclusive magnetized microneedles made from hyaluronic acid and iron oxide NPs for the evaluation of skin permeability and magnetic responsiveness. The study highlighted the development of hyaluronic acid microneedles embedded with superparamagnetic iron oxide NPs, using hyaluronic acid as the matrix material. The results demonstrated that the microneedles exhibited 90% transmittance, a coercivity of 60 Oe, and a magnetic susceptibility of 4×10^{-6} emu/gOe [47]. Furthermore, Li et al. [48] studied the delivery of local anesthetic lidocaine hydrochloride using light-activated, implantable polymeric microneedles for the formation of a patient-operated transdermal pain relief system. In this study, polydopamine was coated over iron oxide NPs, followed by polycaprolactone and polyvinyl pyrrolidone, which was identified to be 9–11 nm in size and spherical in shape. The results showed that the nanoparticle-microneedles possess exceptional penetration capability for painless drug delivery via barrier breaking of the stratum corneum [48].

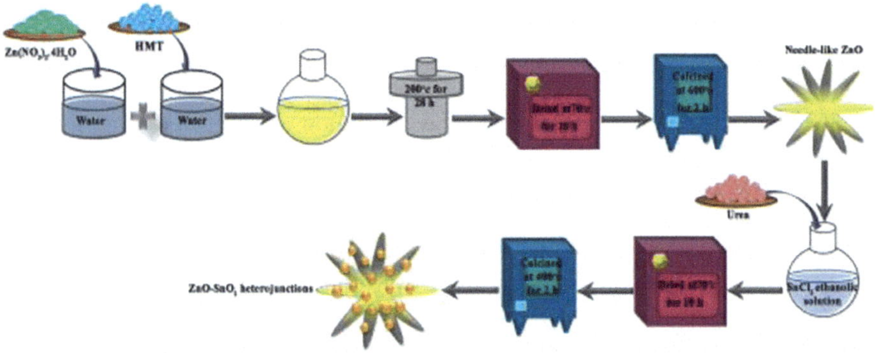

Fig. 13.2 Schematic of actinomorphic zinc oxide microneedle synthesis decorated with tin dioxide nanospheres. Reproduced with permission from Haghighi and Haghighatzadeh [46], ©Springer Nature, 2020

13.4 Drug Delivery Applications of Microneedle-Metal Oxide NPs

In recent times, the formulation of metal oxide NPs into microneedles has shown significant benefits in drug delivery, particularly for applications such as antimicrobial, anticancer, targeted bioimaging, and antidiabetic treatments.

13.4.1 Antimicrobial Application

Li et al. [49] demonstrated the dissolution of microneedles based on zinc oxide NPs and an ionic liquid as synergistic antibacterial agents. The study employed zinc oxide NPs dispersed in imidazolidinyl ionic liquids to fabricate microneedles, resulting in enhanced antibacterial efficacy compared to the individual use of ionic liquids or zinc oxide NPs. The results revealed that these microneedles are highly beneficial as a potential antibacterial agent against gram-positive *Streptococcus aureus* and gram-negative *Escherichia coli* [49]. A photothermal microneedle patch, loaded with hybrid antimicrobial peptide and manganese dioxide NPs was fabricated by Wang et al. [50] for chronic wound healing application as displayed in Fig. 13.3. In this study, a series of double-layered hollow NPs were synthesized using biocompatible methacrylated gelatin through electrostatic assembly and subsequently loaded onto the tips of microneedles. The resultant microneedle-metal oxide nanoparticle formulation was identified to possess photothermal conversion and anti-oxidative nanozyme activity with the ability to inhibit the proliferation of host bacteria. Also, the results revealed that the microneedle formulation possess enhanced wound healing property within 15 days in *S. aureus* infected skin defect model [50]. Recently, Gong et al. [51] developed microneedles loaded with zinc-germanate:copper nanorods for sustained reactive oxygen species (ROS) production to treat wounds infected by methicillin-resistant Staphylococcus aureus (MRSA). In this study, luminescent nanorods were fabricated and incorporated into dissolvable microneedles designed to penetrate biofilms and facilitate minimally invasive healing of wounds caused by multidrug-resistant bacteria. The findings revealed that the microneedle tips dissolved rapidly, releasing nanorods that sustained enhanced ROS production for approximately 48 h, effectively inhibiting bacterial growth while demonstrating biocompatibility and reducing inflammation [51]. Furthermore, Shen et al. [52] demonstrated the synthesis of microneedle patch infused with calcium peroxide NPs for oxygen therapy and biofilm suppression in diabetic wound treatment application. In this study, microneedle-encapsulated calcium dioxide possesses enhanced oxygen release and antimicrobial activity. The results revealed that the microneedle-encapsulated NPs possess the ability to eliminate bacteria, reduce inflammation, promote formation of collagen, stimulate growth of blood vessel, and promote accelerated healing of wounds in diabetic foot ulcer treatment [52].

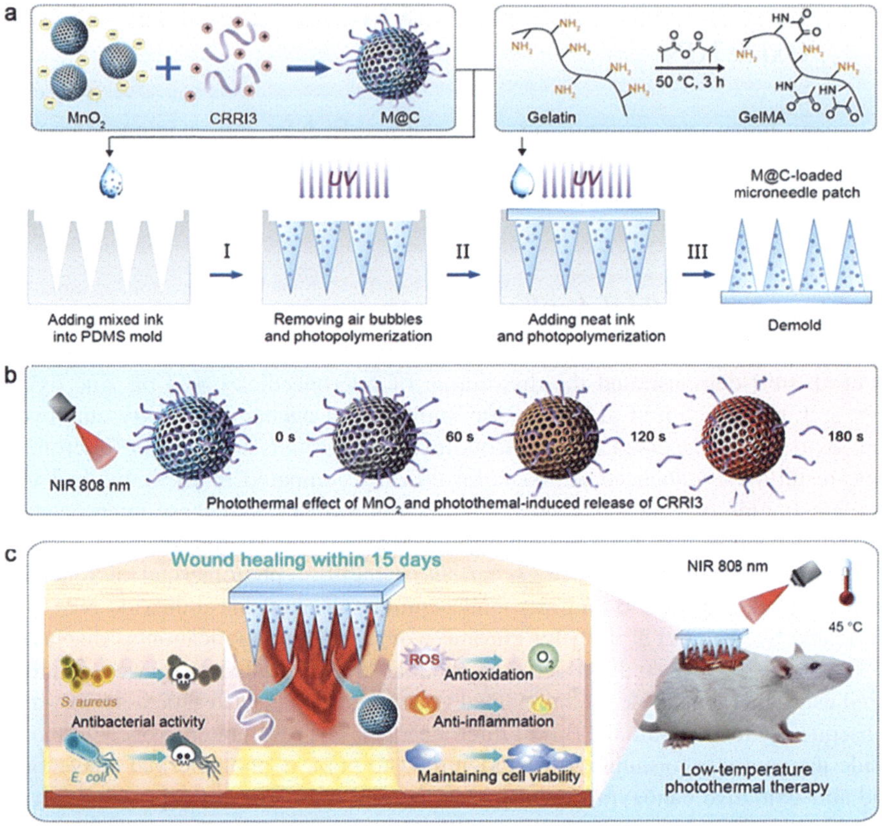

Fig. 13.3 Schematic of double-layered manganese dioxide nanoparticle preparation, **a** hollow manganese dioxide NPs were used for the preparation using UV light-based cross-linkage of hydrogels, **b** photothermal effect of microneedle-NPs at near-infrared region, **c** chronic would healing application of microneedle-nanoparticle with low temperature photothermal therapy. Reproduced with permission from Wang et al. [50], ©Elsevier, 2024

13.4.2 Anticancer Ability

Recent research has highlighted the significant benefits of microneedles in the treatment of melanoma, a type of skin cancer. However, limited studies have specifically demonstrated the effectiveness of microneedle-formulated metal oxide NPs in skin cancer therapy. Chen et al. [53] developed a responsive microneedle system combining copper oxide NPs to enhance chemodynamic and photothermal therapies for efficient anti-melanoma treatment. The microneedle system was fabricated using poly(vinylpyrrolidone) (PVP) as the substrate, with PVP-stabilized copper oxide NPs as the therapeutic agent. This design ensured excellent skin-insertion capability and dissolvability, enabling effective skin penetration to deliver NPs directly to the targeted melanoma site. The results highlighted the microneedle system's ability to self-supply hydrogen peroxide and deplete glutathione, thereby enhancing chemodynamic therapy. The system also demonstrated superior near-infrared photothermal properties for melanoma inhibition [53]. Similarly, Yu et al. [54] designed a subcutaneous rapid-dissolving microneedle patch incorporating copper oxide and disulfiram to enhance antimelanoma efficacy through an integrated multimodal approach combining chemodynamic, photothermal and chemotherapeutic effects. The microneedle system was fabricated using PVP as the substrate and biocompatible Panax notoginseng polysaccharide for the microneedle tips, with copper oxide NPs and F127 micelles containing disulfiram encapsulated as the therapeutic agents for melanoma treatment. This microneedle formulation demonstrated the ability to release copper ions and hydrogen peroxide in the tumor's acidic environment, facilitating self-supplied hydrogen peroxide production to support localized chemodynamic therapy. [54]. Likewise, Zhao et al. [55] prepared a tumor microenvironment-sensitive microneedle patch for chemodynamic treatment of oral squamous cell carcinoma. In this study, iron oxide NPs and vitamin C were loaded onto soluble microneedles, that were prepared using polyvinyl alcohol (PVA). The results emphasized that the microneedle formulation possess enhanced targeted release ability towards oral squamous cell carcinoma tissue and generate intracellular hydrogen peroxide in tumor cells [55] (Fig. 13.4).

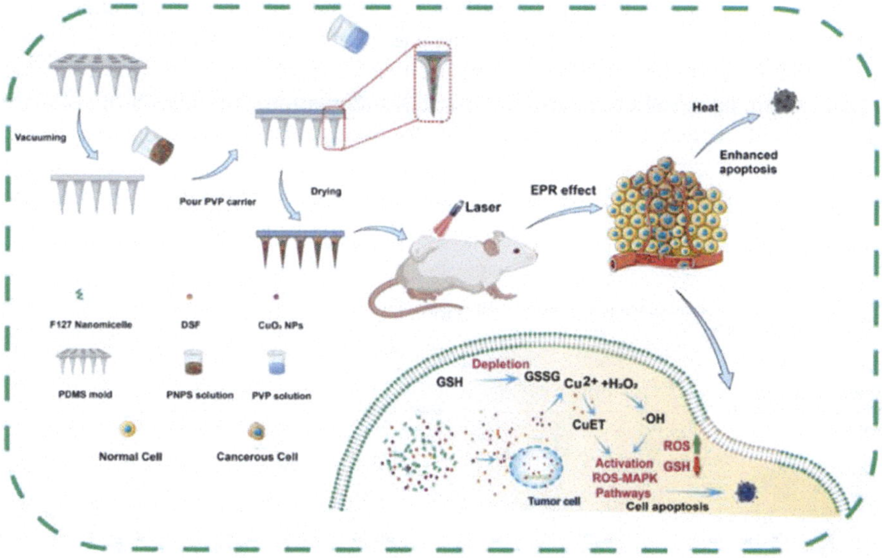

Fig. 13.4 Schematic of copper oxide NPs formulated in microneedles for melanoma treatment. Reproduced with permission from Yu et al. [54], ©American Chemical Society, 2023

13.4.3 Other Applications

Microneedles loaded with metal oxide NPs have been identified to be beneficial for other applications, such as treatment of rheumatoid arthritis, androgenic alopecia, and diabetic wounds. Xia et al. [56] developed microneedles loaded with cerium-manganese oxide NPs to promote macrophage polarization from the M1 to M2 phenotype, alleviate inflammation, and leverage their antioxidant and catalytic-like properties to eliminate ROS, making them an effective drug carrier. In this study, the first-line rheumatoid arthritis medication methotrexate was incorporated into the NPs, which were subsequently modified with bovine serum albumin and embedded into dissolving hyaluronic acid-based microneedles for transdermal delivery [56]. A novel transdermal composite microneedle consisting of mesoporous iron oxide nano-raspberry and polyvinyl alcohol (PVA) was designed by Fang et al. [57] for androgenic alopecia treatment as shown in Fig. 13.5. In this study, microneedles responsive to magnets were prepared via 3D printing-molding approach for improved drug delivery efficiency of Minoxidil. The results showed that the hair density of mice with alopecia exhibited an improvement of 800%, after magnetic microneedle formulation at 10-days post-treatment [57]. Shen et al. [52] developed a microneedle patch infused with calcium peroxide NPs for oxygen release and inhibition of biofilm in diabetic wound treatment. In this study, the microneedles dissolved upon contact with biological fluids, delivering the NPs to the wound site, where they break down to generate oxygen and hydrogen peroxide, enhancing antibacterial efficacy. In vivo experimental results highlighted significant benefits, including induction of blood vessel growth, bacterial elimination, reduced inflammation, stimulation of collagen formation, and accelerated wound repair for the treatment of infected diabetic ulcers [52].

In a study to develop a novel microneedle for glucose-mediated insulin delivery, Xu et al. [58] used zinc oxide quantum dots as pH-sensitive switch to employ

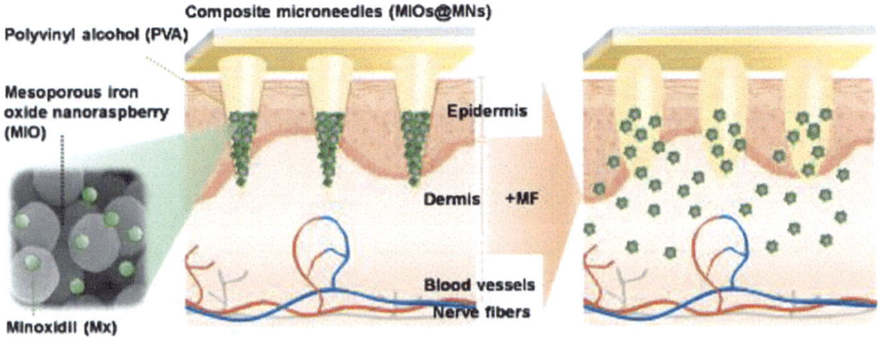

Fig. 13.5 Microneedle-mesoporous iron oxide nano-raspberry composites with polyvinyl alcohol for alopecia treatment. Reproduced with permission from Fang et al. [57], ©MDPI, 2020

nanopores cap of mesoporous bioactive glasses by electrostatic interaction. Additionally, insulin as a drug and glucose oxidase/catalase as a glucose-responsive factor was sealed into the pores of the microneedles. The results demonstrated the potential of microneedle-metal oxide nanoparticle-based transdermal delivery systems for effective diabetes treatment [58]. In another study, Zheng et al. [59] developed innovative photothermal microneedle patches loaded with magnesium oxide-polydopamine NPs, integrated with chitosan gel dressings for the treatment of infected wounds. These nanocomposite-loaded microneedle patches, together with the chitosan gel, were shown to prevent scab formation and facilitate drug delivery into deeper tissue layers. The NPs were released from both the backing layer and the microneedle tips, enabling increased drug loading, sustained magnesium ion release in moist environments, enhanced tissue migration, and the formation of micro-vessels. The findings showed that the microneedle-metal oxide nanocomposite exhibited excellent antioxidant, anti-inflammatory, antibacterial, and pro-angiogenic properties, making it highly effective for treating infectious wounds [59]. Jin et al. [60] demonstrated the creation of a minimally invasive sensing system comprising hydrogel microneedle patches coupled with screen-printed carbon electrodes modified using gold-copper oxide nanospheres. This system was designed for the real-time monitoring of glucose levels in interstitial skin fluid. The study showed that the gold doping into copper oxide led to an active and highly conductive electrocatalyst for glucose oxidation to enhance the sensing performance of glucose sensor. The results emphasized that the microneedle-based sensing system possess 0.02–6 mM of wide linear range, 0.018 mM of low detection limit and 110.38 $\mu A/mM/cm^{-2}$ [60]. Similarly, Li et al. [61] developed swellable colorimetric microneedles for the detection of glucose, utilizing glucose oxidase-like gold NPs. These microneedle patches were capable of rapidly absorbing a significant amount (60.2 mg) of interstitial skin fluid within 10 min. The results indicated that the microneedle patches outperformed natural nanozymes, demonstrating excellent stability and sensitivity in complex environments for glucose detection [61].

13.5 Toxicity of Microneedle-Metal Oxide NPs

The toxicity of microneedle-metal oxide NPs is a critical factor in their suitability for biomedical applications. Zhang et al. [62] conducted an in vivo evaluation of the safety, toxicology, and biodistribution of polyvinyl alcohol microneedles during 160 days of uninterrupted application in mice. In this study, polymeric dissolving microneedles were used as matrix materials to assess their long-term use in managing skin-related complications. Healthy mice received the microneedles for 160 days, with observations indicating that the microneedles dissolved completely and became undetectable in skin tissues by the sixth day after administration. The findings revealed that the microneedles did not cause significant toxicity in healthy mice over the 160-day period [62]. In an investigation into the in vitro cytotoxicity of carbohydrate-based materials for the fabrication of dissolvable microneedle arrays, Yalcintas et al. [63] assessed the cytotoxicity of microneedles at concentrations ranging from 3 to 80 mg/mL, using mouse cell lines, including MPEK-BL6 keratinocytes and NIH-3T3 fibroblasts, as well as human cell lines such as HaCaT keratinocytes, NHDF fibroblasts, and HEK-293 embryonic kidney cells. The results indicated that materials like maltodextrin, hyaluronic acid, carboxymethyl cellulose, and trehalose exhibited negligible cytotoxicity, whereas glucose showed significant cytotoxic effects at concentrations exceeding 50 mg/mL [63]. Yun et al. [64] examined the safety of microneedles coated with influenza antigen for application on rat skin and their ability to elicit an in vivo antigen-specific immune response. In this study, polylactic acid was employed to prepare the microneedle patches, with the protein content on the patches measured at 96.41 ± 3.18 μg. The findings revealed that the microneedles caused no significant skin irritation or systemic toxicity but successfully induced a strong antigen-specific immune response [64]. Several studies have highlighted the significant cytotoxic effects of metal oxide NPs on both human and mouse cell lines [65, 66]. Similarly, toxicological evaluations of metal oxide NPs have demonstrated concentration-dependent toxicity in zebrafish [67]. Collectively, these findings suggest that both microneedles and metal oxide NPs may induce dose- or concentration-dependent toxicity in mouse and human cell lines, animal models, and humans. However, no studies to date have specifically confirmed the toxicity of metal oxide NPs when formulated within microneedles.

13.6 Future Perspectives

Microneedles are extensively utilized for the treatment and diagnosis of skin-related diseases [68]. However, they lack stability in biological fluids with a dose/concentration-dependent toxicity towards cell lines [69]. To address these issues, microneedles are often blended or formulated with metal oxide NPs to enhance their stability and mitigate toxicity [70]. Nonetheless, standalone metal oxide NPs have also been reported to cause toxic effects in biomedical applications [71]. Currently, no studies have specifically reported the toxicity profiles of metal oxide NPs formulated in microneedles. A promising strategy to overcome this limitation involves biosynthesis approaches to reduce toxicity [72]. Microneedles fabricated using biopolymers [73] and metal oxide NPs synthesized via green or biosynthesis methods [20] offer sustainable alternatives to traditional chemical or physical synthesis processes for microneedle-metal oxide nanoparticle systems. Additionally, the integration of advanced computational approaches, such as artificial intelligence, computational modeling, simulation, and machine learning, can significantly improve the design and synthesis of metal oxide NPs formulated in microneedles [74–76]. Furthermore, incorporating computational techniques into the biosynthesis of microneedle-embedded metal oxide NPs holds immense potential for their future application in biomedical fields [77].

13.7 Conclusion

This chapter provides an overview of various metal oxide NPs integrated with microneedles for effective drug delivery applications. The limitations associated with conventional microneedles and metal oxide NPs are addressed through the innovative formulation of microneedle-metal oxide NPs. These formulations have shown promise in treating microbial infections and skin cancers. However, evaluations have demonstrated that microneedles with metal oxide NPs can exhibit dose- or concentration-dependent cytotoxicity. To mitigate these concerns, conventional microneedles can be fabricated using less toxic or nontoxic biopolymers, while metal oxide NPs can be synthesized via biosynthesis approaches. The combination of these strategies enables the development of less toxic microneedle-metal oxide NPs, paving the way for their potential in biomedical applications.

References

1. Yang, D., Chen, M., Sun, Y., Jin, Y., Lu, C., Pan, X., Quan, G., Wu, C.: Microneedle-mediated transdermal drug delivery for treating diverse skin diseases. Acta Biomater. **121**, 119–133 (2021)
2. Priya, S., Singhvi, G.: Microneedles-based drug delivery strategies: a breakthrough approach for the management of pain. Biomed. Pharmacother. **155**, 113717 (2022)
3. Chudzińska, J., Wawrzyńczak, A., Feliczak-Guzik, A.: Microneedles based on a biodegradable polymer—Hyaluronic acid. Polymers **16**(10), 1396 (2024)
4. Chen, H., Wu, B., Zhang, M., Yang, P., Yang, B., Qin, W., Wang, Q., Wen, X., Chen, M., Quan, G.: A novel scalable fabrication process for the production of dissolving microneedle arrays. Drug Deliv. Transl. Res. **9**, 240–248 (2019)
5. Aldawood, F.K., Andar, A., Desai, S.: Investigating laser ablation process parameters for the fabrication of customized microneedle arrays for therapeutic applications. Pharmaceutics **16**(7), 885 (2024)
6. Choi, J.-E., Cha, H.-R., Kim, S., Kim, J.S., Kim, M.-J., Chung, H.W., Baek, S.-K., Lee, J.M., Park, J.-H.: Preparation of particle-attached microneedles using a dry coating process. J. Control. Release **351**, 1003–1016 (2022)
7. Luo, X., Yang, L., Cui, Y.: Microneedles: materials, fabrication, and biomedical applications. Biomed. Microdevice **25**(3), 20 (2023)
8. Traverso, G., Schoellhammer, C.M., Schroeder, A., Maa, R., Lauwers, G.Y., Polat, B.E., Anderson, D.G., Blankschtein, D., Langer, R.: Microneedles for drug delivery via the gastrointestinal tract. J. Pharm. Sci. **104**(2), 362–367 (2015)
9. Khalid, R., Mahmood, S., Mohamed Sofian, Z., Hilles, A.R., Hashim, N.M., Ge, Y.: Microneedles and their application in transdermal delivery of antihypertensive drugs—A review. Pharmaceutics **15**(8), 2029 (2023)
10. Gowda, B.J., Ahmed, M.G., Sanjana, A.: Can microneedles replace hypodermic needles? Painless drug delivery. Resonance **27**(1), 63–85 (2022)
11. Hou, X., Li, J., Hong, Y., Ruan, H., Long, M., Feng, N., Zhang, Y.: Advances and prospects for hydrogel-forming microneedles in transdermal drug delivery. Biomedicines **11**(8), 2119 (2023)
12. Dul, M., Alali, M., Ameri, M., Burke, M.D., Craig, C.M., Creelman, B.P., Dick, L., Donnelly, R.F., Eakins, M.N., Frivold, C.: Assessing the risk of a clinically significant infection from a microneedle array patch (MAP) product. J. Control. Release **361**, 236–245 (2023)

13. McConville, A., Hegarty, C., Davis, J.: Mini-review: assessing the potential impact of microneedle technologies on home healthcare applications. Medicines **5**(2), 50 (2018)
14. Joudeh, N., Linke, D.: Nanoparticle classification, physicochemical properties, characterization, and applications: a comprehensive review for biologists. J. Nanobiotechnol. **20**(1), 262 (2022)
15. Nikam, A., Prasad, B., Kulkarni, A.: Wet chemical synthesis of metal oxide nanoparticles: a review. CrystEngComm **20**(35), 5091–5107 (2018)
16. Tsuzuki, T.: Mechanochemical synthesis of metal oxide nanoparticles. Commun. Chem. **4**(1), 143 (2021)
17. Jeevanandam, J., Chan, Y.S., Danquah, M.K.: Biosynthesis of metal and metal oxide nanoparticles. ChemBioEng Rev. **3**(2), 55–67 (2016)
18. Cele, T.: Preparation of nanoparticles. In: Engineered Nanomaterials-Health and Safety, vol. 12, no. 10, pp. 1–14 (2020)
19. Khan, I., Saeed, K., Khan, I.: Nanoparticles: properties, applications and toxicities. Arab. J. Chem. **12**(7), 908–931 (2019)
20. Jeevanandam, J., Kiew, S.F., Boakye-Ansah, S., Lau, S.Y., Barhoum, A., Danquah, M.K., Rodrigues, J.: Green approaches for the synthesis of metal and metal oxide nanoparticles using microbial and plant extracts. Nanoscale **14**(7), 2534–2571 (2022)
21. Zhong, Y., Zeng, X., Zeng, Y., Yang, L., Peng, J., Zhao, L., Chang, Y.-T.: Nanomaterials-based imaging diagnosis and therapy of cardiovascular diseases. Nano Today **45**, 101554 (2022)
22. Yadav, K.: Nanotechnology in diabetes management: revolutionizing treatment and diagnostics. J. Mol. Liquids 126117 (2024)
23. Cheng, Z., Li, M., Dey, R., Chen, Y.: Nanomaterials for cancer therapy: current progress and perspectives. J. Hematol. Oncol. **14**, 1–27 (2021)
24. Suganthi, K., Rajan, A.S.: Metal oxide nanofluids: review of formulation, thermo-physical properties, mechanisms, and heat transfer performance. Renew. Sustain. Energy Rev. **76**, 226–255 (2017)
25. Nigam, S., Bahadur, D.: Dendrimer-conjugated iron oxide nanoparticles as stimuli-responsive drug carriers for thermally-activated chemotherapy of cancer. Colloids Surf. B **155**, 182–192 (2017)
26. Talelli, M., Rijcken, C.J., Lammers, T., Seevinck, P.R., Storm, G., van Nostrum, C.F., Hennink, W.E.: Superparamagnetic iron oxide nanoparticles encapsulated in biodegradable thermosensitive polymeric micelles: toward a targeted nanomedicine suitable for image-guided drug delivery. Langmuir **25**(4), 2060–2067 (2009)
27. De Leo, V., Maurelli, A.M., Giotta, L., Catucci, L.: Liposomes containing nanoparticles: preparation and applications. Colloids Surf. B **218**, 112737 (2022)
28. Nasrollahzadeh, M., Shafiei, N., Soleimani, F., Nezafat, Z., Bidgoli, N.S.: Physicochemical characterization of biopolymer-based metal nanoparticles. Biopolym. Met. Nanoparticle Chem. Sustain. Appl. **1**, 317–478 (2021)
29. Chenthamara, D., Subramaniam, S., Ramakrishnan, S.G., Krishnaswamy, S., Essa, M.M., Lin, F.-H., Qoronfleh, M.W.: Therapeutic efficacy of nanoparticles and routes of administration. Biomater. Res. **23**(1), 20 (2019)
30. Tort, S., Mutlu Agardan, N.B., Han, D., Steckl, A.J.: In vitro and in vivo evaluation of microneedles coated with electrosprayed micro/nanoparticles for medical skin treatments. J. Microencapsul. **37**(7), 517–527 (2020)
31. Teo, A.L., Shearwood, C., Ng, K.C., Lu, J., Moochhala, S.: Transdermal microneedles for drug delivery applications. Mater. Sci. Eng. B **132**(1–2), 151–154 (2006)
32. Martin, C., Allender, C.J., Brain, K.R., Morrissey, A., Birchall, J.C.: Low temperature fabrication of biodegradable sugar glass microneedles for transdermal drug delivery applications. J. Control. Release **158**(1), 93–101 (2012)
33. Bodhale, D.W., Nisar, A., Afzulpurkar, N.: Structural and microfluidic analysis of hollow side-open polymeric microneedles for transdermal drug delivery applications. Microfluid. Nanofluid. **8**, 373–392 (2010)

34. Tsioris, K., Raja, W.K., Pritchard, E.M., Panilaitis, B., Kaplan, D.L., Omenetto, F.G.: Fabrication of silk microneedles for controlled-release drug delivery. Adv. Func. Mater. **22**(2), 330–335 (2012)
35. Aldawood, F.K., Parupelli, S.K., Andar, A., Desai, S.: 3D printing of biodegradable polymeric microneedles for transdermal drug delivery applications. Pharmaceutics **16**(2), 237 (2024)
36. Nguyen, H.X., Bozorg, B.D., Kim, Y., Wieber, A., Birk, G., Lubda, D., Banga, A.K.: Poly(vinyl alcohol) microneedles: fabrication, characterization, and application for transdermal drug delivery of doxorubicin. Eur. J. Pharm. Biopharm. **129**, 88–103 (2018)
37. Zoudani, E.L., Soltani, M.: A new computational method of modeling and evaluation of dissolving microneedle for drug delivery applications: extension to theoretical modeling of a novel design of microneedle (array in array) for efficient drug delivery. Eur. J. Pharm. Sci. **150**, 105339 (2020)
38. Thakur Singh, R.R., Tekko, I., McAvoy, K., McMillan, H., Jones, D., Donnelly, R.F.: Minimally invasive microneedles for ocular drug delivery. Expert Opin. Drug Deliv. **14**(4), 525–537 (2017)
39. Bodhale, D., Nisar, A., Afzulpurkar, N.: Design, fabrication and analysis of silicon microneedles for transdermal drug delivery applications. In: The Third International Conference on the Development of Biomedical Engineering in Vietnam: BME2010, pp. 84–89. Springer, Ho Chi Minh City, Vietnam (2010)
40. Sawon, M.A., Samad, M.F.: Design and optimization of a microneedle with skin insertion analysis for transdermal drug delivery applications. J. Drug Deliv. Sci. Technol. **63**, 102477 (2021)
41. Larraneta, E., Lutton, R.E., Brady, A.J., Vicente-Pérez, E.M., Woolfson, A.D., Thakur, R.R.S., Donnelly, R.F.: Microwave-assisted preparation of hydrogel-forming microneedle arrays for transdermal drug delivery applications. Macromol. Mater. Eng. **300**(6), 586–595 (2015)
42. Nguyen, H.X., Banga, A.K.: Fabrication, characterization and application of sugar microneedles for transdermal drug delivery. Ther. Deliv. **8**(5), 249–264 (2017)
43. Sabri, A.H., Kim, Y., Marlow, M., Scurr, D.J., Segal, J., Banga, A.K., Kagan, L., Lee, J.B.: Intradermal and transdermal drug delivery using microneedles—Fabrication, performance evaluation and application to lymphatic delivery. Adv. Drug Deliv. Rev. **153**, 195–215 (2020)
44. Seeni, R.Z., Yu, X., Chang, H., Chen, P., Liu, L., Xu, C.: Iron oxide nanoparticle-powered micro-optical coherence tomography for in situ imaging the penetration and swelling of polymeric microneedles in the skin. ACS Appl. Mater. Interfaces **9**(24), 20340–20347 (2017)
45. Cui, J., Huang, J., Yan, Y., Chen, W., Wen, J., Wu, X., Liu, J., Liu, H., Huang, C.: Ferroferric oxide loaded near-infrared triggered photothermal microneedle patch for controlled drug release. J. Colloid Interface Sci. **617**, 718–729 (2022)
46. Haghighi, S., Haghighatzadeh, A.: Actinomorphic ZnO microneedles decorated with SnO2 nanospheres: synthesis, characterization and optical studies. Appl. Phys. A **126**(2), 107 (2020)
47. Choi, J.-G., Hasan, M., Lee, S.-S.: Skin permeability and magnetic susceptibility analysis of hyaluronic acid-based magnetized microneedles containing iron oxide nanoparticles. AIP Adv. **14**(1) (2024)
48. Li, Y., Liao, X., Zheng, B.: Studies on local anesthetic lidocaine hydrochloride delivery via photo-triggered implantable polymeric microneedles as a patient-controlled transdermal analgesia system. J. Biomater. Sci. Polym. Ed. **33**(2), 155–173 (2022)
49. Li, X., Du, W., Xu, W., Ling, G., Zhang, P.: Dissolving microneedles based on ZnO nanoparticles and an ionic liquid as synergistic antibacterial agents. J. Mater. Chem. B **11**(19), 4354–4364 (2023)
50. Wang, G., Wang, W., Chen, Z., Hu, T., Tu, L., Wang, X., Hu, W., Li, S., Wang, Z.: Photothermal microneedle patch loaded with antimicrobial peptide/MnO2 hybrid nanoparticles for chronic wound healing. Chem. Eng. J. **482**, 148938 (2024)
51. Gong, J.-H., Chen, L.-J., Zhao, X., Yan, X.-P.: Persistent production of reactive oxygen species with Zn2GeO4: Cu nanorod-loaded microneedles for methicillin-resistant Staphylococcus aureus infectious wound healing. ACS Appl. Mater. Interfaces **14**(15), 17142–17152 (2022)
52. Shen, H., Ma, Y., Zhang, C., Qiao, Y., Chen, J., Sun, F.: Microneedle patch loaded with calcium peroxide nanoparticles for oxygen healing and biofilm inhibition in diabetic wound healing. ACS Appl. Nano Mater. **7**(13), 15171–15184 (2024)

53. Chen, J., Cao, Y., Lin, S., Niu, H., Zhang, H., Guan, L., Shu, C., Wu, A., Bian, Y., Zhu, Y.: A responsive microneedle system for efficient anti-melanoma by combining self-enhanced chemodynamic therapy with photothermal therapy. Chem. Eng. J. **431**, 133466 (2022)
54. Yu, S., Ni, H., Xu, X., Cai, Y., Feng, J., Zhang, J.: Subcutaneous rapid dissolution microneedle patch integrated with CuO2 and Disulfiram for Augmented Antimelanoma efficacy through multimodal synergy of photothermal therapy, Chemodynamic therapy, and chemotherapy. ACS Biomater. Sci. Eng. **9**(11), 6425–6437 (2023)
55. Zhao, S., Li, Y., Cheng, B.: A tumor microenvironment-responsive microneedle patch for chemodynamic therapy of oral squamous cell carcinoma. Nanoscale Adv. **5**(22), 6162–6169 (2023)
56. Xia, T., Zhu, Y., Li, K., Hao, K., Chai, Y., Jiang, H., Lou, C., Yu, J., Yang, W., Wang, J.: Microneedles loaded with cerium-manganese oxide nanoparticles for targeting macrophages in the treatment of rheumatoid arthritis. J. Nanobiotechnol. **22**(1), 103 (2024)
57. Fang, J.-H., Liu, C.-H., Hsu, R.-S., Chen, Y.-Y., Chiang, W.-H., Wang, H.-M.D., Hu, S.-H.: Transdermal composite microneedle composed of mesoporous iron oxide nanoraspberry and PVA for androgenetic alopecia treatment. Polymers **12**(6), 1392 (2020)
58. Xu, B., Cao, Q., Zhang, Y., Yu, W., Zhu, J., Liu, D., Jiang, G.: Microneedles integrated with ZnO quantum-dot-capped mesoporous bioactive glasses for glucose-mediated insulin delivery. ACS Biomater. Sci. Eng. **4**(7), 2473–2483 (2018)
59. Zheng, G., Xie, J., Yao, Y., Shen, S., Weng, J., Yang, Q., Yan, Q.: MgO@ polydopamine nanoparticle-loaded photothermal microneedle patches combined with chitosan gel dressings for the treatment of infectious wounds. ACS Appl. Mater. Interfaces **16**(10), 12202–12216 (2024)
60. Jin, D., Xu, Z., Zhao, H., Deng, S., Qu, Z., Dou, R., Liu, W.: A minimally invasive sensing system based on hydrogel microneedle patches and Au/Cu2O nanospheres modified screen-printed carbon electrode for glucose monitoring in interstitial skin fluid. Microchem. J. **205**, 111367 (2024)
61. Li, X., Lv, J., Zhao, J., Ling, G., Zhang, P.: Swellable colorimetric microneedles for glucose detection based on glucose oxidase-like gold nanoparticles. Anal. Chim. Acta **1288**, 342152 (2024)
62. Zhang, X.P., Wang, B.B., Li, W.X., Fei, W.M., Cui, Y., Guo, X.D.: In vivo safety assessment, biodistribution and toxicology of polyvinyl alcohol microneedles with 160-day uninterruptedly applications in mice. Eur. J. Pharm. Biopharm. **160**, 1–8 (2021)
63. Yalcintas, E.P., Ackerman, D.S., Korkmaz, E., Telmer, C.A., Jarvik, J.W., Campbell, P.G., Bruchez, M.P., Ozdoganlar, O.B.: Analysis of in vitro cytotoxicity of carbohydrate-based materials used for dissolvable microneedle arrays. Pharm. Res. **37**, 1–18 (2020)
64. Yun, T.-S., Song, B., Hwang, Y.-R., Jin, M., Seonwoo, H., Kim, D., Kim, H.W., Kim, B.C., Kim, D., Park, B.: Safety of applying influenza-antigen-coated microneedles to rat skin and the antigen specific immune response in vivo. J. Pharm. Investig. 1–12 (2024)
65. Sengul, A.B., Asmatulu, E.: Toxicity of metal and metal oxide nanoparticles: a review. Environ. Chem. Lett. **18**(5), 1659–1683 (2020)
66. Attarilar, S., Yang, J., Ebrahimi, M., Wang, Q., Liu, J., Tang, Y., Yang, J.: The toxicity phenomenon and the related occurrence in metal and metal oxide nanoparticles: a brief review from the biomedical perspective. Front. Bioengineering Biotechnol. **8**, 822 (2020)
67. Bai, C., Tang, M.: Toxicological study of metal and metal oxide nanoparticles in zebrafish. J. Appl. Toxicol. **40**(1), 37–63 (2020)
68. Zhang, W., Zhang, W., Li, C., Zhang, J., Qin, L., Lai, Y.: Recent advances of microneedles and their application in disease treatment. Int. J. Mol. Sci. **23**(5), 2401 (2022)
69. Handa, M., Afzal, O., Beg, S., Sanap, S.N., Kaundal, R.K., Verma, R.K., Mishra, A., Shukla, R.: Harnessing personalized tailored medicines to digital-based data-enriched edible pharmaceuticals. Drug Discov. Today **28**(5), 103555 (2023)
70. Chevala, N.T., Jitta, S.R., Marques, S.M., Vaz, V.M., Kumar, L.: Polymeric microneedles for transdermal delivery of nanoparticles: Frontiers of formulation, sterility and stability aspects. J. Drug Deliv. Sci. Technol. **65**, 102711 (2021)

71. Manuja, A., Kumar, B., Kumar, R., Chhabra, D., Ghosh, M., Manuja, M., Brar, B., Pal, Y., Tripathi, B., Prasad, M.: Metal/metal oxide nanoparticles: Toxicity concerns associated with their physical state and remediation for biomedical applications. Toxicol. Rep. **8**, 1970–1978 (2021)
72. Huang, H., Shen, Y., Yang, X., Hou, C., Ke, X., Yang, R.: Dissolvable microneedles loaded with Asiaticoside nanocrystals stabilized by Panax Notoginseng Saponins for hypertrophic scar treatment. J. Drug Deliv. Sci. Technol. 105854 (2024)
73. Koyani, R.D.: Biopolymers for microneedle synthesis: From then to now. Biomanuf. Rev. **4**(1), 1 (2019)
74. He, W., Kong, S., Lin, R., Xie, Y., Zheng, S., Yin, Z., Huang, X., Su, L., Zhang, X.: Machine learning assists in the design and application of microneedles. Biomimetics **9**(8), 469 (2024)
75. Biswas, A.A., Dhondale, M.R., Agrawal, A.K., Serrano, D.R., Mishra, B., Kumar, D.: Advancements in microneedle fabrication techniques: artificial intelligence assisted 3D-printing technology. Drug Deliv. Transl. Res. **14**(6), 1458–1479 (2024)
76. Rezapour Sarabi, M., Alseed, M.M., Karagoz, A.A., Tasoglu, S.: Machine learning-enabled prediction of 3D-printed microneedle features. Biosensors **12**(7), 491 (2022)
77. Johnson, A.R., Caudill, C.L., Tumbleston, J.R., Bloomquist, C.J., Moga, K.A., Ermoshkin, A., Shirvanyants, D., Mecham, S.J., Luft, J.C., DeSimone, J.M.: Single-step fabrication of computationally designed microneedles by continuous liquid interface production. Plos One **11**(9), e0162518 (2016)

Chapter 14
Microneedle-Based Sensor for Soil Quality Management

Nurul Asmak Md. Lazim and Rini Asnida Abdullah

14.1 Introduction

The intricate interplay between the atmosphere, land surface, aquatic bodies, and living organisms determines the earth's climate system. A rise in the average world temperature of more than 0.1 °C every ten years poses a threat to food security and agricultural output. The state of the soil and climate change are closely related. Soil qualities such as resistance and resiliency are impacted by climate change both directly and indirectly. While climate change adaptation options bring on indirect effects, direct effects include changes in organic carbon transformations and nutrient cycling through altered soil moisture and temperature regimes, as well as increased soil erosion rates due to increased frequency of high-intensity rainfall events [1]. At the same time, it is reported that the World's population is expected to reach 8.5 billion people by 2030 [2]. Since the onset of the Industrial Revolution, humankind has advanced to create a deviation from the planet's natural equilibrium. As the World's population grows, more resources will be required to meet everyone's daily needs. This suggests that to meet this necessary food demand, and there is a need to increase global food production in addition to sustainable land cultivation. With 20–30 years of lagged time in our global climate network, the globe is predicted to warm by 1.50 °C between 2020 and 2024, despite international commitments to reduce emissions through the reduction of greenhouse gases at every attempt [3]. To support this expansion, forests have been converted by deforestation to total burning; for example, the Amazon rainforest emits around 2 billion tonnes of carbon dioxide per year, or 5% of global emissions, while the property was put to human use [4]. However, with regard to colder weather in countries, for example, Japan, Sapporo, and many others,

N. A. Md. Lazim
Faculty of Chemical and Energy Engineering, Johor Bahru, Malaysia

R. A. Abdullah (✉)
Faculty of Civil Engineering, Universiti Teknologi Malaysia, Johor Bahru, Johor, Malaysia
e-mail: asnida@utm.my

N. Talreja et al. (eds.), *Microneedles (MNs)-Based Technology*, Series in BioEngineering, https://doi.org/10.1007/978-981-96-3916-8_14

the ground temperature is typically below 0 °C in the coldest month of the year [5, 6]. Slope failures and landslides in colder climates are thought to be primarily caused by differences in climate conditions, such as air temperature and precipitation [7, 8]. As a result, a lot of research has been conducted to examine the trigger factors and failure processes of slope failures in cold climates [9, 10]. According to Ishikawa et al. [11] there are extra elements, such as freeze-thaw action, that cause the failure mechanism of slopes to differ in cold and warm-temperate regions. Owing to this, Ishikawa et al. [11] classified slope failures in cold climates into two primary classes: those that happen during the snow-melting season (March–April–May) and those that happen during the period of high rainfall (August–September). Investigating how climate change affects slope stability in seasonally cold areas is therefore essential as well.

Because of global warming, it is anticipated that many mountain regions may experience an increase in the frequency of severe rainfall events during the heavy rainfall season [12]. One of the primary causes of slope failures or landslides is thought to be the decrease in matrix suction brought on by rainfall infiltration [13–15]. As per the statistical data analyzed by the Japan Meteorological Agency [16] for a sample of 1000 places in Japan, the yearly average of the number of events with rainfall intensity above 50 mm/h rose from 195 in 1987 to 98 in 1998 to 226 in 1999 to 2010. However, because of the recent acceleration of global warming, there have been more documented instances of slope failures in typically cold areas during the snowmelt season. As an illustration, according to a survey conducted in April by the Ministry of Land, Infrastructure, Transport, and Tourism (MLIT), the average number of slope failures in Japan climbed from over 50 times between 2008 and 2012 to approximately 100 times between 2013 and 2017, nearly doubling. In August, however, this figure doubled, rising from roughly 120 times (2008–2012) to almost 240 times (2013–2017). As a result, climate change has an increasing impact on slope stability. The impact of climatic changes on slope failures in colder climates, however, receives less consideration [14, 17, 18]. Many efforts have also been undertaken to evaluate the effects of climate change on slope failures by using downscaled synthetic future climatic scenarios that are derived from GCMs as input for slope stability [12, 19, 20]. However, according to Chiang and Chang [12], Taiwan's average total unstable area will increase by 12.8%, while the average annual maximum rainfall will rise by 15.2% between 2010 and 2099 compared to 1960–2008. Ciabatta et al. [20] evaluated the effects of climate change on landslide danger in the Umbria region, Italy, over three distinct decades (1990–2013 (baseline), 2040–2069, and 2070–2099). They did this by using the PRESSCA system, an operational early warning system. They discovered that a general rise (up to more than 40%) in the occurrence of landslides is anticipated for the Umbria territory, Italy, in the future, primarily during the winter season.

Taking into consideration the event, in urban construction, for example, geotechnical engineering is crucial [21]. Numerous geotechnical issues, including pile settlements, soil nail breaking, and slope instability, are getting worse because of the hard working conditions and extended working hours [22]. Unintentional infrastructure failure may result in fatalities and significant financial damage in the absence of scientific geotechnical monitoring techniques [23]. Under these circumstances, geotechnical monitoring is becoming increasingly important in evaluating the performance and safety state of geotechnical infrastructure [24]. Numerous variables associated with geotechnical engineering, such as those related to soil material qualities, environmental influences, and engineering design and construction, make geotechnical monitoring extremely difficult [25].

Determining the thermal conductivity of geomaterials is crucial for developing energy-related geostructures, such as subterranean heat pumps and underground electric power cable tunnels [31–34], since thermal conduction is the main heat transfer mechanism [26–30]. The volumetric proportions of constituents and their thermal conductivity significantly impact heat transmission in geo-materials like soils and rocks because they are multiphase porous media made up of solid grains and pore fluids. The thermal conductivity of materials can be measured in several ways, ranging from the steady-state method [35] to the transient hot wire approach [36, 37], which uses cold and hot plates. Specifically, for the transient hot wire method, also known as the thermal needle probe method, a temperature sensor (or thermal needle probe) and a long, thin metallic probe are inserted into the target medium. The temperature sensor then records the transient thermal response while the line heat source generates heat at a steady, controlled flux [26, 36–43]. Due to its ease of application in both laboratory and field settings for assessing exposed near-surface soils, this approach is widely applied.

Although the target medium should be in contact with the thermal needle probe method, it is challenging to insert these probes into hard ground, such as consolidated or densely compacted soils and rocks, because the relatively high length-to-thickness ratio of the probes often causes them to buckle during this process. Furthermore, putting a probe into hard ground greatly disturbs the area around it, which makes it difficult for the probe to make good contact with the target medium and produces inaccurate thermal conductivity readings. The transient hot wire method, in a half-space arrangement where the heating wire is positioned on the flat surface of a reference material of known thermal conductivity and good thermal insulation, can be used to measure the thermal conductivity of such hard geo-materials in situ fields [35]. The measuring of thermal conductivity in natural geomaterials is difficult and intimidating due to their uneven and rough surface conditions. This method also demands a good contact between the flat surface and the ground to be measured. Consequently, it is suggested that a pre-drilled hole filled with thermally conductive grease (thermal grease) be used as part of an enhanced pre-drilling thermal needle probe approach. However, to increase accuracy and provide accurate data to improve performance, we recommend adopting a microneedle-based sensor, as the size of the sensor can affect the device's sensitive performance. Afterwards, we introduce a completely integrated microneedle-based sensor and provide a detailed breakdown of its components and functions. By doing so, soil quality management will be able to recommend appropriate decisions for future events and facilitate the referral of highly accurate data for future research.

14.2 Classification and Application of Thermal Needle Sensor

Since thermal conduction is the primary heat transmission mechanism [26–30], determining the thermal conductivity of geomaterials is essential for developing energy-related geostructures, such as underground electric power cable tunnels and subterranean heat pumps [31–34]. Since soils and rocks are multiphase porous media composed of solid grains and pore fluids, the volumetric proportions of constituents and their thermal conductivity majorly affect heat transfer in these materials. There are several techniques for measuring a material's thermal conductivity, from the transient hot wire approach [36, 37], which makes use of hot and cold plates, to the steady-state method [35]. In particular, a temperature sensor (or thermal needle probe) and a long, thin metallic probe are placed into the target medium for the transient hot wire method, also referred to as the thermal needle probe method. Then, while the line heat source produces heat at a constant, regulated flow, the temperature sensor logs the transient thermal response [26, 36–43]. This method is frequently used because it is simple to use for evaluating exposed near-surface soils in both laboratory and field settings (Fig. 14.1a). However, the following study has been improved by Lee et al. [44]. A pre-drilled hole filled with thermally conductive grease (thermal grease) is proposed (Fig. 14.1b). The thermal conductivity derived from the suggested pre-drilling thermal needle probe method was investigated through a series of laboratory experiments and numerical simulation with a finite element code. To investigate their impact, a parametric study on the parameters was carried out, including (1) the thermal conductivity of thermal grease, (2) probe diameter, and (3) pre-drilled hole diameter. Due to slow heat transfer, delays in temperature increase were caused by increases in the diameter of the pre-drilled hole and probe, as well as a decrease in the thermal conductivity of thermal grease. All estimated values agreed well with the reference values, with relative errors of less than 10%. That is, after a certain amount of time, the thermal conductivity of soils mostly controls the increase in temperature. Lees' findings demonstrate that the suggested method may be applied to a variety of materials, including compacted soils, soft rocks, and hard rocks that are impervious to physical penetration. It also yields accurate thermal conductivity readings with a manageable error. Another study by Hukseflux [45] proposed two sensors for soil temperature tests. They were TP01 and STP01 (Fig. 14.1c). A sensor for long-term soil thermal conductivity monitoring is called TP01. To gain a better knowledge of the dynamic (changing heat flux) thermal behaviour of soils, TP01 measurements can also be utilized to estimate the volumic heat capacity and soil thermal diffusivity. TP01 is intended for continuous usage at a single measurement site. When TP01 is used in meteorological surface flux measuring systems, it enhances estimates of both the so-called storage term and soil heat flux. The sensor is a unique Hukseflux design that combines a heater and a temperature-difference sensor with a high sensitivity and a very low thermal mass. Two thermopiles make up the temperature-difference sensor found inside TP01. With unprecedented sensitivity, it gauges the radial temperature differential surrounding a heating wire. A very thin plastic foil contains both the sensor and the heating wire. Soil thermal conductivity is measured with TP01. Buried under the ground, it is intended for long-term on-site operation. Its rated operating range, which encompasses the majority of inorganic soil types, is 0.3–4 W/(m K). Moreover, TP01's low thermal mass makes it appropriate for evaluating the volumic heat capacity and soil thermal diffusivity [45]. On the other hand, the soil's temperature profile is precisely measured by STP01 at five depths in proximity to the surface. Measurements of surface energy balance at the scientific level are done with it. Since the sensor is buried, it is typically impossible to take it to a lab for calibration. The integrated heated wire can be used in the online self-test to confirm the measurement stability of STP01. The STP01 soil temperature profile sensor provides precise temperature difference measurements at five different depths below the soil surface: 0.02, 0.05, 0.1, 0.2, and 0.5 m, respectively. Additionally, the distance between the measurement places is fixed and clearly defined. For the qualities of the soil to become representative of the natural World, it is preferable to leave soil temperature sensors in the soil for as long as feasible. The user can confirm the stable performance of sensors without sending them to a laboratory by using self-testing. Compared to measurements using traditional sensor types, the outcome is substantially better measurement accuracy and quality assurance [46].

14.3 Characterization, Fabrication, and Performance of the Microneedle-Based Sensor

In this chapter, a microneedle-based sensor was proposed for soil quality management. Expanding soil is proposed to be tested for their unique type of soil that is found all over the World that changes size in response to variations in moisture content [47] (Fig. 14.2). Silicate minerals, such as montmorillonite, illite, and kaolin, make up the majority of expansive soil. The expansive soil's potential to expand and contract varies based on the relative amounts of its constituent parts. It was categorized into five categories by Suits et al. [48]: insignificant, low, moderate, high, and very high. An expanding soil's potential to expand and contract is proportional to its montmorillonite content, as demonstrated by X-ray diffraction. Expansive soil slopes with varying slope ratios present a very difficult stability problem. It is widely accepted that frequent ice-water conversion affects the expansive soil slope in seasonally frozen places, causing structural damage to the soil [49] and a major decline in the slope's safety factor [50].

Different slope ratio tests will be set up, and the optimal moisture content responding to the maximum dry density will be adjusted in order to evaluate the slope stability characteristics under various slope ratio settings. Consequently, the slope model's moisture content was 20%, which is the ideal moisture percentage, and its slope ratios were 1:1.5, 1:1, and 1:0.5 [51]. This is because the amount of unfrozen water in frozen soils, the pace of water movement, and other intrinsic

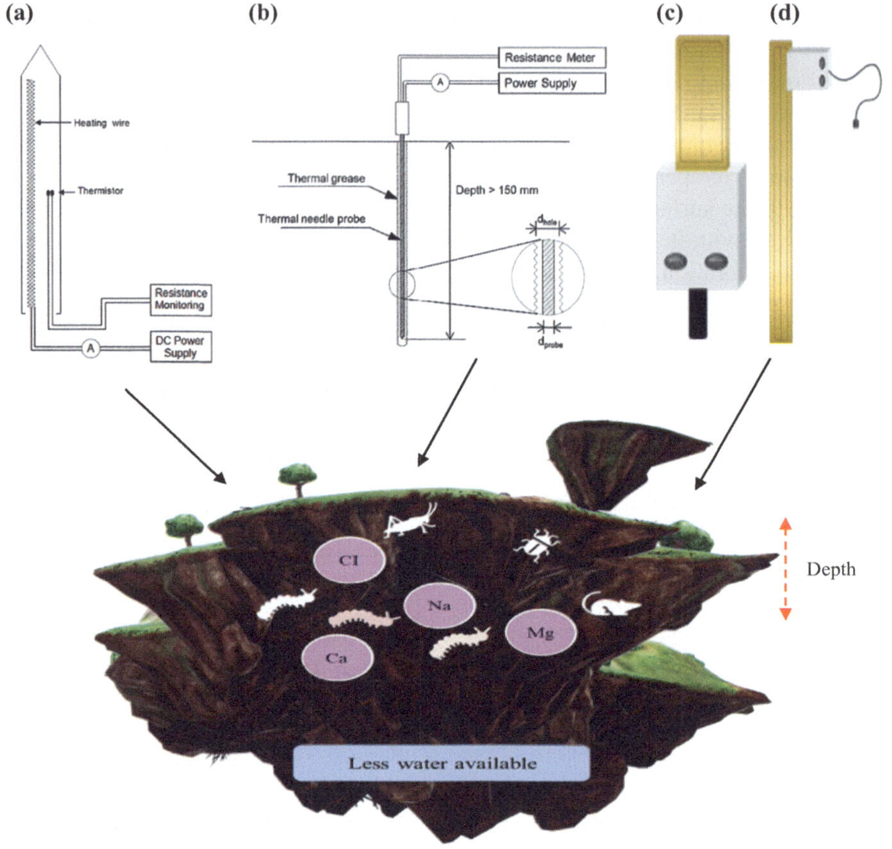

Fig. 14.1 **a** The basic structure of the thermal needle probe (Adopted from [26, 36–43]), **b** pre-drilled hole filled with thermally conductive grease (thermal grease) proposed by Lee et al. figure was reproduced with the permission [44], **c** TP01 and **d** STP01 with a standard cable length of 5 m

qualities of soil water, for example, are all impacted by the magnitude of soil temperature. For instance, the average wintertime soil temperature in the vicinity of Fairbanks, Alaska, is between 0 and −50 °C at significant depths. This suggests different amounts of unfrozen water [52]. Most of Malaysia's slopes have seen different forms of weathering, from grade IV to grade VI weathered granite. Since different soils may require different soil stabilization processes, the following parameters affect the slope angles, stability, and construction ability of the soil: particle size distribution, specific gravity, plastic index, soil-water characteristic curve, and shear strength. The twenty-first century has seen tremendous advances in civilization due to technical improvements [53]. Therefore, our proposed sensor is relevant to the situation in tropical countries, especially Malaysia.

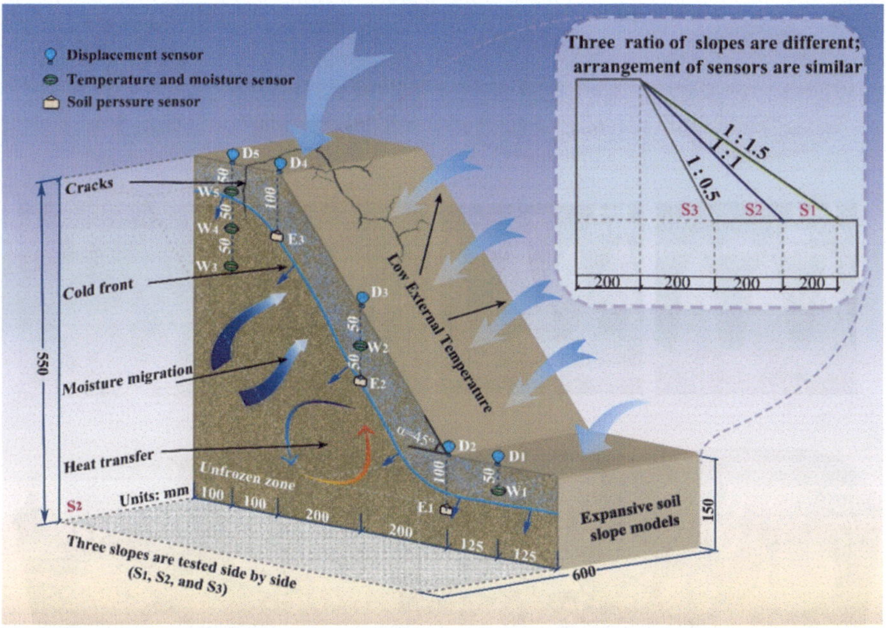

Fig. 14.2 Example of slope freezing model and freeze-thaw monitoring system. The figure was reproduced with the permission [51]

14.3.1 Type of Microneedles for Agriculture

The authors do not find detailed information specifically on MNs types for soil management, however in general, MNs fall into one of four categories: solid, coated, dissolving, or hollow (Fig. 14.3). The mechanisms by which solid, coated, dissolving, and hollow MNs function are "poke-and-patch," "coat-and-poke," "poke-and-release," and "poke-and-flow". Each MN type will be briefly introduced below in terms of shape, materials, and design.

Plant tissues are pre-treated with solid MNs in order to distribute agrochemicals [55]. Applying MNs to plant tissue in two steps, also referred to as "poke-and-patch," results in transient micropores on the surface that facilitate more efficient exposure to agrochemicals [56]. The desired geometric features determine how solid MNs are designed; greater aspect ratios and smaller tip angles call for more complex methods and cutting-edge materials [57].

The composition of solid MNs influences their biocompatibility, reusability, and mechanical strength. Because silicon is more mechanically strong and biocompatible than other materials, it is frequently utilized to fabricate solid MNs [58]. In addition to silicon, other materials that are widely employed as the primary component in the production of solid MNs include metals like titanium (Ti) [59], tungsten (W) [60], and stainless steel [61], as well as polymers like poly (methyl vinyl ether) (PMVE) [62],

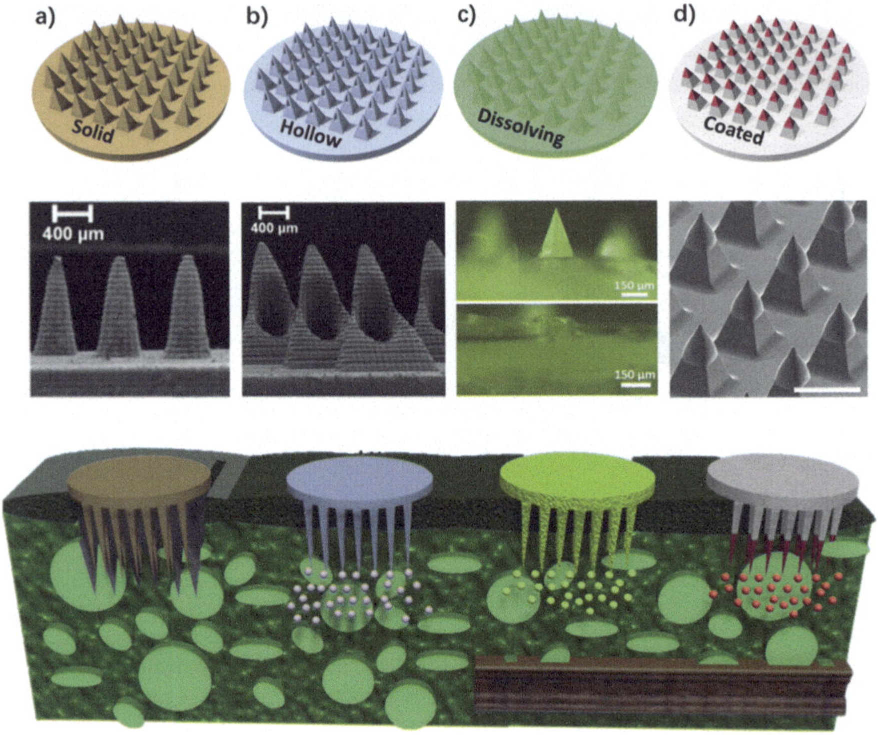

Fig. 14.3 Different types of microneedles for agriculture. The figure was reproduced with permission [54]

and poly(methyl methacrylate) (PMMA) [63]. Moreover, biocompatible substances like chitosan [64] and poly(lactic acid) (PLA) [65] are appropriate for adding to solid MN composition [66, 67]. A specific method of microfabrication is used to process a given material.

The technology known as microelectromechanical systems (MEMS) holds promise for producing solid MNs with ideal design characteristics for use in agriculture [68]. However, the approach is intricate and necessitates thin film deposition, etching, and photolithography in cleanroom settings [69]. While the result is extremely accurate, MN manufacturing can be expensive, time-consuming, and labour-intensive. Thus, techniques for cleanroom-free fabrication have been utilized, including CO_2 laser cutting [57], microinjection molding [70], hot embossing [71], electroplating [72], drawing lithography [73], a mix of spray deposition and micromilling [74], two-photon polymerization [75], and 3D printing methods [76].

The burdensome requirement for several repeated applications in the field limits the use of solid MNs in agriculture. In order to distribute a greater quantity of agrochemicals, one way to solve the problem is to increase the density of MN arrays on one patch [77]. The impact of various densities of polylactic acid (PLA) solid MNs on drug release rates has been studied, for example, by Li et al. [78]. The medication was delivered in the amounts of 152, 210, 284, and 418 ng/cm^2 using MNs with densities of 100, 144, 196, and 256 MNs/cm^2, respectively. Utilizing solid MNs also has the drawback of having rapidly healing perforated pores, which reduces the amount of agricultural chemicals that can be delivered. Furthermore, solid MNs have not often been the chosen MN types for agricultural applications due to the potential of infections in the perforated area [79]. The need for additional MN types has increased in this direction. Notwithstanding the drawbacks above, the design of solid MNs is especially important since coated and dissolving MNs are subsequently manufactured using them as a template.

The production of MNs with hollow structures has increased recently because of advancements in microfabrication processes [80]. With a size reduction to the micron level, these MNs resemble traditional hypodermic needles. The hollow MNs use pressure-driven flow to move the medicine or sample through the reservoir steadily, relying on the "poke-and-flow" method [81]. By either delivering or gathering the necessary drug or samples, Hollow MNs can be used in drug release and detection experiments [82]. Hollow MNs [83] are made in large part from silicon [84], metal [85], polymeric [86], and glass materials [87]. Traditionally, the MEMS method is employed for large-scale MN manufacture, which is followed by deep reactive ion etching (DRIE).

Hollow MNs have also been designed and made using 3D laser lithography [88], micro-stereolithography [89], CNC machining [90], and UV lithographic methods [91]. Hollow MNs are inappropriate for therapy and prevention in agriculture, much like dissolving MNs. Hollow MNs are very rarely used in agricultural products. Such MN systems have been used in the past to electrochemically detect p-cresol, a phenolic substrate necessary to assess a plant's potential for biofuel [92]. These systems have been applied to the surfaces of leaves, stems, and petals. Even though they are not frequently used in agriculture, they might be included as a detecting tool in agricultural products. Hollow MNs have an advantage over coated MNs, which are frequently used in agriculture. Compared to other types of MNs, hollow MNs may extract a volume of sample that is orders of magnitude bigger, which can significantly increase the analysis's accuracy and limit of detection. Beyond simple detection, hollow molecular nanoparticles have a number of uses, including medication administration and precision agriculture.

Recently, there has been a noticeable interest in dissolving MNs because of their excellent qualities, which include high drug loading capacity, ease of use, and simplicity in manufacture [93, 94]. These MNs are typically employed in drug delivery applications as opposed to biosensing ones, and they serve as an adjunct to the "poke-and-release" approach [95]. To put it briefly, medications are incorporated into microneedles (MNs), which dissolve once they enter the target tissues and release the pharmaceuticals under controlled conditions [96]. Typically, there are two steps involved in their manufacturing [97]; a solid MN is first created and utilized as a male template. Subsequently, a polymer, usually polydimethylsiloxane (PDMS), is applied to the male template to create female PDMS molds. These molds are then allowed to cure for about two hours on a hot plate at a temperature of around 80 °C. The cured female PDMS mold is then covered with a polymer solution, which is then polymerized under specific circumstances. These parameters typically include temperature, solvent type, length of polymerization, monomer concentration, and crosslinking agents. For instance, the degree of polymerization is influenced by monomer concentration. To improve the polymer's stability and mechanical qualities, crosslinking agents can be added. The rate of polymerization is directly impacted by temperature. Lower temperature values typically lead to a slower rate of polymerization, which ultimately results in incomplete polymerization. Elevated temperatures speed up the polymerization process, but they can also generate unintended byproducts. The choice of solvent affects the solubility of the monomer and the polymer mixture, changing the final polymer's mechanical properties and shape. The degree of polymerization and the molecular weight distribution are significantly altered by the total amount of time allotted for the polymerization reaction. Certain polymer deterioration may result from longer polymerization times. Because of this, these parameters need to be carefully chosen based on the type of polymer. The cured polymer is then peeled from the PDMS surface to produce the dissolving MNs when the necessary polymerization has taken place. It is possible to utilize photopolymerization to expedite the curing process [98].

For this kind of synthesis, other methods like micro molding [99] and drawing lithography [100] have also been used. Since they are also regarded as polymeric MNs, there are increased interest in studies pertaining to polymers [101]. The most often utilized materials are those that are water soluble, intelligent, and biodegradable polymers (such as gelatin and polyvinyl alcohol (PVA)) [102]. Hollow MNs can accomplish zero-order medication release, disintegrate without leaving behind hazardous components, and are biocompatible [103]. However, because polymers are expensive and complex to synthesize, reusability is not practicable, which limits their usage in agriculture.

Coated MNs are biosensing and delivery gadgets that work on the basis of the "coat-and-poke" method. They are produced by coating MN tips with immobilizing chemicals [104]. Delivery or detection procedures can be completed in a single step as opposed to the poke-and-patch method (solid MNs) [96]. The amount of coating agents that must be immobilized on the surface is influenced by the size and shape of the MN arrays, which makes them fundamental [105]. The coating agent is either released from the MN tip to treat a disease or is captured as a signal (chemical and electrochemical) to detect infections or toxin levels once they interact with the plant's interstitial fluid [106]. Their designs' fabrication techniques are comparable to those applied to solid MNs. Their fabrication entails a second immobilization process, in contrast to solid MN. The literature has reported on a wide range of coating procedures [107–109]. Dip-coating [110], inkjet coating [111], immersion coating [112], drop coating [113], and spray coating [114] are the most often used coating techniques. Materials selection is crucial, given that the interactions between the MN material and coating agent dictate the drug loading capacity and biosensing precision. Materials like the *Francisella novicida* bacterium [115], the Au@SnO_2-vertical graphene [116], the nanocomposite of carbon nanotubes and cellulose nanocrystals

[117], titanium/gold [118], and gold/copper [119] have all been used as coating agents for plant identification up to this point.

Numerous parameters, such as polyphenol concentrations, water transport efficiency, light exposure, bioimpedance, nitrate concentration, and hormone levels, have been evaluated and tracked in plants using coated MNs. All things considered, MN technology is a promising means of overcoming the limitations associated with traditional methods and their application in agriculture. Technologies that are more timely, accurate, and ecologically friendly are essential for a more realistic agricultural future. MN technology is developing, and more and more research is being done on detection and therapeutics. Numerous reviews have addressed the use of MNs in such situations, and the application of MNs in animals as a delivery and detection technique has been extensively studied [120–125].

Comprehending the characteristics of soil is essential for effective soil management. However, managing soil resources sustainably is made more difficult by the varied composition of the soil. Maps containing soil data from specific regions of interest have been compiled and recorded historically. The ancient Egyptians were most likely the first civilization to create measuring tools for mapping soil and land resources and surveying land expanses. Over time, other societies made incremental improvements to the measuring tools and incorporated measures of land slope into soil surveys. The Romans were able to create soil maps with more precision and detail than earlier civilizations thanks to these advancements [126]. Ancient farmers were also astute observers of the performance of their crops. On various soil types, they saw advantages from applying varying quantities of manure and liming ingredients. During the 1620s, colonists saw how Indian farmers applied fertilizer differently, putting fish right at the roots of every corn plant [127]. When applying limestone materials at varying rates, Goering [128] proposed marking a field at predetermined intervals to identify the field location. All of these early studies helped to establish the foundation of soil understanding and improved soil management. Recent advancements in accurate soil survey tools and soil analysis techniques have been crucial to the description of soil. Sophisticated evaluations of soil heterogeneity attributes could be produced using modern information technology. With the use of topographic data and photo-grammatic techniques to analyze aerial pictures, it is now possible to extend soil boundaries outside the sample region [129]. Our understanding of soil is expanding and growing with time.

Our understanding of how to manage soil resources is typically derived from site descriptions, individual profiles, or pit excavations. These data are then converted into qualitative soil maps that are distributed throughout a landscape. Because of this, the traditional method ultimately depends on extrapolating from small sample sites or locations in order to create a generalized soil management plan and suggestion packages. As a result of a complex interaction between soil-forming elements intrinsic to edhaphic and climatic bases, soils are actually very diverse and varied in space, both laterally and vertically [130]. Soil spatial variability directly affects crop yield since cropping is the only and final use of soil resources in arable environments. Managing spatially variable soil qualities has long been a challenging task, largely due to sampling effort, incomplete data, poor information quality, and ambiguity in decision-making [131]. However, the need to switch from a limited-sampling-based soil management approach to an exhaustive-information-based management approach is growing as a result of growing awareness among soil scientists and agriculturists as well as the introduction of new technologies and soil sensors. With the potential to give quantitative, supplementary, and complementary information on soil variability, soil sensors will be able to lead the way in this new era.

In close proximity to the ground, proximal soil sensors were function [132]. The proximate soil sensors gather soil data at a far finer spatial resolution than do remote sensors that are launched into the air or space and are meant for comparable purposes. They lower operating and data acquisition costs while also gathering high resolution soil information. In the past ten years, electrical and electromagnetic induction (EMI) approaches have been the foundation of the most popular proximal sensors [133]. These sensors are ideal for doing in-depth research on arable lands. A geographical information system (GIS) can be used to aggregate the sensor-collected data to make site-specific decisions, which can have a positive impact on crop performance and reduce the environmental impact. The sensors also have the added benefit of being time-efficient because they allow for a mobile survey arrangement. It has been discovered that EMI-based sensors are among the most useful proximal sensors for inferring a variety of soil attributes. A multitude of soil properties, including soil moisture, salinity, clay, organic matter, depth to contrasting soil layers, and bulk density variation, affect the apparent electrical conductivity (ECa) of the soil, which is measured by EMI sensors [134].

14.4 Soil Information and Soil Variability

Typically, a small number of soil samples are taken in order to gather soil information for the assessment of soil attributes. Generally, the samples are combined (without considering their locations), subjected to laboratory analysis, and the averaged results are applied as a single value to the whole research region [135]. A sampling design specifies how many sampling points are used and where they are placed within a field. It is typical practice to overlay a uniform grid with a predetermined size on a field's boundaries or to randomly select sample locations [136]. Sample resolution is mostly determined by expenses and expected return because underlying field variability is sometimes not pre-understood. Ultimately, no single sample technique has been shown to be ideal, and this issue persists because varied fields and soil characteristics exist [137]. Scientists know that soil can change in space over time, either gradually or continuously, or occasionally even suddenly over small distances. Due to variations in temperature, origin, terrain, parent material, time, and historical and contemporary cultural activities, soils differ greatly from one another [138]. Crop productivity is significantly impacted by the physical, chemical, and biological heterogeneity of the soil. Insufficient auger observations are insufficient to obtain the fine-grained data needed to control field variability. On the other hand, sustainable management of soil attributes requires a thorough quantitative understanding of soil variability.

Crop production systems that manage soil spatial variability must make a number of interconnected decisions that ultimately impact crop yield. Crop output is frequently unpredictable because of unmeasured variability in agronomic circumstances as well as random forces, namely the weather [139]. This part of the management system's uncertainty may be decreased by using precise soil management technologies to enhance decisions made about particular site circumstances. According to Yang et al. [140], the performance is contingent upon the interplay between site circumstances and stochastic elements, including climate. Instead of focusing on individual components taken in isolation, site-specific management needs to be based on procedures that allow comprehensive information to be collected to incorporate the simultaneous effects of the most relevant factors determining yield.

Effective techniques, for example, will probably rely on soil properties related to nutrient response curves, meteorological circumstances, and other variables apart from the fertilizer applied. At the start of the crop season, the capacity to adapt to shifting and dynamic circumstances is perhaps just as crucial as comprehending temporal unpredictability. The consequences of temporal variation could be significantly reduced by intelligent soil management that enables the producer to respond to seasons with higher yield potential than usual while initially adjusting for a lower yield objective [141]. It would be especially useful to be able to adapt to temporal variations when managing weather-related risk. Lack of water to support aquatic crops is likely the biggest danger to crop productivity for example in irrigated paddy production areas [142]. Paddy farmers frequently choose to apply fertilizer at lower initial rates due to limited water availability, forgoing the anticipated return. Similar circumstances arise when the quantity of nitrogen that may be supplied in a single application is restricted due to the possibility of water losses through percolation and nutrient leaching in light sandy soils [143].

14.5 Conclusion

In general, the accuracy increases with decreasing scale of measurement. Greater coverage drives this relationship in the right-top direction, the advancement of the technologies covered in this chapter. The chapter provides numerous case studies and examples that demonstrate how geotechnical engineers can enhance their comprehension of soil behavior and soil-structure interactions by utilizing the new dataset derived from new sensor and monitoring technologies. The soil monitoring can be accomplished with technologies like thermal needle sensor, pre-drilled hole filled with thermally conductive grease and many others. However, this chapter reviews the possible design of a microneedle-based sensor and its application to soil health monitoring, with an emphasis on the working principles of the sensor and needle in general, the development of new sensors, and possible implementation methods in different geotechnical engineering. There are still some limitations regarding the proposed idea that limit the application and development of these sensing technologies as a whole in geotechnical health monitoring due to the authors' ongoing study, and the detail design is still in progress. However, the studies took suggestions from experts to properly fabricate, optimize the concept of sensors, explore new installation methods for sensors on geotechnical structures, improve the interrogation technique for higher sensitivity, and manage to fulfil all properties to optimize performance so that the accuracy of the respective sensor is relevant to the situation.

Acknowledgements This research was supported by Universiti Teknologi Malaysia and Faculty Civil Engineering.

References

1. Haokip, I.C., Devi, M.H., Jangir, C., Sunil, B.H., Kundu, S., Surekha, K., Rao, C.S.: Climate Change and Indian Agriculture: challenges and Adaptation Strategies. ICAR-National Academy of Agricultural Research Management, India (2020)
2. Kumar, S.S., Mahale, A.G., Patil, A.C.: Mitigation of climate change through approached agriculture-soil carbon sequestration (a review). Curr. J. Appl. Sci. Technol. 47–64 (2020)
3. World Meteorological Organization.: New climate predictions assess global temperatures in coming five years (2024). https://public.wmo.int/en/media/pressrelease/new-climate-predictions-assess-global-temperatures-coming-five-years
4. Nobre, C.: Deforested parts of Amazon 'emitting more CO2 than they absorb (2024). https://www.bbc.com/news/science-environment-51464694
5. Andersland, O.B., Ladanyi, B.: Frozen Ground Engineering, pp. 1–4. Wiley, Toronto (2003)
6. Li, S., Zhang, M., Pei, W., Lai, Y.: Experimental and numerical simulations on heat-water-mechanics interaction mechanism in a freezing soil. Appl. Therm. Eng. **132**, 209–220 (2018). https://doi.org/10.1016/j.applthermaleng.2017.12.061
7. Dhakal, A.S., Sidle, R.C.: Distributed simulations of landslides for different rainfall conditions. Hydrol. Process. **18**(4), 757–776 (2004). https://doi.org/10.1002/hyp.1365
8. Coe, J.A., Godt, J.W.: Review of approaches for assessing the impact of climate change on landslide hazards, 371–377 (2012)
9. Huang, D., Gu, D.M.: Influence of filling-drawdown cycles of the three gorges reservoir on deformation and failure behaviors of anaclinal rock slopes in the Wu gorge. Geomorphology **295**, 489–506 (2017). https://doi.org/10.1016/j.geomorph.2017.07.028
10. Ortuño, M., Guinau, M., Calvet, J., Furdada, G., Bordonau, J., Ruiz, A., Camafort, M.: Potential of airborne LiDAR data analysis to detect subtle landforms of slope failure: Portainé, Central Pyrenees. Geomorphology **295**, 364–382 (2017). https://doi.org/10.1016/j.geomorph.2017.07.015
11. Ishikawa, T., Tokoro, T., Seiichi, M.: Geohazard at volcanic soil slope in cold regions and its influencing factors. Jpn. Geotech. Soc Spec Publ **1**(1), 1–20 (2015). https://doi.org/10.3208/jgssp.KEY-1
12. Chiang, S.H., Chang, K.T.: The potential impact of climate change on typhoon-triggered landslides in Taiwan, 2010–2099. Geomorphology **133**(3–4), 143–151 (2011). https://doi.org/10.1016/j.geomorph.2010.12.028
13. Huang, C.C., Yuin, S.C.: Experimental investigation of rainfall criteria for shallow slope failures. Geomorphology **120**(3–4), 326–338 (2010). https://doi.org/10.1016/j.geomorph.2010.04.006
14. Gariano, S.L., Guzzetti, F.: Landslides in a changing climate. Earth Sci. Rev. **162**, 227–252 (2016). https://doi.org/10.1016/j.earscirev.2016.08.011
15. Zhu, Y., Ishikawa, T., Subramanian, S.S., Luo, B.: Simultaneous analysis of slope instabilities on a small catchment-scale using coupled surface and subsurface flows. Eng. Geol. **275**, 105750 (2020). https://doi.org/10.1016/j.enggeo.2020.105750

16. Japan Meteorological Agency (JMA), 2020.: Past weather data. http://www.data.jma.go.jp/gmd/risk/obsdl/index.php (in Japanese). Accessed 30 June 2024
17. Dijkstra, T.A., Dixon, N.: Climate change and slope stability in the UK: challenges and approaches. Q. J. Eng. Geol. Hydrogeol. **43**(4), 371–385 (2010). https://doi.org/10.1144/1470-9236/09-036
18. Crozier, M.J.: Deciphering the effect of climate change on landslide activity: a review. Geomorphology **124**(3–4), 260–267 (2010). https://doi.org/10.1016/j.geomorph.2010.04.009
19. Shou, K.J., Yang, C.M.: Predictive analysis of landslide susceptibility under climate change conditions-a study on the Chingshui River watershed of Taiwan. Eng. Geol. **192**, 46–62 (2015). https://doi.org/10.1016/j.enggeo.2015.03.012
20. Ciabatta, L., Camici, S., Brocca, L., Ponziani, F., Stelluti, M., Berni, N., Moramarco, T.: Assessing the impact of climate-change scenarios on landslide occurrence in Umbria region, Italy. J. Hydrol. **541**, 285–295 (2016). https://doi.org/10.1016/j.jhydrol.2016.02.007
21. Song, H.B., Pei, H.F., Zhu, H.H.: Monitoring of tunnel excavation based on the fiber Bragg grating sensing technology. Measurement **169**, 108334 (2021). https://doi.org/10.1016/j.measurement.2020.108334
22. Pei, H.F., Zhang, S.Q., Borana, L., et al.: Slope stability analysis based on real-time displacement measurements. Measurement **131**, 696–693 (2019a). https://doi.org/10.1016/j.measurement.2018.09.019
23. Zhang, S.Q., Pei, H.F.: Determining bound water content of montmorillonite from molecular simulations. Eng. Geol. **294**, 106353 (2021). https://doi.org/10.1016/j.enggeo.2021.106353
24. Zhu, H.H., Garg, A., Yu, X., et al.: Editorial for internet of things (IoT) and artificial intelligence (AI) in geotechnical engineering. J. Rock Mech. Geotech. Eng. **14**, 1025e1027 (2022a). https://doi.org/10.1016/j.jrmge.2022.07.001
25. Ma, J., Pei, H., Zhu, H., Shi, B., Yin, J.: A review of previous studies on the applications of fiber optic sensing technologies in geotechnical monitoring. Rock Mech. Bull. **2**, 100021 (2023)
26. Carslaw, H.S., Jaeger, J.C.: Conduction of Heat in Solids, 2nd edn., pp. 1–455. Oxford University Press, London, UK (1959)
27. Incropera, F.P., Burkhard, S.: Fundamentals of Heat and Mass Transfer, 4th ed., pp. 1–241. Wiley, Hoboken, NJ, USA (1996)
28. Chen, S.X.: Thermal conductivity of sands. Heat Mass Transf. **44**, 1241–1246 (2008)
29. Yun, T.S., Santamarina, J.C.: Fundamental study of thermal conduction in dry soils. Granul. Matter. **10**, 197–207 (2008)
30. Nasirian, A., Cortes, D.D., Dai, S.: The physical nature of thermal conduction in dry granular media. Géotech. Lett. **5**, 1–5 (2015)
31. Kumai, W., Hashimoto, I., Ohsawa, S., Mitani, M., Matsuda, Y.: Completion of high-efficiency water pipe cooling system for underground transmission line. IEEE Trans. Power Deliv. **9**, 585–590 (1994)
32. Brandl, H.: Energy foundations and other thermos-active ground structures. Géotechnique **56**, 81–122 (2006)
33. Johnston, I.W., Narsilio, G.A., Colls, S.: Emerging geothermal energy technologies. KSCE J. Civ. Eng. **15**, 643–653 (2011)
34. Choi, J.C., Lee, S.R., Lee, D.S.: Numerical simulation of vertical ground heat exchangers: intermittent operation in unsaturated soil conditions. Comput. Geotech. **38**, 949–958 (2011)
35. Sass, J.H., Stone, C., Munroe, R.J.: Thermal conductivity determinations on solid rock-A comparison between a steady-state divided-bar apparatus and a commercial transient line-source device. J. Volcanol. Geotherm. Res. **20**, 145–153 (1984)
36. Van Der Held, E.F.M., Van Drunen, F.G.: A method of measuring thermal conductivity of liquids. Physica **15**, 865–881 (1949)
37. Blackwell, J.H.: The axial-flow error in the thermal conductivity probe. Can. J. Phys. **34**, 412–417 (1956)
38. Von Herzen, R., Maxwell, A.E.: The measurement of thermal conductivity of deep-sea sediments by a needle-probe method. J. Geophys. Res. **64**, 1557–1563 (1959)

39. Woodside, W.M.J.H., Messmer, J.H.: Thermal conductivity of porous media I. Unconsolidated sands. J. Appl. Phys. **32**, 1688–1699 (1961)
40. Jones, B.W.: Thermal conductivity probe: development of method and application to a coarse granular medium. J. Phys. E Sci. Instrum. **21**, 832–839 (1988)
41. Gori, F., Marino, C., Pietrafesa, M.: Experimental measurements and theoretical predictions of the thermal conductivity of two phases glass beads. Int. Commun. Heat Mass. **28**, 1091–1102 (2001)
42. Go, G.H., Lee, S.R., Kim, Y.S., Park, H.K., Yoon, S.: A new thermal conductivity estimation model for weathered granite soils in Korea. Geomech. Eng. **6**, 359–376 (2014)
43. Standard Test Method for Determination of Thermal Conductivity by Thermal Needle Probe Procedure: ASTM D5334-14; ASTM International: West Conshohocken, PA, USA (2014)
44. Lee, S.J., Choi, J.C., Baek, S., Kwon, T.H., Ryu, H.H., Song, K.I.: Use of a pre-drilled hole for implementing thermal needle probe method for soils and rocks. Energies **9**, 846 (2016). https://doi.org/10.3390/en9100846
45. Hukseflux: TP01 manual v2028, 1–35, (2020)
46. Hukseflux: STP01 manual v2028, 1–31, (2020)
47. Jones, L.D., Jefferson, I.: Expansive Soils, pp. 413–441. ICE Publishing, London, UK (2012)
48. Suits, L.D., Sheahan, T., Prakash, K., Sridharan, A.: Free swell ratio and clay mineralogy of fine-grained soils. Geotech. Test. J. **27**, 220–225 (2004)
49. Tang, L., Cong, S., Geng, L., Ling, X., Gan, F.: The effect of freeze-thaw cycling on the mechanical properties of expansive soils. Cold Reg. Sci. Technol. **145**, 197–207 (2018)
50. Guo, Y., Shan, W.: Monitoring and experiment on the effect of freeze-thaw on soil cutting slope stability. Procedia Environ. Sci. **10**, 1115–1121 (2011)
51. Yang, Z., Lv, J., Shi, W., Zhang, Q., Lu, Z., Zhang, Y., Ling, X.: Model test study on stability factors of expansive soil slopes with different initial slope ratios under freeze-thaw conditions. Appl. Sci. **11**, 8480 (2021). https://doi.org/10.3390/app11188480
52. Farouki, O.T.: Thermal properties of soils. Cold Regions Research and Engineering Laboratory, Hanover, New Hampshire, U.S.A (1981)
53. Zainal Ariffin, Z.: Slope Stability Analysis in Residual Soils. Bachelor of Civil Engineering. Universiti Teknologi Petronas (2022)
54. Ece, E., Es, I., Inci, F.: Microneedle technology as a new standpoint in agriculture: treatment and sensing. Mater. Today **68**, 275–297 (2023)
55. Kundu, A., Nogueira Campos, M.G., Santra, S., Rajaraman, S.: Precision vascular delivery of agrochemicals with micromilled microneedles (μMMNs). Sci. Rep. **9**, 1–8 (2019)
56. Prausnitz, M.R.: Microneedles for transdermal drug delivery. Adv. Drug Deliv. Rev. **56**, 581–587 (2004)
57. Nejad, H.R., et al.: Low-cost and cleanroom-free fabrication of microneedles. Microsyst. Nanoeng. **4**, 1–7 (2018)
58. Tilli, M., et al.: Handbook of Silicon Based MEMS Materials and Technologies. Elsevier (2020)
59. Parker, E.R., et al.: Bulk micromachined titanium microneedles. J. Microelectromech. Syst. **16**, 289–295 (2007)
60. Yoshida, K., et al.: Implantation mechanics of tungsten microneedles into peripheral nerve trunks. Med. Biol. Eng. Comput. **45**, 413–420 (2007)
61. Hoang, M., Ita, K., Bair, D.: Solid microneedles for transdermal delivery of amantadine hydrochloride and Pramipexole Dihydrochloride. Pharmaceutics **7**, 379–396 (2015)
62. Turner, J.G., et al.: Hydrogel-forming microneedles: current advancements and future Trends. Macromol. Biosci. **21**, 2000307 (2021)
63. Choi, S.-O., et al.: An electrically active microneedle array for electroporation. Biomed. Microdevices **12**, 263–273 (2010)
64. Sadeqi, A., et al.: Hard polymeric porous microneedles on stretchable substrate for transdermal drug delivery. In: 40th Annual International Conference IEEE Engineering in Medicine and Biology Society IEEE 2018, pp. 5737–5740 (2018)

65. Wang, Q.L., Zhu, D.D., Chen, Y., Guo, X.D.: A fabrication method of microneedle molds with controlled microstructures. Mater. Sci. Eng. C **65**, 135–142 (2016)
66. Kim, D.-S., et al.: Microneedle array patch (MAP) consisting of crosslinked hyaluronic acid nanoparticles for processability and sustained release. Pharm. Res. **37**, 50 (2020)
67. Cha, K.J., et al.: Simple and cost-effective fabrication of solid biodegradable polymer microneedle arrays with adjustable aspect ratio for transdermal drug delivery using acupuncture microneedles. J. Micromech. Microeng. **24**, 115015 (2014)
68. Baek, S., Jeon, E., Park, K.S., Yeo, K.-H., Lee, J.: Monitoring of water transportation in plant stem with microneedle sap flow sensor. J. Microelectromech. Syst. **27**, 440–447 (2018)
69. Dragicevic N., Maibach, H.I.: Percutaneous Penetration Enhancers Physical Methods in Penetration Enhancement. Springer (2017)
70. Gülçür, M., et al.: A cost-effective process chain for thermoplastic microneedle manufacture combining laser micro-machining and micro-injection moulding. J. Manuf. Sci. Technol. **32**, 311–321 (2021)
71. Li, J., Zhou, Y., Yang, J., et al.: Fabrication of gradient porous microneedle array by modified hot embossing for transdermal drug delivery. Mater. Sci. Eng. C **96**, 576–582 (2019)
72. Miller, P.R., et al.: Fabrication of hollow metal microneedle arrays using a molding and electroplating method. MRS Adv. **4**, 1417–1426 (2019)
73. Lee, K., et al.: Drawing lithography: three-dimensional fabrication of an ultrahigh-aspect-ratio microneedle. Adv. Mater. **22**, 483–486 (2010)
74. Kim, M.J., et al.: Fabrication of circular obelisk-type multilayer microneedles using micro-milling and spray deposition. Front. Bioeng. Biotechnol. **6**, 54 (2018)
75. FarajiRad, Z., Prewett, P.D., Davies, G.J.: High-resolution two-photon polymerization: the most versatile technique for the fabrication of microneedle arrays. Microsyst. Nanoeng. **7**, 71 (2021)
76. Krieger, K.J., et al.: Simple and customizable method for fabrication of high-aspect ratio microneedle molds using low-cost 3D printing. Microsyst. Nanoeng. **5**, 1–14 (2019)
77. Yan, G., et al.: Evaluation needle length and density of microneedle arrays in the pretreatment of skin for transdermal drug delivery. Int. J. Pharm. **391**, 7–12 (2010)
78. Li, Q.Y., et al.: A solid polymer microneedle patch pretreatment enhances the permeation of drug molecules into the skin. Rsc Adv. **7**, 15408–15415 (2017)
79. Lee, K.J., et al.: A practical guide to the development of microneedle systems—In clinical trials or on the marketInt. J. Pharm. **573**, 118778 (2020)
80. Park, K.: Nanotechnology: what it can do for drug delivery. J. Control. Release Off. J. Control. Release Soc. **120**, 1 (2007)
81. Ita, K.: Ceramic microneedles and hollow microneedles for transdermal drug delivery: two decades of research. J. Drug Deliv. Sci. Technol. **44**, 314–322 (2018)
82. Gardeniers, H.J.G.E., et al.: Silicon micromachined hollow microneedles for transdermal liquid transport. J. Microelectromech. Syst. **12**, 855–862 (2003)
83. Bariya, S.H., et al.: Microneedles: an emerging transdermal drug delivery system. J. Pharm. Pharmacol. **64**, 11–29 (2012)
84. Ahn, B.: Optimal microneedle design for drug delivery based on insertion force experiments with variable geometry. Int. J. Control. Autom. Syst. **18**, 143–149 (2020)
85. Davis, S.P., Prausnitz, M.R., Allen, M.G.: In: TRANSDUCERS '03. 12th Int. Conf. Solid-State Sensors, Actuators Microsystems. Dig. Tech. Pap. (Cat. No.03TH8664), IEEE, n.d., pp. 1435–1438
86. Dardano, P., et al.: One-shot fabrication of polymeric hollow microneedles by standard photolithography. Polymers (Basel) **13**, 520 (2021)
87. Jiang, J., et al.: Intrascleral drug delivery to the eye using hollow microneedles. Pharm. Res. **26**, 395–403 (2009)
88. Suzuki, M., Takahashi, T., Aoyagi, S.: 3D laser lithographic fabrication of hollow microneedle mimicking mosquitos and its characterization. Int. J. Nanotechnol. **15**, 157–173 (2018)
89. Kim, H., Moon, S.K., Seo, M.: Hybrid layering scanning-projection micro-stereolithography for fabrication of conical microlens array and hollow microneedle array. Microelectron. Eng. **153**, 15–19 (2016)

90. Le Thanh, H., et al.: Fabrication of monolithic SU-8 microneedle arrays having different needle geometries using a simplified process. J. Microelectromech. Syst. **24**, 1583–1593 (2015)
91. Zhu, M.W., et al.: Silica needle template fabrication of metal hollow microneedle arrays. J. Micromech. Microeng. **19**, 115010 (2009)
92. Harper, J.C., Brozik, S.M., Flemming, J.H., McClain, J.L., Polsky, R., Raj, D., et al.: Fabrication and testing of a microneedles sensor array for p-cresol detection with potential biofuel applications. ACS Appl. Mater. Interfaces **1**, 1591–1598 (2009)
93. Erdem, Ö., et al.: Recent advances in microneedle-based sensors for sampling, diagnosis and monitoring of chronic diseases. Biosensors **11**, 296 (2021)
94. Nagarkar, R., et al.: A review of recent advances in microneedle technology for transdermal drug delivery. J. Drug Deliv. Sci. Technol. **59**, 101923 (2020)
95. Ita, K.: Dissolving microneedles for transdermal drug delivery: advances and challenges. Biomed. Pharmacother. **93**, 1116–1127 (2017)
96. Jin, X., et al.: Insulin delivery systems combined with microneedle technology. Adv. Drug Deliv. Rev. **127**, 119–137 (2018)
97. Lin, S., et al.: Novel strategy for immunomodulation: dissolving microneedle array encapsulating thymopentin fabricated by modified two-step molding technology. Eur. J. Pharm. Biopharm. **122**, 104–112 (2018)
98. Lim, S.H., et al.: High resolution photopolymer for 3D printing of personalized microneedle for transdermal delivery of anti-wrinkle small peptide. J. Control. Release **329**, 907–918 (2021)
99. Rodgers, A.M., Courtenay, A.J., Donnelly, R.F.: Dissolving microneedles for intradermal vaccination: manufacture, formulation, and stakeholder considerations. Expert Opin. Drug Deliv. **15**, 1039–1043 (2018)
100. Lee, K., Jung, H.: Drawing lithography for microneedles: a review of fundamentals and biomedical applications. Biomaterials **33**, 7309–7326 (2012)
101. Li, J., et al.: Microneedle patches as drug and vaccine delivery platform. Curr. Med. Chem. **24**, 2413–2422 (2017)
102. Vora, L.K., et al.: Pullulan-based dissolving microneedle arrays for enhanced transdermal delivery of small and large biomolecules. Int. J. Biol. Macromol. **146**, 290–298 (2020)
103. Amodwala, S., Kumar, P., Thakkar, H.P.: Statistically optimized fast dissolving microneedle transdermal patch of meloxicam: a patient friendly approach to manage arthritis. Eur. J. Pharm. Sci. **104**, 114–123 (2017)
104. Sharma, S., et al.: A pilot study in humans of microneedle sensor arrays for continuous glucose monitoring. Anal. Methods **10**, 2088–2095 (2018)
105. Al-Qallaf, B., Das, D.B., Davidson, A.: Transdermal drug delivery by coated microneedles: geometry effects on drug concentration in blood. Asia-Pacific J. Chem. Eng. **4**, 845–857 (2009)
106. Andrianov, A.K., Marin, A., DeCollibus, D.P.: Microneedles with intrinsic immunoadjuvant properties: microfabrication, protein stability, and modulated release. Pharm. Res. **28**, 58–65 (2011)
107. Haj-Ahmad, R., et al.: Microneedle coating techniques for transdermal drug delivery. Pharmaceutics **7**, 486–502 (2015)
108. Ingrole, R.S.J., Gill, H.S.: Microneedle coating methods: a review with a perspective. J. Pharmacol Exp. Ther. **370**, 555–569 (2019)
109. Gill, H.S., Prausnitz, M.R.: Coated microneedles for transdermal delivery. J. Control. Release **117**, 227–237 (2007)
110. Zhang, Y., et al.: Development of lidocaine-coated microneedle product for rapid, safe, and prolonged local analgesic action. Pharm. Res. **29**, 170–177 (2012)
111. Uddin, M.J., et al.: Inkjet printing of transdermal microneedles for the delivery of anticancer agents. Int. J. Pharm. **494**, 593–602 (2015)
112. Matriano, J.A., et al.: Macroflux microprojection array patch technology: a new and efficient approach for intracutaneous immunization. Pharm. Res. **19**, 63–70 (2002)

113. Chen, X., et al.: Dry-coated microprojection array patches for targeted delivery of immunotherapeutics to the skin. J. Control. Release **139**, 212–220 (2009)
114. McGrath, M.G., Vrdoljak, A., O'Mahony, C., Oliveira, J.C., Moore, A.C., Crean, A.M.: Determination of parameters for successful spray coating of silicon microneedle arrays. Int. J. Pharm. **415**, 140–149 (2011)
115. Chandler, C.E., et al.: In vivo intradermal delivery of bacteria by using microneedle arrays. Infect. Immun. **86**, e00406–e00418 (2018)
116. Wang, Z., et al.: Au@SnO2-vertical graphene-based microneedle sensor for in-situ determination of abscisic acid in plants. Mater. Sci. Eng. C **127**, 112237 (2021)
117. Mugo, S.M., Lu, W.: Flexible molecularly imprinted electrochemical sensor for cortisol monitoring in sweat. Anal. Bioanal. Chem. **412**, 7063–7072 (2020)
118. Bukhamsin, A., et al.: Robust, long-term, and exceptionally sensitive microneedle-based bioimpedance sensor for precision farming. Adv. Sci. **8**, 2101261 (2021)
119. O'Flynn, B. et al.: Novel smart sensor glove for arthritis rehabilitation. In: IEEE Sensors, pp. 1–3. IEEE (2017)
120. Park, J.-H., et al.: A microneedle roller for transdermal drug delivery. Eur. J. Pharm. Biopharm. **76**, 282–289 (2010)
121. Coulman, S.A., et al.: Microneedle mediated delivery of nanoparticles into human skin. Int. J. Pharm. **366**, 190–200 (2009)
122. Chu, L.Y., Choi, S.-O., Prausnitz, M.R.: Fabrication of dissolving polymer microneedles for controlled drug encapsulation and delivery: bubble and pedestal microneedle designs. J. Pharm. Sci. **99**, 4228–4238 (2010)
123. Mohan, A.M.V., et al.: Continuous minimally-invasive alcohol monitoring using microneedle sensor arrays. Biosens. Bioelectron. **91**, 574–579 (2017)
124. Windmiller, J.R., et al.: Bicomponent microneedle array biosensor for minimally-invasive glutamate monitoring. Electroanalysis **23**, 2302–2309 (2011)
125. Ciui, B., et al.: Wearable wireless tyrosinase bandage and microneedle sensors: toward melanoma screening. Adv. Healthc. Mater. **7**, 1701264 (2018)
126. Dorling, D., Fairbairn, D.: Mapping: ways of Representing the World. Addison Wesley Longman Ltd., Harlow (1997)
127. Kellogg, C.E.: We seek; we learn. In: Stefferud, A. (ed.) The Yearbook of Agriculture Soil, pp. 1–11. US Government Printing Office, USDA. Washington, DC (1957)
128. Goering. C.E.: Recycling a concept. Agricultural Engineering Magazine. November. St. Joseph, Mich., American Society of Agricultural Engineering (1993)
129. Jensen, J.R.: Remote Sensing of the Environment: an Earth Resource Perspective. Prentice Hall, Upper Saddle River, New Zealand (2007)
130. Jenny, H.: Factors of Soil Formation. McGraw-Hill, New York (1941)
131. McBratney, A.B.: On variation, uncertainty and informatics in environmental soil-management. Aust. J. Soil Res. **30**, 913–935 (1992)
132. Viscarra Rossel, R.A., Fouad, Y., Walter, C.: Using a digital camera to measure soil organic carbon and iron contents. Biosyst. Eng. **100**, 149–159 (2008)
133. Adamchuk, V.I., Hummel, J.W., Morgan, M.T., Upadhyaya, S.K.: On-the-go soil sensors for precision agriculture. Comput. Electron. Agric. **44**, 71–91 (2004)
134. Saey, T., Van Meirvenne, M., De Smedt, P., Cockx, L., Meerschman, E., Islam, M.M., Meeuws, F.: Mapping depth-to-clay using fitted multiple depth response curves of a proximal EMI sensor. Geoderma **162**, 151–158 (2011)
135. Wollenhaupt, N.C., Mulla, D.J., Crawford, C.A.G.: Soil sampling and interpolation techniques for mapping spatial variability of soil properties. In: Pierce, F.J., Sadler, E.J. (eds.) The State of Site Specific Management for Agriculture, pp. 19–54. ASA-CSSA-SSSA, Madison (1997)
136. Earl, R., Taylor, J.C., Wood, G.A., Bradley, I., James, I.T., Waine, T., Welsh, J.P., Godwin, R.J., Knight, S.M.: Soil factors and their influence on within-field crop variability, part 1: field observation of soil variation. Biosyst. Eng. **84**, 425–440 (2003)

137. Hollands, K.R.: Relationship of nitrogen and topology. In: Robert, P.C., Rust, R.H., Larson, W.E. (eds.) Precision Agriculture: proceedings of the Third International Conference on Precision Agriculture, pp. 3–12. American Society of Agronomy, Crop Science Society of America and Soil Science Society of America, Madison, WISC (1996)
138. Sommer, M.: Influence of soil pattern on matter transport in and from terrestrial biogeosystems-a new concept for landscape pedology. Geoderma **133**, 107–123 (2006)
139. Lark, R.M.: Exploring scale-dependent correlation of soil properties by nested sampling. Eur. J. Soil Sci. **56**, 307–317 (2005)
140. Yang, C., Peterson, C.L., Shropshire, G.J., Otawa, T.: Spatial variability of field topography and wheat yield in the Palouse Region of the Pacific Northwest. Trans. ASAE **41**, 17–27 (1998)
141. Ferguson, R.B., Hergert, G.W., Schepers, J.S., Gotway, C.A., Cahoon, J.E., Peterson, T.A.: Site-specific nitrogen management of irrigated maize: yield and soil residual nitrate effects. Soil Sci. Soc. Am. J. **66**, 544–553 (2002)
142. De Datta, S.K.: Principles and Practices of Rice Production, p. 618. Wiley-Interscience, New York (1981)
143. Booltink, H.W.G., Verhagen, J., Bouma, J., Thornton, P.K.: Application of simulation models and weather generators to optimize farm management strategies. In: Robert, P.C., Rust, R.H., Larson, W.E. (eds.) Precision Agriculture: proceedings of the Third International Conference on Precision Agriculture, pp. 343–360. American Society of Agronomy, Crop Science Society of America, and Soil Science Society of America, Madison, WIS (1996)

Chapter 15
Microneedle-Based Technology for Agriculture and Safety

Mazura Jusoh, Nurul Asmak Md Lazim, and Zaki Yamani Zakaria

15.1 Introduction

Global concerns about crop health and food supply are a result of population growth and climate change. On the one hand, extreme weather events and plant diseases brought on by climate change can stress plants, which has a significant impact on crop productivity [1]. However, in order to satisfy the growing demand for food safety and quality, society must improve agricultural production, raise crop yields, and decrease crop losses [2]. In order to maintain food security and agricultural sustainability, these concerns are propelling the development of new crop management technologies toward the identification of plant diseases and the ongoing monitoring of plant stress [3].

The world's population is predicted to reach 9.7 billion by 2050 and 10.9 billion by 2100, from an anticipated 8.5 billion in 2030 [4]. This suggests that there is a rising market for agricultural products, which are the most important raw materials used to produce a wide range of food ingredients [5]. The Food and Agriculture Organization (FAO) and the Food and Drug Administration (FDA) have been forced to swiftly enact stringent regulations to help farmers and stockholders to ensure the safety of agricultural products due to the extraordinary rise in concerns about the quality and safety of crop products [6]. Agricultural products have consistently faced illnesses and pollutants. Despite these strong guidelines, the result is a decrease in crop output and the raising of concerns about consumer safety [7]. Stated that plants are either exposed to high concentrations of agrochemicals to minimize crop loss, or they are

M. Jusoh (✉) · N. A. Md Lazim · Z. Y. Zakaria
Faculty of Chemical and Energy Engineering, Universiti Teknologi Malaysia, Johor Bahru, Johor, Malaysia
e-mail: r-mazura@utm.my

M. Jusoh
Centre for Lipid Engineering and Applied Research (CLEAR), Universiti Teknologi Malaysia, Johor Bahru, Johor, Malaysia

N. Talreja et al. (eds.), *Microneedles (MNs)-Based Technology*, Series in BioEngineering, https://doi.org/10.1007/978-981-96-3916-8_15

given pharmaceutical substances to improve their health status against external causes [8]. Overuse of agrochemicals usually causes trace levels of potentially dangerous substances to accumulate in crops, which negatively impacts postharvest procedures (packing, storing, and transportation) [9]. To address this problem, agrochemicals are used in the recommended Codex Alimentarius limits, as established by the World Health Organization (WHO) and the Food and Agriculture Organization (FAO) [10]. As a result, agricultural products undergo extensive testing in laboratories utilizing detection techniques to guarantee that they adhere to legal limitations for contaminants and meet quality and safety standards [11]. In recent years, analytical chemistry has advanced significantly, enabling the detection of pollutants in agricultural products at trace levels [12]. The complexity in screening such compounds persists despite the development of a variety of rapid and sensitive screening techniques because the limits allowed by Codex Alimentarius for particular agrochemicals are incredibly low, and the effects of agrochemicals on public health and the environment are verifiable [13].

When compared to traditional agricultural methods, "smart agriculture," or "precision agriculture," has the potential to use less resources to produce more, which has drawn increasing attention. It also enhances the social aspects and working conditions of farming, ranching, and other related vocations [14]. In order to optimize production while taking into consideration the uncertainties and variabilities present in sustainable agricultural systems, smart agriculture is a collection of technology that includes sensors, information systems, improved machinery, and knowledgeable management [14, 15]. Advanced sensing systems are among the technologies that are most important because they gather and assess vital information for management and decision-making, particularly in situations where crop growth conditions vary greatly over time and space. These systems monitor crop developments and soil health and conditions. The characteristics of the soil, illnesses, weeds, pests, and past land management practices can all cause spatial heterogeneity. Specifically, long-term spatial patterns may be formed by certain soil characteristics (such as moisture, pH, nutrients, and so on) and plant diseases. The causes of temporal variability are weather patterns and management techniques [15].

Therefore, this chapter focused on microneedle technology and its various agricultural uses. Additionally, a summary of the components and manufacturing processes for sophisticated agricultural sensor types is given. In keeping with the bioeconomic strategy of Industrial Revolution (IR) 5.0, this chapter demonstrated the significant benefits of this new instrument in agriculture.

15.2 Types and Materials of Microneedles (MNs)

MNs can be broadly classified into five types: hydrogel-forming polymer **MNs**, hollow (or open-channel), dissolving, coated, and solid **MNs** [46] (see Fig. 15.1). Therapeutic chemicals can be injected into the tissue and reach its deeper layers by way of the micropores created on the tissue's surface by a solid microneedle array.

A coated microneedle is a solid needle that has been pre-coated with a desired medication prior to insertion. The microneedle surface area and coating layer thickness determine how much medication may be loaded [47]. Hollow **MNs**, like hypodermic needles, have an interior channel that allows fluid to pass between the skin and microneedle under pressure. The concept of open-channel microneedle design is relatively new [48, 49]. Similar to hollow **MNs**, the open-channel design offers two-dimensional (2D) fluid flows that can be utilized for biological fluid extraction and medication delivery. The main difficulty with hollow **MNs** is that trapped dermal tissue can clog the channels. Because of this, the open-channel design minimizes or completely eliminates tissue blockage inside the microneedle lumen because of its side-opened design [50]. Furthermore, it is easier to make open-channel **MNs** than hollow ones [51]. Biodegradable polymer materials are used to create dissolving **MNs**, and the medicine of interest is enclosed within the polymer. When the **MNs** are inserted into the skin, they dissolve and release the medication [52]. Polymer **MNs** that create hydrogels are composed of materials that super swell, namely crosslinked hydrogels. Because of its hydrophilic characteristics, this kind of microneedle may absorb a lot of water into its polymeric network. If the medication is incorporated within the hydrogel-forming **MNs**' polymeric structure, they can be utilized for drug delivery applications in addition to sampling [53]. Although there are many materials used to make **MNs**, silicon, metals, ceramics, and polymers are the most often utilized materials. The material that has been utilized the most frequently to create **MNs** is silicon. Furthermore, silicon was used to create the first **MNs**. However, the lengthy and intricate multistep manufacturing process needed to create silicon **MNs** limits their applicability in practical settings. Another issue is that silicon is a fragile material that might fracture during insertion and leave residues in the skin, which could cause reactions to foreign bodies. Additionally, metals, including titanium, stainless steel, nickel, and palladium, are used to make **MNs**.

Metals have superior mechanical and biocompatibility qualities than silicon. However, creating intricate microneedle structures out of metals is difficult. MNs have also been made from ceramics such as alumina, organically modified ceramics, calcium phosphate dihydrate, and calcium sulfate dihydrate because of their strong chemical characteristics and great compression resistance [54, 55]. The medical industry is becoming more interested in polymer materials because of their superior mechanical qualities, high biocompatibility, low toxicity, and less expensive production than alternative materials. Polymers that dissolve and produce hydrogels and don't dissolve are used to make polymer MNs. Microneedle fabrication involves the use of a wide range of polymers, including polylactic acid (PLA), poly (vinyl alcohol) (PVA), poly (methyl methacrylate) (PMMA), polyglycolic acid (PGA), poly (lactic-co-glycolic acid) (PLGA), poly (vinylpyrrolidone) (PVP), poly (carbonate), cyclic-olefin copolymer, polystyrene (PS), SU-8 photoresist, and poly (methyl vinyl ether-co-maleic anhydride) [54, 56]. These polymers are typically used to fabricate microneedle arrays that dissolve and create hydrogels. When a drug is inserted into dissolving MNs, it is often coated on the polymer material's surface or enclosed within the polymer structure. The drug will either dissolve entirely or partially from the surface. Microneedle architectures can be loaded with functional nanoparticles to

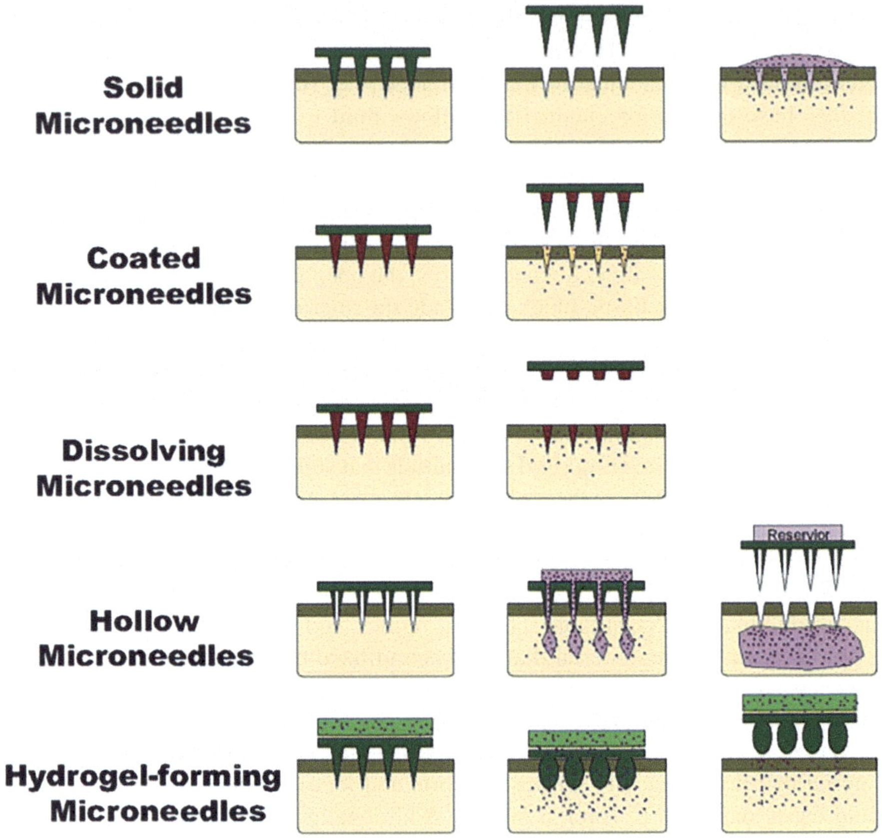

Fig. 15.1 Different types of MNs. The figure was reproduced with permission [59]

enable sample extraction, or they can be built with nonporous structures for sample collecting. Hydrogel MNs are a relatively new type of microneedle; upon insertion, the materials absorb the sample from the tissue and enlarge due to the hydrogel's swelling capabilities [53, 57, 58]. It is possible to extract the sample from the polymer structure using techniques like centrifuging or rinsing. Additional uses for the sample include illness monitoring and diagnosis.

15.3 Microneedle Technology and Its Applications in Agriculture

By 2050, crop output must double in order to meet the growing demand brought on by population growth [16]. However, given the effects of declining agricultural land availability and climate change, it seems doubtful that this objective will be

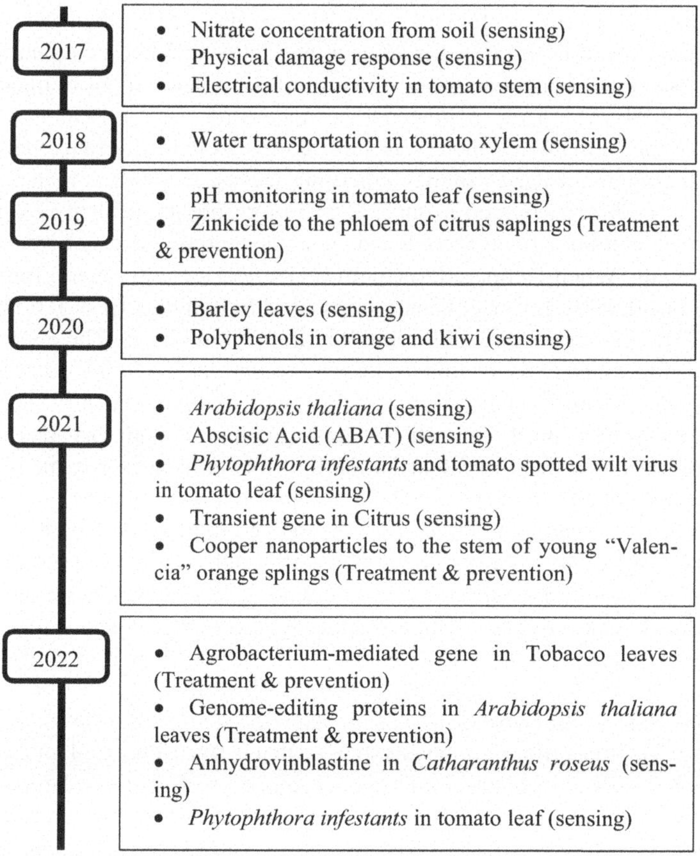

Fig. 15.2 Different types of microneedle-based applications in agriculture

met. Therefore, quick action is required to increase agricultural productivity, such as the use of sensors. Among the various sensor designs, MN-based sensors are favoured because of their capacity to efficiently gather important information by barely penetrating the plant's waxy skin. The prompt identification of plant diseases with the application of MN-based sensors can greatly increase agricultural production [17] (see Fig. 15.2).

15.3.1 Vegetables

Cao et al. studied the application of GA_3 to several vegetable plant types, such as spinach, lettuce, and tomatoes. The size of the MNs was adjusted to fit the anatomy of the plants, demonstrating the adaptability of the delivery device. The researchers employed MNs with a height of 1006 ± 65 μm and a base diameter of 259 ±

8 μm for injection into the tomato stem. The identical MNs that were used for ft-10 were employed in the trial for spinach and lettuce. When compared to other treatments, such as GA_3 foliar spray (GA_S), tomato plants ($n = 7$) injected with GA_3-loaded MNs (GA_INJ) displayed a substantial increase in height and similar diameters of the stem. While there was a discernible colour difference between the 4th and 6th compound leaves (newly emerging leaves, length less than 2 cm after treatment) in both GA_3-treated groups, it was more pronounced in GA_INJ than in GA_S. The amount of carotenoids was not substantially affected by any of the three treatments. When compared to Control, GA_INJ's freshly emerging leaf (4th leaf) had considerably lower concentrations of chlorophyll a ($p = 0.0049$) and b ($p = 0.0025$); in contrast, only chlorophyll b dropped in GA_S. The application of GA_3 reduced the sensitivity of mature leaves (second leaf). On the other hand, the chlorophyll b content was effectively reduced by GA_3 injection ($p = 0.039$). GA_3 had an impact on the third leaf, which was expanding in both GA_INJ and GA_S. These results show that GA_3 distribution by MNs influences growth at different stages in tomato plants and is more efficient than foliar spraying [18].

Following GA_3 treatment, lettuce ($n = 8$) showed morphological changes in the developing leaves as well as an extended stem. Although this type of treated lettuce is unsuitable for commercialization, it demonstrated that GA3 may be delivered using MNs and could be useful in breeding for rapid maturation. Furthermore, these phenotypes are associated with an overabundance of GA_3, which is easily adjusted to a range that is appropriate for vegetative growth [19]. Following injection with GA_3-loaded MNs, spinach likewise displayed an extended stem, demonstrating the MNs' efficacy ($n = 5$). Transcriptome analysis showed that in ft-10 mutant plants, wounding responses were reduced 24 h after microneedle insertion. The researchers used wild-type *L. sativa* var. Parris Island for time-course quantitative reverse transcription-polymerase chain reaction (RT-qPCR) tests to illustrate the wounding effects of MNs in a plant of great agricultural relevance. After injecting lettuce plants with GA_3-loaded MNs (GA_INJ) or MNs devoid of GA (Mock), samples were taken for RT-qPCR analysis either one, three, or twenty-four hours later. The closest lettuce homologs of the Arabidopsis wounding DEGs identified in our ft-10 transcriptome are the wounding response marker genes that were employed. Data on gene expression were standardized to control (untreated) plants and gathered at the appropriate intervals (1, 3, or 24 h). One to three hours following microneedle injection, the researchers saw an increase in gene expression for every gene evaluated in the Mock and GA_INJ groups. After the microneedle injection, 24 h later, the expression levels of injury indicators went back to normal. When compared to the expression levels at 24 h, the relative expression levels at 1 or 3 h considerably decreased ($p < 0.05$ or $p < 0.01$). The 24-h gene expression levels were similar to those seen in the Control samples that were taken at the same time. The data obtained from ft-10 *A. thaliana* mutant plants indicates that the wounding effects caused by silk-based MNs were reduced to a minimum within a day, implying that MNs are a very effective and less invasive method of delivering payloads [18].

15.3.2 Cereals

Cereals are considered vital for the delivery of their payload because of their importance to global food security and agriculture. As a result, the researchers investigated the possibility of using MNs technology in monocotyledonous cereals [20]. Monocot plant species, such as rice (*Oryza sativa*), maize (*Zea mays*), and barley (*Hordeum vulgare*), differ from dicot plant species in that their vasculature is dispersed. This makes it difficult to target the xylem and phloem precisely in these plants [21]. However, GA_3's previously mentioned tiny size may allow for both apoplast and symplast delivery of the hormone after injection, obviating the requirement for a precisely timed vascular injection. Due to size similarities between the ft-10 petiole and the stem of the cereals, rice, maize, and barley were chosen as the model monocots. The same MNs used for ft-10 were applied to these crops. Because of its significance in agriculture, soybean, a dicot legume, was also investigated [18].

Rice seedlings that were 7 days old were given either empty (Mock) or GA_3-loaded (GA_INJ) MNs. In comparison to the Mock ($p = 5.32 \times 10 - 13$) and Control ($p = 2.79 \times 10 - 13$) groups, GA_INJ plants displayed a greater plant height. In addition, the length of the leaf blades in GA_INJ plants was longer than in the Mock ($p = 9.16 \times 10 - 8$) and control ($p = 1.61 \times 10 - 7$) groups. These findings demonstrated the successful transport and functionality of GA_3 in rice, indicating the efficacy of distribution by MNs [18].

Two treatments were applied to 13-day-old maize seedlings: GA_3 solution foliar spray (GA_S) and GA_3-loaded MNs (GA_INJ) ($n = 8$). The seedlings had a well-developed first internode (between the first and second leaf) and an emerging second internode. In comparison to control (no treatment) ($p = 5.98 \times 10 - 5$ and $p = 3.52 \times 10 - 5$) and GA_S ($p = 4.19 \times 10 - 5$ and $p = 1.36 \times 10 - 5$), GA_INJ demonstrated an increase in plant height and second internodal length, but there was no significant difference in first internodal length. These findings show the great efficiency of GA_3 microneedle administration overspray, and they also show that GA_3 was transported to and functioned in maize [22], especially effectively in the developing tissues. Even with the same dosage of GA_3 as MNs, spray administration was unable to increase maize's height. This is most likely caused by the high length–width ratio of the leaf morphology, which leads to significant off-target loss and inefficient leaf GA_3 uptake. Similar to maize, barley showed enhanced plant height and first internodal length when compared to GA_S ($p = 2.39 \times 10 - 6$ and $p = 1.02 \times 10 - 4$) ($n = 8$) and Control ($p = 8.13 \times 10 - 7$ and $p = 5.19 \times 10 - 5$). The reason the internode is different is that when treated, 13-day-old barley seedlings showed signs of emerging their first internode. Taller soybean seedlings and higher internodal length were also the results of microneedle-mediated GA_3 delivery [23] ($n = 5$) 3 days posttreatment compared to control ($p = 2.15 \times 10 - 5$), and the GA_3-induced difference was more pronounced 6 days after treatment. These findings confirmed the adaptability of MNs as a payload delivery technique and showed their efficacy in staple food crops [18].

15.4 Soil

15.4.1 Temperature

The temperature of the soil, which can range from −10 to 50 °C [24], is a crucial factor in agriculture as it affects various plant growth processes such as germination, blooming, and compositing. It also has a major impact on the physical, chemical, and microbiological processes in soils, all of which are essential to plant growth [25, 26]. Properties of the soil, such as bulk density, specific heat capacity, thermal conductivity, texture, water content, and surface covering materials, all have a significant impact on soil temperature [27, 28]. A soil temperature sensor is made up of bias and read-out circuits to interpret the electrical signal into digital data and temperature probes to convert temperature variation into an electrical signal [28].

15.4.2 PH Value

The pH of soil represents the cumulative impacts of the soil-forming elements, such as parent material, organisms, and climate, and indicates how basic or alkaline the soil is. In addition to affecting microbial activity, soil pH influences plant nutrition availability by dictating the chemical forms of certain nutrients. Most plants can grow in soil that has an ideal pH between 5.5 and 7.5 [29]. The pH value of the soil can give important information for smart and precision agriculture, helping to maintain the health of the soil within an appropriate range for particular farm crops.

15.4.3 Nutrient

In soil organic materials (SOMs), the three most crucial nutrients for crop productivity are nitrogen (N), phosphorus (P), and potassium (K) [30, 31]. Fertilizers, including N, P, and K, are utilized in agriculture engineering to increase crop yield. However, there are substantial negative consequences, both environmentally and economically, of excessive anthropogenic nutrient inputs. Crops use very little fertilizer (such as N and P), and a large portion of it is lost to erosion, volatilization, and leaching. Efficient and precise identification of nutrient molecules in soil is crucial for precision agricultural development and environmental sustainability, enabling plants to adapt to settings deficient in nutrients [32–37].

15.5 Material and Fabrication Techniques for Advanced Sensor Types in Agriculture

The application of nanotechnology in several food sectors, such as agriculture and farming (e.g., pesticides, fertilizers, and vaccinations), food processing, food packaging (preservation and protection), and nutritional supplements, has been acknowledged by researchers and industry partners. The most dynamic assessment reports in food science are built on the sophisticated manufacture of nanoscale sensor devices, which have a wide range of applications related to food safety. In this way, food development, creation, handling, and packaging all make use of assessment methodologies and tools [38] (see Fig. 15.153).

Chemical sensor approaches are widely employed for food safety and quality assessment to measure or differentiate physical parameters and changes in reactive species, agents, and contaminant targets into read-out signals by means of instrumentation or observation. Sensors have the potential to revolutionize food safety research by enabling high-sensitivity and selectivity real-time investigations. This includes basic considerations such as agriculture and farming, food processing, food packaging, and nutritional supplementation. The evaluation of intrinsic properties (selectivity, sensitivity, sensory detection range, limitation, and lifetime), economic manufacturability, new research methods with quick response times, straightforward sensing strategies, and real applicability are some of the keys that can be used to identify progress in sensors [39, 40]. For instance, the engineering and construction of biosensors rely largely on the integration of two elements: a transducer and a bioreceptor. Transducers translate the physiochemical change that results from the interaction between food-reactive targets and biological elements into a quantified signal whose intensity is related to target concentrations. Bioreceptors immobilize sensitive biological elements (such as enzymes, proteins, antibodies, and whole cells) that function as active interfacial surfaces [41, 42]. Transducing elements, chemical sensors, and biosensors can be divided into optical (including reflectometry or absorptiometry, fluorescence, and surface plasmon resonance [SPR]) and surface-enhanced Raman spectroscopy (SERS) assay sensors, piezoelectric sensors, and thermal sensors. These sensors can convert reactions or binding with target species into a digital signal [39, 41, 43–45] (see Fig. 15.3).

As regards sensor-based safety, actuators are used to respond to feedback and take control of situations, while sensors gather data about environmental and physical characteristics. Context is the collection of data from sensors that describes an object or environment and is used to identify people, places, things, and their states [60, 61]. When modelling scenarios of domains with a range of temporal variation features, context acquisition makes a significant contribution. One such area is agriculture. The agriculture domain has the following requirements:

1. Gathering data on the weather, crops, and soil.
2. Keeping an eye on the allocated land.
3. Several crops on one plot of land.

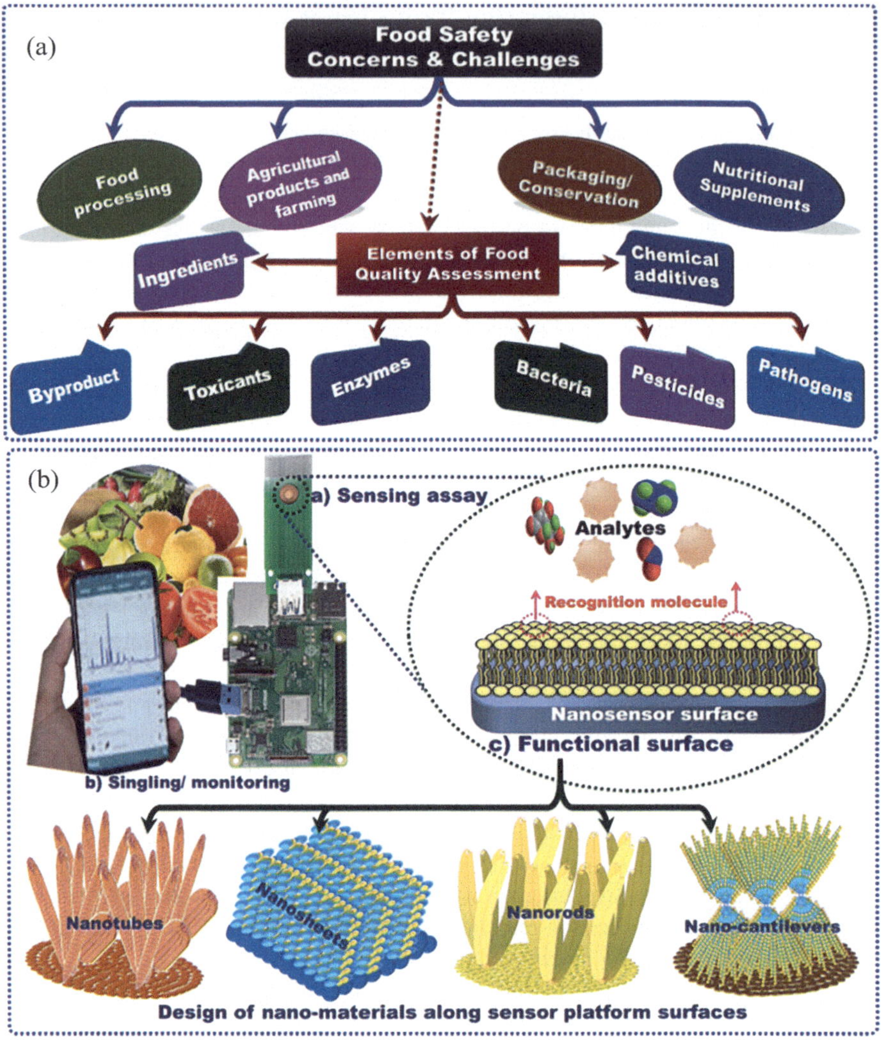

Fig. 15.3 Applications of nanotechnology (**a**), sensor fabrication (**b**). The figure was reproduced with permission [38]

4. Distinct portions of uneven ground require various amounts of water and fertilizer.
5. Varying crop needs according to various soil and climatic conditions.
6. Preventive measures as opposed to reactive ones.

The aforementioned specifications call for distributed and parallel application and processing. Furthermore, wireless sensors and actuators are needed to gather the necessary data and respond to various circumstances. Processed information is required instead of raw sensor data for decision support. In order to meet these needs, wireless sensors, actuators, and associated networks offer a compelling option for the creation of a system for context acquisition. This system would provide acquired data to distant decision support systems, enabling the provision of a controlled environment based on decision-making [62].

A wireless sensor network (WSN) is made up of a number of units known as "nodes." The smart devices known as nodes are employed to gather the data requirements that are specific to applications. A sensor network uses hardware, software, and algorithms to carry out three fundamental tasks: (i) sensing, (ii) communication, and (iii) computation. The nodes carry out many tasks. The scattered nodes that gather data are referred to as source nodes, while the node that compiles data from all source nodes is referred to as the sink node or, occasionally, gateway node. There is a chance the sink node has a lot of processing power. Because multiple-hop routing is necessary, a source node also functions as a routing node. WSAN stands for Wireless Sensor and Actuator Network, a variation of WSN that includes actuators as an extra component type. The addition of an actuator expands WSN's capabilities beyond monitoring to include Control [62].

Numerous sensor network-based research projects include wireless communication technologies such as WiFi, Bluetooth, Wibree, and ZigBee. These technologies complement each other and have various capacities and attributes. Because of its low cost and low power consumption, ZigBee wireless communication technology (IEEE 802.15.4) is chosen over other technologies for the construction of wireless sensor networks. It was unveiled in May 2003 and runs worldwide on the 2.4 GHz Industrial, Scientific, and Medical (ISM) band. In the 2.4 GHz range, there are 16 ZigBee channels with a 5 MHz bandwidth each. A wireless sensor network's fundamental component is the wireless sensor node. It is made up of four fundamental modules: the power module, the communication module, the processing/computation module, and the sensor/actuator module. An optional module called external memory can be required if local decision-making requires the storage of data. Many factors, including size, housing, scalability, and energy saving, must be taken into account during design. The sensor/actuator module provides interfaces to transducers and actuators. The main application domain, issue, and distribution pattern requirements are taken into account while choosing a sensor mode. Some of the main characteristics of a sensor node that set it apart from others are its size, transmission range, memory, functioning frequency band, processor/microcontroller, and available suitable sensors [62].

The majority of domain concerns are closely related to weather qualities, licensing-free ISM bands, and specialized attributes of microcontrollers. Microcontrollers are essential because they offer the processing capability needed for data aggregation and local decision-making, as well as energy management features like sleep and wake modes. Due to its low power consumption, several sleeping modes, flash memory, effective byte-oriented storage, and compatibility with almost all TinyOS programs, the ATmega128L is the most widely used microcontroller. The requirements specific to the agriculture domain are outlined in the next section. Owing to these unique requirements, sensor nodes such as the Lofar node and SPWAS are also carefully developed to maintain features like housing, robustness,

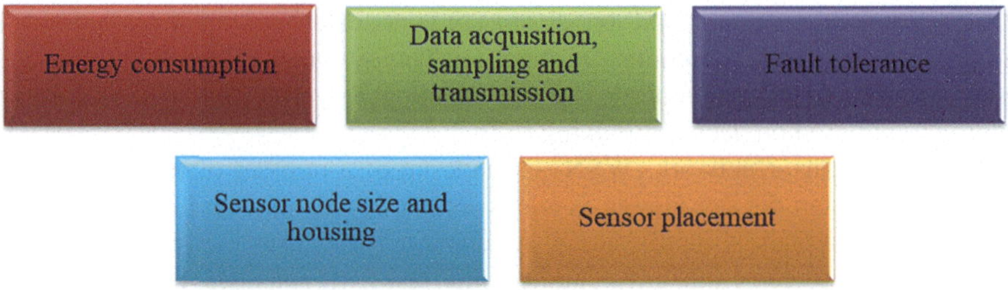

Fig. 15.4 Sensor nodes design

and compatibility for different energy sources. Wireless sensor network problems have been described in the literature that has already been written [63, 64], and other people have tackled these problems [65, 66] with varying approaches. However, farming is sometimes done on farms that span several hundred acres and may entail a variety of terrain types, resources, and microclimates. The placement of sensor nodes in open, uncontrolled environments presents a number of design and implementation challenges, some of which are critical and are listed below (Fig. 15.4).

(a) Energy consumption

In a wireless sensor network, each sensor node is in charge of data processing, event detection, and transmission. In a network with multiple hops, a node additionally has to do data routing duties. All the aforementioned actions necessitate the consumption of energy. Typically, a node has a finite and restricted energy supply, such as lithium cells or alkaline batteries. As a result, the lifespan of a sensor node is heavily dependent on the battery life. Battery life may be increased by many extra months by implementing an appropriate energy management approach in hardware and software. An alternative would be to employ renewable energy sources, such solar power or kinetic energy, where a little more costly solution would be given up in exchange for a longer lifespan for the sensor nodes. Furthermore, as their placement and access are typically well-defined, these sensors might be periodically recharged by changing their batteries in the case of agricultural lands [62].

(b) Data acquisition, sampling and transmission

Energy is used in the sampling, processing, and transmission of each piece of data. In order to ensure that energy is not wasted and that pertinent and valuable data is recorded, an effective and optimized data collection and sampling rate must be set. Conversely, acquiring data often will result in a large number of packets being broadcast, which rapidly depletes the batteries' energy. The sample rate used to collect data in agriculture is typically quite low. On the other hand, it might be modified according to the resources, crop kind, and environment. There are numerous wireless transmission systems at one's disposal. It is also possible to plan a data transmission strategy to minimize the number of transmissions, thus preserving energy. This could involve storing data locally at the sensor. Nodes and intelligent data transfer, such as the combined values or just the modified ones, etc. Additionally, the application makes it possible to use features like sleep mode to awaken the transmits only when required. Additionally, dependability, sensor nodes and communication lines need to be positioned close enough to mutually guarantee dependable multi-hop communication [62].

(c) Fault tolerance

When situated in an open, hostile environment, sensor nodes are vulnerable to interference, blockages, and physical damage. The entire function of the network must not be impacted by the failure of a sensor node in order for a WSN to remain reliable. Rearranging the sensor network, using redundant sensor nodes, overlapping sensing zones, and other strategies are some of the methods used to improve the network's fault tolerance or dependability [62].

(d) Sensor node size and housing

The sensor node size needs to be deployable and compact. The node ought to be housed in a protective casing that can withstand extreme weather conditions, such as heat and rain, as well as physical abuse from people or animals [62].

(e) Sensor placement

Sensor node: when it comes to the network's architecture, algorithms, topologies, and parameters being sensed, placement is crucial. In order to create a WSN that can function dependably and autonomously, it must be thoughtfully planned and cleverly executed. Sensors should be positioned such that they cover the entire area of interest and are at the right heights and positions to measure the parameters without obstruction. Light sensors, for instance, need to be positioned higher to prevent obstruction by plant leaves. On the other hand, for precise measurements, water level and moisture sensors need to be positioned close enough to the ground. The sensors may be moved from their intended locations by strong winds and water currents, so the nodes must be supported by the appropriate fittings [62]. In the realm of agriculture, the use of sensors, actuators, and their networks is currently advanced. Different technologies, including context-aware computing and grid computing, make use of WSN and WSAN [67, 68]. This section reviews agricultural services such as pesticide spraying, fertilizer, irrigation, and pasture and animal monitoring. Because of its significance, horticulture is also taken into consideration (Table 15.1).

(i) Irrigation

One of the most crucial elements of agriculture is irrigation, which is the deliberate use of water on agricultural land. Due to water scarcity in many countries, water must be used properly, meaning that it should only be supplied to those locations where it is required and in the necessary amount. In order to address the issue of water waste in traditional irrigation techniques like flood irrigation and furrow irrigation, several irrigation techniques are being used, such as drip irrigation and spray irrigation. A remote-controlled, autonomous irrigation system was created and tested by Damas et al. [69] for an irrigated area in Spain. There were seven sub-regions within

Table 15.1 Agriculture works

Activities	Description
(i) Irrigation	Drip irrigation and spray irrigation
(ii) Fertilization	Spraying, broadcasting, and manual spreading
(iii) Pest control	Pest management techniques and product development
(iv) Animal and pastures monitoring	GPS receiver, PDA, WiFi flash card for WLAN, audio amplifier, and speaker
(v) Horticulture	Low-intensity or small-scale farming

the larger territory. A control sector watched over and contested each sub-region. Through a WLAN network, the seven control sectors were linked to the central controller and to each other. Significant water conservation, or up to 30–60%, was demonstrated by the results. Precision irrigation control for self-propelled, linear-move, and centre-pivot irrigation systems was the focus of Evans and Bergman's [70] research. The technology employed meteorological and remotely sensed data to schedule watering with the help of wireless sensors. In order to gather soil moisture and climate data for smart irrigation in Portugal, Morais et al. [71] developed a wireless data-collecting network. A number of SPWAS (Solar Powered Wireless Data-Acquisition Stations) have been installed to measure soil moisture in an effort to increase irrigation efficiency. Several methods have been employed to detect soil moisture.

A number of sensor-based projects have been created with the understanding that greater yield production requires control over irrigation water. Designing and developing a computer-based drip irrigation control system with the capability of remote data collecting [72]. Additionally, they discussed the advantages of storing sensor data for statistical analysis to determine which crops demand how much irrigation. A sensor-based irrigation system was created by Yunseop Kim et al. [73]. Bluetooth and GPS technology were used to remotely monitor the temperature and moisture levels of the soil, meteorological data, and sprinkler location. The project was developed with the goal of maximizing output while conserving water. Kim and Evans created a real-time wireless in-field sensor and control software for site-specific sprinkler irrigation [74]. They used Bluetooth wireless radio communication to facilitate decision-making and real-time irrigation operation monitoring, and they combined an irrigation controller tailored to the individual site with feedback from infield data.

(ii) Fertilization

Fertilizers are used to improve soil fertility, which has a direct impact on plant development and food quality. There are several methods of fertilization, including spraying, broadcasting, and manual spreading. It is challenging to provide fertilizer to the right location at the right time, and it calls for sensing abilities. Researchers have proposed various approaches to suitable fertilization. Cugati et al. [75] used GPS, real-time sensors, and Bluetooth to build an automated fertilizer applicator with input, decision support, and output modules. The Decision Support System (DSS), which determines the ideal amount and application pattern of fertilizer based on real-time sensor data acquisition through Bluetooth communication modules, receives GPS and sensor data values from the input module. The application rate for fertilizer regulations was based on the DSS calculations. For site-specific fertilization, Ehlert et al. [76] employed a mechanical sensor they had constructed themselves, a Pendulum meter. A pendulum meter was installed on the front of the tractor to gauge the density of the crops. To estimate the application of nitrogen in the field, the sensor was used in conjunction with a task calculator, a customized CAN-bus onboard computer, and a fertilizer spreader. Using wireless sensors, the LAN and GPS analysis server, He et al. [77] created and integrated an optimal fertilization decision support system utilizing IEEE 802.11 protocol (WiFi). Real-time data on soil moisture, conductivity, temperature, PH value, air temperature, humidity, CO_2 concentration, light, and other factors were collected using sensors.

In order to enable maximum interaction, the system was created using the B/S (browser/server) structure method. In order to take advantage of data reduction for energy conservation, the data from small experiment plots were interpolated for bigger plots using a GIS analysis server. In their fertilization decision-making system, they employed the "3414" test [78], which refers to fertilization trials using three variables (nitrogen, phosphorus, and potassium), four levels, and fourteen applications of fertilizer. The "3414" fertilizer efficiency test is mostly utilized in systems that make decisions about fertilizing. It is a crop yield and fertilization statistical model. Theoretically, the ideal yields of this model crop and the appropriate amounts of fertilizer may be computed [79]. The plan is made to take advantage of the low treatment count and high efficiency of optimal regression design. Fertilizer is used at four levels in the 3414 test: 0, 1, 2, and 3. Level 2 is the ideal fertilizer level for the local estimate, while Level 0 indicates no fertilizer at all. Level 2 refers to the regular level. These are the definitions of other levels: 2level $\times$ 0:5 = 1level, and 2level $\times$ 1:5 = 3level. Level 3 indicates overfertilization. On 14 treatment locations of the test field, several fertilization schemes (based on various NPK fertilization rates) are applied. These are then utilized to determine the ideal fertilization application rate and how it relates to crop output [62].

(iii) Pest control

For better "pest" control, agricultural pest management techniques and product development have long dominated. Rather than application tactics, the focus remained primarily on the development of new and effective products to replace outdated and hazardous products. When farming methods are more in line with ecological systems, pest management may become more sustainable. Additionally, sensor networks have been utilized to combat pest and fungal issues. Aline Baggio [80] created a scheme to address the *Phytophthora* disease that affects potato crops. Sensors detected temperature and humidity. They were able to lessen the illness by keeping an eye on these two facts. Plant disease frequently affects isolated crop patches, necessitating the administration of fungicides at varying rates as opposed to evenly throughout the field. Dammer suggested the use of CROP-meter-controlled variable rate field sprayers

[81]. Information regarding the spatial distribution of plant leaf areas to be sprayed with pesticides could be obtained by using a CROP-meter, a real-time sensor to assess crop biomass density. Their algorithm uses sensor data to regulate the fungicide dosage that the field sprayer will apply.

(iv) Animal and pastures monitoring

A moving virtual fence algorithm was created by Butler et al. [82] to manage the cow herd. Every cow in the herd has a smart collar that includes a GPS receiver, PDA, WiFi flash card for WLAN, audio amplifier, and speaker. GPS is used to confirm the animal's location in relation to the virtual fence. When an animal approaches a virtual fence, a sound cue is produced to eject it off the fence line. Large-scale cattle monitoring comes with a number of requirements and difficulties. A peer-to-peer distributed architecture for statewide livestock monitoring based on a mobile ad hoc wireless sensor network was described by Radenkovic and Wietrzyk [83]. Their main focus is on using Distributed Hash Tables (DHT) for new data storage and routing, which lessens the significant issue of node disconnections. Farming cattle has a close relationship with the agricultural sector. Cattle use green meadows to graze on. A pervasive, self-configurable sensor-based system was developed by Tim Wark and his colleagues [84] at the CSIRO ICT Centre in Australia to evaluate animal behaviour and control as well as pasture assessment. With the use of camera sensors, the development of the grass was examined in order to direct animals into verdant pastures. The main goal of their effort was to create durable technology that could be used outside to simulate the behaviour of both individual and herd animals. They monitored animal behaviours, including sleeping, grazing, ruminating, etc., using a specially made sensor. They believe that this type of behaviour analysis when combined with their movement behaviour, can be utilized to analyze patterns in herd behavior as well as the link between cows and calves. While keeping an eye on cattle is a crucial responsibility, there are a number of obstacles to overcome, such as animal mobility and radio attenuation. In order to maximize the radio coverage of the collar, Andonovic et al. [85] built an enhanced, low-cost, low-power collar with a solar power relay router and two antennas positioned at strategic locations. Additionally, they suggested the Implicit Routing Protocol (IRP) to address the high rate of packet loss brought on by animal movement.

(v) Horticulture

The production, distribution, and usage of flowers, fruits, greenhouse plants, and other ornamentals are all covered under horticulture. Low-intensity or small-scale farming are other names for it. Due to the significant interest in greenhouse and vineyard research from the research community, these topics are covered in this section. In order to analyze the current state of plant nurseries, Zhang et al. [86] used sensor networks to monitor air temperature, humidity, ambient light, soil moisture, and temperature. Plant diseases may also be found with the aid of such a network. The authors of [67] discussed how the approach conserves water compared to conventional watering techniques. They used Ech2o soil moisture sensors and TelosB sensor motes to create a context-aware irrigation control system. The watering system for university green spaces was tested. According to field testing, 30–50% of the water conservation. A lower-cost irrigation control system was developed by creating a sensor/actuator node, and the WSAN protocol on-site is described in [87]. Five sensor nodes equipped with light, humidity, and soil moisture sensors were used to control the irrigation of the university garden area. Because of the Atmel ATMega64L AVR microcontroller's capability for six different sleep modes Idle, ADC Noise Reduction, Power-save, Power-down, Standby, and Extended Standby and the XBee/X-Bee-Pro OEM RF module's low power consumption features, the node was developed with this microcontroller in mind. The cheap cost and power consumption characteristics of the design were prioritized over the sensor mote's size.

The agriculture area lacks a framework that might aid in the development of an entire system, from context collection to modeling and from a decision support system to actuation. One of these frameworks, the context-aware sensor grid framework, is composed of three layers: the sensor network layer, the grid computing layer, and the context-aware application layer [67, 88]. The necessity of several building pieces for every layer, from sensor-based data collecting to context-aware applications for timely decision-making, was highlighted by the authors as a way to improve interactivity. The idea of mixed communities of communicative plants and artefacts was introduced by Goumopoulos et al. [68, 89]. ePlants and eGadgets are the two main concepts covered by their PLANTS ontology. The authors created a system architecture for precision agriculture applications using the PLANTS ontology. They contend that the system will be proactive and scalable thanks to local decision-making at each node. There is not a single organization that offers a set process for creating sensor-based agricultural systems, from collecting data to modeling and making decisions. Under the umbrella of precision agriculture, several research groups are working independently to construct such systems utilizing common technologies. One of the standardization initiatives being carried out by the Food and Agriculture Organization of the United Nations (http://www.fao.org) is the creation of a multilingual Agriculture Vocabulary (AGROVOC) to facilitate the ontology-based development of context modelling. The idea of the Agricultural Ontology Service/ Concept Server (AOS/CS) [90] has also been offered in this regard with the aim of standardizing the agricultural lexicon and offering rich and semantically sound terminology.

15.6 Bioeconomic Plan in Industrial Revolution (IR) 5.0

An additional Industry 5.0 vision is the bioeconomy. One can achieve a balance between the industrial sector, the economy, and the environment by strategically using biological resources for industrial purposes. The European Commission defines the bioeconomy as "the production of renewable biological resources and the conversion of these resources and waste streams into value-added products, such as food, feed, bio-based products, and bioenergy." Aside from agriculture, forestry, fishing, food, and pulp and paper industries, it also covers parts of the chemical, biotechnology, and energy industries. Its sectors have a strong potential for innovation due to the utilization of a wide range of sciences (life sciences, agronomy, ecology, food science, and social sciences), industrial and enabling technologies (biotechnology, nanotechnology, information and communication technologies (ICT), and engineering), as well as local and tacit knowledge. Biologization, the central idea of the bioeconomy, has the potential to fundamentally change a wide range of industries. Therefore, the next industrial revolution may centre on the bioeconomy or at least include its elements. Furthermore, other problems like space industries, space mining, and space life might either be the next revolution or a part of it. We already have advice from scientists to use space resources carefully. Space mining could be the next big item to hit it big.

15.7 Conclusion

MNs-based technologies are essential in many domains because of their adaptable geometries and special qualities that may be adjusted and maximized according to the intended use. Even with these benefits, there are still certain issues that need to be resolved, such as improving the sensitivity, scalability, reversibility, and reproducibility of measurement and production. Notwithstanding these difficulties, MNs have already brought about a significant revolution in a number of industries, including the biomedical sector, agriculture, food sector, food quality control, environmental procedures, and more. Further advancements in MNs-based devices could include the merging of sensing and therapy into a single closed-loop apparatus, thereby utilizing the unique qualities of each MN type independently and enhancing their combined benefits. As a result, this will allow for a more sophisticated and useful application of the technology in customized sensing and therapy. This chapter has provided some combined sensing and therapeutic MN techniques; nevertheless, more optimizations and validations need to be addressed. In conclusion, because of their quick development, simplicity in manufacturing, and special geometry in agricultural and environmental applications, MNs have enormous promise and potential in a variety of disciplines.

15.8 Challenges and Future Perspective

Agriculture and plant health have benefited recently from the application and transfer of new technologies that were first created in the field of human medicine. Innovations in precision agriculture and plant treatment are essential to addressing the problems facing the agriculture sector because they boost crop output, increase plant tolerance to stressors and diseases, and improve the sustainability of crop production. Growing advancements in nanotechnology and micro-nanofabrication methods have made major strides in the creation of novel platforms for crop and plant health protection. The creation of plant-specific microneedle devices opens up new avenues for improving crop health and biosecurity, developing novel diagnostic tools, and facilitating advances in plant engineering. For instance, standard methods for determining the ionic content of plants through chemical analysis are harmful and necessitate the death of the plant. On the other hand, microneedle technology reduces the time required for sample extraction from hours to minutes or allows for noninvasive plant health monitoring. The extracted samples can be put onto a microsensor for in situ pathogen detection, or they can be used for PCR amplification without purification. This can solve the issues with traditional nucleic acid amplification-based molecular diagnostic techniques, allowing farmers and field workers to identify plant diseases in the field and apply treatment plans there instead of having to send samples to labs for analysis. Furthermore, the biosensor's modular architecture guarantees that the apparatus can be upgraded and adjusted as needed by the industry. The standard, costly, and time-consuming laboratory test techniques can be replaced with agricultural microsensors without sacrificing sensitivity or obstructing crop growth. The combination of miniaturized amplification assays and microneedle patches will enable quick, affordable, straightforward, and field-deployable plant pathogen identification and amplification.

Studies show that agricultural sensors are capable of precisely identifying pathogens that cause different types of plant and crop illnesses. The development of a low-cost product is made possible by improvements in nanotechnology production and intelligent material usage. These benefits have prompted advancements in agricultural biosensor research in recent years. In nanomedicine, this technique has been widely used to detect various biomarkers. Nevertheless, there are not many studies on POC diagnosis of plant and food illnesses. Farmers and other field personnel may identify plant diseases in the field and carry out on-site treatments with the use of a POC plant pathogen detection system, all without having to send samples to labs for analysis. With the help of this technology, the rapidly developing sector of precision farming will be able to use resources more effectively and have a smaller negative environmental impact. A few benefits of micro-nanotechnology application in point-of-care (POC) plant diagnostics include reduced sample usage without sacrificing assay sensitivity, quick turnaround of data, cost-effectiveness, quick analysis time, and the ability to produce disposable devices. Nonetheless, research into creating dependable little tools for diagnosing, treating, and tracking plant illnesses is still expanding.

References

1. Singh, B.K., Delgado-Baquerizo, M., Egidi, E., Guirado, E., Leach, J.E., Liu, H., Trivedi, P.: Climate change impacts on plant pathogens, food security and paths forward. Nat. Rev. Microbiol. **21**, 640–656 (2023). https://doi.org/10.1038/s41579-023-00900-7

2. Chakraborty, S., Newton, A.C.: Climate change, plant diseases and food security: an overview. Plant Pathol. **60**, 2–14 (2011). https://doi.org/10.1111/j.1365-3059.2010.02411.x
3. Li, Z., Yu, T., Paul, R., Fan, J., Yang, Y., Wei, Q.: Agricultural nanodiagnostics for plant diseases: recent advances and challenges. Nanoscale Adv. **2**, 3083–3094 (2020). https://doi.org/10.1039/c9na00724e
4. United Nations Department of Economic and Social Affairs.: Global Population Growth and Sustainable Development (2021)
5. Jumhawan, U., Bamba, T.: In: The Application of Green Solvents in Separation Processes, pp. 483–516. Elsevier (2017)
6. Eskola, M., et al.: Worldwide contamination of food-crops with mycotoxins: validity of the widely cited 'FAO estimate' of 25. Crit. Rev. Food Sci. Nutr. **60**, 2773–2789 (2020)
7. Liliane, T.N., Charles, M.S., Agron. Chang: factors affecting yield of crops. Food Secur. **9** (2020)
8. Swaroopa, R.T., et al.: Agrochem, pp. 437–464. Detect. Treat. Remediat, Butterworth-Heinemann (2020)
9. Alengebawy, A., et al.: Heavy metals and pesticides toxicity in agricultural soil and plants: ecological risks and human health implications. Toxics **9**, 42 (2021)
10. Zikankuba, V.L., et al.: Pesticide regulations and their malpractice implications on food and environment safety. Cogent Food Agric. **5**, 1601544 (2019)
11. Senyuva, H.: Encyclopedia of Food Safety, pp. 152–158. Academic Press, Waltham (2014)
12. Kaur, N., et al.: Advanced green analytical chemistry for environmental pesticide detection. Curr. Opin. Green Sustain. Chem. **30**, 100488 (2021)
13. Ece, E., Eş, E., Inci, F.: Microneedle technology as a new standpoint in agriculture: treatment and sensing. Mater. Today **68**, 275–297 (2023)
14. Stafford, J.V.: Implementing precision agriculture in the 21st century. J. Agric. Eng. Res. **76**, 267–275 (2000)
15. Zhang, N., Wang, M., Wang, N.: Precision agriculture—A worldwide overview. Comput. Electron. Agric. **36**, 113–132 (2002)
16. Ray, D.K., Mueller, N.D., West, P.C., Foley, J.A.: Yield trends are insufficient to double global crop production by 2050. Plos One **8**, 6 (2013)
17. Omar, R., Zheng, Y., Wang, J., Haick, H.: Microneedle sensors for multiplex applications: toward advanced biomedical and environmental analysis. Adv. Sens. Res. **2**, 2200032 (2023)
18. Cao, Y., Koh, S.S., Han, Y., Tan, J.J., Kim, D., Chua, N.-H., Urano, D., Marelli, B.: Drug delivery in plants using silk MNs. Adv. Mater. **35**, 2205794 (2023)
19. Miceli, A., Moncada, A., Sabatino, L., Vetrano, F.: Effect of gibberellic acid on growth, yield, and quality of leaf lettuce and rocket grown in a floating system. Agronomy **9**, 382 (2019)
20. FAO; IFAD; UNICEF; WFP; WHO.: The State of Food Security and Nutrition in the World 2022: repurposing Food and Agricultural Policies to Make Healthy Diets More Affordable, FAO, Rome, Italy (2022). https://doi.org/10.4060/cc0639en
21. Cao, Y., Lim, E., Xu, M., Weng, J.K., Marelli, B.: Precision delivery of multiscale payloads to tissue-specific targets in plants. Adv. Sci. **7**, 1903551 (2020)
22. Shahzad, K., Hussain, S., Arfan, M., Hussain, S., Waraich, E.A., Zamir, S., et al.: Exogenously applied gibberellic acid enhances growth and salinity stress tolerance of maize through modulating the morpho-physiological, biochemical and molecular attributes. Biomolecules **11**, 1005 (2021)
23. Shan, F., Zhang, R., Zhang, J., Wang, C., Lyu, X., Xin, T., et al.: Study on the regulatory effects of GA3 on soybean internode elongation. Plants **10**, 1737 (2021)
24. Taylor, S.A., Jackson, R.D.: Heat Capacity and Specific Heat. C, pp. 941–944 (1986)
25. Viets, F.G.: Soil fertility and organic matter as critical components of production systems. Agric. Ecosyst. Environ. **30** (1990)
26. Brusseau, M.L., Pepper, I.L., Gerba, C.: Environmental and Pollution Science. Academic Press (2019)
27. Stuart Chapin, F., Matson, P.A., Vitousek, P.M.: Principles of terrestrial ecosystem ecology. Princ. Terr. Ecosyst. Ecol. 1–529 (2012). https://doi.org/10.1007/978-1-4419-9504-9

28. Kevin, J.M.: Chapter 5 Soil Heat. in 1183–1199 (2002)
29. SS, N. Food Analysis. 4th edn. Springer Science and Business Media, New York (2010)
30. Sharma, L.K., Zaeen, A.A., Bali, S.K., Dwyer, J.D.: Improving nitrogen and phosphorus efficiency for optimal plant growth and yield. Intech I, **13** (2016)
31. Scarnecchia, D.L., Magdoff, F.: Building soils for better crops. Organic Matter Management. J. Range Manag. **47** (1994)
32. Guignard, M.S., et al.: Impacts of nitrogen and phosphorus: from genomes to natural ecosystems and agriculture. Front. Ecol. Evol. **5** (2017)
33. Zhu, L., et al.: A novel method for soil organic matter determination by using an artificial olfactory system. Sensors (Switzerland) **19** (2019)
34. Lvova, L., Nadporozhskaya, M.: Chemical sensors for soil analysis: principles and applications. In: New Pesticides and Soil Sensors. Elsevier Inc., (2017). https://doi.org/10.1016/B978-0-12-804299-1/00018-7
35. Patkar, R.S., Ashwin, M., Rao, V.R.: Piezoresistive microcantilever based lab-on-a-chip system for detection of macronutrients in the soil. Solid State Electron. **138**, 94–100 (2017)
36. Razaq, M., Zhang, P., Shen, H.L., Salahuddin.: Influence of nitrogen and phosphorous on the growth and root morphology of Acer mono. Plos One **12**, 1–13 (2017)
37. Schachtman, D.P., Shin, R.: Nutrient sensing and signaling: NPKS. Annu. Rev. Plant Biol. **58**, 47–69 (2007)
38. Shenashen, M.A., Emran, M.Y., El Sabagh, A., Selim, M.M., Elmarakbi, A., El-Safty, S.A.: Progress in sensory devices of pesticides, pathogens, coronavirus, and chemical additives and hazards in food assessment: Food safety concerns. Progr. Mater. Sci. **124**, 100866 (2022)
39. Lu, X.: Sensing Techniques for Food Safety and Quality Control: sensing Techniques for Food Safety and Quality Control. Royal Society of Chemistry (2017)
40. Hosnedlova, B., Sochor, J., Baron, M., Bjørklund, G., Kizek, R.: Application of nanotechnology based-biosensors in analysis of wine compounds and control of wine quality and safety: a critical review. Crit. Rev. Food Sci. Nutr. **60**, 3271–3289 (2020)
41. Dincer, C., Bruch, R., Costa-Rama, E., Fernández-Abedul, M.T., Merkoçi, A., Manz, A., et al.: Disposable sensors in diagnostics, food, and environmental monitoring. Adv. Mater. **31**, 1806739 (2019)
42. Anshori, I., Suzuki, H.: Microfluidic device for high-sensitivity coulometric detection of proteins. Sens Actuat. B: Chem. **256**, 835–838 (2018)
43. Allen, J.B., Larry, R.F.: Electrochemical Methods Fundamentals and Applications. Wiley (2001)
44. Guo, X., Kuhlmann, J., Heineman, W.R.: Biosensors on enzymes, tissues, and cells. In: Environmental Analysis by Electrochemical Sensors and Biosensors, pp. 283–312. Springer (2014)
45. Muslim, N.Z.M., Ahmad, M., Heng, L.Y., Saad, B.: Optical biosensor test strip for the screening and direct determination of l-glutamate in food samples. Sens. Actuat. B: Chem. **161**, 493–497 (2012)
46. Faraji Rad, Z., Prewett, P.D., Davies, G.J.: Parametric optimization of two-photon direct laser writing process for manufacturing polymeric microneedles. Addit. Manuf. **56**, 102953 (2022)
47. Makvandi, P., Kirkby, M., Hutton, A.R.J., Shabani, M., Yiu, C.K.Y., Baghbantaraghdari, Z., et al.: Engineering microneedle patches for improved penetration: analysis, skin models and factors affecting needle insertion. Nano-Micro Lett. **13**, 93 (2021)
48. Faraji Rad, Z., Nordon, R.E., Davies, G.J., Anthony, C.J., Prewett, P.: US Patent 1,085,0082 (2020)
49. Ebrahiminejad, V., Faraji Rad, Z.: Design, development, and testing of polymeric microblades: a novel design of microneedles for biomedical applications. Adv. Mater. Interfaces 2201115 (2022)
50. Faraji Rad, Z., Nordon, R.E., Anthony, C.J., Bilston, L., Prewett, P.D., Arns, J.-Y., et al.: High-fidelity replication of thermoplastic microneedles with open microfluidic channels. Microsyst. Nanoeng. **3**, 17034 (2017)

51. Faraji Rad, Z.: Microneedles Fabrication For Subcutaneous Fluid Sampling And Drug Delivery, Ph.D. thesis, University of Birmingham, College of Engineering and Physical Sciences, Birmingham (2016)
52. Ita, K.: Transdermal delivery of drugs with microneedles-potential and challenges. Pharmaceutics **7**, 90 (2015)
53. Turner, J.G., White, L.R., Estrela, P., Leese, H.S.: Hydrogel-forming microneedles: current advancements and future trends. Macromol. Biosci. **21**, 2000307 (2021)
54. Faraji Rad, Z., Prewett, P.D., Davies, G.J., Beilstein, J.: An overview of microneedle applications, materials, and fabrication methods. Nanotechnol. **12**, 1034 (2021)
55. Waghule, T., Singhvi, G., Dubey, S.K., Pandey, M.M., Gupta, G., Singh, M., et al.: Microneedles: a smart approach and increasing potential for transdermal drug delivery system. Biomed. Pharmacother. **109**, 1249 (2019)
56. Aldawood, F.K., Andar, A., Desai, S.: A comprehensive review of microneedles: types, materials, processes, characterizations and applications. Polymers **13**, 2815 (2021)
57. Peng, K., Vora, L.K., Domínguez-Robles, J., Naser, Y.A., Li, M., Larraneta, E., et al.: Hydrogel-forming MNs for rapid and efficient skin deposition of controlled release tip-implants. Mater. Sci. Eng. C **127** (2021)
58. Faraji Rad, Z.: Microneedle technologies for food and crop health: recent advances and future perspectives. Adv. Eng. Mater. **25**(4), 2201194 (2023)
59. Abowd, G., Dey, A.K., Brown, P., Davies, N., Smith, M., Steggles, P.: Towards a better understanding of context and context-awareness. In: The Workshop on The What, Who, Where, When, and How of Context-Awareness as part of the 2000 Conference on Human Factors in Computing Systems (CHI 2000), pp. 304–307, Springer, The Netherlands (1999)
60. Schilit, B.N., Theimer, M.M.: Disseminating active map information to mobile hosts. IEEE Netw. **8**(5), 22–32 (1994)
61. Aqeel-ur-Rehman, Abbasi, A.Z., Islam, N., Shaikh, Z.A.: A review of wireless sensors and networks' applications in agriculture. Comput. Standards Interfaces **36**, 263–270 (2014)
62. Akyildiz, I.F., Su, W., Sankarasubramaniam, Y., Cayirci, E.: A survey on sensor networks. IEEE Commun. Mag. **40**(8), 102–114 (2002)
63. Willig, A.: Wireless sensor networks: concept, challenges and approaches. e & i Elektrotechnik und Informationstechnik **123**(6), 224–231 (2006)
64. Anastasi, G., Conti, M., Di Francesco, M., Passarella, A.: Energy conservation in wireless sensor networks: a survey. Ad Hoc Netw. **7**(3), 537–568 (2009)
65. Ozdemir, S., Xiao, Y.: Secure data aggregation in wireless sensor networks: a comprehensive overview. Comput. Netw. **53**(12), 2022–2037 (2009)
66. Aqeel-ur-Rehman, Z.A., Shaikh, N.A., Shaikh, N.: An integrated framework to develop context-aware sensor grid for agriculture. Aust. J. Basic Appl. Sci. **4**(5), 922–931 (2010)
67. Goumopoulos, C., Kameas, A.D., Cassells, A.: An ontology-driven system architecture for precision agriculture applications. Int. J. Metadata Semant. Ontol. **4**(1), 72–84 (2009)
68. Damas, M., Prados, A.M., Gómez, F., Olivares, G.: HidroBus system: fieldbus for integrated management of extensive areas of irrigated land. Microprocess. Microsyst. **25**(3), 177–184 (2001)
69. Evans, R., Bergman, J.: Relationships between cropping sequences and irrigation frequency under self-propelled irrigation systems in the Northern Great Plains (NGP), USDA Annual Report. Project, 003–002 (2003)
70. Morais, R., Valente, A., Serôdio, C.: A Wireless Sensor Network for Smart Irrigation and Environmental Monitoring, pp. 845–850. EFITA/WCCA Joint Congress on IT in Agriculture, Portugal (2005)
71. Basu, T., Thool, M.V.R., Thool, R.C., Birajdar, A.C.: Computer based drip irrigation control system with remote data acquisition system. In: 4th World Congress of Computers in Agriculture and Natural Resources, USA (2006)
72. Kim, Y., Evans, R.G., Iversen, W.M.: Remote sensing and control of an irrigation system using a distributed wireless sensor network. IEEE Trans. Instrum. Meas. **57**(7), 1379–1387 (2008)

73. Kim, Y., Evans, R.G.: Software design for wireless sensor-based site-specific irrigation. Comput. Electron. Agric. **66**(2), 159–165 (2009)
74. Cugati, S., Miller, W., Schueller, J.: Automation concepts for the variable rate fertilizer applicator for tree farming. In: The Proceedings of the 4th European Conference in Precision Agriculture, pp. 14–19, Berlin, Germany (2003)
75. Ehlert, D., Schmerler, J., Voelker, U.: Variable rate nitrogen fertilization of winter wheat based on a crop density sensor. Precision Agric. **5**(3), 263–273 (2004)
76. He, J., Wang, J., He, D., Dong, J., Wang, Y.: The design and implementation of a integrated optimal fertilization decision support system. Math. Comput. Model. (in press)
77. Chen, X., Zhang, F.: The establishment of fertilization technology index system based on "3414" fertilizer experiment. China Agric. Technol. Extension **22**(4), 36–39 (2006)
78. Yanlin, H., Shoulun, C.: Summarization of fertilization model research. Chin. J. Soil Sci. **35**(4), 493–501 (2004)
79. Baggio, A.: Wireless sensor networks in precision agriculture, ACM Workshop Real-World Wireless Sensor Networks, Stockholm, Sweden (2005)
80. Dammer, K.H.: Variable rate application of fungicides. In: Precision Crop Protection the Challenge and Use of Heterogeneity, pp. 351–365. Springer Science and Business Media (2010)
81. Butler, Z., Corke, P., Peterson, R., Rus, D.: Virtual fences for controlling cows. In: The 2004 IEEE International Conference on Robotics and Automation (ICRA) , pp. 4429–4436, New Orleans, LA (2004)
82. Radenkovic, M., Wietrzyk, B.: Wireless mobile ad-hoc sensor networks for very large scale cattle monitoring. In: 6th Int'l Workshop Applications and Services in Wireless Networks (ASWN 06), pp. 47–58 (2006)
83. Wark, T., Corke, P., Sikka, P., Klingbeil, L., Guo, Y., Crossman, C., et al.: Transforming agriculture through pervasive wireless sensor networks. In: IEEE Pervasive Computing, pp. 50–57 (2007)
84. Andonovic, I., Michie, C., Gilroy, M., Goh, H.G., Kwong, K.H., Sasloglou, K., Wu, T.: Wireless Sensor Networks for Cattle Health Monitoring, ICT Innovations 2009, pp. 21–31. Springer, Berlin, Heidelberg (2010)
85. Zhang, W., Kantor, G., Singh, S.: Integrated wireless sensor/actuator networks in an agricultural application. In: 2nd ACM International Conference on Embedded Networked Sensor Systems, p. 317 (2004)
86. Aqeel-ur-Rehman, Shaikh, Z.A., Yousuf, H., Nawaz, F., Kirmani, M., Kiran, S.: Crop irrigation control using wireless sensor and actuator network (WSAN). In: 2nd IEEE International Conference on Information and Emerging Technologies (ICIET 2010), pp. 1–5, Karachi, Pakistan (2010)
87. Aqeel-ur-Rehman, Shaikh, Z.A.: Towards design of context-aware sensor grid framework for agriculture. In: Fifth International Conference on Information Technology, XXVIII-WASET Conference, pp. 244–247, Rome, Italy (2008)
88. Goumopoulos, C., Christopoulou, E., Drossos, N., Kameas, A.: The PLANTS system: enabling mixed societies of communicating plants and artefacts. In: Ambient Intelligence. Springer Berlin/Heidelberg, pp. 184–195 (2004)
89. Lauser, B., Sini, M.: From AGROVOC to the agricultural ontology service/concept server: an OWL model for creating ontologies in the agricultural domain. In: Proceedings of the 2006 International Conference on Dublin Core and Metadata Applications: Metadata for Knowledge and Learning, Dublin Core Metadata Initiative, pp. 76–88, Manzanillo, Colima, Mexico (2006)

The manufacturer's authorised representative in the EU is Springer Nature Customer Service Centre GmbH, Europaplatz 3, 69115 Heidelberg, Germany. If you have any concerns regarding our products, please contact ProductSafety@springernature.com

Printed and bound by CPI Group (UK) Ltd, Croydon, CR0 4YY
01/10/2026
02225928-0012